BOUNDARY VALUE PROBLEMS

ADIWES INTERNATIONAL SERIES

IN MATHEMATICS

A. J. Lohwater, Consulting Editor

BOUNDARY VALUE PROBLEMS

F. D. GAKHOV

Translation edited by

I. N. SNEDDON

Simson Professor in Mathematics
University of Glasgow

PERGAMON PRESS

OXFORD · LONDON · EDINBURGH · NEW YORK
PARIS · FRANKFURT

ADDISON-WESLEY PUBLISHING COMPANY, INC.

READING, MASSACHUSETTS · PALO ALTO · LONDON

Sole distributors in the U.S.A.
ADDISON-WESLEY PUBLISHING COMPANY, INC.
Reading, Massachusetts · Palo Alto · London

———————————

PERGAMON PRESS
International Series of Monographs in
Pure and Applied Mathematics

Volume 85

Library of Congress Catalog Card No. 62-10263

This edited translation has been made from the
second revised and enlarged edition of
Краевые задачи (Krayevye zadachi)
published in 1963 by Fizmatgiz, Moscow

CONTENTS

IV. HILBERT BOUNDARY VALUE PROBLEM
AND SINGULAR INTEGRAL EQUATIONS WITH HILBERT KERNEL

V. VARIOUS GENERALIZED BOUNDARY VALUE PROBLEMS

VI. BOUNDARY VALUE PROBLEMS AND SINGULAR INTEGRAL EQUATIONS WITH DISCONTINUOUS COEFFICIENTS AND OPEN CONTOURS

VII. INTEGRAL EQUATIONS SOLUBLE IN CLOSED FORM

FOREWORD TO THE FIRST EDITION

THE present book is based on a course of lectures given by the author from 1947 onwards to students first of Kazan and then of Rostov Universities. This has had a great deal to do with the arrangement of the subject matter in the book, the style of which is closer to that of a manual than to a monograph. Having delivered lectures for many years to students whose interest and response could be observed both during lectures themselves and later at examinations, I have quite understandably taken great pains to ensure a clarity and simplicity of style, which in many places has taken preference over mathematical rigour, general validity of the methods used, and strict mathematical consistency. Sometimes the treatment of a problem commences not from the most general formulation of it, but from an examination of a simple case, in order to bring out more clearly the principles of the problem; in such cases the treatment of the problem is sufficiently detailed. In going on to the general treatment of problems I have limited myself to indicating essentially new facts and to a presentation of formulae. There are many examples of this, two of which are as follows: the consideration of the Riemann boundary value problems starts from the simplest case of a simply-connected domain (Chapter II), and the investigation of the theory of integral representations of analytic functions from the Cauchy type integral with real density (§ 34). Quite apart from considerations of expediency from a pedagogical viewpoint, I took into account those readers whose needs would be satisfied by a presentation of the simplest case only. Hence it would have been doubly unjustified to have commenced the exposition of the problem from a general formulation of it.

As a rule, a detailed treatment of a problem is preceded by brief remarks of a heuristic nature. In the most important cases there are also concluding sections. This sometimes results in repetitions which could have been avoided if a different method of exposition had been used.

The method I have used might arouse criticism, but in a desire to express my views on the subject I would like to quote the words of the well known mathematician and teacher, F. Klein, with which I am fully in agreement. ". . . One will always come across people, who, following the example of medieval Scholastics, begin their teaching from the most general concepts, and who defend this method as being the only true scientific method of approach. However, this reasoning is far from correct: to teach scientifically means to teach a person to think scientifically, and not to dim his powers of thought from the very begining with cold, hard systematics."†

For this very reason I have tried, where possible, to avoid special terminology and symbols, replacing them by descriptive phrases. The use of symbols to abbreviate formulation is very convenient indeed for people who are well acquainted with such symbols, i.e. for people who are constantly dealing with problems of this kind. But there can be no doubt too that it constitutes an obstacle, frequently a serious one, for the large circle of people who are not specialists in a given field, but who wish to become acquainted with it. On the subject of this listen now to the words of one of the most inspired mathematicians of our era, N. G. Chebotarev: ". . . If a science (any science) instead of making itself unnecessarily inaccessible (the use of obscure scientific terminology, special requirements with regard to quotations, etc.), tried by means of popular works to make itself more accessible to the general reader, then it would progress incomparably quicker. We mathematicians are less guilty than others in this respect."‡ Although the last sentence of this quotation might sound like something of a consolation for us mathematicians, every mathematician knows from personal experience just how difficult it is to make use of the results of any mathematical field removed from his own, because of the excessive use of special terminology and symbols.

Owing to lack of space I have sometimes been forced to depart from the generally accepted style of presentation and to present the material concisely; however, the problems concerned in this are, for the most part, not fundamental, and appear in small print. I can only say once again that my main task has been to bring the text

† F. Klein. *Elementary mathematics from the viewpoint of higher mathematics* (Elementary matematika s tochki zreniya vysshei matematiki). ONTI, Vol. I, p. 426 (1935).

‡ N. G. Chebotarev. Mathematical autobiography (Matematicheskaya avtobiografiya). *Usp. matem. nauk,* **3**, No. 4 (26), p. 65 (1948).

as a whole within the understanding of as wide a range of readers as possible. For this very reason I have omitted problems which require a knowledge of the theory of functions of a real variable, or of functional analysis. The book makes use only of the fundamentals of the theory of functions of a complex variable and of differential and integral equations. There is an exception to this at the end of the book (§§ 51, 52), where a knowledge of the fundamentals of the theory of automorphic functions is required. Readers who wish to become more fully acquainted with this theory, are advised to have recourse to specialized works on the subject.

The possibility of qualitative investigation is most important in the theory presented here. Many practically important problems, (for example, the mixed problem for analytic functions, the Dirichlet problem for a plane with slits, etc.) were solved by various particular methods long before the general theory was founded. However, only the general theory, which gives the simple relations between the number of solutions, the conditions of solubility and the index, and between the latter and the admissible class of functions, has succeeded in clarifying the problem.

In those cases when the solution is given in a closed form, the appropriate formulae are presented in full, but when the problem is reduced to solving integral equations, the kernels of the latter, as a rule, are not written down. This occurs for example in Chapters III and VII, in the regularization of singular equations, and in Chapter V in the reduction of the general boundary value problems to integral equations. We only establish the reducibility of the problem to the equation and indicate the procedure of deriving the latter. The reader who has mastered the principles of the theory will easily construct the appropriate equation in solving any practical problems.

The theory treated here is closely connected with applications and, moreover, these are responsible for its creation. Applications, however, are outside the scope of the present book; they are given in the books of such qualified authors as N. I. Muskhelishvili [17], [17*], [18], M. A. Lavrent'ev and B. V. Shabat [14], L. A. Galin [5]†. I have preferred to use the space for topics, in the discussion of which new material can be introduced.

† See References at the end of the book. The numbers in square brackets denote monographs and surveys (Sec. A). The numbers in the parenthesis denote original papers in Russian (Sec. B) and in non-Russian languages (Sec. C).

In those cases where there is a variety of proofs, I have attempted to select the simplest; other methods, if they are of any theoretical interest, are presented as problems at the end of each chapter. Some problems are formulated in such a way that the reader who does not solve them independently, can use them as additional theoretical material. The most difficult problems are marked by asterisks.

All chapters end with historical notes which together give an idea of the history of the problem. The author feels that historical facts brighten and increase interest in the text.

In editing individual sections of the book I was assisted by Dozent V. S. Rogozhin and the postgraduates: T. A. Batchurina, A. A. Govorukhina, E. G. Khasabov, G. S. Litvinchuk, I. M. Mel'nik, L. G. Mikhailov, I. A. Paradoksova, Iv. I. Cherskii, S. V. Yanovski, R. Kh. Zaripov. In writing the book I was greatly helped by my collaborator in the Department, Dozent L. A. Chikin. Professor D. I. Sherman read the manuscript and made a number of valuable comments. A large number of observations which considerably improved the presentation were made by editor of the book I. G. Aramanovich. I express my gratitude to all these people.

THE AUTHOR

FOREWORD TO THE SECOND EDITION

THE present edition of the book has been somewhat enlarged. Sections §§ 48–50, which were originally in Chapter VII, have now been transferred to Chapter VI and §§ 52, 53 have been added dealing with new types of integral equations soluble in closed form. Accordingly Chapters VI and VII have new chapter headings. Chapters IV and V also have two extra sections (33 and 36*) concerning inverse boundary value problems. Other additions and revisions have been made in the light of results which have appeared since the first edition went to press. For the most part these changes are only given without detailed proof.

The original text remains basically unchanged. Only in a few places have defects of exposition had to be adjusted.

In preparing this edition great assistance was rendered to me by my faculty colleague V. A. Kakichev. In editing various parts other faculty colleagues N. I. Borovskaya, A. A. Govorukhina, I. M. Melnik, I. A. Paradoksova, N. A. Rysyuk, V. I. Smagina, I. Kh. Kairullin, E'. G. Kahsabov, Yu. I. Cherskii and F. V. Chumakov participated. I wish to express my gratitude to all these persons.

THE AUTHOR

INTRODUCTION

THE subject of this book is the theory of boundary value problems for analytic functions and its applications to the solution of singular integral equations with Cauchy and Hilbert kernels. We confine ourselves to linear problems for one unknown function. The detailed table of contents is sufficient to give the reader a fairly good idea of the type of material contained in the book.

The theories treated originated at the beginning of the present century, in the works of Hilbert and Poincaré; they advanced considerably at the beginning of twenties, owing to the works of Noether and Carleman and have undergone rapid development during the last twenty years. This latter period is almost exclusively connected with the works of Soviet mathematicians among which first of all the following should be mentioned: B. V. Khvedelidze, V. D. Kupradze, D. A. Kveselava, L. G. Magnaradze, S. G. Mikhlin, N. I. Muskhelishvili, I. I. Privalov, D. I. Sherman, I. N. Vekua, N. P. Vekua. The author has also taken part in the development of the theory.

Let us indicate some aspects of the terminology.

As far as the names of boundary value problems are concerned, we use those which were in use up to 1944.

For analytic functions we do not use at all such terms as "holomorphic" and "regular"; only the term "analytic" being used. The basis is analyticity at a point, i.e. the expandibility of a function into a series in the vicinity of a point. By a function analytic in a domain we understand a function analytic at all points of the domain, obtained by the analytic continuation from any of the elements. In the case of a multiply-connected domain it may turn out to be many-valued. Single-valuedness is regarded as an additional condition. In the case when the analytic function may possess some singularities, this fact will always be specially mentioned with a precise indication as to what singularities are admissible.

CHAPTER I

INTEGRALS OF THE CAUCHY TYPE

IT IS well known from books on the equations of mathematical physics that the basic boundary value problems for the Laplace equation—the Dirichlet and Neumann problems—are solved by means of the so-called potentials of simple and double layers.

In solving boundary value problems connected with other differential equations, generalized potentials of various types are employed. For the solution of the boundary value problems of the theory of analytical functions of complex variable, the analogous device is constituted by the integral of the Cauchy type and its various generalizations.

Prior to proceeding directly to the subject of the present book, i. e. to the solution of boundary value problems, it is necessary to make a preliminary examination of the properties of the mathematical apparatus used for it.

The present chapter is auxiliary. Its material will not be used all at once, but gradually; consequently, in order to come more quickly to the fundamental contents of the book we recommend the reader to study this chapter by parts. First it is necessary to master the contents of the first five sections, and then one may proceed to study Chapter II devoted to the Riemann boundary value problem. Before proceeding to Chapter III containing the theory of singular integral equations, one has to assimilate § 7 dealing with the change of the integration order in double singular integrals. To understand Chapter IV the knowledge of § 6 is required, and for Chapter V that of § 10. Only then is it expedient to study the remaining sections of this chapter.

§ 1. Definition of the Cauchy type integral and examples

Let L be a smooth closed contour† in the plane of the complex variable z. The domain within the contour L is called the interior domain

---

† By a smooth contour we understand here and hereafter a simple (i.e. without points of intersection with itself), closed or open line with continuously varying tangent, having no recurrent points (cusps).

and denoted as D^+, whilst the complementary domain to $D^+ + L$ is called the exterior domain and denoted by D^-.

If $f(z)$ is an analytic function in D^+ and continuous in $D^+ + L$, then according to the Cauchy formula in the theory of functions of a complex variable,

$$\frac{1}{2\pi i} \int\limits_L \frac{f(\tau)}{\tau - z}\, d\tau = \begin{cases} f(z), & z \in D^+, \\ 0, & z \in D^-. \end{cases} \tag{1.1}$$

If, however, $f(z)$ is analytic in D^- and continuous in $D^- + L$, then

$$\frac{1}{2\pi i} \int\limits_L \frac{f(\tau)}{\tau - z}\, d\tau = \begin{cases} f(\infty), & z \in D^+, \\ -f(z) + f(\infty), & z \in D^-. \end{cases} \tag{1.2}$$

The positive direction of the contour L is usually that for which D^+ is on the left.

Cauchy's formula enables the value of a function to be calculated at any point within a domain if values on the boundary are known; one may thus say that Cauchy's formula solves the boundary value problem in analytic functions. The integral on the left-hand side of the formulae (1.1) and (1.2) is known as *Cauchy's integral*.

Let now L be a smooth closed or open contour situated wholly in a finite part of the plane; τ is the complex coordinate of its points and $\varphi(\tau)$ is a continuous function of position on the contour. Then the integral

$$\Phi(z) = \frac{1}{2\pi i} \int\limits_L \frac{\varphi(\tau)}{\tau - z}\, d\tau, \tag{1.3}$$

constructed in the same way as the Cauchy integral, is called the Cauchy type integral. The function $\varphi(\tau)$ is called its density and $1/(\tau - z)$ the kernel.

It is readily observed that the Cauchy type integral constitutes a function analytic in the entire plane of the complex variable, except for the points of the contour of integration L. The proof of analyticity of $\Phi(z)$ consists in establishing the possibility of differentiation with respect to the variable z (parameter) in the integrand, and is given in all textbooks on the theory of functions of a complex variable.

We shall carry out an analogous reasoning for a somewhat more general case which we shall encounter later on. This reasoning will imply the analyticity of the Cauchy type integral as a particular case.

THEOREM. *Let L be a smooth contour (closed or open), $f(\tau, z)$ a function continuous with respect to the variable $\tau \in L$ and analytic in z in some domain for all values τ. Then the function represented by the curvilinear integral*

$$F(z) = \int_L f(\tau, z)\, d\tau, \tag{1.4}$$

is a function analytic in the variable z.

To prove the theorem let us consider the difference

$$\frac{F(z + \Delta z) - F(z)}{\Delta z} - \int_L f_z'(\tau, z)\, d\tau$$

$$= \int_L \left[\frac{f(\tau, z + \Delta z) - f(\tau, z)}{\Delta z} - f_z'(\tau, z) \right] d\tau.$$

In view of the analyticity in z the quantity

$$\left| \frac{f(\tau, z + \Delta z) - f(\tau, z)}{\Delta z} - f_z'(\tau, z) \right|$$

can be made arbitrarily small for all τ for a sufficiently small $|\Delta z|$. Since, by assumption, the contour L is of finite length, by using an estimate of the integral and passing to the limit we obtain

$$F'(z) = \lim_{\Delta z \to 0} \frac{F(z + \Delta z) - F(z)}{\Delta z} = \int_L f_z'(\tau, z)\, d\tau. \tag{1.5}$$

Thus, $F(z)$ is a function analytic in z everywhere in the domain of analyticity of $f(\tau, z)$. Those values of z at which $f(\tau, z)$ ceases to be analytic are singular points of the function $F(z)$.

For the Cauchy type integral with a continuous density $\varphi(\tau)$ the only points for which the integrand ceases to be analytic with respect to z, are the points of the line L. This line is the singular line for the function $\Phi(z)$ given by (1.3). Important problems concerning the behaviour of the Cauchy type integral as the point z approaches the contour L and meets the contour, will be dealt with below, in §§ 4 and 5.

If L is an open contour $\Phi(z)$ is a function analytic in the entire plane, with the singular line L. Now let L be a closed contour. By definition the function $\Phi(z)$ splits into two independent functions.

viz. $\Phi^+(z)$ defined in D^+, and $\Phi^-(z)$ defined for points of D^-. These functions are not generally an analytic continuation of each other.

An analytic function $\Phi(z)$ defined in two domains D^+, D^-, complementary to each other in the entire plane by two independent expressions $\Phi^+(z)$, $\Phi^-(z)$, will frequently be called a sectionally analytic function.

We now note an important property of the Cauchy type integral. The function $\Phi(z)$ represented by the Cauchy type integral (1.3) vanishes at infinity.

In fact, let us expand $\Phi(z)$ in the vicinity of infinity into series with respect to the powers of $1/z$. Since

$$\frac{1}{\tau - z} = -\frac{1}{z} - \frac{\tau}{z^2} - \cdots - \frac{\tau^{n-1}}{z^n} + \cdots,$$

multiplying the series by $\dfrac{1}{2\pi i}\varphi(\tau)$ and integrating term by term we obtain

$$\Phi^-(z) = \sum_{k=1}^{\infty} \frac{c_k}{z^k},$$

where

$$c_k = -\frac{1}{2\pi i}\int_L \tau^{k-1}\varphi(\tau)\,d\tau.$$

In the series the zero power is absent, whence it follows that $\Phi^-(\infty) = 0$.

EXAMPLE. Let us compute the Cauchy type integral over the unit circle $|z| = 1$ with the density $\varphi(\tau) = \dfrac{2}{\tau(\tau - 2)}$, i.e.

$$\Phi(z) = \frac{1}{2\pi i}\int_L \frac{1}{\tau - 2}\frac{d\tau}{\tau - z} - \frac{1}{2\pi i}\int_L \frac{1}{\tau}\frac{d\tau}{\tau - z}.$$

The function $1/(z - 2)$ is analytic in D^+, and $1/z$ is analytic in D^-, and vanishes at infinity. The first integral from formula (1.1) equals $1/(z - 2)$ for $z \in D^+$ and zero for $z \in D^-$. The second integral from formula (1.2) equals $-1/z$ for $z \in D^-$ and vanishes for $z \in D^+$. Hence

$$\Phi^+(z) = \frac{1}{z - 2}; \quad \Phi^-(z) = \frac{1}{z}.$$

§ 2. Functions satisfying the Hölder condition

2.1. Definition and properties

Before proceeding to the basic problem, the investigation of the behaviour of the Cauchy type integral on the line of integration, we consider an auxiliary problem concerning the classes of functions.

Let $\varphi(t)$ be a function which with its argument t, may take both real and complex values. It is known that the continuity of the function $\varphi(t)$ consists in the fact that $|\varphi(t_2) - \varphi(t_1)|$ can be made arbitrarily small when $|t_2 - t_1|$ is sufficiently small, i.e. the increments of the function and the argument together tend to zero.

The problem of the order of smallness of the increment of a function with respect to the increment of the argument is not considered in this case; this order may be arbitrary. Many properties of functions, however, for instance expansions into series of various types and the rapidity of their convergence, integral representations, etc., are closely connected with the order of smallness of the continuity modulus † of the function. Consequently, it is expedient to divide the extensive set of continuous functions into classes, depending on the order of smallness of the continuity modulus. A very important class is constituted by the functions, the continuity modulus of which is a power function in the increment of the argument. We now proceed to the investigation of this class which is most interesting for us.

Let L be a smooth contour and $\varphi(t)$ a function of position on it. *The function $\varphi(t)$ is said to satisfy on the curve the Hölder condition (H condition), if for two arbitrary points of this curve*

$$|\varphi(t_2) - \varphi(t_1)| < A\,|t_2 - t_1|^\lambda, \qquad (2.1)$$

where A and λ are positive numbers. A is called the Hölder constant and λ the Hölder index. If λ were greater than unity, then the condition (2.1) would imply that the derivative $\varphi'(t)$ vanishes everywhere, the function $\varphi(t)$, therefore, would identically be equal to a constant. Consequently we assume that

$$0 < \lambda \leqq 1.$$

If $\lambda = 1$ the Hölder condition is identical to the familiar Lipschitz condition. (Observe that in many books the Hölder condition is called the Lipschitz condition of order λ.)

---

† The continuity modulus of a function $\varphi(t)$ is the function $\omega(\delta) = \sup |\varphi(t_2) - \varphi(t_1)|$ where the points t_1 and t_2 belong to L and $|t_2 - t_1| < \delta$.

If t_1, t_2 are sufficiently close and the Hölder condition is satisfied for some value of the index λ_1, then it is evident that it is satisfied for every index $\lambda < \lambda_1$. The converse, in general, is not true. Thus, smaller λ is associated with a wider class of functions. The narrowest class is the class of functions obeying the Lipschitz condition.

On the basis of the latter property it is easy to find that if the functions $\varphi_1(t)$, $\varphi_2(t)$ satisfy the Hölder condition with the indices λ_1, λ_2, respectively, then their sum, product and quotient, under the condition that the denominator does not vanish, satisfy the Hölder condition with the index $\lambda = \min(\lambda_1, \lambda_2)$.

If $\varphi(t)$ is differentiable and has a finite derivative, it satisfies the Lipschitz condition. This statement follows from the theorem of finite increment. The converse, in general, is not true; this is shown by the following example of a function given on the real axis:

$$\varphi(x) = |x|.$$

The function $\varphi(x)$ satisfies the Lipschitz condition, but its derivative does not exist at the origin of the coordinate system, since the left and right derivatives there are equal to -1, $+1$, respectively.

For a complex function, with superposition of functional relationships of different classes, the resultant function as a rule belongs to the widest (worst) class of all circuit elements. One often comes across analytic functions of functions which satisfy Hölder's condition. Such functions satisfy Hölder's condition with the same indices. Proof is obtained with direct verification having regard to the fact that the analytic function satisfies the Lipschitz condition.

Let us present two more examples.

1. The function $\varphi(x) = \sqrt{x}$ satisfies the H condition with the index $\lambda = \frac{1}{2}$ on any interval of the real axis; if now this interval does not contain the origin, this function is analytic on it and, consequently, it satisfies the Lipschitz condition.

2. Consider the function $\varphi(x) = 1/\ln x$ for $0 < x \leqq \frac{1}{2}$, and $\varphi(0) = 0$. It is readily observed that this function is continuous on the whole closed interval (segment) $0 \leqq x \leqq \frac{1}{2}$. However, since $\lim\limits_{x \to 0} x^\lambda \ln x = 0$ for any $\lambda > 0$, no matter what are the values of A and λ, there can always be found a value of x such that $|\varphi(x) - \varphi(0)| = \left|\dfrac{1}{\ln x}\right| > A x^\lambda$. Consequently, the function $\varphi(x)$ on the segment under consideration does not satisfy the Hölder condition.

In what follows we shall encounter problems in which the Hölder condition is satisfied on the contour everywhere, except for some given isolated points. This will be understood as obeying the Hölder condition on every closed part of the contour not containing the above indicated points. In the same sense we understand this condition as being satisfied on an open contour, if the exceptional points are its ends. In the second of the examples presented above the function $\varphi(x)$ satisfies the Hölder condition on the semi-open interval $0 < x \leqq \frac{1}{2}$ and does not satisfy it on the segment $0 \leqq x \leqq \frac{1}{2}$.

2.2. Functions of many variables

The concept of the Hölder condition can be extended to functions of an arbitrary number of variables. Let us consider specifically the case of a function of two variables. The function $\varphi(t, \tau)$ is said to satisfy the Hölder condition, if for any pair values (t_1, t_2), (τ_1, τ_2) belonging to the specified sets,

$$|\varphi(t_2, \tau_2) - \varphi(t_1, \tau_1)| < A|t_2 - t_1|^\mu + B|\tau_2 - \tau_1|^\nu,$$

where A, B, μ, ν are positive constants, and μ, $\nu \leqq 1$.

If λ is the smallest of the numbers μ, ν, a constant C can be found, such that the following inequality is satisfied:

$$|\varphi(t_2, \tau_2) - \varphi(t_1, \tau_1)| < C[|t_2 - t_1|^\lambda + |\tau_2 - \tau_1|^\lambda].$$

It is evident that if $\varphi(t, \tau)$ satisfies the Hölder condition with respect to the set of variables t, τ, then it satisfies this condition with respect to the variable t uniformly in τ, and with respect to τ uniformly in t.

§ 3. Principal value of the Cauchy type integral

3.1. Improper integral

Consider a definite integral of a real function. In accordance with the original definition as the limit of a sum, it has a meaning only for bounded functions. If the integrand is infinite at a point c of the integration range, $a \leqq c \leqq b$, then the integral of such a function is given a meaning by means of a second passing to the limit. Thus, a neighbourhood of the point c is cut out, the integral is taken over the remaining part and passing to the limit is carried out, when the length of the segment cut out tends to zero. Let c be an interior point of the interval. The limit

$$\lim_{\substack{\varepsilon_1 \to 0 \\ \varepsilon_2 \to 0}} \left[\int_a^{c-\varepsilon_1} f(x)\, dx + \int_{c+\varepsilon_2}^b f(x)\, dx \right].$$

if it exists, is called the *improper integral* of the unbounded function $f(x)$ in the interval (a, b).

In this definition it is significant that the neighbourhood cut out is entirely arbitrary, if only its length tends to zero. Consequently, ε_1 and ε_2 tend to zero according to an arbitrary law, independently of each other.

It is shown in courses of Analysis that an improper integral exists if the order of infinity of $f(x)$ is smaller than unity, i.e.

$$|f(x)| < \frac{M}{|x - c|^\alpha} \quad (\alpha < 1).$$

If $f(x)$ tends to infinity of order one or greater, the improper integral does not exist or, in other words, *it diverges*.

3.2. Principal value of singular integral

Consider the integral

$$\int_b^a \frac{dx}{x - c} \quad (a < c < b).$$

Computing it as an improper integral we have

$$\int_a^b \frac{dx}{x - c} = \lim_{\substack{\varepsilon_1 \to 0 \\ \varepsilon_2 \to 0}} \left[-\int_a^{c - \varepsilon_1} \frac{dx}{c - x} + \int_{c + \varepsilon_2}^b \frac{dx}{x - c} \right]$$

$$= \ln \frac{b - c}{c - a} + \lim_{\substack{\varepsilon_1 \to 0 \\ \varepsilon_2 \to 0}} \ln \frac{\varepsilon_1}{\varepsilon_2}. \tag{3.1}$$

It is evident that the limit of the last expression depends on the manner in which ε_1 and ε_2 tend to zero. Consequently, the integral regarded as improper does not exist. It is called a *singular integral*. However, it can be given a meaning if a dependence between ε_1 and ε_2 be established. Assuming that the interval cut out is situated symmetrically with respect to the point c, i.e.

$$\varepsilon_1 = \varepsilon_2 = \varepsilon, \tag{3.2}$$

we are led to the concept of the Cauchy principal value of the improper integral.

DEFINITION. The Cauchy principal value of the singular integral

$$\int_a^b \frac{dx}{x-c} \quad (a < c < b)$$

is the expression

$$\lim_{\varepsilon \to 0} \left[\int_a^{c-\varepsilon} \frac{dx}{x-c} + \int_{c+\varepsilon}^b \frac{dx}{x-c} \right].$$

Taking into account formula (3.1) and condition (3.2) we have

$$\int_a^b \frac{dx}{x-c} = \ln \frac{b-c}{c-a}. \qquad (3.3)$$

Consider now the more general integral

$$\int_a^b \frac{\varphi(x)}{x-c} dx,$$

where $\varphi(x)$ is a function which obeys in the interval (a, b) the Hölder condition. This case can easily be reduced to the preceding simpler one by representing the integral in the form

$$\int_a^b \frac{\varphi(x)}{x-c} dx = \int_a^b \frac{\varphi(x) - \varphi(c)}{x-c} dx + \varphi(c) \int_a^b \frac{dx}{x-c}.$$

In view of the Hölder condition

$$\left| \frac{\varphi(x) - \varphi(c)}{x-c} \right| < \frac{A}{|x-c|^{1-\lambda}},$$

the first integral exists as improper, and the second is identical with (3.3).

Consequently, we obtain the following result: *the singular integral*

$$\int_a^b \frac{\varphi(x)}{x-c} dx,$$

where $\varphi(x)$ satisfies the Hölder condition, exists in the sense of the Cauchy principal value, and it is equal to

$$\int_a^b \frac{\varphi(x)}{x-c}\,dx = \int_a^b \frac{\varphi(x)-\varphi(c)}{x-c}\,dx + \varphi(c)\ln\frac{b-c}{c-a}.$$

To denote the singular integral some authors use special symbols, e.g. $\int'$ or v.p. $\int$ (valeur principale). It is however by no means necessary because, on one hand, if the integral

$$\int_a^b \frac{\varphi(x)}{x-c}\,dx$$

exists as an ordinary or an improper integral, it also exists in the sense of its principal value, the corresponding values being identical; on the other hand, the singular integral will always be understood in the sense of the principal value. We shall therefore denote the singular integral by the ordinary integral sign.

Observe that the concept of the principal value is in some cases valid in the presence of singularities of the integrand of order higher than one (see Problems 2 and 3 in the end of the chapter).

3.3. Many-valued functions

In the next section we shall come across a logarithmic function which is many-valued. Later we shall time and time again have to deal with both power and logarithmic elementary many-valued analytic functions. Functions of this kind present considerable difficulty to beginners and not enough attention is paid to them in the usual courses on the theory of functions. It is therefore advisable to lay the foundations here for what will be required later.

In defining the complex number $z = \varrho e^{i\theta}$ the modulus ϱ is single-valued, but the argument θ is determined by neglecting a term with multiplicity 2π. This fails to lead to a many-valued representation of the number since the argument enters via the function $e^{i\theta}$ which has a period 2π. If, however, the analytic function is such that the argument (angle) enters by means of a non-periodic function, it becomes many-valued. Amongst such elementary functions are logarithmic and power functions with a non-integer index. We will consider them in parallel:

$$\ln(z - z_0) = \ln|z - z_0| + i \arg(z - z_0) = \ln \varrho + i\Theta,$$

$$(z - z_0)^\gamma = \varrho^\gamma e^{i\gamma\Theta}.$$

The general representation of the argument θ is:

$$\Theta = \theta + 2k\pi,$$

where k may assume any integer value ($k = 0, \pm 1, \pm 2, ...$), whilst θ is the smallest absolute value of the argument shown in Fig. 1.

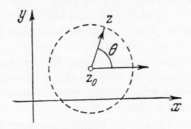

FIG. 1.

To each selected value of k corresponds a definite branch of the many-valued function. With the logarithmic function the number of branches is infinite; the number of branches is also infinite for a power function with an irrational or complex index. If, however, the index is rational, $\gamma = p/q$, then it has q branches. The branches of the logarithmic function differ from each other by a term of the type $2m\pi i$, but those of power functions differ by the factor $e^{2m\gamma\pi i}$ (m is an integer). To define a many-valued function, it is obviously necessary to indicate the selected branch. But, as distinct from functions of a real variable, this is still insufficient for a complete definition of a many-valued function of a complex variable. For such a function, points exist in a plane which possess the property that, as the variable varies along the closed contour surrounding a point and returns to its original value, the selected branch of the function does not remain the same, but becomes some other branch. These points are called *branch points* of a many-valued function. For these functions the branch points are z_0 and the infinitely remote point. If the variable describes a contour round point z_0 (the broken line in Fig. 1) in the positive (negative) direction, then the argument θ changes by $\pm 2\pi$; the logarithmic function acquires the addendum term $\pm 2\pi i$, but the power function acquires the factor $e^{2\gamma\pi i}$; the branch corresponding to the selected value $k = n$ therefore changes to the adjacent branch corresponding to $k = n \pm 1$. The investigation at the infinitely remote point is carried out in the usual way by substituting $z = 1/\zeta$ and considering the point $\zeta = 0$.

The selected branch of a function may only be preserved if traversal of the branch points is made impossible. This is achieved by making cuts in the plane which connect the branch points. For the logarithmic and power function under consideration, it is necessary to cut along a line starting from the point z_0 and passing to infinity. For the purpose in hand (to fix on a selected branch) the shape of the cut is immaterial. But if other circumstances have to be taken into account as well, the shape of the cut is sometimes pre-determined. If there are no special considerations, the simplest cut is selected, namely a rectilinear cut preferably parallel to the axis of the abscissae. A many-valued function may be regarded as fully defined if its branch is selected and its cut is prescribed.

The bounds of the variation of the argument θ are determined by the position of the cut. If, for example, the cut is along a line at an angle θ_0 to the axis of the abscissae, then for the principal branch† $(k = 0)$ $\theta_0 \leqq \theta \leqq \theta_0 + 2\pi$. In particular, $0 \leqq \theta \leqq 2\pi$ for a cut along a positive semi-axis of the abscissae, and $-\pi \leqq \theta \leqq \pi$ along a negative semi-axis. If the cut is curvilinear, the bounds of variation of the argument are functions of a point. The initial value of the argument corresponds to the left-hand approach (seen from point z_0) of the cut, and the final value to the right-hand approach. Suppose we put Θ^+ and Θ^- for the value of the argument on the left and right approaches respectively. We then get

$$\Theta^- - \Theta^+ = 2\pi.$$

The cut for the selected branch will be a line of discontinuity. On the approaches of the cut:

$$\ln(z^- - z_0) = \ln(z^+ - z_0) + 2\pi i, \qquad (z^- - z_0)^\gamma = e^{i2\pi\gamma}]\ (z^+ - z_0)^\gamma.$$

The property of discontinuity of the branches of many-valued functions on the approaches of a cut is widely used in the solution of boundary value problems with discontinuous boundary conditions.

---

† If, by definition, the approaches of the cut are excluded from the surface, the equality signs are discarded.

The logarithm is used if the discontinuous function enters the boundary condition as an addendum, and the power function if as a multiplier (see §§ 41.2, 41.3).

Each branch of a many-valued function in an appropriately cut plane is thus single-valued and discontinuous. Preservation of continuity is only possible if the argument of the many-valued function can be regarded as varying in a multi-sheet plane—a Riemann surface, the number of sheets of which is equal to the number of branches of the many-valued function and the sheets are fixed on the approaches to the cuts in such a way that transition from one sheet to another rigorously corresponds to the replacement of the branches on traversal of the branch points. Questions of this kind do not arise later and so no further comment is required. For a good introduction to the representation of many-valued functions on Riemann surfaces, reference could well be made to V. F. Kagan's paper in the supplement to vol. 1 of F. Klein's book "Elementary mathematics from the advanced point of view" (Elementarnaya matematika s tochki zreniya vysshei).

3.4. Principal value of singular curvilinear integral

Let L be a smooth contour and τ, t complex coordinates of its points. Consider the singular curvilinear integral

$$\int_L \frac{\varphi(\tau)}{\tau - t}\, d\tau. \tag{3.4}$$

If σ, s are the lengths of the arcs from a fixed origin of integration to the points τ, t, and $\tau = \tau(\sigma)$ is the equation of the contour in the complex form, then substituting in the integral $\tau = \tau(\sigma)$, $t = t(s)$, $d\tau = \tau'(\sigma)\, d\sigma$ we would reduce it to two real singular integrals, as will be done in § 10. Subsequently, we could make full use of the results of the preceding article, without resorting to new considerations. However, since in what follows we shall constantly solve problems directly in complex variables, without passing to real ones, it is expedient to consider the singular integral independently as a function of a complex variable.

Let us describe about the point t of L as centre a circle of radius ϱ, and let t_1, t_2 be the points of intersection of this circle with the curve (Fig. 2). The radius is assumed to be sufficiently small, so that the circle has no other points of intersection with the curve L besides

t_1, t_2. Denote by l the part of the contour L cut out by the circle and take the integral over the remaining arc

$$\int_{L-l} \frac{\varphi(\tau)}{\tau - t}\, d\tau.$$

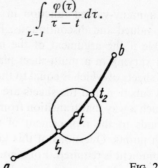

FIG. 2.

DEFINITION. *The limit of the integral*

$$\int_{L-l} \frac{\varphi(\tau)}{\tau - t}\, d\tau$$

as $\varrho \to 0$ is called the principal value of the singular integral

$$\int_{L} \frac{\varphi(\tau)}{\tau - t}\, d\tau.$$

We commence the investigation of the problem of existence of the singular integral whith the simplest case

$$\int_{L} \frac{d\tau}{\tau - t}.$$

The integrand has the primitive $\ln(\tau - t)$; the investigation therefore can be carried out as in § 3.2. The only difficulty is due to the fact that the primitive is a multi-valued function. We assume that $\ln(\tau - t)$ is the contour value of the analytic function $\ln(z - t)$ which is single-valued in the plane cut along a curve connecting the branch points t and ∞. Let us agree for definiteness that the cut is made on the right of the curve L. Then

$$\int_{L-l} \frac{d\tau}{\tau - t} = \ln(\tau - t)\big|_a^{t_1} - \ln(\tau - t)\big|_{t_2}^b = \ln\frac{b-t}{a-t} + \ln\frac{t_1 - t}{t_2 - t},$$

where a and b are the ends of the contour L.

Now

$$\ln\frac{t_1 - t}{t_2 - t} = \ln\left|\frac{t_2 - t}{t_1 - t}\right| + i[\arg(t_2 - t) - \arg(t_1 - t)].$$

By definition $|t_2 - t| = |t_1 - t|$, the first term therefore vanishes. The expression in the square brackets is equal to the angle α between the vectors $\overrightarrow{tt_1}$, $\overrightarrow{tt_2}$ (Fig. 3), and in view of the choice of the cut this angle should be measured on the left of the curve.

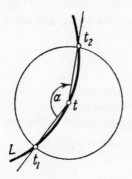

FIG. 3.

Hence

$$\lim_{\varrho \to 0} \ln\frac{t_1 - t}{t_2 - t} = i\pi$$

and consequently

$$\int_L \frac{d\tau}{\tau - t} = \ln\frac{b - t}{a - t} + i\pi. \qquad (3.5)$$

The latter integral can also be represented in the form

$$\int_L \frac{d\tau}{\tau - t} = \ln\frac{b - t}{t - a} \qquad (3.6)$$

(selecting the branch of the logarithmic function in such a way that $\ln(-1) = \pi i$).

If the contour L is closed, then setting $a = b$ we observe that the first term of the right-hand side of formula (3.5) vanishes, and the integral assumes the simple form

$$\int_L \frac{d\tau}{\tau - t} = i\pi. \qquad (3.7)$$

2*

Let us now examine the singular integral

$$\int_L \frac{\varphi(\tau)}{\tau - t}\, d\tau,$$

where $\varphi(\tau)$ satisfies the Hölder condition. Representing it in the form

$$\int_L \frac{\varphi(\tau)}{\tau - t}\, d\tau = \int_L \frac{\varphi(\tau) - \varphi(t)}{\tau - t}\, d\tau + \varphi(t) \int_L \frac{d\tau}{\tau - t} \qquad (3.8)$$

and repeating the reasoning of the preceding article we find that *the singular integral*

$$\int_L \frac{\varphi(\tau)}{\tau - t}\, d\tau$$

where the function $\varphi(\tau)$ satisfies the Hölder condition, exists in the sense of the Cauchy principal value.

This integral can be represented in the two following forms:

$$\int_L \frac{\varphi(\tau)}{\tau - t}\, d\tau = \int_L \frac{\varphi(\tau) - \varphi(t)}{\tau - t}\, d\tau + \varphi(t) \left[\ln \frac{b - t}{a - t} + \pi i \right]; \quad (3.9)$$

$$\int_L \frac{\varphi(\tau)}{\tau - t}\, d\tau = \int_L \frac{\varphi(\tau) - \varphi(t)}{\tau - t}\, d\tau + \varphi(t) \ln \frac{b - t}{t - a}. \qquad (3.10)$$

In particular, for a closed contour, setting $a = b$ in (3.9) we obtain

$$\int_L \frac{\varphi(\tau)}{\tau - t}\, d\tau = \int_L \frac{\varphi(\tau) - \varphi(t)}{\tau - t}\, d\tau + i\pi\varphi(t). \qquad (3.11)$$

In all subsequent references to a singular integral we shall always have in mind its principal value.

3.5. Properties of the singular integral

It is quite obvious that a singular integral possesses some of the properties of an ordinary integral, namely an integral of a sum is equal to the sum of integrals and a constant factor may be taken outside the integral sign. Before proceeding to the derivation of less trivial properties—the rules of change of variable and integration by parts, let us make an important remark.

In introducing the concept of the principal value it was emphasized as a most significant fact that the vicinity cut out is taken symmetrically with respect to the point under investigation. However, if the definition of the principal value be subject to a more precise examination, it will turn out that there is no need for rigorous symmetry. In fact, in § 3.2 the significant requirement was not that $\varepsilon_1 = \varepsilon_2$, but that $\lim\limits_{\varepsilon_1, \varepsilon_2 \to 0} \dfrac{\varepsilon_1}{\varepsilon_2} = 1$, exactly as in § 3.4 where the significant assumption is not that the points t_1, t_2 lie on the same circle ($|t_2 - t| = |t_1 - t|$) but that the following condition is satisfied:

$$\lim_{\substack{t_1 \to t \\ t_2 \to t}} \left| \frac{t_2 - t}{t_1 - t} \right| = 1. \tag{3.12}$$

Thus, if the symmetry requirement be discarded and only condition (3.12) preserved, then the singular integral defined from this more general standpoint coincides with its principal value introduced above.

THEOREM (the rule of change of variable). *If the function $\tau = \alpha(\zeta)$ has a continuous first derivative $\alpha'(\zeta)$ which does not vanish anywhere, and constitutes a one to one mapping of the contour L onto a contour L', then*

$$\int_L \frac{\varphi(\tau)}{\tau - t}\, d\tau = \int_{L'} \frac{\varphi[\alpha(\zeta)]\alpha'(\zeta)}{\alpha(\zeta) - \alpha(\xi)}\, d\zeta, \tag{3.13}$$

where

$$t = \alpha(\xi).$$

Proof. Let us cut out of the contour L' by means of a sufficiently small circle with centre at the point ξ, an arc l'. Let its ends be ξ_1 and ξ_2 and the corresponding points of the contour L be t_1 and t_2.

We define the principal value of the transformed integral as follows:

$$\int_{L'} \frac{\varphi[\alpha(\zeta)]\alpha'(\zeta)}{\alpha(\zeta) - \alpha(\xi)}\, d\zeta = \lim_{\xi_1, \xi_2 \to \xi} \int_{L'-l'} \frac{\varphi[\alpha(\zeta)]\alpha'(\zeta)}{\alpha(\zeta) - \alpha(\xi)}\, d\zeta.$$

Making the substitution $\zeta = \beta(\tau)$, where $\beta(\tau)$ is the function inverse to $\alpha(\zeta)$ (in view of the condition of the theorem the inverse function exists and is unique), we find that the right-hand side of the relation coincides with the expression

$$\lim_{t_1, t_2 \to t} \int_{L-l} \frac{\varphi(\tau)}{\tau - t}\, d\tau.$$

The points t_1 and t_2 are not situated symmetrically with respect to the point t on L any more; however, as we shall now prove, they satisfy condition (3.12).

Expanding the function $\alpha\,(\zeta)$ into a Taylor series at the point ξ and taking only the first two terms we obtain

$$t_2 = \alpha(\xi_2) = t + [\alpha'(\xi) + \varepsilon_2(\xi_2,\,\xi)]\,(\xi_2 - \xi),$$
$$t_1 = \alpha(\xi_1) = t + [\alpha'(\xi) + \varepsilon_1(\xi_1,\,\xi)]\,(\xi_1 - \xi),$$

where in view of the assumed continuity of $\alpha'(\zeta)$, ε_1 and ε_2 tend to zero as $\xi_1, \xi_2 \to \xi$.

Thus

$$\left| \frac{t_2 - t}{t_1 - t} \right| = \left| \frac{\alpha'(\xi) + \varepsilon_2}{\alpha'(\xi) + \varepsilon_1} \right| \left| \frac{\xi_2 - \xi}{\xi_1 - \xi} \right|$$

and consequently

$$\lim_{t_1,\,t_2 \to t} \left| \frac{t_2 - t}{t_1 - t} \right| = \lim_{\xi_1,\,\xi_2 \to \xi} \left| \frac{\xi_2 - \xi}{\xi_1 - \xi} \right| = 1.$$

In accordance with the remark made at the beginning of this article, this completes the proof of the theorem.

The condition that the transformation is one-to-one is essential, although there is no particular need to emphasize it, since it is essential for an ordinary integral as well.

THEOREM (integration by parts). *If $\varphi(\tau)$ is a continuously differentiable function and the point t does not coincide with an end of the contour L (a or b), then the following formula of integration by parts is valid:*

$$\int_L \frac{\varphi(\tau)}{\tau - t}\,d\tau = \pm\, i\pi\varphi(t) + \varphi(b)\ln(b - t) - \varphi(a)\ln(a - t) -$$
$$- \int_L \varphi'(\tau)\ln(\tau - t)\,d\tau. \qquad (3.14)$$

The first term is taken with positive sign if the cut connecting the point t with the point at infinity, by means of which the single-valued branch of $\ln(\tau - t)$ is distinguished (see § 3.4), is made on the right of L; the negative sign is taken in the opposite case.

To prove the theorem we start from the integral entering the right-hand side

$$\int_L \varphi'(\tau)\ln(\tau - t)\,d\tau$$

which is taken as improper. To compute it we may cut out the vicinity of the point t in an arbitrary way. Let us perform it as in determining the principal value; thus

$$\int_L \varphi'(\tau) \ln(\tau - t) \, d\tau$$

$$= \lim_{\varrho \to 0} \left[\int_a^{t_1} \varphi'(\tau) \ln(\tau - t) \, d\tau + \int_{t_2}^b \varphi'(\tau) \ln(\tau - t) \, d\tau \right].$$

Carrying out the integration by parts on the ordinary integrals in the square brackets, we have

$$\int_a^{t_1} \varphi'(\tau) \ln(\tau - t) \, d\tau + \int_{t_2}^b \varphi'(\tau) \ln(\tau - t) \, d\tau$$

$$= \varphi(b) \ln(b - t) - \varphi(a) \ln(a - t) + \varphi(t_1) \ln(t_1 - t) -$$

$$- \varphi(t_2) \ln(t_2 - t) - \int_a^{t_1} \frac{\varphi(\tau)}{\tau - t} \, d\tau - \int_{t_2}^b \frac{\varphi(\tau)}{\tau - t} \, d\tau.$$

Passing to the limit, $\varrho \to 0$, we find that the first two terms are unaltered; the last two give together the integral

$$- \int_L \frac{\varphi(\tau)}{\tau - t} \, d\tau,$$

in the sense of the principal value. To determine the limit of the remaining terms let us carry out the transformation

$$\varphi(t_1) \ln(t_1 - t) - \varphi(t_2) \ln(t_2 - t) = \varphi(t) \left[\ln(t_1 - t) - \ln(t_2 - t) \right] +$$

$$+ \left[\varphi(t_1) - \varphi(t) \right] \ln(t_1 - t) - \left[\varphi(t_2) - \varphi(t) \right] \ln(t_2 - t).$$

The function $\varphi(t)$, as a continuously differentiable function, satisfies the Lipschitz condition. Hence, on the basis of the result $\lim_{x \to 0} x \ln x = 0$ we find that the limits of the last two terms of the above expression are zero.

The limit of $\ln(t_1 - t) - \ln(t_2 - t)$ has already been examined in § 3.4. It is equal to $+ i\pi$ if the cut is made on the right of L, and to $- i\pi$ in the other case. This completes the proof.

If L is a closed contour ($a = b$) the formula (3.14) is simplified†:

$$\int_L \frac{\varphi(\tau)}{\tau - t}\, d\tau = + i\pi\varphi(t) - \int_L \varphi'(\tau) \ln(\tau - t)\, d\tau. \qquad (3.15)$$

The present article is based on the paper of S. G. Mikhlin [16].

§ 4. Limiting values of the Cauchy type integral. Integrals over the real axis

4.1. The basic lemma

It has been indicated above that for an analytic function defined by an integral of the Cauchy type, the contour of integration is a singular line. We now proceed to investigate the principal problem— the behaviour of the Cauchy type integral on the integration contour. The fundamental result which will be deduced below is the fact that an integral of the Cauchy type with the density satisfying the Hölder condition, has the same properties as the potential of a double layer with continuous density, i.e. it has continuous limiting values on approaching the contour from both sides, but these limiting values are distinct, so that on passing through the contour a jump takes place.

THE BASIC LEMMA. *If the density $\varphi(\tau)$ satisfies the Hölder condition and the point t does not coincide with an end of the contour, then the function*

$$\psi(z) = \int_L \frac{\varphi(\tau) - \varphi(t)}{\tau - z}\, d\tau$$

on passing through the point $z = t$ of the contour behaves as a continuous function, i.e. this function has a definite limiting value on approaching the point t by z from any side of the contour, along any path:

$$\lim_{z \to t} \psi(z) = \int_L \frac{\varphi(\tau) - \varphi(t)}{\tau - t}\, d\tau = \psi(t).$$

To prove the assertion we shall estimate the difference

$$\psi(z) - \psi(t) = \int_L (z - t) \frac{\varphi(\tau) - \varphi(t)}{(\tau - z)(\tau - t)}\, d\tau.$$

† The second sign does not occur, since the cut can be made only on the right of L.

Let us split the integral into the two following terms: I_1 taken over the segment L_δ of the contour L, lying inside a circle of a sufficiently small radius δ with centre at the point t, and I_2 taken over the remaining part $L - L_\delta$ (Fig. 4). Let us estimate I_1. Assume first that z approaches t along a path which is not tangent to the contour. Then

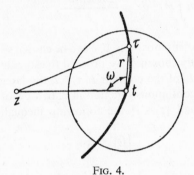

FIG. 4.

for a sufficiently small δ the non-obtuse angle ω at the point t has a lower bound $\omega_0 > 0$. Applying the sine theorem to the triangle $zt\tau$ we obtain

$$\frac{|z - t|}{|\tau - z|} = \frac{\sin \beta}{\sin \omega} \leqq \frac{1}{\sin \omega_0} = K, \tag{4.1}$$

where K is a positive number.

By virtue of the Hölder condition we have

$$\left| \frac{\varphi(\tau) - \varphi(t)}{\tau - t} \right| < A |\tau - t|^{\lambda - 1} = A r^{\lambda - 1}, \tag{4.2}$$

where $r = |\tau - t|$.

Furthermore, let us make use of the following property of smoothness of the contour L: for a smooth contour the ratio ds/dr where s is the length of an arc of the contour and r is the length of the corresponding chord, is a bounded quantity, i.e.

$$\left| \frac{ds}{dr} \right| \leqq m,$$

where m is a positive constant. Hence,

$$|d\tau| = |ds| \leqq m |dr|. \tag{4.3}$$

The inequality (4.3) will frequently be employed below.

Applying now the estimates (4.1) – (4.3) we obtain

$$|I_1| \leqq \int\limits_{L_\delta} \frac{|z - t|}{|\tau - z|} \left| \frac{\varphi(\tau) - \varphi(t)}{\tau - t} \right| |d\tau| < KAm \int\limits_{L_\delta} r^{\lambda-1} |dr|$$

$$= 2KAm \int\limits_0^\delta r^{\lambda-1} dr = \frac{2KAm\delta^\lambda}{\lambda}.$$

Selecting an arbitrary number $\varepsilon > 0$, δ can be chosen so that $|I_1| < \varepsilon/2$. Now, having already chosen δ, we proceed to estimate the integral I_2. On the segment $L–L_\delta$ of the contour L, $\tau \neq t$, the integral I_2 therefore is at the point t a continuous function in z. In view of the continuity, for sufficiently small $|z - t|$ the following inequality is valid:

$$|I_2| < \frac{\varepsilon}{2},$$

whence

$$|\psi(z) - \psi(t)| \leqq |I_1| + |I_2| < \varepsilon.$$

It still remains to eliminate the condition that z tends to t along a path non-tangent to the contour.

Let us first observe that the estimate of the difference $|\psi(z) - \psi(t)|$ is valid independently of t and, consequently, the passage of $\psi(z)$ to its limit occurs uniformly. It follows that the limiting value of $\psi(z)$ on the contour L, the function $\psi(t)$, is continuous. In fact, if t and t_1 are two points of the contour L, then

$$|\psi(t) - \psi(t_1)| \leqq |\psi(t) - \psi(z)| + |\psi(z) - \psi(t_1)|.$$

In view of the uniform passage of $\psi(z)$ to the limit, both terms in the right-hand side of the inequality can be made arbitrarily small, for sufficiently close points t, t_1 and z.

Assume now that z tends to t along a curve γ tangent to L. Let us take on the curve γ a point z sufficiently close to t and draw through it a curve γ_1, such that it intersects the contour L at a point t_1 sufficiently close to t, and it is not tangential to L. The line γ_1 can always be so chosen that the lengths of the chords zt and zt_1 are simultaneously arbitrarily small† (Fig. 5).

Applying first the statement proved above concerning the existence of the limit along a non-tangential path, and next the properties of con-

---

† The reader may find the proof of this not quite self-evident statement in the book of N. I. Muskhelishvili [17*], p. 41.

tinuity of the limiting values, we find that

$$|\psi(z) - \psi(t_1)| \quad \text{and} \quad |\psi(t) - \psi(t_1)|$$

are arbitrarily small, whence

$$|\psi(z) - \psi(t)| \leq |\psi(z) - \psi(t_1)| + |\psi(t) - \psi(t_1)|$$

is also arbitrarily small. This establishes the existence of the limit along any path.

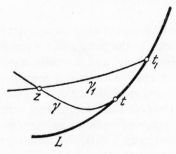

FIG. 5.

Let us note that the existence of the limit of the function $\psi(z)$ is a local property, i.e. its validity at a prescribed point t follows from the properties of the density $\varphi(\tau)$ in the vicinity of this point. In fact, in our proof we estimated directly the integral taken over an arbitrarily small arc of the contour containing the point t, and for this purpose the Hölder condition was used. It is not required, for the continuity of the function $\psi(z)$ at the point t, that $\varphi(\tau)$ should satisfy the Hölder condition on the remaining part of the contour as well; there it may simply be continuous and even possess discontinuities, only the condition of integrability being preserved.

4.2. The Sokhotski formulae

We are now in a position to examine the fundamental problem of the existence of the limiting values of an integral of the Cauchy type on the contour of integration, and to establish the connection between the limiting values and the singular integral.

Consider the function

$$\Phi(z) = \frac{1}{2\pi i} \int_L \frac{\varphi(\tau)}{\tau - z} \, d\tau, \tag{4.4}$$

where $\varphi(\tau)$ satisfies the Hölder condition.

2 a*

We assume also that the contour L is closed. In the case of an open contour we may supplement it by an arbitrary curve so that it becomes a closed one, setting on the additional curve $\varphi(\tau) = 0$.

To investigate the limiting values of $\Phi(z)$ at a point t of the contour, let us take the function examined in the preceding article

$$\psi(z) = \frac{1}{2\pi i} \int_L \frac{\varphi(\tau) - \varphi(t)}{\tau - z} \, d\tau. \tag{4.5}$$

We shall denote the limiting values of the analytic functions $\Phi(z), \psi(z)$ when the point z tends to the point t of the contour from the inside, by $\Phi^+(t), \psi^+(t)$, respectively, and from the outside by $\Phi^-(t), \psi^-(t)$, respectively. (For an open contour this corresponds to the limiting values from the left and from the right.) To emphasize the direction of passing to the limit we shall accordingly write $z \to t^+$ or $z \to t^-$. The values of the corresponding functions at the point t of the contour will simply be denoted by $\Phi(t), \psi(t)$; $\Phi(t)$ denoting the singular integral in the sense of the principal value.

$$\frac{1}{2\pi i} \int_L \frac{\varphi(\tau)}{\tau - t} \, d\tau$$

Starting from the relations

$$\int_L \frac{d\tau}{\tau - z} = \begin{cases} 2\pi i, & z \in D^+, \\ 0, & z \in D^-, \\ \pi i, & z \in L, \end{cases} \tag{4.6}$$

we obtain

$$\psi^+(t) = \lim_{z \to t^+} \left[\frac{1}{2\pi i} \int_L \frac{\varphi(\tau)}{\tau - z} \, d\tau - \frac{\varphi(t)}{2\pi i} \int_L \frac{d\tau}{\tau - z} \right] = \Phi^+(t) - \varphi(t),$$

$$\psi^-(t) = \lim_{z \to t^-} \left[\frac{1}{2\pi i} \int_L \frac{\varphi(\tau)}{\tau - z} \, d\tau - \frac{\varphi(t)}{2\pi i} \int_L \frac{d\tau}{\tau - z} \right] = \Phi^-(t),$$

$$\psi(t) = \frac{1}{2\pi i} \int_L \frac{\varphi(\tau)}{\tau - t} \, d\tau - \frac{\varphi(t)}{2\pi i} \int_L \frac{d\tau}{\tau - t} = \Phi(t) - \frac{1}{2} \varphi(t).$$

Since, according to the basic lemma, the function $\psi(t)$ is continuous, the right-hand sides of the above relations are identical, i.e.

$$\Phi^+(t) - \varphi(t) = \Phi^-(t) = \Phi(t) - \tfrac{1}{2}\varphi(t). \tag{4.7}$$

Thus, we obtain finally

$$\Phi^+(t) = \frac{1}{2}\,\varphi(t) + \frac{1}{2\pi i}\int\limits_L \frac{\varphi(\tau)}{\tau - t}\,d\tau,$$

$$\Phi^-(t) = -\frac{1}{2}\,\varphi(t) + \frac{1}{2\pi i}\int\limits_L \frac{\varphi(\tau)}{\tau - t}\,d\tau, \qquad (4.8)$$

the singular integral

$$\int\limits_L \frac{\varphi(\tau)}{\tau - t}\,d\tau$$

being understood in the sense of the principal value.

The formulae (4.8), derived first in 1873 by the Russian mathematician Yu. V. Sokhotski [25] will be called the *Sokhotski formulae*. They are fundamental for all following considerations.

Let us now state the deduced result.

THEOREM. *Let L be a smooth contour (closed or open) and $\varphi(\tau)$ a function of position on the contour, which satisfies the Hölder condition. Then the Cauchy type integral*

$$\Phi(z) = \frac{1}{2\pi i}\int\limits_L \frac{\varphi(\tau)}{\tau - z}\,d\tau$$

has the limiting values $\Phi^+(t)$, $\Phi^-(t)$ at all points of the contour L not coinciding with its ends, on approaching the contour from the left or from the right along an arbitrary path, and these limiting values are expressed by the density of the integral $\varphi(t)$ and the singular integral $\Phi(t)$ in accordance with the Sokhotski formulae (4.8).

Subtracting and adding the formulae (4.8) we obtain the following two equivalent formulae:

$$\Phi^+(t) - \Phi^-(t) = \varphi(t), \qquad (4.9)$$

$$\Phi^+(t) + \Phi^-(t) = \frac{1}{\pi i}\int\limits_L \frac{\varphi(\tau)}{\tau - t}\,d\tau, \qquad (4.10)$$

which will be frequently employed hereafter.

4.3. The conditions ensuring that an arbitrary complex function is the boundary value of a function analytic in the domain

Suppose that on a smooth closed contour L a continuous complex function of position $\varphi(t)$ is prescribed, and let $t = t(s) = t_1(s) + it_2(s)$

be the equation of the contour in the complex form, $t(s)$ being a function of the arc s measured from an arbitrary point of the contour.

Substituting into the expression for the function $\varphi(t)$ the complex coordinate, and separating the real and imaginary parts we have

$$\varphi(t) = \varphi[t(s)] = \varphi_1(s) + i\varphi_2(s).$$

It is easy to prove that in general the following question has a negative answer: does there exist a function analytic in the domain $D^+(D^-)$, such that the prescribed complex function $\varphi(t)$ is its limiting value on the contour? In fact, having the values of the real part $\varphi_1(s)$ we can construct a function $u(x, y)$ which is harmonic in the domain D^+ (or D^-) and the limiting values of which on the contour are identical with the prescribed function $\varphi_1(s)$ (the Dirichlet problem). The function $u(x, y)$ being found, we can determine the conjugate harmonic function $v(x, y)$ to within an arbitrary constant term. We thus obtain the analytic function $f(z) = u(x, y) + iv(x, y)$ for which $\varphi_1(s)$ is the boundary value of its real part. Calculating now the limiting value of the imaginary part $v(x, y)$ we find that, in general, it does not coincide with the given function $\varphi_2(s)$ and, consequently, the prescribed complex function $\varphi(t)$ may not be the limiting value of a function analytic in D^+ (or D^-). The above reasoning indicates that only one part of a complex function may be prescribed in an arbitrary way—either the real or the imaginary part; then the second part is determined to within a constant.

Thus, if a given complex function constitutes the limiting value of an analytic function, it must satisfy certain relationships. We now proceed to their derivation.

In what follows we assume that $\varphi(t)$ satisfies the Hölder condition. Consider the Cauchy type integral with the density $\varphi(t)$:

$$\Phi(z) = \frac{1}{2\pi i} \int_L \frac{\varphi(\tau)}{\tau - z} \, d\tau.$$

If $z \in D^+$ and $\varphi(t)$ is the boundary value of a function analytic in D^+, then

$$\Phi^+(z) = \varphi(z);$$

If $z \in D^-$ and $\varphi(t)$ is the boundary value of a function analytic in D^-, then according to the Cauchy formula for the infinite domain

$$\Phi^-(z) = -\varphi(z) + \varphi(\infty).$$

Taking now the limiting values on the contour L and making use of the Sokhotski formulae (4.8) we obtain

$$\varphi(t) = \frac{1}{2}\varphi(t) + \frac{1}{2\pi i}\int_L \frac{\varphi(\tau)}{\tau - t}\,d\tau$$

if $\varphi(z)$ is analytic in D^+, and

$$-\varphi(t) + \varphi(\infty) = -\frac{1}{2}\varphi(t) + \frac{1}{2\pi i}\int_L \frac{\varphi(\tau)}{\tau - t}\,d\tau$$

if it is analytic in D^-. Hence

$$-\frac{1}{2}\varphi(t) + \frac{1}{2\pi i}\int_L \frac{\varphi(\tau)}{\tau - t}\,d\tau = 0; \tag{4.11}$$

$$\frac{1}{2}\varphi(t) + \frac{1}{2\pi i}\int_L \frac{\varphi(\tau)}{\tau - t}\,d\tau - \Gamma = 0 \quad (\Gamma = \varphi(\infty)). \tag{4.12}$$

In view of the above reasoning, we see that the conditions (4.11), (4.12) are necessary for a function $\varphi(t)$ to be the boundary value of a function analytic in the domains D^+, D^-, respectively. It is easy to prove that these conditions are also sufficient.

In fact, suppose that $\varphi(t)$ satisfies for instance condition (4.11). It means, that for the Cauchy type integral,

$$\Phi(z) = \frac{1}{2\pi i}\int_L \frac{\varphi(\tau)}{\tau - z}\,d\tau,$$

in accordance with the Sokhotski formulae (4.8), $\Phi^-(t) = 0$. Consequently, on the basis of the relation (4.9)

$$\varphi(t) = \Phi^+(t) - \Phi^-(t) = \Phi^+(t).$$

Let us formulate the result deduced above.

THEOREM. *Suppose that on a smooth closed contour L a complex function $\varphi(t)$ is given, and it satisfies the Hölder condition. In order that this function be the boundary value of a function analytic in the interior domain D^+, it is necessary and sufficient that the conditions (4.11) are satisfied. In order that $\varphi(t)$ be the boundary value of a function analytic in the exterior domain D^- and takes at infinity the given value Γ, it is necessary and sufficient that the condition (4.12) is satisfied.*

A more general problem may be formulated, namely: to find the condition that a complex function given on the contour is the boundary value of a function analytic in the domain, except at some isolated points where it has prescribed singularities. For subsequent considerations it is of interest to examine the case when $\varphi(t)$ is the boundary value of a function analytic in the domain D^-, except at infinity where it has a pole of order n, the principal part being known:

$$\gamma(z) = a_0 z^n + a_1 z^{n-1} + \cdots + a_{n-1} z + a_n.$$

An argument similar to the preceding one readily leads to the conclusion that the condition has exactly the same form as (4.12), the only difference being that the constant Γ is replaced by the expression $\gamma(t)$:

$$\frac{1}{2}\,\varphi(t) + \frac{1}{2\pi i} \int\limits_L \frac{\varphi(\tau)}{\tau - t}\,d\tau - \gamma(t) = 0. \tag{4.13}$$

Some authors call the relations (4.11), (4.12) the condition of analytic continuability of a complex function prescribed on the contour, into the domain. The term analytic continuation, however, is frequently applied not only to functions analytic at all points of the domain, but also to functions which have singularities at some points. For such functions the conditions (4.11), (4.12) should be replaced by certain more complicated ones (see Problem 12 at the end of this chapter).

The formulae (4.11), (4.12) can be used for computing singular integrals in the case when the density of the integral is the boundary value of an analytic function (see Problem 10).

4.4. Limiting values of the derivatives. Derivatives of limiting values. Derivatives of a singular integral

1. Let $\varphi(t)$ be a function of position on a closed contour L, the mth derivative of which satisfies the Hölder condition.

We shall prove that the mth derivative of the function $\Phi(z)$ defined by the Cauchy type integral

$$\Phi(z) = \frac{1}{2\pi i} \int\limits_L \frac{\varphi(\tau)}{\tau - z}\,d\tau,$$

possesses limiting values on the contour and that these limiting values satisfy relations analogous to the Sokhotski formulae (4.8)

$$\Phi^{(m)+}(t) = \frac{1}{2}\,\varphi^{(m)}(t) + \frac{1}{2\pi i}\int_L \frac{\varphi^{(m)}(\tau)}{\tau - t}\,d\tau, \qquad (4.14)$$

$$\Phi^{(m)-}(t) = -\frac{1}{2}\,\varphi^{(m)}(t) + \frac{1}{2\pi i}\int_L \frac{\varphi^{(m)}(\tau)}{\tau - t}\,d\tau. \qquad (4.15)$$

It follows from the relation (1.3) that the mth derivative of a Cauchy type integral has the form

$$\Phi^{(m)}(z) = \frac{m!}{2\pi i}\int_L \frac{\varphi(\tau)}{(\tau - z)^{m+1}}\,d\tau.$$

Let us integrate the right-hand side m times by parts. Since the contour is closed, the integrated part vanishes every time. We have therefore

$$\Phi^{(m)}(z) = \frac{1}{2\pi i}\int_L \frac{\varphi^{(m)}(\tau)}{\tau - z}\,d\tau.$$

Applying to the derived Cauchy type integral the Sokhotski formulae (4.8) we arrive at the relations (4.14), (4.15).

2. We shall now show that in formulae (4.14) and (4.15) the operations of differentiation and passing to the limit on the contour are interchangeable, i.e. that the limit values on a contour of the derivatives of a Cauchy integral coincide with the derivatives of its limit values. We commence with the first derivative. It is required to prove that

$$[\Phi'(t)]^{\pm} = [\Phi^{\pm}(t)]'.$$

We put

$$\Psi(z) = \frac{1}{2\pi i}\int_L \frac{\varphi'(\tau)}{\tau - z}\,d\tau.$$

It has been proved that

$$[\Phi'(t)]^{\pm} = \Psi^{\pm}(t);$$

we now also prove that

$$[\Phi^{\pm}(t)]' = \Psi^{\pm}(t).$$

We shall consider $[\Phi'(t)]^+$. Since $[\Phi^+(z)]'$ is continuous right up to the contour and satisfies the H condition on the contour, $[\Phi^+(t)]'$

also exists and satisfies the H condition. We have:

$$[\Phi^+(t)]' = \lim_{\Delta t \to 0} \frac{\Phi^+(t + \Delta t) - \Phi^+(t)}{\Delta t}.$$

But by virtue of Cauchy's theorem

$$\Phi^+(t + \Delta t) - \Phi^+(t) = \int_t^{t+\Delta t} [\Phi^+(t)]' \, dt = \int_C [\Phi^+(z)]' \, dz,$$

where C is a contour lying entirely within D^+. Furthermore,

$$\int_C [\Phi^+(z)]' \, dz = \int_C \Psi^+(z) \, dz = \Psi^+(t)\Delta t + \int_C [\Psi^+(z) - \Psi^+(t)] \, dz.$$

We select a contour C such that its length is $2|\Delta t|$ with the following condition fulfilled for a specified $\varepsilon > 0$

$$|\Psi^+(z) - \Psi^+(t)| < \frac{\varepsilon}{2}$$

This is obviously possible owing to the continuity of $\Psi^+(z)$ right up to the contour. We therefore have

$$\left| \frac{\Phi^+(t + \Delta t) - \Phi^+(t)}{\Delta t} - \Psi^+(t) \right| < \varepsilon,$$

and hence

$$[\Phi^+(t)]' = \Psi^+(t),$$

as required. Likewise for $\Phi^-(z)$. Repeating the reasoning, we obtain

$$[\Phi^{\pm}(t)]^{(m)} = [\Phi^{(m)}(t)]^{\pm}, \tag{4.16}$$

provided that the density $\varphi^{(m)}(t) \in H$.

3. The rule for the differentiation of a singular integral can now be deduced

$$\Phi(t) = \frac{1}{2\pi i} \int_L \frac{\varphi(\tau)}{\tau - t} \, d\tau.$$

We write Sokhotski's formulae for $\Phi^+(z)$ and $[\Phi^{(m)}(z)]^+$:

$$\Phi^+(t) = \tfrac{1}{2}\varphi(t) + \Phi(t),$$

$$[\Phi^{(m)}(t)]^+ = \frac{1}{2} \varphi(t)^{(m)} + \frac{1}{2\pi i} \int_L \frac{\varphi^{(m)}(\tau)}{\tau - t} \, d\tau.$$

Since $\varphi(t)$, by assumption, and $\Phi^+(t)$, as proved above, have up to mth order derivatives, it follows that $\Phi^+(t)$, also has derivatives of the same order. Differentiating the first equality m times and comparing it with the second, we obtain, having regard to formula (4.16):

$$\Phi^{(m)}(t) = \left[\frac{1}{2\pi i}\int_L \frac{\varphi(\tau)}{\tau - t} d\tau\right]^{(m)} = \frac{1}{2\pi i}\int_L \frac{\varphi^{(m)}(\tau)}{\tau - t} d\tau. \qquad (4.17)$$

The results given under items 2 and 3 are due to S. Ya. Al'per and Yu. I. Cherskii, being published here for the first time. Note that (4.17) could also have been obtained from formula (3.15) by successive differentiation and integration by parts.

4.5. The Sokhotski formulae for corner points of a contour

In investigating problems of existence of the a singular integral and limiting values of a Cauchy type integral, we everywhere assumed the condition that the integration contour is a smooth line. It is easily seen that this condition is not necessary.

Observing carefully the reasoning of § 3.4 and § 4.1 we find that the property of smoothness of the contour was employed twice, the first time in § 3.4 in computing the value of $\lim \ln (t_2 - t)/(t_1 - t)$, and again in the formula (4.3) in replacing the quantity $|d\tau| = ds$ by $m\, dr$; in both cases the smoothness in the immediate vicinity of the investigated point was relevant. In the investigations of § 4.1 the property of the contour which was in fact employed can be stated as follows: the ratio of the small arc of the contour to the corresponding chord is a bounded quantity. This property, however, is valid not only for contours with a continuously varying tangent, but also for contours having sharp corners. Thus, all considerations of § 4.1 are also valid in this case, i.e. when the point under investigation is a corner point of the contour.

In § 3.4 the property of smoothness was essential; however it is easy to see what alterations should be made in the reasoning, so that it is valid for a corner point of the contour. Let α be the angle between the two tangents to the contour L at the point t, measured on the left of the contour (Fig. 6). Then, obviously, we have

$$\lim_{t_1,\, t_2 \to t} \ln \frac{t_2 - t}{t_1 - t} = -i\alpha$$

whence

$$\int_L \frac{d\tau}{\tau - t} = i\alpha. \tag{4.16}$$

The relations (4.7) take the form

$$\Phi^+(t) - \varphi(t) = \Phi^-(t) = \Phi(t) - \frac{\alpha}{2\pi} \varphi(t), \tag{4.17}$$

and, consequently,

$$\Phi^+(t) = \left(1 - \frac{\alpha}{2\pi}\right) \varphi(t) + \frac{1}{2\pi i} \int_L \frac{\varphi(\tau)}{\tau - t} d\tau, \tag{4.18}$$

$$\Phi^-(t) = -\frac{\alpha}{2\pi} \varphi(t) + \frac{1}{2\pi i} \int_L \frac{\varphi(\tau)}{\tau - t} d\tau. \tag{4.19}$$

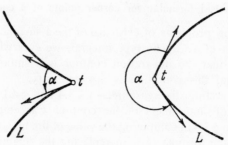

FIG. 6.

This result may be formulated as follows.

THEOREM. *If an integral of the Cauchy type is taken over a contour having a finite number of corner points, the limiting values of the integral exist, and for the non-corner points the ordinary Sokhotski formulae (4.8) remain valid, whereas for the corner points the formulae (4.18), (4.19) take their place.*

The proof given above cannot be directly extended to the case of cusps. Nevertheless, with the help of additional considerations it can be proved that formulae (4.18), (4.19) remain valid in this case as well. It is only necessary to set $\alpha = 0$ or $\alpha = 2\pi$, depending on whether the point of the cusp is directed to the right or to the left of the contour.

The proof of this assertion can be found in the book of N. I. Muskhelishvili [17*], pp. 562–7.

4.6. Integrals of the Cauchy type over the real axis

Let $\varphi(\tau)$ be a complex function of the real variable τ, which obeys the Hölder condition for all finite τ and tends to a definite limit $\varphi(\infty)$ as $\tau \to \pm \infty$. We shall require that for large τ the following inequality holds:

$$|\varphi(\tau) - \varphi(\infty)| < \frac{A}{|\tau|^\mu}, \quad \mu > 0, \quad A > 0. \tag{4.20}$$

Consider the Cauchy type integral

$$\Phi(z) = \frac{1}{2\pi i} \int_{-\infty}^{\infty} \frac{\varphi(\tau)}{\tau - z} d\tau, \tag{4.21}$$

assuming that z does not lie on the real axis.

If $\varphi(\infty) \neq 0$ the improper integral (4.21) is divergent, i.e. the expression

$$\int_{N'}^{N''} \frac{\varphi(\tau)}{\tau - z} d\tau$$

does not tend to a limit when N' and N'' tend to $-\infty$ and $+\infty$, respectively, independently of each other. In fact,

$$\int_{N'}^{N''} \frac{\varphi(\tau) \, d\tau}{\tau - z} = \int_{N'}^{N''} \frac{\varphi(\tau) - \varphi(\infty)}{\tau - z} d\tau + \varphi(\infty) \int_{N'}^{N''} \frac{d\tau}{\tau - z}. \tag{4.22}$$

In the first integral of the right-hand side, in view of (4.20) the integrand for large $|\tau|$ is of order $|\tau|^{-1-\mu}$, whence the corresponding integral with infinite limits is convergent, in accordance with the well-known criterion for convergence of improper integrals.

The second integral can easily be computed, namely

$$\int_{N'}^{N''} \frac{d\tau}{\tau - z} = \ln(N'' - z) - \ln(N' - z) = \ln \frac{|N'' - z|}{|N' - z|} \pm i\alpha.$$

Here α is the angle between the straight lines connecting the point z with the points N' and N'', and the sign of the second term is positive when z lies in the upper semi-plane and negative when it lies in the lower semi-plane (the reader is advised to draw a figure).

If N' and N'' tend (independently of each other) to $-\infty$ and $+\infty$, respectively, then α tends to π but $\ln |N'' - z|/|N' - z|$ does not tend to any limit, whence it follows that also the left-hand side of the relation (4.22) does not tend to a limit.

We now assume that the limits of integration N' and N'' tend to infinity symmetrically, i.e. $-N' = N'' = N$. Then

$$\lim_{N \to \infty} \ln \frac{|N - z|}{|-N - z|} = \ln 1 = 0,$$

Thus we obtain

$$\lim_{N \to \infty} \int_{-N}^{N} \frac{\varphi(\tau)\, d\tau}{\tau - z} = \int_{-\infty}^{\infty} \frac{\varphi(\tau) - \varphi(\infty)}{\tau - z}\, d\tau \pm \pi i \varphi(\infty).$$

the sign being chosen in the manner indicated above.

The expression appearing in the left-hand side is called the principal value of the integral (4.21) with infinite limits. In what follows integrals with infinite limits are always understood in the sense of their principal values. If the corresponding integral exists as improper, it is evident that the principal value is identical with its value as improper integral.

Thus, we have established the existence of the principal value of the integral (4.21), and

$$\frac{1}{2\pi i} \int_{-\infty}^{\infty} \frac{\varphi(\tau)\, d\tau}{\tau - z} = \frac{1}{2\pi i} \int_{-\infty}^{\infty} \frac{\varphi(\tau) - \varphi(\infty)}{\tau - z}\, d\tau \pm \frac{1}{2} \varphi(\infty), \qquad (4.23)$$

the integral on the right-hand side having ordinary sense. If $\varphi(\infty) = 0$ the integral (4.21) is convergent.

Since the integral

$$\int_{-\infty}^{\infty} \frac{\varphi(\tau)\, d\tau}{(\tau - z)^2}$$

is absolutely convergent for any z not lying on the real axis, the integral (4.21) may be differentiated with respect to the parameter z, and, consequently, $\Phi(z)$ is a function analytic in the upper and lower semi-planes. These functions will be denoted $\Phi^+(z)$ and $\Phi^-(z)$, respectively.

Suppose now that $\text{Im}\, z = 0$, i.e. the point $z = t$ is situated on the integration line. Then by the integral

$$\int_{-\infty}^{\infty} \frac{\varphi(\tau)}{\tau - z}\, d\tau \qquad (4.24)$$

we understand the principal value defined in the following way:

$$\int_{-\infty}^{\infty} \frac{\varphi(\tau)}{\tau - z}\, d\tau = \lim_{\substack{N \to \infty \\ \varepsilon \to 0}} \left\{ \int_{-N}^{t-\varepsilon} \frac{\varphi(\tau)}{\tau - t}\, d\tau + \int_{t+\varepsilon}^{N} \frac{\varphi(\tau)}{\tau - t}\, d\tau \right\}.$$

In particular

$$\int_{-\infty}^{\infty} \frac{d\tau}{\tau - t} = \lim_{\substack{N \to \infty \\ \varepsilon \to 0}} \left\{ \int_{-N}^{t-\varepsilon} \frac{d\tau}{\tau - t} + \int_{t+\varepsilon}^{N} \frac{d\tau}{\tau - t} \right\}$$

$$= \lim_{\substack{N \to \infty \\ \varepsilon \to 0}} \ln \frac{-\varepsilon(N - t)}{(-N - t)\,\varepsilon} = 0.$$

Making use of this relation we may write

$$\int_{-\infty}^{\infty} \frac{\varphi(\tau)\, d\tau}{\tau - t} = \lim_{\substack{N \to \infty \\ \varepsilon \to 0}} \left\{ \int_{-N}^{t-\varepsilon} \frac{\varphi(\tau) - \varphi(\infty)}{\tau - t}\, d\tau + \int_{t+\varepsilon}^{N} \frac{\varphi(\tau) - \varphi(\infty)}{\tau - t}\, d\tau \right\}$$

$$= \lim_{\varepsilon \to 0} \left\{ \int_{-\infty}^{t-\varepsilon} \frac{\varphi(\tau) - \varphi(\infty)}{\tau - t}\, d\tau + \int_{t+\varepsilon}^{\infty} \frac{\varphi(\tau) - \varphi(\infty)}{\tau - t}\, d\tau \right\},$$

the last two integrals converging in the ordinary sense. To prove the existence of the limit as $\varepsilon \to 0$, the reasoning of § 3.4 may be repeated. The fact that in our case the length of the integration contour is infinite has no essential importance, since we can return to the case of a finite contour by carrying out the transformation

$$\int_{-\infty}^{t-\varepsilon} + \int_{t+\varepsilon}^{\infty} = \left(\int_{-\infty}^{t-A} + \int_{t+A}^{\infty} \right) + \int_{t-A}^{t-\varepsilon} + \int_{t+\varepsilon}^{t+A}.$$

The integrals in parenthesis are independent of ε, and the limit of the sum of the last two integrals, as it was established in § 3.4, exists. Thus, if $\varphi(\tau)$ satisfies the conditions formulated above, the principal value of the integral (4.24) exists.

Repeating the reasoning of § 4.2 we can prove the validity of the Yu. V. Sokhotski formulae for the infinite contour:

$$
\left.
\begin{aligned}
\Phi^+(t) &= \frac{1}{2}\,\varphi(t) + \frac{1}{2\pi i}\int\limits_{-\infty}^{\infty}\frac{\varphi(\tau)}{\tau - t}\,d\tau; \\[2em]
\Phi^-(t) &= -\frac{1}{2}\,\varphi(t) + \frac{1}{2\pi i}\int\limits_{-\infty}^{\infty}\frac{\varphi(t)}{\tau - t}\,d\tau.
\end{aligned}
\right\}
\tag{4.25}
$$

Here $\Phi^+(t)$ and $\Phi^-(t)$ are the limits of $\Phi(z)$ as z tends to t from the upper and lower semi-plane, respectively.

Let us also investigate the behaviour of the function $\Phi(z)$ in the vicinity of infinity, which is now a boundary point. To this end let us make the change of variable

$$
z = -\frac{1}{\zeta}.
$$

When the point $z = \tau$ traverses the real axis in the positive direction, the corresponding point $\sigma = -1/\tau$ also traverses the real axis in the positive direction.

Performing in the integral (4.21) the change of variables and introducing the notation

$$
\Phi(z) = \Phi\left(-\frac{1}{\zeta}\right) = \Phi^*(\zeta), \quad \varphi(\tau) = \varphi\left(-\frac{1}{\sigma}\right) - \varphi^*(\sigma),
$$

we obtain

$$
\Phi^*(\zeta) = \frac{\zeta}{2\pi i}\int\limits_{-\infty}^{\infty}\frac{\varphi^*(\sigma)\,d\sigma}{\sigma(\sigma - \zeta)} = \frac{1}{2\pi i}\int\limits_{-\infty}^{\infty}\frac{\varphi^*(\pi)\,d\sigma}{\sigma - \zeta} - \frac{1}{2\pi i}\int\limits_{-\infty}^{\infty}\frac{\varphi^*(\sigma)\,d\sigma}{\sigma}.
\tag{4.26}
$$

All the integrals are understood in the sense of the principal values. The second integral of the right-hand side of the last formula is a constant quantity; therefore, the investigation of the function $\Phi(z)$ in the vicinity of the point $z = \infty$ is reduced to the investigation of the integral

$$\frac{1}{2\pi i} \int_{-\infty}^{\infty} \frac{\varphi^*(\sigma)\,d\sigma}{\sigma - \zeta}$$

in the vicinity of the point $\zeta = 0$, i.e. to the problem with which we are already familiar. Assume that the function $\varphi^*(\sigma)$ satisfies the Hölder condition in the vicinity of the point $\sigma = 0$, i.e.

$$|\varphi^*(\sigma_2) - \varphi^*(\sigma_1)| \leqq B|\sigma_2 - \sigma_1|^\mu \quad (0 < \mu \leqq 1).$$

(this condition does not follow from (4.20)). Thus, we have for the function $\varphi(\tau)$

$$|\varphi(\tau_2) - \varphi(\tau_1)| \leqq B \left| \frac{1}{\tau_2} - \frac{1}{\tau_1} \right|^\mu \quad (0 < \mu \leqq 1). \tag{4.27}$$

It is readily observed that the last inequality implies the condition (4.20).

If z tends to infinity along an arbitrary path remaining in the upper or lower semi-plane, then $\zeta \to 0$ also remains in the upper or lower semi-plane. Consequently, applying to the first integral of the right-hand side of (4.26) the first formula (4.25) we obtain

$$\Phi^+(\infty) = \Phi^{*+}(0)$$

$$= \frac{1}{2}\varphi^*(0) + \frac{1}{2\pi i} \int_{-\infty}^{\infty} \frac{\varphi^*(\sigma)\,d\sigma}{\sigma} - \frac{1}{2\pi i} \int_{-\infty}^{\infty} \frac{\varphi^*(\sigma)\,d\sigma}{\sigma} = \frac{1}{2}\varphi(\infty).$$

Reasoning in the same way for Φ^- we are led to the two formulae

$$\Phi^+(\infty) = \tfrac{1}{2}\varphi(\infty), \quad \Phi^-(\infty) = -\tfrac{1}{2}\varphi(\infty). \tag{4.28}$$

A similar investigation can also be performed for the Cauchy type integral over an arbitrary smooth curve going to infinity.

Let us observe that the change of variables

$$\zeta = \frac{z - i}{z + i}, \quad \sigma = \frac{\tau - i}{\tau + i}$$

reduces the investigation of a Cauchy type integral over the real axis to that of an integral over a unit circle in the ζ plane.

§ 5. Properties of the limiting values of the Cauchy type integral

5.1. The limiting values satisfy the Hölder condition

It was proved in § 4.1 that the limiting values of a Cauchy type integral—the functions $\Phi^+(t)$ and $\Phi^-(t)$ are continuous on the contour L. These functions, however, possess a more profound property than continuity, namely they obey the Hölder condition.

This property is established by the following theorem.

THEOREM. *If L is a smooth closed contour and $\varphi(t)$ satisfies on L the Hölder condition with an index λ, then the limiting values of the Cauchy type integral, $\Phi^+(t)$ and $\Phi^-(t)$, also satisfy this condition, the index being the same when $\lambda < 1$, and arbitrarily close to λ when $\lambda = 1$.*

It follows from the relation (3.8) that it is sufficient to prove the theorem stated above for the function

$$\psi(t) = \frac{1}{2\pi i} \int_L \frac{\varphi(\tau) - \varphi(t)}{\tau - t} \, d\tau.$$

To this end let us estimate the expression

$$|\psi(t_2) - \psi(t_1)| = \left| \frac{1}{2\pi i} \int_L \left\{ \frac{\varphi(\tau) - \varphi(t_2)}{\tau - t_2} - \frac{\varphi(\tau) - \varphi(t_1)}{\tau - t_1} \right\} d\tau \right| \quad (5.1)$$

for two arbitrary, sufficiently close points t_1 and t_2.

From point t_1 we describe a circle of radius δ such that it intersects L at two points a and b. The part of the contour L lying within this circle is denoted by l. Let t_2 be an arbitrary fixed point on the arc l other than points a or b. We set $\delta = k|t_2 - t_1|$; it is obvious that $k > 1$.

Denote by $s = s(t, \tau)$ the length of the smaller of the two arcs of the contour L, with the ends t and τ.

It follows from the property of smoothness of the contour L, (4.3), that for any two points t_1 and t_2 we may write

$$s(t_1, t_2) \leqq m|t_1 - t_1|, \quad (5.2)$$

where m is a positive constant.

Let us cut off an arc l of the contour L, taking on both sides of the point t_1 equal arcs of lengths $2s(t_1, t_2)$. The ends of this arc will be denoted by a and b. The integral we are interested in, (5.1), can be represented as follows:

$$\psi(t_2) - \psi(t_1) = \frac{1}{2\pi i} \int_l \frac{\varphi(\tau) - \varphi(t_2)}{\tau - t_2} d\tau - \frac{1}{2\pi i} \int_l \frac{\varphi(\tau) - \varphi(t_1)}{\tau - t_1} d\tau +$$

$$+ \frac{1}{2\pi i} \int_{L-l} \left\{ \frac{\varphi(\tau) - \varphi(t_2)}{\tau - t_2} - \frac{\varphi(\tau) - \varphi(t_1)}{\tau - t_1} \right\} d\tau$$

$$= \frac{1}{2\pi i} \int_l \frac{\varphi(\tau) - \varphi(t_2)}{\tau - t_2} d\tau - \frac{1}{2\pi i} \int_l \frac{\varphi(\tau) - \varphi(t_2)}{\tau - t_1} d\tau +$$

$$+ \frac{1}{2\pi i} \int_{L-l} \frac{\varphi(t_1) - \varphi(t_2)}{\tau - t_1} d\tau +$$

$$+ \frac{1}{2\pi i} \int_{L-l} \frac{[\varphi(\tau) - \varphi(t_2)](t_2 - t_1)}{(\tau - t_1)(\tau - t_2)} d\tau = I_1 + I_2 + I_3 + I_4.$$

In accordance with § 4.1, an estimate for the integral I_2 is

$$|I_2| \leq \frac{1}{2\pi} \int_l \left| \frac{\varphi(\tau) - \varphi(t_1)}{\tau - t_1} \right| |d\tau| \leq C' \int_l \frac{|d\tau|}{|\tau - t_1|^{1-\lambda}}$$

$$\leq C'' \int_0^{C|t_2-t_1|} \frac{ds}{s^{1-\lambda}} \leq C''' s^\lambda(t_1, t_2) \leq A_1 |t_2 - t_1|^\lambda.$$

Here, and throughout, all numerical coefficients are positive constants. In an entirely similar way

$$|I_1| \leq A_2 |t_2 - t_1|^\lambda.$$

For the integral I_3 we have

$$|I_3| \leq \frac{|\varphi(t_1) - \varphi(t_2)|}{2\pi} \left| \int_{L-l} \frac{d\tau}{\tau - t_1} \right| \leq \frac{A |t_2 - t_1|^\lambda}{2\pi} \left| \int_{L-l} \frac{d\tau}{\tau - t_1} \right|.$$

The last integral can be directly computed, namely

$$\int_{L-l} \frac{d\tau}{\tau - t_1} = \ln \frac{a - t_1}{b - t_1}.$$

Consequently, it is bounded for any t_1 on L. Therefore, we have the estimate
$$|I_3| \leqq A_3 |t_2 - t_1|^\lambda.$$

Let us now proceed to estimate the most complicated integral I_4. Again, making use of the Hölder condition and the inequality (4.3) we obtain

$$|I_4| \leqq A \frac{|t_2 - t_1|}{2\pi} \int\limits_{L-l} \frac{ds}{|\tau - t_1| |\tau - t_2|^{1-\lambda}}$$

$$\leqq A' |t_2 - t_1| \int\limits_{L-l} |\tau - t_1|^{\lambda-2} \left| \frac{\tau - t_1}{\tau - t_2} \right|^{1-\lambda} |d\tau|.$$

But
$$|\tau - t_1| - |t_1 - t_2| \leqq |\tau - t_2| \quad \text{and} \quad |\tau - t_2| \geqq \delta = k |t_2 - t_1|,$$
so that
$$|\tau - t_1| \leqq \frac{k-1}{k} |\tau - t_2|.$$

It follows that
$$|I_4| \leqq A'' \left(\frac{k-1}{k} \right)^{1-\lambda} |t_1 - t_2| \int\limits_R^\delta r^{\lambda-2} \, dr,$$

where
$$R = \max_{\tau \in L-l} |\tau - t_1|.$$

If $\lambda < 1$, computing the latter integral we find
$$|I_4| \leqq A_4 |t_2 - t_1|^\lambda.$$

If $\lambda = 1$, in an analogous way we obtain the estimate
$$|I_4| \leqq A'_4 |t_2 - t_1| \left| \ln |t_2 - t_1| \right|.$$

Bearing in mind that as $x \to 0$, $\ln x$ increases slower than any negative power $|x|^{-\varepsilon}$ $(\varepsilon > 0)$ we obtain
$$|I_4| \leqq A'_4 |t_2 - t_1|^{1-\varepsilon}.$$

Collecting the derived estimates for I_1, I_2, I_3 and I_4 and noting that when $\lambda = 1$ the index λ in the estimates of I_1, I_2, I_3 may be replaced by $1 - \varepsilon$, we verify the validity of the theorem.

The proof has been given for the case of a closed contour. However, the remark in § 4.2, concerning the transition from an open contour to a closed one, implies that the theorem holds also for an open contour, except on its ends.

It follows directly from the proof of the theorem on the property of the limiting values of a Cauchy type integral, that the following property of the integral in the sense of the Cauchy principal value is true:

If $\varphi(t)$ satisfies the Hölder condition with an index λ on a smooth closed contour L, then

$$\Phi(t) = \frac{1}{2\pi i} \int_L \frac{\varphi(\tau)\,d\tau}{\tau - t}$$

also satisfies this condition, the index being the same when $\lambda < 1$, and the index being $1 - \varepsilon$, where ε is an arbitrarily small number, when $\lambda = 1$.

This statement, as well as the preceding theorem are easily generalized (see N. I. Muskhekishvili [17*], p. 57) to the case when the density of the integral $\varphi(\tau, \zeta)$ depends on a parameter ζ, satisfying the Hölder condition with respect to this parameter as well. In particular, the case $\zeta = t$ is possible.

5.2. Extension of the assumptions

As has already been indicated, the fundamental assumptions under which the properties of the limiting values of the Cauchy type integral have been derived (§§ 4, 5)—the smoothness of the contour and the Hölder condition for the density—are only sufficient, but not necessary.

For instance, in § 4.5 the Sokhotski formulae were generalized to the case of a contour with a finite number of corner points. We shall now indicate some other possible generalizations.

The extension of the conditions imposed on the contour and the density may follow two trends. (1) The trend in which we preserve the property of uniform passing to the limit at all points of the contour, in approaching it along any path, and as a result of this—preserving the condition that the limiting values are continuous functions on the contour. (2) The trend in which the limiting values exist only almost everywhere (with exception of a set of points of the contour, of measure zero) and only along paths non-tangential to the contour.

First let us consider the generalizations of the first type, thus remaining within the range of the classical definition of the Riemann integral. As we have seen, the problem consists in investigating the integral

$$\int_L \frac{\varphi(\tau) - \varphi(t)}{\tau - t}\,d\tau.$$

Let us denote the arc of the contour L between the points t_1 and t_2 by $s(t_1, t_2)$, and the length of its chord by $r(t_1, t_2) = |t_1 - t_2|$.

If we preserve the condition that the ratio of the length of the arc to the chord is a bounded quantity (the class of curves satisfying this condition includes the class of piecewise smooth curves), then in order that the expression $[\varphi(\tau) - \varphi(t)]/(\tau - t)$ be integrable it is sufficient to require, as we did, that the function $\varphi(t)$ satisfies the Hölder condition.

The integrability condition will also be satisfied if we impose on the function $\varphi(t)$ weaker restrictions: for instance we may require that the following inequality holds:

$$|\varphi(\tau) - \varphi(t)| < \frac{A}{\left|\ln|\tau - t|\right|^p} \quad (p > 1)$$

or the even more general condition

$$|\varphi(\tau) - \varphi(t)| <$$
$$< \frac{A}{\left|\ln|\tau - t|\left|\ln|\ln|\tau - t|\right|\right| \cdots \left|\ln\left|\ln\cdots|\ln|\tau - t|\cdots\right|\right|\right|^p} \quad (p > 1).$$

The quantities appearing in the right-hand sides of the latter inequalities tend to zero as $|\tau - t| \to 0$ slower than $|\tau - t|^\lambda$; hence the class of functions satisfying the last conditions contains the class of functions satisfying the Hölder condition.

If the class of contours under consideration be extended, it is necessary to narrow accordingly the class of allowable functions $\varphi(t)$. A clear confirmation of this assertion is provided by the following theorem†.

(1) *If the contour consists of a rectifiable Jordan curve, and the density of the Cauchy type integral satisfy the conditions*

$$s(t_1, t_2) < C|t_1 - t_2|^\alpha \qquad (0 < \alpha \leq 1),$$
$$|\varphi(t_1) - \varphi(t_2)| < A|t_1 - t_2|^\beta \qquad (1 - \alpha < \beta \leq 1),$$

then the integral of the Cauchy type has continuous limiting values everywhere on the contour except perhaps at its ends.

(2) *If the contour L is a rectifiable Jordan curve and the density $\varphi(t)$ satisfies the Lipschitz condition* $\qquad |\varphi(t_1) - \varphi(t_2)| < A|t_1 - t_2|,$

then the Cauchy type integral has continuous limiting values on the contour L, except perhaps at its ends.

Generalizations of the second type, connected with introducing into consideration Lebesgue and Stieltjes integrals, have at present essentially theoretical value, and have so far been little connected with practical applications.

Without going into details, we shall state one most general result, given by I. I. Privalov‡.

THEOREM. *Let L be a closed rectifiable Jordan curve of length l, θ the angle between the positive direction of the x-axis and the tangent to L, and $F(s)$ a complex function of the arc s of bounded variation on the segment $[0, l]$. Then the integral*

---

† See for instance I. I. Privalov [21], p. 197.

‡ See I. I. Privalov [21], Ch. III, or G. M. Goluzin [6], Ch. X.

of the Cauchy–Stieltjes type

$$\Phi(z) = \frac{1}{2\pi i} \int_0^l \frac{e^{i\theta(\sigma)} dF(\sigma)}{\tau(\sigma) - z}$$

has finite limiting values along all non-tangent paths almost everywhere on the contour L, and these values satisfy the relations

$$\Phi^{\pm}(t) = \pm \frac{1}{2} F'(s) + \frac{1}{2\pi i} \int_L \frac{e^{i\theta(\sigma)} dF(\sigma)}{\tau(\sigma) - t},$$

where the integral in the right-hand side is defined as singular.

In particular, if $F(s)$ is absolutely continuous, the Cauchy–Stieltjes integral is reduced to the Cauchy–Lebesgue integral

$$\frac{1}{2\pi i} \int \frac{\varphi(\tau)}{\tau - z} d\tau$$

where $\varphi(t) = \varphi[t(s)] = F'(s)$ almost everywhere on the contour L, and the preceding formulae are reduced to the ordinary Sokhotski formulae (4.8).

5.3. Some new results

Of the new works, those of most interest for application to boundary value problems are due to Khvedelidze [11] using the theory of Cauchy integrals with density of class L_p, $p > 1$ (Lebesgue-integrable in power p). We formulate two statements which we shall need to use later (see § 16.3).

1. If $\varphi(t) \in L_p$, then the singular integral

$$S\varphi = \frac{1}{\pi i} \int_L \frac{\varphi(\tau)}{\tau - t} d\tau$$

also belongs to the same class L_p. Limit values of the Cauchy integral exist almost everywhere along every non-tangential path and Sokhotski's formulae (4.8) are applicable.

2. The singular integral S in the norm of the space $L_p(p > 1)$ is a bounded operator, i.e.

$$\|S\varphi\|_{L_p} = \left(\int_L \left| \frac{1}{\pi} \int_L \frac{\varphi(\tau)}{\tau - t} d\tau \right|^p ds \right)^{\frac{1}{p}}$$

$$\leq M_p \left(\int_L |\varphi(\tau)|^p ds \right)^{\frac{1}{p}} = M_p \|\varphi\|_{L_p},$$

where M_p is a constant independent of φ. The contour L is regarded as a Lyapounov curve (the angle $\alpha(s)$ formed by the tangent to the contour satisfies Hölder's condition).

The proof uses the corresponding result of Riesz for integrals with a Hilbert kernel and an investigation of the following integral

$$\int_L \varphi(\tau) \left[\frac{d\tau}{\tau - t} - \cot \frac{\sigma - s}{2} d\sigma \right],$$

which has a singularity of order less than unity under the assumptions made about the contour.

§ 6. The Hilbert formulae for the limiting values of the real and imaginary parts of an analytic function

6.1. The Cauchy and Schwarz kernels

Let $f(z) = u(x, y) + iv(x, y)$ be a function analytic in the unit circle. We denote by s and σ the lengths of the arcs of the circle, measured from the point of intersection of the arc with the positive direction of the x-axis.

Further, suppose that the continuous function $u(s)$ is the limiting value of the real part of a function $f(z)$, on the contour of the circle.

It is known that in this case the Schwarz formula

$$f(z) = \frac{1}{2\pi} \int_0^{2\pi} u(\sigma) \frac{e^{i\sigma} + z}{e^{i\sigma} - z} d\sigma + iv_0 \tag{6.1}$$

makes it possible to express the function $f(z)$ analytic in the circle $|z| < 1$ by the values of its real part on the contour, to within a constant imaginary term iv_0. Setting $z = 0$ in (6.1) and making use of the theorem of mean value, we obtain

$$v_0 = v(0, 0) = \frac{1}{2\pi} \int_0^{2\pi} v(\sigma) d\sigma. \tag{6.2}$$

The expression $(e^{i\sigma} + z)/(e^{i\sigma} - z)$ is called the Schwarz kernel.

There exists a simple relation between the Schwarz and Cauchy kernels. Denoting by τ the complex coordinate of a point of the circle $(\tau = e^{i\sigma})$ we have

$$\frac{e^{i\sigma} + z}{e^{i\sigma} - z} d\sigma = \left(-1 + \frac{2e^{i\sigma}}{e^{i\sigma} - z} \right) d\sigma = \frac{2}{i} \frac{d\tau}{\tau - z} - d\sigma. \tag{6.3}$$

This implies the relation

$$\frac{1}{2\pi} \int_0^{2\pi} u(\sigma) \frac{e^{i\sigma} + z}{e^{i\sigma} - z} \, d\sigma = \frac{1}{2\pi i} \int_L \frac{2u(\sigma)}{\tau - z} \, d\tau - \frac{1}{2\pi} \int_0^{2\pi} u(\sigma) \, d\sigma. \quad (6.4)$$

This formula, which establishes the relation between the Schwarz integral and the Cauchy type integral with a real density taken over the unit circle, will be employed in future considerations.

6.2. The Hilbert formulae

We proceed to derive relations between the limiting values on the contour of the real and imaginary parts of a function analytic in the unit circle.

Let us pass to the limit in the formula (6.4), i.e. let z tend to the point $t = e^{is}$ of the unit circle; making use of the Sokhotski formulae (4.8) we obtain

$$u(s) + iv(s) = \frac{1}{2} \, 2u(s) + \frac{1}{2\pi i} \int_L \frac{2u(\sigma)}{\tau - t} \, d\tau - \frac{1}{2\pi} \int_0^{2\pi} u(\sigma) \, d\sigma + iv_0.$$

After a transformation, combining the two integrals on the right-hand side we have

$$v(s) = \frac{1}{2\pi i} \int_0^{2\pi} u(\sigma) \frac{e^{i\sigma} + e^{is}}{e^{i\sigma} - e^{is}} \, d\sigma + v_0. \quad (6.5)$$

Since

$$\frac{e^{i\sigma} + e^{is}}{e^{i\sigma} - e^{is}} = \frac{e^{i\frac{\sigma-s}{2}} + e^{-i\frac{\sigma-s}{2}}}{e^{i\frac{\sigma-s}{2}} - e^{-i\frac{\sigma-s}{2}}} = \frac{1}{i} \cot \frac{\sigma - s}{2},$$

we finally obtain

$$v(s) = -\frac{1}{2\pi} \int_0^{2\pi} u(\sigma) \cot \frac{\sigma - s}{2} \, d\sigma + v_0. \quad (6.6)$$

This formula gives the boundary value of the imaginary part of an analytic function, in terms of the real part.

To deduce the expression for $u(s)$ in terms of $v(s)$ observe that for the function

$$-if(z) = v - iu$$

the real part is v and the imaginary $-u$. Applying the formula (6.6) to the latter function we obtain

$$u(s) = \frac{1}{2\pi} \int\limits_{0}^{2\pi} v(\sigma) \cot \frac{\sigma - s}{2} \, d\sigma + u_0. \tag{6.7}$$

The symmetric formulae (6.6), (6.7) are called the *Hilbert inversion formulae*, and the expression $\cot \frac{1}{2}(\sigma - s)$ the *Hilbert kernel*.

For $\sigma = s$ the kernel is infinite of first order; hence, the integrals appearing in the Hilbert formulae are singular and should be understood in the sense of the principal value. Inversion formulae (6.6), (6.7) may be interpreted as the solution of the integral equation (see § 31.3):

$$\frac{1}{2\pi} \int\limits_{0}^{2\pi} u(\sigma) \, \mathrm{ctg} \frac{\sigma - s}{2} \, d\sigma = c(s).$$

We shall not pursue here the complicated problem of the least limiting conditions which should be imposed upon the functions $u(s)$ and $v(s)$ in order that the formulae (6.6) and (6.7) hold. We confine ourselves to the obvious statement that it is sufficient that they satisfy the Hölder condition.

Problems 14, 16 and 25 at the end of the chapter are complementary to this section.

§ 7. The change of the order of integration in a repeated singular integral

7.1. The case when one integral is ordinary

Before proceeding to the investigation of the problem of the change of the order of integration in a repeated singular integral, let us first consider the case when in the repeated integral only one integral is singular, while the second is ordinary.

Suppose that in the repeated integral

$$I = \int\limits_{L} \omega(\tau, z) \, d\tau \int\limits_{L} \frac{\varphi(\tau, \tau_1)}{\tau_1 - \tau} \, d\tau_1 \tag{7.1}$$

the function $\varphi(\tau, \tau_1)$ satisfies the Hölder condition with respect to both variables and $\omega(\tau, z)$ is integrable with respect to τ for all z belonging to a prescribed set of values; L is a smooth contour.

Denote the integral obtained by the change of the order of integration as follows:

$$I' = \int_L d\tau_1 \int_L \frac{\omega(\tau, z)\, \varphi(\tau, \tau_1)}{\tau_1 - \tau}\, d\tau. \qquad (7.2)$$

LEMMA. *Under the assumptions made above*† *with respect to the functions* φ, ω *and the contour* L, *the integral* (7.1) *may undergo a change of the order of integration, i.e.*

$$\int_L \omega(\tau, z)\, d\tau \int_L \frac{\varphi(\tau, \tau_1)}{\tau_1 - \tau}\, d\tau_1 = \int_L d\tau_1 \int_L \frac{\omega(\tau, z)\, \varphi(\tau, \tau_1)}{\tau_1 - \tau}\, d\tau. \qquad (7.3)$$

To carry out the proof let us divide the contour of integration in the integral I with respect to the variable τ, into two parts: the arc l cut out of L by a circle of radius δ with centre at the point $\tau_1 = \tau$, and the rest of the contour $L - l$. The same is done for the integral I', taking the centre of the circle at the point $\tau = \tau_1$.

Introduce the notation

$$I = I_0 + I_\delta,$$

where

$$I_0 = \int_L \omega(\tau, z)\, d\tau \int_{L-l} \frac{\varphi(\tau, \tau_1)}{\tau_1 - \tau}\, d\tau_1$$

and

$$I_\delta = \int_L \omega(\tau, z)\, d\tau \int_l \frac{\varphi(\tau, \tau_1)}{\tau_1 - \tau}\, d\tau_1.$$

The integral I' is split in an analogous way:

$$I' = I_0' + I_\delta'.$$

In the repeated integrals I_0, I_0' all integrals are ordinary; the change of the order of integration is therefore legitimate. Consequently‡,

$$I_0 = I_0'.$$

---

† This result is true under the condition that $\varphi(\tau, \tau_1)$ is simply integrable. The additional limitation is imposed to simplify the proof. In what follows the lemma will be required only under the limitations imposed above.

‡ This relation will become particularly clear if the positions of the points τ and τ_1 on the contour L are determined by the arc coordinates s and $s_1 (0 \le s \le l,$ $0 \le s_1 \le l)$ where l is the length of the contour L, which are regarded as rectilinear coordinates of a point in an auxiliary plane $O s s_1$. Then for both integrals I_0 and I_0' the domain of integration is a square with side l, a narrow strip along a diagonal being excluded.

3*

Hence

$$|I - I'| = |I_\delta - I'_\delta| \leqq |I_\delta| + |I'_\delta|.$$

Let us estimate the last two integrals.

In the inner integral of the repeated integral

$$I_\delta = \int\limits_L \omega(\tau, z)\, d\tau \int\limits_l \frac{\varphi(\tau, \tau_1)}{\tau_1 - \tau}\, d\tau_1$$

let us make the usual transformation

$$\int\limits_l \frac{\varphi(\tau, \tau_1)}{\tau_1 - \tau}\, d\tau_1 = \int\limits_l \frac{\varphi(\tau, \tau_1) - \varphi(\tau, \tau)}{\tau_1 - \tau}\, d\tau_1 + \varphi(\tau, \tau) \int \frac{d\tau_1}{\tau_1 - \tau}.$$

The first integral of the right-hand side was already examined in § 4.1. For a sufficiently small δ its modulus can be made smaller than the positive number $\varepsilon/4M$ where

$$M = \max \left| \int\limits_L \omega(\tau, z)\, d\tau \right|.$$

Next, denoting the ends of the arc l by τ', τ'' and applying the formula (3.6) we obtain

$$\int\limits_l \frac{d\tau}{\tau_1 - \tau} = \ln \frac{\tau'' - \tau}{\tau - \tau'} = i\left[\arg(\tau'' - \tau) - \arg(\tau - \tau')\right] = i\alpha,$$

where α is the angle between the secants (Fig. 7).

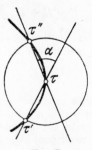

FIG. 7.

For a sufficiently small δ, in view of the smoothness of the curve this angle can be made arbitrarily small. Consequently, we can obtain the estimate

$$\left| \varphi(\tau, \tau) \int_l \frac{d\tau_1}{\tau_1 - \tau} \right| < \frac{\varepsilon}{4M}.$$

Then

$$|I_\delta| < 2 \frac{\varepsilon}{4M} \left| \int_L \omega(\tau, z)\, d\tau \right| \leqq \frac{\varepsilon}{2}.$$

Analogous reasoning leads to the estimate

$$|I_\delta'| < \frac{\varepsilon}{2},$$

whence

$$|I - I'| < \varepsilon.$$

On the other hand, however, it is evident that the difference $I - I'$ is independent of the quantity δ; consequently,

$$I - I' = 0.$$

This completes the proof.

7.2. The transposition formula

Consider the pair of repeated singular integrals, differing in the order of integration

$$I(t) = \frac{1}{\pi i} \int_L \frac{d\tau}{\tau - t} \frac{1}{\pi i} \int_L \frac{\varphi(\tau, \tau_1)}{\tau_1 - \tau}\, d\tau_1, \qquad (7.4)$$

$$I_1(t) = \frac{1}{\pi i} \int_L d\tau_1 \frac{1}{\pi i} \int_L \frac{\varphi(\tau, \tau_1)}{(\tau - t)(\tau_1 - \tau)}\, d\tau. \qquad (7.5)$$

Both integrals make sense. In fact, the first,

$$\frac{1}{\pi i} \int_L \frac{\chi(\tau)\, d\tau}{\tau - t}, \quad \text{where} \quad \chi(\tau) = \frac{1}{\pi i} \int_L \frac{\varphi(\tau, \tau_1)}{\tau_1 - \tau}\, d\tau_1,$$

exists in the sense of the Cauchy principal value, since on the basis of §5.1 the density $\chi(\tau)$ satisfies the Hölder condition. In the second integral we carry out the following transformations.

Let us expand the function $1/[(\tau - t)(\tau_1 - t)]$ into partial fractions:

$$\frac{1}{(\tau - t)(\tau_1 - \tau)} = \frac{1}{\tau_1 - t} \left[\frac{1}{\tau - t} - \frac{1}{\tau - \tau_1} \right]. \qquad (7.6)$$

Then, letting

$$\omega(\zeta, \tau_1) = \frac{1}{\pi i} \int\limits_L \frac{\varphi(\tau, \tau_1)}{\tau - \zeta} d\tau,$$

we obtain

$$I_1 = \frac{1}{\pi i} \int\limits_L \frac{\omega(t, \tau_1) - \omega(\tau_1, \tau_1)}{\tau_1 - t} d\tau_1.$$

By virtue of the theorem of § 5.1 $\omega(\zeta, \tau_1)$ satisfies the Hölder condition; consequently, I_1 exists as an improper integral.

This reasoning is valid for all points t which do not coincide with an end of the contour. In the latter case an additional investigation is necessary. We shall not, however, pursue this problem here.

Although the integrals I and I_1 differ only in the order of integration, they are not equal. This fact is shown by the Poincaré–Bertrand transposition formula[†]:

$$\frac{1}{\pi i} \int\limits_L \frac{d\tau}{\tau - t} \frac{1}{\pi i} \int\limits_L \frac{\varphi(\tau, \tau_1)}{\tau_1 - \tau} d\tau_1 = \varphi(t, t) +$$

$$+ \frac{1}{\pi i} \int\limits_L d\tau_1 \frac{1}{\pi i} \int\limits_L \frac{\varphi(\tau, \tau_1)}{(\tau - t)(\tau_1 - \tau)} d\tau. \quad (7.7)$$

To prove the assertion let us construct the functions $I(z)$, $I_1(z)$ replacing in the integrals $I(t)$, $I_1(t)$ the variable t by the variable z varying over the entire plane:

$$I(z) = \frac{1}{\pi i} \int\limits_L \frac{d\tau}{\tau - z} \frac{1}{\pi i} \int\limits_L \frac{\varphi(\tau, \tau_1)}{\tau_1 - \tau} d\tau_1; \quad (7.8)$$

$$I_1(z) = \frac{1}{\pi i} \int\limits_L d\tau_1 \frac{1}{\pi i} \int\limits_L \frac{\varphi(\tau, \tau_1)}{(\tau - z)(\tau_1 - \tau)} d\tau. \quad (7.9)$$

For all z not lying on L, in view of the last lemma we may change the order of integration. Hence

$$I(z) = I_1(z).$$

Denoting the limiting values of these functions as before, we have

$$I^+(t) = I_1^+(t); \quad I^-(t) = I_1^-(t).$$

---

† Poincaré [20], Bertrand (1).

But, in accordance with the formula (4.10)

$$I(t) = \tfrac{1}{2}\left[I^+(t) + I^-(t)\right],$$

or, making use of the preceding relations,

$$I(t) = \tfrac{1}{2}\left[I_1^+(t) + I^-(t)\right]. \tag{7.10}$$

Using the expansion (7.6), let us represent $I_1(z)$ in the form

$$I_1(z) = \frac{1}{\pi i}\int\limits_L \frac{d\tau_1}{\tau_1 - z}\,\frac{1}{\pi i}\left[\int\limits_L \frac{\varphi(\tau,\tau_1)}{\tau - z}\,d\tau - \int\limits_L \frac{\varphi(\tau,\tau_1)}{\tau - \tau_1}\,d\tau\right]$$

$$= \frac{1}{\pi i}\int\limits_L \frac{\psi(z,\tau_1)}{\tau_1 - z}\,d\tau_1.$$

The last expression proves that $I_1(z)$ is a Cauchy type integral with the density

$$\psi(z,\tau_1) = \frac{1}{\pi i}\int\limits_L \frac{\varphi(\tau,\tau_1)}{\tau - z}\,d\tau - \frac{1}{\pi i}\int\limits_L \frac{\varphi(\tau,\tau_1)}{\tau - \tau_1}\,d\tau,$$

depending on the parameter z.

Employing the Sokhotski formulae (4.7), (4.8), we find the values of $I_1^+(t)$ and $I_1^-(t)$, taking into account† that as the value of the density $\psi(z,t)$ at the point t we should take $\lim\limits_{z\to t}\psi(z,t)$ equal to $\psi^+(t,t)$ as $z \to t$ remaining on the left of L, and $\psi^-(t,t)$ as $z \to t$ on the right of L.

Thus,

$$I_1^+(t) = \psi^+(t,\,t) + \frac{1}{\pi i}\int\limits_L \frac{\psi^+(t,\tau_1)}{\tau_1 - t}\,d\tau_1;$$

$$I_1^-(t) = -\,\psi^-(t,\,t) + \frac{1}{\pi i}\int\limits_L \frac{\psi^-(t,\tau_1)}{\tau_1 - t}\,d\tau_1,$$

whence

$$\tfrac{1}{2}\left[I_1^+(t) + I_1^-(t)\right] = \tfrac{1}{2}\left[\psi^+(t,\,t) - \psi^-(t,\,t)\right] +$$

$$+ \frac{1}{2\pi i}\int\limits_L \frac{\psi^+(t,\tau_1) + \psi^-(t,\tau_1)}{\tau_1 - t}\,d\tau. \tag{7.11}$$

† We do not quote here the justification of the applicability of the Sokhotski formulae in the case when the density of the Cauchy type integral depends on z (see N. I. Muskhelishvili [17*], p. 128).

Now, according to the formulae (4.9), (4.10)

$$\tfrac{1}{2}\left[\psi^+(t,\,t) - \psi^-(t,\,t)\right] = \varphi(t,\,t),$$

$$\tfrac{1}{2}\left[\psi^+(t,\,\tau_1) + \psi^-(t,\,\tau_1)\right] = \frac{1}{\pi i}\int\limits_L \frac{\varphi(\tau,\,\tau_1)}{\tau - t}\,d\tau - \frac{1}{\pi i}\int\limits_L \frac{\varphi(\tau,\,\tau_1)}{\tau - \tau_1}\,d\tau$$

$$= \frac{t - \tau_1}{\pi i}\int\limits_L \frac{\varphi(\tau,\,\tau_1)}{(\tau - t)(\tau - \tau_1)}\,d\tau.$$

Introducing these relations into the formula (7.11) and bearing in mind (7.10) we arrive at the required formula (7.7).

The transposition formula (7.7) can also be written in the following form:

$$\int\limits_L \frac{d\tau}{\tau - t}\int\limits_L \frac{\varphi(\tau,\,\tau_1)}{\tau_1 - \tau}\,d\tau_1 = -\pi^2\varphi(t,\,t) +$$

$$+ \int\limits_L d\tau_1 \int\limits_L \frac{\varphi(\tau,\,\tau_1)}{(\tau - t)(\tau_1 - \tau)}\,d\tau. \tag{7.7'}$$

The proof outlined here was given by N. I. Muskhelishvili [17*]. There also exists another derivation of the transposition formula (see Problems 27, 28 at the end of this chapter).

7.3. Inversion of the singular integral with the Cauchy kernel for the case of a closed contour

We shall now present an application of the transposition formula (7.7) to the derivation of the formula of inversion of the singular integral with the Cauchy kernel, for the case of a closed contour.

Let there be given the singular integral

$$\psi(t) = \frac{1}{\pi i}\int\limits_L \frac{\varphi(\tau)}{\tau - t}\,d\tau. \tag{7.12}$$

We shall now try to express the density $\varphi(t)$ by the values of the integral itself, i.e. by the function $\psi(t)$. Let us replace in both sides the variable t by τ_1, multiply by $(1/\pi i)[d\tau_1/(\tau_1 - t)]$ and integrate over the contour L. Changing the order of integration we obtain

$$\frac{1}{\pi i}\int\limits_L \frac{\psi(\tau_1)}{\tau_1 - \tau}\,d\tau_1 = \varphi(t) + \frac{1}{\pi i}\int\limits_L \varphi(\tau)\,d\tau\,\frac{1}{\pi i}\int\limits_L \frac{d\tau_1}{(\tau_1 - t)(\tau - \tau_1)}.$$

Now

$$\int\limits_L \frac{d\tau_1}{(\tau_1 - t)(\tau - \tau_1)} = \frac{1}{\tau - t}\left[\int\limits_L \frac{d\tau_1}{\tau_1 - t} - \int\limits_L \frac{d\tau_1}{\tau_1 - \tau}\right]$$

$$= \frac{1}{\tau - t}(i\pi - i\pi) = 0,$$

so that

$$\varphi(t) = \frac{1}{\pi i}\int\limits_L \frac{\psi(\tau)}{\tau - t}\,d\tau. \tag{7.13}$$

This formula is the formula for the inversion of the singular integral (7.12).

Let us observe that the formulae (7.12), (7.13) are similar to the Hilbert inversion formulae (6.6), (6.7). The latter may be regarded as the formulae of inversion of the singular integral with the Hilbert kernel $\cot[(\sigma - s)/2]$ (see § 6.2).

§ 8. Behaviour of the Cauchy type integral at the ends of the contour of integration and at the points of density discontinuities

8.1. The case of density satisfying the Hölder condition on L, including the ends

In the investigations of §5 on the behaviour of the limiting values of the Cauchy type integral on the contour, we always excluded from the considerations the ends of the contour. The fact that an end constitutes an exceptional point is clear, since it is impossible to include it into a two-sided interval and, consequently, it is impossible to define the singular integral as the principal value. Now we proceed to the examination of the behaviour of the Cauchy type integral on the ends of the contour.

Let L be an open contour with ends a and b, and let $\varphi(t)$ be a function satisfying the Hölder condition on the whole curve, including the ends.

Envisage the Cauchy type integral

$$\Phi(z) = \frac{1}{2\pi i}\int\limits_L \frac{\varphi(\tau)}{\tau - z}\,d\tau.$$

Proceeding as in § 3.4 let us represent $\Phi(z)$ in the form

$$\Phi(z) = \frac{\varphi(t)}{2\pi i} \int_L \frac{d\tau}{\tau - z} + \frac{1}{2\pi i} \int_L \frac{\varphi(\tau) - \varphi(t)}{\tau - z} \, d\tau$$

$$= \frac{\varphi(t)}{2\pi i} \ln \frac{b - z}{a - z} + \frac{1}{2\pi i} \int_L \frac{\varphi(\tau) - \varphi(t)}{\tau - z} \, d\tau,$$

where t is an arbitrary point of the contour L.

For $z = t$ this integral exists as improper. Therefore, it constitutes a function which is bounded and tends to a definite limit as the point z approaches any point of the contour, including its ends (see § 4.1). The principal singularity of $\Phi(z)$ on the ends is due to the first term.

Setting in the last expression successively $t = a$, $t = b$ we obtain

$$\Phi(z) = -\frac{\varphi(a)}{2\pi i} \ln(z - a) + \Phi_1(z), \tag{8.1}$$

$$\Phi(z) = \frac{\varphi(b)}{2\pi i} \ln(z - b) + \Phi_2(z), \tag{8.2}$$

where $\Phi_1(z)$, $\Phi_2(z)$ are functions bounded in the vicinities of the respective ends and tending to definite limits as z tends to a or b respectively.

Thus, at the ends of the contour of integration the Cauchy type integral possesses singularities of logarithmic type, completely determined by the values of $\varphi(a)$ and $\varphi(b)$.

8.2. The case of discontinuity of the first kind

Consider now the related problem of the behaviour of the Cauchy type integral at the points at which the density has a discontinuity of the first kind. Let c be the point of discontinuity. We shall call the points preceding c in moving along the contour in the positive direction the left vicinity of the point c, and the points following c the right vicinity. The limits from the right and from the left will as before be denoted $\varphi(c \pm 0)$.

Let us split the integral over the contour ab into two integrals over the sections ac and cb as follows:

$$\int_{ab} = \int_{ac} + \int_{cb}.$$

Then c is the right and the left end of the partial contours, respectively, and in the first integral we have $\varphi(c) = \varphi(c - 0)$, whereas in the second $\varphi(c) = \varphi(c + 0)$.

Applying the formulae (8.1) and (8.2) we obtain the representation

$$\Phi(z) = \frac{\varphi(c - 0) - \varphi(c + 0)}{2\pi i} \ln(z - c) + \Phi_0(z), \qquad (8.3)$$

where $\Phi_0(z)$ has in the vicinity of the point c the same properties as the functions $\Phi_1(z)$, $\Phi_2(z)$ in the formulae (8.1), (8.2).

The relations (8.1)–(8.3) will be employed below.

8.3. The particular case of a power singularity

Before proceeding to a general investigation of the behaviour of the Cauchy type integral at the points where the density has an infinity of the power type, let us consider the Cauchy type integral of a particular kind

$$\Omega(z) = \frac{1}{2\pi i} \int\limits_{ab} \frac{d\tau}{(\tau - t_0)^\gamma (\tau - z)} \left(\begin{matrix} \gamma = \alpha + i\beta, \\ 0 \leqq \alpha < 1 \end{matrix} \right), \qquad (8.4)$$

where t_0 is a prescribed point of the contour.

In examining this integral there arise specific difficulties, due to the many-valuedness of both the integrand and the integral itself. It is necessary to observe carefully the cut in the plane of complex variable which singles out the single-valued branches of the functions under consideration. Problems of this type have already been dealt with in §§ 3.4, 3.5, and will be encountered in Chapter VI.

We proceed to investigate the integral $\Omega(z)$ for various positions of the point t_0.

1. *Point t_0 coincides with the left end* $(t_0 = a)$. The many-valued function $(t - a)^{-\gamma}$ is regarded as the contour value of the analytic function $(z - a)^{-\gamma}$. The latter is considered as single-valued in the plane cut along a line connecting the branch points a and ∞. We agree to make the cut through the contour L. Let us select on the left-hand side of the cut a branch of this function, and take it as the value $(t - a)^{-\gamma}$ on the contour of integration and as the density of the integral (8.4); thus according to the definition

$$(t - a)^{-\gamma} = [(t - a)^{-\gamma}]^+.$$

Its value on the right-hand side of the contour will be obtained by making a circle about the point a in the positive direction (Fig. 8), taking into account that then the power function $(t - a)^{-\gamma}$ acquires the factor $e^{-2\pi i \gamma}$.

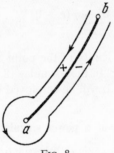

FIG. 8.

For an arbitrary point t of the contour we have

$$[(t - a)^{-\gamma}]^+ - [(t - a)^{-\gamma}]^- = (1 - e^{-2\pi i \gamma})(t - a)^{-\gamma}. \qquad (8.5)$$

We now prove that, in the vicinity of the point $z = a$, $\Omega(z)$ has the representation

$$\Omega(z) = \frac{e^{i\pi\gamma}}{2i \sin\gamma\pi}(z - a)^{-\gamma} + \Omega_0(z), \qquad (8.6)$$

where $\Omega_0(z)$ is a function analytic in the vicinity of a.

To prove this let us first observe that according to the Sokhotski formulae

$$\Omega^+(t) - \Omega^-(t) = (t - a)^{-\gamma}. \qquad (8.6')$$

Let us denote

$$\omega(z) = \frac{(z - a)^{-\gamma}}{1 - e^{-2\pi i \gamma}} = \frac{e^{i\gamma\pi}}{2i \sin\gamma\pi}(z - a)^{-\gamma}$$

and consider the function $\Omega(z) - \omega(z)$.

In view of (8.5) and (8.6') we have

$$[\Omega(t) - \omega(t)]^+ = [\Omega(t) - \omega(t)]^-. \qquad (*)$$

We shall now show that in the neighbourhood of the point a the function $\Omega(z) - \omega(z)$ satisfies the inequality

$$|\Omega(z) - \omega(z)| \leqq \frac{\text{const}}{|z - a|^\nu}, \qquad (**)$$

where ν is a real number satisfying conditions $0 \leqq \alpha < \nu < 1$.

Putting $z - a = \varrho e^{i\theta}$, we find that

$$(z - a)^{-\gamma} = e^{-\gamma \ln(z-a)} = \varrho^{-\alpha} e^{\beta \theta} e^{-i(\beta \ln \varrho + \alpha \theta)} = \frac{h(z)}{|z - a|^\alpha}.$$

The function $h(z)$ is bounded if $z \to a$ and therefore

$$|\omega(z)| \leqq \frac{C_1}{|z - a|^\alpha} \qquad (0 \leqq a < 1).$$

Proceeding to the estimate $|\Omega(z)|$ and putting $|\tau - a| = r$ and $0 \leqq \alpha < \nu < 1$, we have

$$\Omega(z) = \frac{1}{\varrho^\nu} \left[\frac{1}{2\pi i} \int\limits_{ab} \frac{(\varrho^\nu - r^\nu) h(\tau)}{r^\alpha (\tau - z)} \, d\tau + \frac{1}{2\pi i} \int\limits_{ab} \frac{r^{\nu - \alpha} h(\tau)}{\tau - z} \, d\tau \right]$$

$$= \frac{1}{\varrho^\nu} (I_1 + I_2).$$

We satisfy ourselves as to the boundedness of the integrals I_1 and I_2. Having regard to the inequalities

$$|\varrho - r| = ||z - a| - |\tau - a|| \leqq |z - \tau|$$

and†
$$|\varrho^\nu - r^\nu| \leqq |\varrho - r|^\nu \leqq |z - \tau|^\nu$$

and the boundedness of $h(\tau)$, we get

$$|I_1| \leqq C_2 \int\limits_{ab} \frac{dr}{r^\alpha |r - \varrho|^{1-\nu}}.$$

The last integral exists as non-singular for all values of ϱ and it is therefore bounded. The boundedness of the integral I_2 follows from the fact that $r^{\nu - \alpha} h(\tau) \in H(\nu - \alpha)$. Thus

$$|\Omega(z)| \leqq \frac{1}{\varrho^\nu} (|I_1| + |I_2|) \leqq \frac{C_3}{|z - a|^\nu} \qquad (0 \leqq a < \nu < 1).$$

Note that the equality (8.6), proved below, provides the exact estimate ($\nu = \alpha$). But to obtain the latter we have to employ the rougher estimate which has just been obtained.

The inequality (**) follows from the estimates for $|\omega(z)|$ and $|\Omega(z)|$.

---

† The inequality $|\varrho^\nu - r^\nu| \leqq |\varrho - r|^\nu$ is established by definition of the maximum of the function $f(x) = (1 - x^\nu)(1 - x)^{-\nu}$ $(0 \leqq x \leqq 1)$, where $x = \varrho$ $\left(x = \dfrac{r}{\varrho} \right)$ (sic), if $r > \varrho (r < \varrho)$.

The equality (*) shows that the functions $[\Omega(z) - \omega(z)]^+$, $[\Omega(z) - \omega(z)]^-$ are analytically-continuable through the contour ab and that they therefore form a single analytic function $\Omega(z) - \omega(z)$ over the entire plane. This function may only have an isolated singularity at point a. By virtue of the estimate (**) the point a cannot be a pole or an essentially singular point. Neither can it be a branch point since a single-valued function with a branch point must have lines of discontinuity (cuts), which is absurd, owing to continuity of the function $\Omega(z) - \omega(z)$ in the domains D^+ and D^- and the condition (*) on the contour. The function $\Omega(z) - \omega(z)$ thus has a removable singularity at point a and it is therefore an analytic function in the neighbourhood of point a, which leads to the representation (8.6).

To deduce the value of $\Omega(z)$ on the contour we again apply the Sokhotski formula (4.10)

$$\Omega(t) = \frac{1}{2\pi i} \int_L \frac{d\tau}{(\tau - a)^\gamma (\tau - t)} = \frac{1}{2}[\Omega^+(t) + \Omega^-(t)].$$

Hence

$$\Omega(t) = \frac{1}{2i}(t - a)^{-\gamma} \cot \gamma \pi + \Omega_0(t), \tag{8.7}$$

where $\Omega_0(t)$ is a function analytic in the vicinity of the point a.

2. *Point t_0 coincides with the right end ($t_0 = b$)*. In this case, under the condition that the cut for the function $(z - b)^{-\gamma}$ which connects the points b and ∞ passes through the contour L, analogous reasoning leads to the relations

$$\Omega(z) = -\frac{e^{-i\gamma\pi}}{2i \sin \gamma \pi}(z - b)^{-\gamma} + \Omega_1(z), \tag{8.8}$$

$$\Omega(t) = -\frac{1}{2i}(t - b)^{-\gamma} \cot \gamma \pi + \Omega_1(t). \tag{8.9}$$

Let us in particular note that if $\gamma = \frac{1}{2}$ and, consequently, $\cot \gamma \pi = 0$ the formulae (8.7), (8.9) yield

$$\Omega(t) = \Omega_i(t) \quad (i = 0, 1),$$

i.e. the singular integral (8.4) is a function bounded near the ends.

3. *Point t_0 is an interior point of the contour ($t_0 = c$)*.

$$\Omega(z) = \frac{1}{2\pi i} \int_{ab} \frac{d\tau}{(\tau - c)^\gamma (\tau - z)}. \tag{8.10}$$

We suppose that the cut for the function $(z - c)^{-\gamma}$ connecting the points c and ∞ passes through the section cb of the contour (Fig. 9).

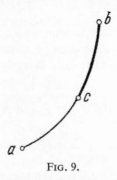

FIG. 9.

Splitting the integral into the two fol owing integrals

$$\int_{ab} = \int_{ac} + \int_{cb},$$

we reduce our investigation to the case where the branch point c coincides with an end of the contour of integration. The second integral can be directly computed with the help of the formulae (8.6), (8.7). To compute the first integral, additional considerations are necessary, since, as distinct from the case 1, the contour ac is not the cut line. Hence, the function $(z - c)^{-\gamma}$ varies continuously in passing through the contour ac:

$$[(t - c)^{-\gamma}]^{+} = [(t - c)^{-\gamma}]^{-} \quad \text{on } ac. \tag{8.11}$$

On the other hand, in accordance with the Sokhotski formulae,

$$\Omega^{+}(t) - \Omega^{-}(t) = (t - c)^{-\gamma} \quad \text{on } L. \tag{8.12}$$

It follows from the above formulae that:

1. The representation of the function $\Omega(z)$ in the vicinity of the point c involves the function $(z - c)^{-\gamma}$.

2. The analytic expressions representing the function $\Omega(z)$ are different on iffe s ides of the contour.

We shall now seek the representation of the first term of the function $\Omega(z)$ in the vicinity of the point c in the form

$$\int_{ac} = A(z - c)^{-\gamma} + \Omega_{2}(z)$$

on the left-hand side of the contour,

$$\int_{ac} = B(z - c)^{-\gamma} + \Omega_2(z)$$

on the right-hand side of the contour, where A and B are unknown coefficients and $\Omega_2(z)$ is a function analytic in the vicinity of the point c. The formulae (8.11) and (8.12) imply that

$$A - B = 1. \tag{8.13}$$

To derive the second relation for A and B let us observe that for the integral

$$\int_{ac} \frac{d\tau}{(\tau - c)^{\gamma} (\tau - t)}$$

the formula (8.9) holds, independently of the manner of selection of the branch of the function $(z - c)^{-\gamma}$. Applying again the formula (4.10) we obtain

$$A + B = -\frac{1}{i} \cot \gamma \pi. \tag{8.14}$$

From (8.13) and (8.14)

$$A = -\frac{e^{-\gamma \pi i}}{2i \sin \gamma \pi}; \quad B = -\frac{e^{\gamma \pi i}}{2i \sin \gamma \pi}.$$

Consequently, $\Omega(z)$ has in the vicinity of the point c the representation

$$\Omega(z) = (z - c)^{-\gamma} + \Omega_3(z) \quad \text{on the left of } L, \tag{8.15}$$

$$\Omega(z) = \Omega_4(z) \quad \quad \text{on the right of } L, \tag{8.16}$$

$$\Omega(t) = \tfrac{1}{2}(t - c)^{-\gamma} + \Omega_5(t) \text{ on } L. \tag{8.16'}$$

All $\Omega_i(z)$ are functions analytic in the vicinity of the point c.

8.4. The general case of a power singularity

Let $\varphi^*(t)$ be a function which satisfies the Hölder condition on the closed arc ab, the index being λ.

Consider the integral

$$\Phi(z) = \frac{1}{2\pi i} \int_{ab} \frac{\varphi^*(\tau)}{(\tau - c)^{\gamma} (\tau - z)} d\tau \quad \begin{pmatrix} \gamma = \alpha + i\beta, \\ 0 \leq \alpha < 1 \end{pmatrix} \tag{8.17}$$

which can be written in the form

$$\Phi(z) = \frac{\varphi^*(c)}{2\pi i} \int_{ab} \frac{d\tau}{(\tau - c)^{\gamma} (\tau - z)} + \frac{1}{2\pi i} \int_{ab} \frac{\varphi^*(\tau) - \varphi^*(c)}{(\tau - c)^{\gamma}} \frac{d\tau}{\tau - z}. \tag{8.18}$$

The first term is the integral examined in the preceding article. We proceed to examine the behaviour of the second integral, as $z \to c$. The function

$$(t-c)^{i\beta} = e^{i\beta[\ln|t-c|+i\theta]} = e^{\beta\theta}[\cos(\beta\ln|t-c|) + i\sin(\beta\ln|t-c|)]$$

at the point c is discontinuous but bounded. Taking into account this result it can be proved, which we shall not do here, that the function $(t-c)^{\alpha+i\beta}$ satisfies for $\alpha > 0$ the Hölder condition with the index α.

Thus

$$\left|\frac{\varphi^*(t) - \varphi^*(c)}{(t-c)^\gamma}\right| < A|t-c|^{\lambda-\alpha}. \tag{8.19}$$

If $\alpha = 0$ the following estimate holds:

$$\left|\frac{\varphi^*(t) - \varphi^*(c)}{(t-c)^{i\beta}}\right| < A|t-c|^{\gamma-\varepsilon}, \tag{8.19'}$$

where ε is an arbitrarily small number.

The above estimates imply the following results.

1. If $\alpha < \lambda$ (including $\alpha = 0$) the integrand in the formula (8.18) satisfies, in view of (8.19), the Hölder condition and, consequently, the integral constitutes a function $\Phi_0(z)$ bounded and tending to a definite limit as $z \to c$.

2. If $\alpha \geq \lambda$, $\Phi_0(z)$ constitutes a function unbounded in the vicinity of the point c, the order of infinity being lower than α.

A full proof of the assertions quoted above involves rather cumbersome calculations, which we shall not carry out here. The reader may find a detailed treatment in the book of N. I. Muskhelishvili [16*].

Without going into details let us consider the case when the function $\varphi^*(t)$ has at the point c a discontinuity of the first kind. If the contour L be divided into two sections ac and cb, then $\varphi^*(t)$ on both closed arcs satisfies the Hölder condition.

Let us split the integral (8.17) into two integrals over the sections ac, cb. Let us subject both integrals to the transformation leading to the form (8.18). To the integrals

$$\frac{\varphi(c-0)}{2\pi i} \int_{ac} \frac{d\tau}{(\tau-c)^\gamma(\tau-z)}, \qquad \frac{\varphi(c+0)}{2\pi i} \int_{cb} \frac{d\tau}{(\tau-c)^\gamma(\tau-z)}$$

we apply the reasoning of the preceding article, and in examining the remaining integrals we make use of the properties 1,2 of the present article.

Finally we find that the function $\Phi(z)$ has in the vicinity of the point c the representation

$$\left.\begin{aligned}
\Phi(z) &= \frac{e^{\gamma\pi i}\varphi^*(c+0) - e^{-\gamma\pi i}\varphi^*(c-0)}{2i\sin\gamma\pi}(z-c)^{-\gamma} \\
&\qquad\qquad + \Phi_0(z) \text{ on the left of } L, \\
\Phi(z) &= \frac{e^{\gamma\pi i}\varphi^*(c+0) - e^{-\gamma\pi i}\varphi^*(c-0)}{2i\sin\gamma\pi}(z-c)^{-\gamma} \\
&\qquad\qquad + \Phi_0(z) \text{ on the right of } L,
\end{aligned}\right\} \tag{8.20}$$

and near the point c, $\Phi_0(z)$ is given by the inequality

$$\Phi_0(z) < \frac{c}{|z - c|^{\alpha_0}} \quad (\alpha_0 < \alpha). \tag{8.21}$$

In particular, if the Hölder index $\lambda > \alpha$, then $\Phi_0(z)$ is bounded.

For the value of the integral (8.17) on the contour the following representations hold:

$$\Phi(t) = \left[\frac{e^{\gamma \pi i}}{2i \sin \gamma \pi} \varphi^*(c + 0) - \frac{\cot \gamma \pi}{2i} \varphi^*(c - 0) \right] (t - c)^{-\gamma} + \Phi^*(t) \quad \text{on } ac, \tag{8.22}$$

$$\Phi(t) = \left[\frac{\cot \gamma \pi}{2i} \varphi^*(c + 0) - \frac{e^{-\gamma \pi i}}{2i \sin \gamma \pi} \varphi^*(c - 0) \right] (t - c)^{-\gamma} + \Phi^*(t) \quad \text{on } cb, \tag{8.23}$$

$\Phi^*(t)$ being given by (8.21).

8.5. Singularity of logarithmic type

Consider the Cauchy type integral

$$\Omega(z) = \frac{1}{2\pi i} \int\limits_{ab} \frac{\ln^{p-1}(\tau - a)}{\tau - z} d\tau, \tag{8.24}$$

where p is a positive integer.

In order to separate the principal part of the expansion of $\Omega(z)$ near the point a we introduce the functions $\omega_p(z, a)$ by the following recursive relations:

$$\omega_1(z, a) = -\frac{1}{2\pi i} \ln (z - a),$$

$$\omega_2(z, a) = -\frac{1}{4\pi i} \ln^2 (z - a) + \frac{1}{2} \ln (z - a),$$

$$\omega_p(z, a) = -\frac{1}{2p\pi i} \ln^p(z - a) - \sum_{k=1}^{p-1} C_{p-1}^k \frac{(2\pi i)^k}{k + 1} \omega_{p-k}(z, a) \quad (p > 1) \tag{8.25}$$

where C_{p-1}^k are the binomial coefficients.

We assume that the contour ab coincides with line of intersection for the function $\omega_1(z, a)$ whence

$$\omega_1^+(t, a) - \omega_1^-(t, a) = 1.$$

It can be proved by a direct calculation that the functions $\omega_p(z, a)$ satisfy for an arbitrary p the relations

$$\omega_p^+(t, a) - \omega_p^-(t, a) = \ln^{p-1}(t - a). \tag{8.26}$$

On the basis of these relations and the Sokhotski formulae for $\Omega(z)$, and applying reasoning analogous to that of § 8.3 we obtain

$$\Omega(z) = \omega_p(z, a) + \Omega_0(z), \tag{8.27}$$

where $\Omega_0(z)$ is a function analytic at the point a.

For points of the contour close to the point a we have

$$\Omega(t) = \frac{1}{2\pi i} \int\limits_{ab} \frac{\ln^{p-1}(\tau - a)}{\tau - t} d\tau = \frac{1}{2} [\omega_p^+(t, a) + \omega_p^-(t, a)] + \Omega_0(t). \quad (8.28)$$

To conclude, let us present somewhat more general relations. Let

$$\Phi(z) = \frac{1}{2\pi i} \int\limits_{ab} \frac{\varphi^*(\pi) \ln^{p-1}(\tau - a)}{\tau - z} d\tau, \quad (8.29)$$

where $\varphi^*(\tau)$ is a function which satisfies the Hölder condition. Then near the point a the following representation is valid:

$$\Phi(z) = \varphi^*(a) \omega_p(z, a) + \Phi_0(z) \quad (8.30)$$

$$\Phi(t) = \tfrac{1}{2} \varphi^*(a) [\omega_p^+(t, a) + \omega_p^-(t, a)] + \Phi_0(t). \quad (8.31)$$

If in the expression for the density of the integral (8.29) a be replaced by b, then the representation of $\Phi(z)$ in the vicinity of the point b has the form (8.30), (8.31), $\varphi^*(a)$ being replaced by $\varphi^*(b)$ and the functions $\omega_p(z, a)$ by $\omega_p^*(z, b)$. Namely we have

$$\omega_1^*(z, b) = \frac{1}{2\pi i} \ln(z - b),$$

$$\omega_2^*(z, b) = \frac{1}{4\pi i} \ln^2(z - b) + \frac{1}{2} \ln(z - b),$$

$$\omega_p^*(z, b) = \frac{1}{2p\pi i} \ln^p(z - b) - \sum_{k=1}^{p-1} (-1)^k C_{p-1}^k \frac{(2\pi i)^k}{k+1} \omega_{p-k}^*(z, b). \quad (8.32)$$

The formulae for points lying inside the contour ab are rather complicated and we do not quote them.

8.6. Singularities of power-logarithmic type

Generalizing the results of § 8.4 and § 8.5 consider the integral

$$\Phi(z) = \frac{1}{2\pi i} \int\limits_{ab} \frac{\varphi^*(\tau) \ln^p(\tau - a)}{(\tau - a)^\gamma (\tau - z)} d\tau. \quad (8.33)$$

Introduce the functions

$$S_0(z, a) = \frac{e^{\gamma\pi i}}{2i \sin \gamma\pi} (z - a)^{-\gamma},$$

$$S_p(z, a) = \frac{e^{\gamma\pi i}}{2i \sin \gamma\pi} \frac{\ln^p(z - a)}{(z - a)^\gamma} + \frac{e^{-\gamma\pi i}}{2i \sin \gamma\pi} \sum_{k=1}^{p} C_p^k (2\pi i)^k S_{p-k}(z, a)$$

$$(p = 1, 2, \ldots). \quad (8.34)$$

It is easy to prove that the functions $S_p(z, a)$ satisfy the conditions

$$S_p^+(t, a) - S_p^-(t, a) = \frac{\ln^p(t - a)}{(t - a)^\gamma} \quad (p = 0, 1, 2, \ldots),$$

This result implies that in the vicinity of the point a the function $\Phi(z)$ has the representation

$$\Phi(z) = \varphi^*(a)\, S_p\,(z,\,a) + \Phi^*(z),\tag{8.35}$$

$$\Phi(t) = \tfrac{1}{2}\,\varphi^*(a)\,[S_p^+(t,\,a) + S_p^-(t,\,a)] + \Phi^*(t).\tag{8.36}$$

where $\Phi^*(z)$ is a function analytic on the plane cut along the contour ab, having in the vicinity of the point a the estimate

$$|\Phi^*(z)| < \frac{A}{|z-a)|^{\alpha_0}}\,;\quad \alpha_0 < \operatorname{Re}\gamma.$$

Similarly, for the integral

$$\Phi(z) = \frac{1}{2\pi i} \int\limits_{ab} \frac{\varphi^*(\tau)\ln^p(\tau-b)}{(\tau-b)^\gamma\,(\tau-z)}\,d\tau\tag{8.37}$$

we obtain in the vicinity of the end b a representation of the same type containing the functions

$$S_0^*(z,\,b) = -\frac{e^{-\gamma\pi i}}{2i\sin\gamma\pi}\,(z-b)^{-\gamma},$$

$$S_p^*(z,\,b) = -\frac{e^{-\gamma\pi i}}{2i\sin\gamma\pi}\,\frac{\ln^p(z-b)}{(z-b)^\gamma} - \frac{e^{\gamma\pi i}}{2i\sin\gamma\pi}\sum_{k=1}^{p} C_p^k\,(-1)^k\,(2\pi i)^k\, S_{p-k}^*(z,\,b).\tag{8.38}$$

The contents of § 8.5 and § 8.6 have been based on the paper of I. M. Melnik (2).

8.7. Integral of the Cauchy type over a complicated contour

So far for simplicity we have been assuming that the Cauchy type integral is taken over a contour consisting of one simple curve, closed or open. There is nothing, however, to prevent us from extending all the results derived above to the case when the contour of integration consists of a finite number of curves which are entirely separate or have a number of points of intersection. Let us at once observe that without loss of generality all curves may be regarded as simple, since if any of the curves were self-intersecting we could divide it into separate sections which would not be self-intersecting, and consider these sections as independent curves.

Thus, let there be given the Cauchy type integral

$$\Phi(z) = \frac{1}{2\pi i} \int\limits_{L} \frac{\varphi(\tau)}{\tau-z}\,d\tau,\tag{8.39}$$

taken over the contour L consisting of a finite number of simple curves L_k, i.e.

$$L = L_1 + L_2 + \cdots + L_m.$$

The set of curves may contain both closed and open curves. It is evident that $\Phi(z)$ can be represented in form of the sum

$$\Phi(z) = \frac{1}{2\pi i} \int\limits_{L_1} \frac{\varphi(\tau)}{\tau-z}\,d\tau + \cdots + \frac{1}{2\pi i} \int\limits_{L_m} \frac{\varphi(\tau)}{\tau-z}\,d\tau,\tag{8.40}$$

Every term of the sum is a Cauchy type integral taken over a simple curve and, consequently, the results of the whole theory obtained above are applicable to it. This fact implies immediately the following theorem.

THEOREM. *The integral of the Cauchy type* (8.39) *taken over the complex L constitutes a function analytic in the entire plane of the complex variable, except for the points of the contour L, and vanishing at infinity.*

Let us examine the limiting values of the integral. Let the point z approach a point t of the contour, which is located on a curve L_j and is not a point of intersection of the curves L_k. All terms of the sum (8.40), except the jth term, are analytic functions and hence they are continuous at the point t. Now, the integral taken over the contour L_j obeys all conditions under which the results of the foregoing theory have been deduced. It follows therefore that this theory is also applicable to the case of complex contour considered now. Thus, in all theorems concerning the limiting values, including the Sokhotski formulae, the formulae for the change of the order of integration, etc., we may regard the contour L as complex, i.e. consisting of a finite number of simple curves, under the condition that the point t at which the limiting values are being investigated, is not a point of intersection of the simple contours L_k.

The case is much more complicated for a point of intersection of simple contours. Here a number of terms of the sum (8.40) undergo a discontinuity simultaneously. The limiting value of the integral (8.39) depends on which of the sectors, into which the plane is divided by simple curves intersecting at the point t, contains the path along which we approach the point t. An investigation of the limiting values at the point of intersection of simple curves offers considerable difficulties, if we consider the integral taken over the whole bundle of simple curves issuing from the point under consideration (as it was done by Trjizinsky(1) who first investigated this problem). However, the difficulties disappear entirely if, as it was proposed by D. A. Kveselava (3), the limiting value at the point of intersection of the curves be taken as the sum of the limiting values of all terms entering the sum (8.40). If the sector in which there lies the path γ along which we approach the point t, is given (Fig. 10), then for each of the integrals over the curves L_k intersecting at the point t, we can exactly determine the side (right or left) on which the approach to the contour occurs; consequently, we can select the Sokhotski formula (4.7) or (4.8) which corresponds to the selected path of approach to the contour.

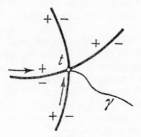

FIG. 10.

We agree in what follows to give no definite value to the integral (8.39) at the point of intersection. The limiting values will be determined for every path of approach to the limit, on the basis of the above formulated principle of splitting the integral into a sum of integrals taken over simple curves. If this method of investigation is adopted, then the case when the point of intersection is an end of the curve passing through this point offers no difficulties, since the examination of the behaviour of the integral is performed by the ordinary method of § 8.2.

§ 9. Limiting values of generalized integrals and double Cauchy integrals

In §§ 9.1–9.3 some of the foregoing results are extended to generalized integrals, and in §§ 9.4–9.6 to double Cauchy integrals.

9.1. Formulation of the problem

Let L be a smooth closed or open contour. Consider the curvilinear integral

$$\Psi(z) = \frac{1}{2\pi i} \int_L \frac{P(\tau, z)}{Q(\tau, z)} d\tau \tag{9.1}$$

under the following assumptions concerning the functions P and Q.

1. P and Q are entire analytic functions with respect to the complex variable z for all values $\tau \in L$.

2. P satisfies the Hölder condition with respect to the variable τ.

3. Q has continuous derivative with respect to τ, which satisfies the Hölder condition.

4. At the points where $Q(\tau, z) = 0$, $Q'_\tau(\tau, z)$ and $Q'_z(\tau, z)$ are distinct from zero.

In accordance with § 1 the integral (9.1) constitutes a function analytic in the entire plane, except for the values of z which satisfy the equation

$$Q(\tau, z) = 0 \quad (\tau \in L), \tag{9.2}$$

and possibly the point at infinity.

The behaviour of $\Phi(z)$ at infinity can be investigated only if certain conditions be imposed upon the behaviour of the entire functions P and Q at infinity. This problem will be left aside.

Equation (9.2) determines a finite or infinite number of curves Γ. Solving this equation for z and taking the value of τ from the equation of the contour L, $\tau = \tau(\sigma)$, we arrive at the equations of the curves Γ in parametric form. In view of the conditions 3 and 4 equation (9.2) determines a one to one continuous correspondence between the contour L and each of the curves Γ. Consequently, all curves Γ are at the same time closed or open, depending on whether the contour L is closed or open. In subsequent considerations we shall, for definiteness, consider the contour L to be closed.

The curves Γ are singular lines for the function $\Phi(z)$.

It turns out that under the conditions 1–4 imposed upon the functions P and Q, the function $\Phi(z)$ behaves on the singular lines Γ in the same manner as the Cauchy type integral on the contour of integration, i.e. it has continuous limiting values on approaching the contour from both sides; these values do not coincide, the curves Γ therefore are the lines of discontinuities for the function $\Phi(z)$.

The method of investigation consists in reducing the involved integral under consideration taken over the contour L, to the Cauchy type integral over the curve Γ. Without explaining the details of the investigation we shall confine ourselves to indicating the principal features.

9.2. Formulae analogous to the Sokhotski formulae for the Cauchy type integral

Let us select a definite curve Γ denoting the complex coordinates of its points by ζ, so that we have the relation

$$Q(\tau, \zeta) = 0. \tag{9.2'}$$

Let

$$\zeta = \psi(\tau) \tag{9.3}$$

be the equation of the curve Γ obtained by solving equation (9.2) with respect to ζ.

According to the Weierstrass theorem the analytic function $Q(\tau, z)$ of two complex variables τ and z, which satisfies the condition $\partial Q/\partial z \neq 0$ is representable in the form

$$Q(\tau, z) = [\psi(\tau) - z]\,\Omega(\tau, z), \tag{9.4}$$

where $\psi(\tau)$, $\Omega(\tau, z)$ are analytic functions in their variables, and $\Omega(\tau, z) \neq 0$ in the vicinities of investigated points.

In our case $Q(\tau, z)$ is not analytic in τ; simple reasoning, however, proves that the expansion (9.4) holds also in this case, the only difference being that ψ and Ω are not analytic in τ, but are functions of the same nature as the function Q itself (condition 3). It is easy to see that the function $\psi(\tau)$ appearing in the expansion (9.4), is the right-hand side of the equation (9.3).

Taking into account the expansion (9.4) we represent the integral (9.1) in the form

$$\Psi(z) = \frac{1}{2\pi i} \int\limits_{L} \frac{P(\tau, z)}{\Omega(\tau, z)}\,\frac{d\tau}{\psi(\tau) - z}. \tag{9.5}$$

In order to reduce the integration over the contour L to the integration over the investigated curve Γ, let us make the substitution (9.3). Denoting by ω the function inverse to ψ,

$$\tau = \omega(\zeta),$$

and introducing the function

$$\varphi(\zeta, z) = \frac{P[\omega(\zeta), z]\,\omega'(\zeta)}{\Omega[\omega(\zeta), z]}, \tag{9.6}$$

we reduce the investigated integral to the following one:

$$\Psi(z) = \frac{1}{2\pi i} \int\limits_{\Gamma} \frac{\varphi(\zeta, z)}{\zeta - z}\,d\zeta. \tag{9.7}$$

The integral (9.7) differs from the ordinary Cauchy type integral only in the fact that its density $\varphi(\zeta, z)$ itself depends on the variable z. However, this fact does not lead to any alterations of the behaviour of the integral near the contour Γ. It is known (see the footnote to § 7.2) that for this integral the ordinary Sokhotski formulae are valid.

Thus, we have

$$\Psi^\pm(w) = \pm \frac{1}{2} \varphi(w, w) + \frac{1}{2\pi i} \int_\Gamma \frac{\varphi(\zeta, w)}{\zeta - w} d\zeta.$$

Setting $\omega(w) = t$ and taking into account that $\omega'(w) = \dfrac{1}{\psi'(t)}$, $\psi'(t)\Omega(t, w) = Q'_\tau(t, w)$, we obtain

$$\Psi^\pm(w) = \pm \frac{1}{2} \frac{P(t, w)}{Q'_\tau(t, w)} + \frac{1}{2\pi i} \int_L \frac{P(\tau, w)}{Q(\tau, w)} d\tau, \tag{9.8}$$

the last integral being understood in the sense of the principal value.

9.3. The formula for change of the order for integration

For the double singular integral of the type considered above, a formula for change of the order of integration is valid, which is similar to the formula (7.7) (the Poincaré–Bertrand formula).

Its derivation is based on the following statement.

LEMMA. *Let there be given on the contour L a function $\varphi(t, \tau)$ which satisfies the conditions:*

1. *It is continuous with respect to both the variables, and its first derivative with respect to τ exists and satisfies the Hölder condition with respect to τ and t.*

2. $\varphi(t, t) = 0$ *but* $\varphi'_\tau(t, t) \neq 0$.

Then $\varphi(t, \tau)$ is representable in the form

$$\varphi(t, \tau) = (\tau - t) s(t, \tau), \tag{9.9}$$

where $s(t, \tau)$ satisfies the Hölder condition and $s(t, t) = \varphi'_\tau(t, t)$.

To prove the lemma we use the obvious relation

$$\varphi(t, \tau) - \varphi(t, t) = \int_t^\tau \varphi'_\tau(t, u)\, du.$$

Taking into account that $\varphi(t, t) = 0$ and substituting

$$u = t + v(\tau - t), \tag{9.10}$$

we obtain the required representation (9.9) in which

$$s(t, \tau) = \int_{L'} \varphi'_\tau[t, t + v(\tau - t)]\, dv,$$

and the integral is taken over the arc L' (with ends 0, 1) into which the substitution (9.10) maps the arc $t\tau$ of the contour L, the mapping being one to one and continuous.

The proof that $s(t,\tau)$ obeys the Hölder condition and that $s(t, t) = \varphi'_\tau(t, t)$ is carried out by direct calculation, and we do not quote it here.

We now proceed to derive the transposition formula. Suppose that

(1) The function $P(\tau, \sigma)$ given on L satisfies the Hölder condition.

(2) The function $\omega(\tau, \sigma)$ satisfies the conditions of the lemma.

(3) $Q(\tau, w)$ is a function which satisfies conditions 3, 4 of § 9.1, w being an arbitrary point of the curve Γ.

Under these conditions the following transposition formula is valid:

$$\int\limits_L \frac{d\tau}{Q(\tau, w)} \int\limits_L \frac{P(\tau, \sigma)}{\omega(\tau, \sigma)}\, d\sigma$$

$$= -\pi^2 \frac{P(t, t)}{Q'_\tau(t, w)\, \omega'_\tau(t, t)} + \int\limits_L d\sigma \int\limits_L \frac{P(\tau, \sigma)}{Q(\tau, w)\, \omega(\tau, \sigma)}\, d\tau. \qquad (9.11)$$

The proof is carried out by reduction to integrals of the Cauchy type. On the basis of the lemma

$$\omega(\tau, \sigma) = (\sigma - \tau)\, s(\tau, \sigma). \qquad (9.9')$$

Since $w = \psi(t)$, the function $Q(\tau, w) = Q[\tau, \psi(t)]$ satisfies the conditions of the lemma. Hence

$$Q[\tau, \psi(t)] = (\tau - t)\, s_1(\tau, t). \qquad (9.9'')$$

This implies that

$$\int\limits_L \frac{d\tau}{Q(\tau, w)} \int\limits_L \frac{P(\pi, \sigma)}{\omega(\tau, \sigma)}\, d\sigma = \int\limits_L \frac{d\tau}{\tau - t} \int\limits_L \frac{\varphi(\tau, \sigma, t)}{\sigma - t}\, d\sigma, \qquad (9.12)$$

where the function

$$\varphi(\tau, \sigma, t) = \frac{P(\tau, \sigma)}{s_1(\tau, t)\, s(\tau, \sigma)}$$

satisfies the Hölder condition with respect to τ and σ, for $t \in L$. Applying to the integral (9.12) the transposition formula (7.7) and bearing in mind that

$$s(t, t) = \omega'_\tau(t, t); \quad s_1(t, t) = Q'_\tau(t, w),$$

i.e.

$$\varphi(t, t, t) = \frac{P(t, t)}{Q'_\tau(t, t)\, \omega'_\tau(t, t)},$$

we arrive at the required relation (9.11).

The problems of these sections were first, not rigorously and not quite precisely, considered in the paper of L. G. Mikhailov (1). The present treatment follows the paper of I. M. Melnik (1).

9.4. Multiple Cauchy integrals. Formulation of the problem

Let the plane of the complex variable $z_k (k = 1, 2, \ldots, n)$ be divided by a given smooth closed contour L_k into two parts—into the interior part D_k^+ and the exterior part D_k^-. In this case the topological products

$$D^{\pm \pm \cdots \pm} = D_1^\pm \times D_2^\pm \times \ldots \times D_n^\pm$$

are regular semi-cylindrical domains†. The boundaries of all these domains have a common part, viz. $L = L_1 \times L_2 \times \ldots \times L_n$, called the seam of the boundaries of the respective domains. The integral

$$\Phi(z_1, \ldots, z_n) = \frac{1}{(2\pi i)^n} \int\limits_L \frac{\varphi(\tau_1, \ldots, \tau_n)\, d\tau_1 \ldots d\tau_n}{(\tau_1 - z_1) \ldots (\tau_n - z_n),}, \tag{9.13}$$

the density of which $\varphi(\tau_1, \ldots, \tau_n)$ is a continuous function of the points of the seam L, is a multiple Cauchy integral.

Below we treat various topics relating to these integrals, the case of two variables being considered for the sake of simplicity. As we shall see, the integral (9.13) largely possesses analogous properties to those of an ordinary single Cauchy integral.

9.5. Singular double integral. Poincaré–Bertrand formula

Suppose that a function $\varphi(t_1, t_2)$ is defined on the seam $L = L_1 \times L_2$ which satisfies Holder's condition with indices λ_1 and λ_2, i.e. that $\varphi \in H(\lambda_1, \lambda_2)$. We set:

$$\varphi_1 = \varphi(\tau_1, t_2) - \varphi(t_1, t_2), \quad \varphi_2 = \varphi(t_1, t_2) - \varphi(t_1, t_2),$$
$$\varphi_{12} = \varphi(\tau_1, \tau_2) - \varphi(\tau_1, t_2) - \varphi(t_1, \tau_2) + \varphi(t_1, t_2).$$

Using the results of 2.2 it is found that

$$\left.\begin{array}{ll} |\varphi_1| \leq A_1 |\tau_1 - t_1|^{\lambda_1}, & |\varphi_2| \leq A_2 |\tau_2 - t_2|^{\lambda_2}, \\ |\varphi_{12}| \leq 2A_1 |\tau_1 - t_1|^{\lambda_1}, & |\varphi_{12}| \leq 2A_2 |\tau_2 - t_2|^{\lambda_2}. \end{array}\right\} \tag{9.14}$$

From the last two inequalities, we have

$$|\varphi_{12}| \leq 2\sqrt{(A_1 A_2)} |\tau_1 - t_1|^{\frac{\lambda_1}{2}} |\tau_2 - t_1|^{\frac{\lambda_2}{2}}. \tag{9.15}$$

Also note the identity:

$$\varphi(\tau_1, \tau_2) = \varphi_1 + \varphi_2 + \varphi_{12} + \varphi(t_1, t_2). \tag{9.16}$$

We consider the singular integrals:

$$\tilde{\varphi}_1 = S_1\varphi = \frac{1}{\pi i} \int\limits_{L_1} \frac{\varphi(\tau_1, t_2)\, d\tau_1}{\tau_1 - t_1,}, \quad \tilde{\varphi}_2 = S_2\varphi = \frac{1}{\pi i} \int\limits_{L_1} \frac{\varphi(t_1, \tau_2)\, d\tau_2}{\tau_2 - t_2,},$$

$$\tilde{\varphi}_{12} = S_{12}\varphi = -\frac{1}{\pi^2} \int\limits_L \frac{\varphi(\tau_1, \tau_2)\, d\tau_1\, d\tau_2}{(\tau_1 - t_1)(\tau_2 - t_2)}.$$

Obviously, the integrals $S_1\varphi$ and $S_2\varphi$ exist in the sense of the principal value (see § 3.4).

---

† The point $(z_1, \ldots, z_n)$ belongs to the topological product $D_1 \times D_2 \times \ldots \times D_n$ of the sets $D_k (k = 1, 2, \ldots, n)$ if $z_k \in D_k$.

We describe the circle $|z_k - t_k| = \varrho$ in the plane z_k with its centre at a point $t_k \in K_k$ and put l_k for the part of L_k within the circle. The principal value of the singular integral $S_{12}\varphi$ is

$$\lim_{\varrho \to 0} \int_{L - l_1 \times l_2} \frac{\varphi(\tau_1, \tau_2)(d\tau_1 \, d\tau_2)}{(\tau_1 - t_1)(\tau_2 - t_2)}. \tag{9.17}$$

If $\varphi \in H(\lambda_1, \lambda_2)$, then by substituting the right-hand side of (9.16) for $\varphi(\tau_1, \tau_2)$ in (9.17) and using the inequalities (9.14), (9.15) and the equality (4.6), it can be shown (as in § 3.4) that the integral $S_{12}\varphi$ also exists in the sense of the principal value.

Note also the following operator equalities, either proved above, or easily proved,

$$S_1 S_2 = S_2 S_1 = S_{12}, \quad S_j^2 = S_0, \quad S_j S_0 = S_0 S_j = S_j, \quad j = 1, 2, \tag{9.18}$$

where S_0 is an identity operator, i.e. $S_0\varphi = \varphi$.

By using Kupradse's method [13] of proving the Poincaré–Bertrand formula, the appropriate Poincaré–Bertrand formula can be obtained for two variables. In operator form this may be written as:

$$S_{12}(a\tilde{\varphi}_{12}) - S_{12}(\varphi\tilde{a}_{12}) = a\varphi + \tilde{a}_1\tilde{\varphi}_1 + \tilde{a}_2\tilde{\varphi}_2 + \tilde{a}_{12}\tilde{\varphi}_{12} - $$
$$- S_1(\varphi_1\tilde{a}_1 + \tilde{\varphi}_2\tilde{a}_{12}) - S_2(\varphi_2\tilde{a}_2 + \tilde{\varphi}_1\tilde{a}_{12}),$$

where $\alpha(t_1, t_2), \varphi(t_1, t_2)$ are functions of points on the seam which satisfy Hölder's condition there.

If $a \equiv 1$ this formula yields a formula for the convolution of the operator S_{12}:

$$S_{12}^2\tilde{\varphi} = S_{12}\varphi_{12} = \varphi,$$

from which the following inversion formulae follow in turn:

if $\psi = S_{12}\varphi$ and $\varphi \in H$, then $\varphi = S_{12}\psi$ and vice versa.

9.6. Sokhotski's formulae

If $\varphi(t_1, t_2) \in H(\lambda_1, \lambda_2)$, then, as in the proof of the lemma in § 4.1, it may be shown that in crossing the contours L_1, L_2 and the seam L, the integrals

$$\psi_1 = \frac{1}{2\pi i} \int_{L_1} \frac{\varphi_1 \, d\tau_1}{\tau_1 - z_1}, \quad \psi_2 = \frac{1}{2\pi i} \int_{L_2} \frac{\varphi_2 \, d\tau_2}{\tau_2 - z_2},$$

$$\psi_{12} = -\frac{1}{4\pi^2} \int_{L} \frac{\varphi_{12} \, d\tau_1 \, d\tau_2}{(\tau_1 - z_1)(\tau_2 - z_2)}$$

behave as continuous functions.

We put $\Phi^{\pm\pm}(t_1, t_2)$ for the limit values of the integral (9.13) when the point $(z_1, z_2) \in D^{\pm\pm}$ tends to the point $(t_1, t_2) \in L$.

From the obvious equality

$$\Phi(z_1, z_2) = \psi_{12} + \frac{\psi_1}{2\pi i} \int_{L_2} \frac{d\tau_2}{\tau_2 - z_2} + \frac{\psi_2}{2\pi i} \int_{L_1} \frac{d\tau_1}{\tau_1 - z_1} -$$
$$- \frac{\varphi(t_1, t_2)}{4\pi^2} \int_L \frac{d\tau_1 \, d\tau_2}{(\tau_1 - z_1)(\tau_2 - z_2)},$$

as in the derivation of Sokhotski's formulae for a one-dimensional Cauchy integral in § 4.2, we have:

$$\left.\begin{matrix} \Phi^{++} \\ \Phi^{--} \end{matrix}\right\} = \frac{1}{4}(\varphi \pm S_1\varphi \pm S_2\varphi + S_{12}\varphi),$$
$$\left.\begin{matrix} \Phi^{+-} \\ \Phi^{-+} \end{matrix}\right\} = \frac{1}{4}(-\varphi \mp S_1\varphi \pm S_2\varphi + S_{12}\varphi). \tag{9.19}$$

Hence

$$\Phi^{++} \pm \Phi^{+-} \pm \Phi^{-+} + \Phi^{--} = \begin{cases} S_{12}\varphi, \\ \varphi, \end{cases}$$
$$\Phi^{++} \mp \Phi^{+-} \pm \Phi^{-+} - \Phi^{--} = \begin{cases} S_1\varphi, \\ S_2\varphi. \end{cases} \tag{9.20}$$

Sokhotski's formulae (9.19) can be compactly written as follows:
$$4\Phi^{\pm\pm} = \{(\pm S_0 + S_1)(\pm S_0 + S_2)\}\,\varphi,$$

where it is first necessary to find the corresponding operators from formulae (9.18) and then apply them to the function $\varphi(t_1, t_2)$.

For arbitrary n Sokhotski's formulae become
$$2^n \Phi^{\pm\pm \cdots \pm} = \{(\pm S_0 + S_1)(\pm S_0 + S_2) \ldots (\pm S_0 + S_n)\}\,\varphi.$$

Using Muskhelishvili's results (Ref. 17*, § 18), we have: if $\varphi(t_1, t_2) \in H(\lambda_1, \lambda_2)$, then the singular integral $S_{12}\varphi = \Phi(t_1, t_2)$ satisfies the condition:
$$|\Phi(t_1, t_2) - \Phi(\tau_1, \tau_2)| \leq \sum_{j=1}^{2} A_j |\tau_j - t_j|^{\lambda_j} |\ln|\tau_j - t_j||.$$

The same condition is also satisfied by the limit values Φ^{++}, implying that $\Phi^{++} \in H(\lambda_1 - \varepsilon, \lambda_2 - \varepsilon)$, where ε is an arbitrarily small positive number. In §§ 9.4–9.6 we have closely followed Kakichev's articles (1 and 2). As regards Sokhotski's formulae, see also the articles of Gagua (1) and Bittner (1).

§ 10. Integral of the Cauchy type and potentials

It was already indicated before that there exists a close connection between the Cauchy type integral and (logarithmic) potentials. We shall now explicitly state this relation.

Assuming that the contour L is closed we set

$$\Phi(z) = u(x, y) + iv(x, y) = \frac{1}{2\pi i} \int_L \frac{\varphi(\tau)}{\tau - z} \, d\tau.$$

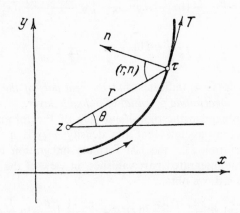

FIG. 11.

For simplicity we assume that the density $\varphi(\tau)$ is a real function. If it were complex it would be necessary to split the Cauchy type integral into two, separating the real and imaginary parts of the density.

Introducing the notation (Fig. 11)

$$\tau - z = re^{i\theta},$$

$$r = |\tau - z|; \quad \theta = \arg(\tau - z), \tag{10.1}$$

we have

$$\frac{d\tau}{\tau - z} = d\ln(\tau - z) = d(\ln r + i\theta) = \frac{dr}{r} + id\theta. \tag{10.2}$$

Hence

$$u(x, y) + iv(x, y) = \frac{1}{2\pi} \int_L \varphi \frac{d\theta}{ds} \, ds - \frac{i}{2\pi} \int_L \varphi \frac{dr}{r}. \tag{10.3}$$

In accordance with the Cauchy–Riemann equations applied to the function $\ln(\tau - z) = \ln r + i\theta$ we have

$$\frac{d\theta}{ds} = -\frac{\partial \ln r}{\partial n} = \frac{\partial}{\partial n} \ln \frac{1}{r}.$$

Consequently,

$$u(x, y) = \frac{1}{2\pi} \int_L \varphi \frac{d\theta}{ds} ds$$

$$= \frac{1}{2\pi} \int_L \varphi \frac{\partial}{\partial n}\left(\ln \frac{1}{r}\right) ds = \frac{1}{2\pi} \int_L \varphi \frac{\cos(r, n)}{r} ds, \qquad (10.4)$$

$$v(x, y) = \frac{1}{2\pi} \int_L \varphi \, d\ln \frac{1}{r} = -\frac{1}{2\pi} \int_L \varphi \frac{dr}{r}. \qquad (10.5)$$

The formula (10.4) indicates that *the real part of the Cauchy type integral is the logarithmic potential of a double layer*.

In order to express the right-hand side of the formula (10.5) by potentials, let us assume that the density $\varphi(t)$ is a differentiable function. Performing in the last formula integration by parts and noting that the integrated part vanishes, in view of the fact that the contour is closed, we obtain

$$v(x, y) = \frac{1}{2\pi} \int_L \frac{d\varphi}{ds} \ln r \, ds = -\frac{1}{2\pi} \int_L \frac{d\varphi}{ds} \ln \frac{1}{r} ds. \qquad (10.6)$$

Comparing the latter expression with the potential of a simple layer in ordinary notation

$$\int_L \mu(s) \ln \frac{1}{r} ds,$$

we observe that *the imaginary part of the Cauchy type integral is the logarithmic potential of a simple layer* with the density

$$\mu(s) = -\frac{1}{2\pi} \frac{d\varphi}{ds}.$$

Since the total charge (mass) of the potential is

$$m = \int_L \mu(s) \, ds = -\frac{1}{2\pi} \int_L \frac{d\varphi}{ds} ds = 0, \qquad (10.7)$$

the formula (10.6) represents a potential of not entirely general form, the condition being implicit that the total charge is zero.

Observe that even if we do not assume that the density $\varphi(t)$ is differentiable, we could integrate by parts, but the resulting integral $\int_L \ln \frac{1}{r} \, d\varphi$ should be understood in the Stieltjes sense.

Making use of the representation of the Cauchy type integral by the potentials it is possible to derive the Sokhotski formulae for its limiting values. The first term—the potential of a double layer—constitutes a function possessing a discontinuity on passing through the contour L. The limiting values are given by the expression

$$\pm \frac{1}{2} \varphi + \frac{1}{2\pi} \int_L \varphi \frac{d\theta}{ds} \, ds . \tag{10.8}$$

The potential of a simple layer behaves as a continuous function in passing through the contour; it is necessary to regard the integral (10.6) as singular in the sense of the principal value. Summing the two terms we arrive at the Sokhotski formulae. Let us note that for curves which satisfy the so-called Lyapounov condition (the function $\alpha(s)$ where α is the angle between the tangent and the x-axis, satisfies the Hölder condition), derivation of the Sokhotski formulae by the method indicated above offers no difficulties (provided that the formulae (10.6) for the limiting values of the potential of a double layer have already been derived). However, for the case of an arbitrary smooth contour considerable difficulties would be encountered, which are due to the questionable existence of the integral

$$\int_L \varphi \frac{d\theta}{ds} \, ds$$

on the contour.

§ 11. Historical notes

It is difficult to name the author who first drew attention to the Cauchy type integral, i.e. who proposed to consider the density of the Cauchy integral as an arbitrary function. It is however possible to indicate exactly the time and the author who first presented a most important investigation on the theory of this integral. In 1873, in his doctor's thesis, Y. V. Sokhotski [25], investigated the behaviour of the Cauchy type integral on the contour and derived the formulae (4.7), (4.8), under the same conditions (the Hölder condition) as ours, in § 4. It is remarkable that he also considered the cases of discontinuities of the density.

These investigations are not rigorous from the modern viewpoint, but they entirely conformed to the requirements of rigor of that time. Nevertheless, the fundamental investigations of Sokhotski were unjustly forgotten and did not influence the subsequent development of the theory†.

The next investigation of the limiting values of the Cauchy type integral is due to Harnack (1), in 1885. This author splits the Cauchy type integral into the sum of two potentials (see § 10) and derives the Sokhotski formulae, on the basis of the formulae for the limiting values of potential of a double layer. Strong limitations are imposed on the density and the contour. The paper of Harnack had no significant influence on the subsequent investigations; although independent of Sokhotski's papers, they were in accordance with the programme outlined by him—the investigation of the Cauchy type integral as an entity.

The two papers of Plemelj (1), (2), are of great importance. A new derivation of the Sokhotski formulae was given, under the same assumptions of the Hölder condition, but already sufficiently rigorously. An outline of the proof of the theorem on the behaviour of the limiting values of the Cauchy type integral, was presented (see § 5.1); the most important result, however, is the fact that here for the first time the Cauchy type integral was employed as a mathematical device for solving a definite boundary value problem of the theory of analytic functions.

The further development of the theory and applications of the Cauchy type integral is mainly connected with the names of Soviet mathematicians—I. I. Privalov and N. I. Muskhelishvili. In the papers of Privalov, the first of which appeared in 1918, the Sokhotski formulae for functions satisfying the Hölder condition were derived with utmost completeness and rigour, and far-reaching generalizations were carried out with respect to the density of the integral and the contour of integration. Some information on these generalizations have been quoted in § 5.2. In the papers of N. I. Muskhelishvili, beginning in 1922, the Cauchy type integral has systematically been applied to solution of problems of the theory of elasticity.

These papers, which have become widely known through his book [18], have been of primary importance in introducing the Cauchy type integral as a mathematical device to various fields of pure and applied mathematics. In 1941 N. I. Muskhelishvili published a paper (1) in which a profound investigation of the behaviour of the Cauchy type integral at the ends of the contour of integration and at the points of discontinuity of density, was presented. This paper immediately found a wide application in the theory of boundary value problems and singular integral equations.

Among the recent papers on the theory of the Cauchy type integral the most valuable are those of L. G. Magnaradze (1), S. G. Mikhlin and B. V. Khvedelidze. In Mikhlin's paper [16] the Cauchy type integral is considered as an operator in the space of functions with integrable square. This makes it possible to use functional analysis in some cases as a tool of investigation. A similar investigation was carried out by B. V. Khvedelidze (2) for functions with integrable power $p > 1$, and with a weight.

The most complete exposition of the Cauchy type integral, under the classical assumptions ensuring existence of continuous limiting values everywhere on the contour, is given in the book of N. I. Muskelishvili [17*]. This exposition constitutes the basis of many sections of this chapter (§ 4, § 7, § 8.3, § 8.4).

† The importance of the papers of Yu. V. Sokhotski in the theory of the Cauchy type integral, was first indicated by A. I. Markushevitch.

Problems on Chapter I

1. Compute the singular integral $\int\limits_a^b \dfrac{dx}{x-c}$ defining it as

$$\lim_{\varepsilon \to 0} \left[\int\limits_a^{c-\varepsilon} \frac{-dx}{c-x} + \int\limits_{c+k\varepsilon}^b \frac{dx}{x-c} \right].$$

2. Prove that if $n > 0$ is an odd number, then the singular integral

$$\int\limits_a^b \frac{dx}{(x-c)^n}$$

defined as the principal value, exists.

3. Prove that if n is odd and $\varphi(x)$ satisfies the condition $|\varphi(x) - \varphi(c)| \leq$ $\leq A|x - c|^\lambda$, where $\lambda > n - 1$, then the integral

$$\int\limits_a^b \frac{\varphi(x)}{(x-c)^n}\, dx$$

exists.

4. Prove that if the singular integral

$$\int\limits_L \frac{d\tau}{\tau - t}$$

be defined by the condition

$$\lim_{t_1, t_2 \to t} \left| \frac{t_2 - t}{t_1 - t} \right| = k$$

(notation of § 3.3), then the limiting values of the Cauchy type integral

$$\Phi(z) = \frac{1}{2\pi i} \int\limits_L \frac{\varphi(\tau)}{\tau - z}\, d\tau$$

are given by the formulae

$$\Phi^+(t) = \frac{1}{2}\left[1 + \frac{\ln k}{\pi i}\right] \varphi(t) + \frac{1}{2\pi i} \int\limits_L \frac{\varphi(\tau)}{\tau - t}\, d\tau,$$

$$\Phi^-(t) = \frac{1}{2}\left[-1 + \frac{\ln k}{\pi i}\right] \varphi(t) + \frac{1}{2\pi i} \int\limits_L \frac{\varphi(\tau)}{\tau - t}\, d\tau.$$

5*. Prove that if $\varphi(t)$ satisfies the condition $H(\lambda)$, then for the derivative of the Cauchy type integral

$$\Phi(z) = \frac{1}{2\pi i} \int\limits_L \frac{\varphi(\tau)}{\tau - z}\, d\tau$$

at the points of the contour, the following estimate holds:

$$|\Phi'(z)| < A|z-t|^{\lambda-1} \quad \text{for} \quad \lambda < 1,$$
$$|\Phi'(z)| < A \ln|z-t| \quad \text{for} \quad \lambda = 1.$$

(N. I. Muskhelishvili [17*], § 20).

6. Compute the singular integral

$$\frac{1}{\pi i} \int_{|\tau|=1} \frac{\tau^2}{\tau^3 - 1} d\tau.$$

Answer 1.

7. Let L be a simple closed contour dividing the plane into the regions D and D^-. Making use of the Cauchy theorem and the Cauchy formulae prove that if the density of the Cauchy type integral

$$\Phi(z) = \frac{1}{2\pi i} \int_L \frac{\varphi(\tau)}{\tau - z} d\tau$$

may be represented in the form

$$\varphi(\tau) = \varphi^+(\tau) + \varphi^-(\tau),$$

where φ^+ and φ^- are the boundary values of functions analytic in D^+ and D^- respectively, then the following formulae are valid:

$$\Phi^+(z) = \varphi^+(z), \quad \Phi^-(z) = -\varphi^-(z) + \varphi^-(\infty).$$

Derive the same on the basis of the Sokhotski formulae.

8. Making use of the result of the preceding problem compute the Cauchy type integral for the following instances:

(a) $\varphi(t)$ is the boundary value of a function analytic in D^+, except at a finite number of points a_k where it has poles.

(b) $\varphi(t)$ is a function of the same type in D^-.

Answer.

(a) $\Phi^+(z) = \varphi(z) - \sum g_k\left(\dfrac{1}{z - a_k}\right); \quad \Phi^-(z) = -\sum g_k\left(\dfrac{1}{z - a_k}\right);$

(b) $\Phi^+(z) = \sum g_k\left(\dfrac{1}{z - a_k}\right) + g(z),$

$\Phi^-(z) = -\varphi(z) + \sum g_k\left(\dfrac{1}{z - a_k}\right) + g(z),$

where $g_k(z), g(z)$ are polynomials representing the principal parts of the expansion of $\varphi(t)$ in the vicinity of the finite poles and the infinity, respectively ($g_k(0) = 0$).

9. Compute the Cauchy type integral

$$\Phi(z) = \frac{1}{2\pi i} \left[\int_L \frac{\tau^3 + i\tau^2 + \tau - 4i}{\tau^4 - 3\tau^2 - 4} + \frac{\ln\dfrac{\tau-2}{\tau-3}}{\tau^2 - 4} - \frac{e^{1/\tau} + \sin\dfrac{1}{\tau-1}}{\tau^2 + 1} \right] \frac{d\tau}{\tau - z},$$

where L is the circle $|z| = \frac{3}{2}$.

Answer.

$$\Phi^+(z) = \frac{z + \ln \dfrac{z-2}{z-3}}{z^2 - 4}, \quad \Phi^-(z) = \frac{e^{1/z} + \sin \dfrac{1}{z-1} - i}{z^2 + 1}.$$

10. Making use of the formulae (4.11), (4.12) compute the singular integral

$$\frac{1}{\pi i} \int_L \left[\frac{\tau^3 + i\tau^2 + \tau - 4i}{\tau^4 - 3\tau^2 - 4} + \frac{\ln \dfrac{\tau-2}{\tau-3}}{\tau^2 - 4} - \frac{e^{1/\tau} + \sin \dfrac{1}{\tau-1}}{\tau^2 + 1} \right] \frac{d\tau}{\tau - t},$$

where L is the contour of Problem 9.

Answer.

$$\frac{t - \ln \dfrac{t-2}{t-3}}{t^2 - 4} + \frac{e^{1/t} + \sin \dfrac{1}{t-1} - i}{t^2 + 1}.$$

11*. Derive the formula for the difference between the limiting values of the Cauchy type integral

$$\Phi^+(t) = \Phi^-(t) - \varphi(t),$$

without assuming that $\varphi(t)$ satisfies the Hölder condition, but only that it is continuous (N. I. Muskhelishvili [17*], § 17).

12. Prove that if $\varphi(t)$ is the boundary value on the contour L of a function $\Phi(z)$ analytic in the domain D^+, except at a finite number of points where it has poles, the sum of the principal parts of the expansion of $\Phi(z)$ in the vicinity of the poles being $R(z)$, then it should satisfy the condition

$$-\frac{1}{2}\varphi(t) + \frac{1}{2\pi i} \int_L \frac{\varphi(\tau)}{\tau - t} d\tau + R(t) = 0.$$

13. Derive the formulae for the limiting values of the mth derivative of the Cauchy type integral

$$\Phi(z) = \frac{1}{2\pi i} \int_L \frac{\varphi(\tau)}{\tau - z} d\tau,$$

assuming that the contour is open, its ends being a and b. The density $\varphi(t)$ is assumed to be differentiable the required number of times.

14. Compute the singular integrals with the Hilbert kernels:

(a) $\dfrac{1}{2\pi} \displaystyle\int_0^{2\pi} \cos n\theta \, \cot \dfrac{\theta - \varphi}{2} \, d\theta;$

(b) $\dfrac{1}{2\pi} \displaystyle\int_0^{2\pi} \sin n\theta \, \cot \dfrac{\theta - \varphi}{2} \, d\theta;$

(c) $\dfrac{1}{2\pi} \displaystyle\int_0^{2\pi} \ln 2 \left| \sin \dfrac{\theta}{2} \right| \cot \dfrac{\theta - \varphi}{2} \, d\ldots$

Answer.

(a) $-\sin n\varphi$;   (b) $\cos n\varphi$;   (c) $\dfrac{\pi - \varphi}{2}$.

N.B. From (a) if $n = 0$ we get

$$\int\limits_0^{2\pi} \cot \frac{\theta - \varphi}{2}\, d\theta = 0.$$

Obtain this result by direct evaluation of the singular integral as the principal value.

15. Compute the singular integral

$$\frac{1}{\pi} \int\limits_{-1}^1 \frac{\ln\,[(1 - \tau)^p\,(1 + \tau)^q]}{(\tau - t)\,\sqrt{(1 - \tau^2)}}\, d\tau .$$

Hint. Perform the change of variables $\tau = \cos\theta$, $t = \cos\varphi$, and take into account the identity

$$\frac{1}{\cos\theta - \cos\varphi} = \frac{1}{2\sin\varphi}\left[\cot \frac{\varphi + \theta}{2} + \cot \frac{\varphi - \theta}{2}\right],$$

making use of the result of the preceding example.

Answer.

$$\frac{p\,(\operatorname{arc\,cos} t - \pi) + q\,\operatorname{arc\,cos} t}{\sqrt{(1 - t^2)}} .$$

16. Prove that if $\varphi(s)$ is a periodic differentiable function, then for the singular integral with the Hilbert kernel the following formula for integration by parts is valid:

$$\frac{1}{2\pi} \int\limits_{-\pi}^{\pi} \varphi(\sigma) \cot \frac{\sigma - s}{2}\, d\sigma = -\frac{1}{2\pi} \int\limits_{-\pi}^{\pi} \varphi'(\sigma) \ln\left[\sin^2 \frac{\sigma - s}{2}\right] d\sigma .$$

17. Compute the Cauchy type integral

$$\Phi(z) = \frac{1}{2\pi i} \int\limits_L \frac{\ln t}{t - z}\, dt,$$

if the origin of the coordinate system belongs to D^+ and $\ln t$ is determined by the condition

$$0 \leq \operatorname{Im} \ln t < 2\pi.$$

Hint. Apply the Cauchy formula or the Cauchy theorem to the integral taken over the contour consisting of the curve L, the edges of the cut connecting the point t_0 with the origin, and a small circle about the origin.

Answer.

$$\Phi^+(z) = \ln(z - t_0), \quad \Phi^-(z) = \ln\left(1 - \frac{t_0}{z}\right).$$

18. Let $L = \sum\limits_{k=1}^{m} L_k$ be a contour consisting of m simple open curves having no common ends. The coordinates of the ends, taken in certain order, are denoted by $c_1, c_2, \ldots, c_{2m}$. Let p be an integer, $0 \leq p \leq 2m$, and $P(z)$ a polynomial. Compute the singular integral

$$I(t) = \frac{1}{\pi i} \int\limits_{L} \frac{\prod\limits_{k=1}^{p} (\tau - c_k)^{\frac{1}{2}}}{\prod\limits_{j=p+1}^{2m} (\tau - c_j)^{\frac{1}{2}}} \cdot \frac{P(\tau)}{\tau - t} \, d\tau .$$

Hint. Consider the Cauchy type integral

$$\Phi(z) = \frac{1}{4\pi i} \int \frac{\prod\limits_{k=1}^{p} (\tau - c_k)^{\frac{1}{2}}}{\prod\limits_{j=p+1}^{2m} (\tau - c_j)^{\frac{1}{2}}} \frac{P(\tau)}{\tau - z} \, d\tau ,$$

taken over a union of loops enclosing the curves L_k. Pass to the limit, the loops tending to the cuts along the curves L_k, and make use of the Sokhotski formula $I = \Phi^+ + \Phi^-$.

Answer. $I(t) = -P^*(t)$, where $P^*(z)$ is a polynomial representing the principal part of the expansion of the density of the integral of the Cauchy type, in the vicinity of infinity, i.e. a polynomial which obeys the condition

$$\lim_{z \to \infty} \left[\frac{\prod\limits_{k=1}^{p} (\tau - c_k)^{\frac{1}{2}}}{\prod\limits_{j=p+1}^{2m} (z - c_j)^{\frac{1}{2}}} P(z) - P^*(z) \right] = 0 .$$

19. Making use of the result of the preceding problem compute the following singular integrals:

(a) $\dfrac{1}{\pi} \int\limits_{-1}^{1} \sqrt{\left(\dfrac{1-t}{1+t}\right)} \dfrac{t^2 + 3}{t - x} \, dt$; (b) $\dfrac{1}{\pi} \int\limits_{-a}^{a} \dfrac{t^n}{\sqrt{(a^2 - t^2)}} \dfrac{dt}{t - x}$ ($n > 0$, an integer)

(c) $\dfrac{1}{\pi i} \int\limits_{a}^{b} \dfrac{\tau^3 - 2\tau^2 + 3\tau + 1}{\sqrt{[(\tau - a)(\tau - b)]}} \dfrac{d\tau}{\tau - t}$.

Answer. (a) $-x^2 + x - 3{\cdot}5$; (b) $x^{n-1} + \sum\limits_{k=1}^{m} \dfrac{1.3 \ldots (2k-1)}{2.4 \ldots 2k} \cdot a^{2k} x^{n-2k-1}$,

where $m = \dfrac{n}{2} - 1$ for n even, and $m = \dfrac{n-1}{2}$ for n odd; (c) $-t^2 + \frac{1}{2}(4 - a - b)t + \frac{1}{8}(8a + 8b - 2ab - 3a^2 - 3b^2)$.

20. Compute the singular integrals

(a) $I_p(t) = \dfrac{1}{\pi i} \displaystyle\int_a^b \dfrac{\left[\ln \dfrac{\tau - a}{\tau - b}\right]^p}{\sqrt{[(\tau - a)(\tau - b)]}} \dfrac{d\tau}{\tau - t};$

(b) $I_p(t) = \dfrac{1}{\pi i} \displaystyle\int_a^b \dfrac{\left[\ln \dfrac{\tau - a}{\tau - b}\right]^p}{\tau - t} d\tau$

(p is a non-negative integer).

Answer.

(a) $I_p(t) = -\dfrac{1}{2} \displaystyle\sum_{k=1}^p C_p^k (2\pi i)^k \left[\dfrac{\ln^{p-k} \dfrac{t-a}{t-b}}{\sqrt{[(t-a)(t-b)]}} + I_{p-k}(t)\right], \quad I_0(t) = 0,$

(b) $I_p(t) = -\dfrac{1}{(p+1)\,2\pi i} \left[2 \ln^{p+1} \dfrac{t-a}{t-b} + \right.$

$\left. + \displaystyle\sum_{k=1}^{p+1} C_{p+1}^k (2\pi i)^k \ln^{p-k+1} \dfrac{t-a}{t-b} + \sum_{k=1}^{p+1} C_{p+1}^k (2\pi i)^k I_{p-k+1}(t)\right],$

where C_p^k are the binomial coefficients.

21. Prove that

$$\Phi(z) = \dfrac{1}{2\pi i} \int_{-a}^a v(t) \coth(t - z)\, dt$$

is a periodic function, the period being πi, and analytic outside the sections $-a \leq x \leq a, y = k\pi i\,(k = 0, \pm 1, \pm 2, \ldots)$.

22. Prove that for the function $\Phi(z)$ of the preceding problem the following relation holds:

$$\Phi^+(\varkappa) + \Phi^-(\varkappa) = \dfrac{1}{\pi i} \int_{-\tanh a}^{\tanh a} v_1(s) \dfrac{ds}{s - \sigma} + \dfrac{1}{\pi i} \int_{-\tanh a}^{\tanh a} v_1(s) \dfrac{s\,ds}{1 - s^2}.$$

Here $v_1(s) = v(\operatorname{arctanh} s)$, $\sigma = \tanh x\,(-a < x < a)$.

23*. Let L be a smooth closed contour, the length of the arc being l, and $G^+(z,\zeta)$, $G^-(z,\zeta)$ the Green functions of the Neumann problem for the domains D^+ and D^- respectively. Prove that the following inversion formulae of Hilbert are valid:

$$v^+(s) = \dfrac{1}{l} \int_0^l \dfrac{\partial G^+(s,\sigma)}{\partial \sigma} u^+(\sigma)\, d\sigma + \dfrac{1}{l} \int_0^l v^+(\sigma)\, d\sigma,$$

$$u^+(s) = -\dfrac{1}{l} \int_0^l \dfrac{\partial G^+(s,\sigma)}{\partial \sigma} v^+(\sigma)\, d\sigma + \dfrac{1}{l} \int_0^l u^+(\sigma)\, d\sigma,$$

$$v^-(s) = -\frac{1}{l} \int_0^l \frac{\partial G^-(s,\sigma)}{\partial \sigma} u^-(\sigma) \, d\sigma + \frac{1}{l} \int_0^l v^-(\sigma) \, d\sigma,$$

$$u^-(s) = \frac{1}{l} \int_0^l \frac{\partial G^-(s,\sigma)}{\partial \sigma} v^-(\sigma) \, d\sigma + \frac{1}{l} \int_0^l u^-(\sigma) \, d\sigma,$$

where

$$G^\pm(s,\sigma) = \cot \frac{\pi}{l}(s-\sigma) + \text{regular function}$$

(Hilbert [9], pp. 84–88).

Hint. Make use of the expression for the solution of the Neumann problem, in terms of the Green function, the Cauchy–Riemann equations in the form

$$\frac{\partial u}{\partial n} = -\frac{\partial v}{\partial s}; \quad \frac{\partial v}{\partial n} = \frac{\partial u}{\partial s},$$

and integrate by parts.

24*. Prove that the necessary and sufficient condition that $f^+(s) = a(s) + ib(s)$ is the boundary value of a function analytic in the domain D^+, is constituted by the relation

$$f^+(s) - \frac{i}{2\pi} \int_0^l \frac{\partial G^+(s,\sigma)}{\partial \sigma} f^+(\sigma) \, d\sigma - \frac{1}{l} \int_0^l f^+(\sigma) \, d\sigma = 0.$$

Similarly, for the domain D^- the condition has the form

$$f^-(s) + \frac{i}{2\pi} \int_0^l \frac{\partial G^-(s,\sigma)}{\partial \sigma} f^-(\sigma) \, d\sigma - \frac{1}{l} \int_0^l f^-(\sigma) \, d\sigma = 0$$

(Hilbert [9]).

Hint. Make use of the formulae of the preceding example.

25. Let $f(z) = u(x,y) + iv(x,y)$ be a function analytic in the upper semi-plane. Denote by $u(x)$, $v(x)$ the limiting values of its real and imaginary parts on the real axis. Prove that the Hilbert inversion formulae are valid:

$$v(x) = -\frac{1}{\pi} \int_{-\infty}^\infty \frac{u(\xi)}{\xi - x} \, d\xi + v_\infty; \quad u(x) = \frac{1}{\pi} \int_{-\infty}^\infty \frac{v(\xi)}{\xi - x} \, d\xi + u_\infty.$$

Here u_∞, v_∞ are the values of the real and imaginary parts at infinity.

26. (a) Compute the Cauchy type integral

$$\Phi(z) = \frac{1}{2\pi i} \int_{-\infty}^\infty \frac{\varphi(\tau)}{\tau - z} \, d\tau,$$

if $\varphi(\tau) = \varphi^+(\tau) + \varphi^-(\tau)$ where $\varphi^\pm(\tau)$ satisfy the Hölder condition everywhere on the axis, and constitute the boundary values of functions analytic in the upper and lower semi-planes, respectively.

Answer.

$$\Phi^+(z) = \varphi^+(z) - \tfrac{1}{2}\varphi^+(\infty) + \tfrac{1}{2}\varphi^-(\infty),$$
$$\Phi^-(z) = -\varphi^-(z) - \tfrac{1}{2}\varphi^+(\infty) - \tfrac{1}{2}\varphi^-(\infty).$$

N.B. We may assume

$$\varphi^+(\infty) = \varphi(\infty), \quad \varphi^-(\infty) = 0 \quad \text{or} \quad \varphi^-(\infty) = \varphi(\infty), \quad \varphi^+(\infty) = 0.$$

(b) Making use of the solution of (a), compute the Cauchy type integral

$$\Phi(z) = \frac{1}{\pi i} \int_{-\infty}^{\infty} \frac{\tau^2}{\tau^2+1} \frac{d\tau}{\tau-z}.$$

Answer.

$$\Phi^+(z) = \frac{z}{z+i}, \quad \Phi^-(z) = \frac{-z}{z-i}.$$

27*. Prove that

$$\int_{ab} \frac{d\tau}{\tau-t} \int_{ab} \frac{d\tau_1}{\tau_1-\tau} = \frac{1}{2}\left[\ln^2 \frac{b-t}{t-a} - \pi^2\right]$$

(V. D. Kupradze [13], p. 222).

28*. Derive the transposition formula (7.7) by means of the transformation

$$\int_L \frac{d\tau}{\tau-t} \int_L \frac{\varphi(\tau, \tau_1)}{\tau_1-\tau} d\tau_1 = \int_L \frac{d\tau}{\tau-t} \int_L \frac{\varphi(\tau, \tau_1) - \varphi(\tau, \tau)}{\tau_1-\tau} d\tau_1 +$$

$$+ \int_L \frac{\varphi(\tau, \tau) - \varphi(t, t)}{\tau-t} d\tau \int_L \frac{d\tau_1}{\tau_1-\tau} + \varphi(t, t) \int_L \frac{d\tau}{\tau-t} \int_L \frac{d\tau_1}{\tau_1-\tau}$$

(V. D. Kupradze [13], p. 220).

Hint. Perform the change of the order of integration in the integrals in which one only integral is singular, and make use of the result of the preceding example.

CHAPTER II

RIEMANN BOUNDARY VALUE PROBLEM

In this chapter we shall consider the principal problem of the theory of boundary value problems of analytic functions, around which the whole material of the book is grouped, namely the Riemann boundary value problem. We shall frequently return to it in the sequel, in connection with various generalizations and applications.

First, we shall carry out a preliminary investigation of the concept of the index of a function, which is of a great value as an auxiliary tool.

§ 12. The index

12.1. Definition and basic properties

Let L be a smooth closed contour and $G(t)$ a continuous non-vanishing function given on L.

DEFINITION. *By the index of the function $G(t)$ with respect to the contour L we understand the increment of its argument, in traversing the curve in the positive direction, divided by 2π.*

If the increment of a quantity ω in traversing the contour L be denoted by $[\omega]_L$, the index of $G(t)$ can be written in the form

$$\varkappa = \operatorname{Ind} G(t) = \frac{1}{2\pi} [\arg G(t)]_L. \tag{12.1}$$

The index can easily be expressed by the change of the logarithm of the function; namely we have

$$\ln G(t) = \ln |G(t)| + i \arg G(t).$$

After having traversed the contour L, $|G(t)|$ returns to its original value. Hence

$$[\arg G(t)]_L = \frac{1}{i} [\ln G(t)]_L.$$

so that

$$\varkappa = \frac{1}{2\pi i} \, [\ln G(t)]_L. \tag{12.2}$$

The index can be expressed by the integral

$$\varkappa = \text{Ind}\, G(t) = \frac{1}{2\pi} \int_L d \arg G(t) = \frac{1}{2\pi i} \int_L d \ln G(t), \tag{12.3}$$

which is understood in the Stieltjes sense. In view of the continuity of $G(t)$ the increment of its argument in traversing the contour should be a multiple of 2π; consequently, we may make the following assertions.

1. *The index of a function which is continuous on a closed contour and does not vanish anywhere on it, is an integer or zero.*

The definition of the index implies immediately the following statement.

2. *The index of a product of functions is equal to the sum of the indices of the factors. The index of a quotient is equal to the difference of the indices of the dividend and the divisor.*

Suppose now that $G(t)$ is differentiable and constitutes the boundary value of a function analytic inside or outside of the contour L. Then

$$\varkappa = \frac{1}{2\pi i} \int_L d \ln G(t) = \frac{1}{2\pi i} \int_L \frac{G'(t)}{G(t)} \, dt \tag{12.4}$$

turns out to be equal to the logarithmic residue of the function $G(t)$. The theorems on the logarithmic residue imply the following properties of the index.

3. *If $G(t)$ is the boundary value of a function analytic inside or outside the contour, then its index is equal to the number of zeros inside or outside the contour, respectively, taken with negative sign.*

4. *If the function $G(z)$ is analytic inside the contour except at a finite number of points where it may have poles, then the number of zeros is to be replaced by the difference between the number of zeros and the number of poles.*

The zeros and poles should be counted the number of times equal to their multiplicity.

Observe that the indices of complex conjugate functions have opposite signs.

12.2. Computation of the index

Let
$$t = t_1(s) + i t_2(s) \quad (0 \le s \le l)$$

be the equation of the contour L. Substituting the expression of the complex coordinate t into the function $G(t)$ we obtain

$$G(t) = G[t_1(s) - i t_2(s)] = \xi(s) + i\eta(s). \tag{12.5}$$

We regard ξ, η as Cartesian coordinates. Thus

$$\xi = \xi(s), \quad \eta = \eta(s)$$

constitutes the parametric equation of a curve Γ. In view of the continuity of the function $G(t)$ and the fact that the contour L is closed, the curve Γ is also closed.

The number of spires of the curve Γ about the origin, i.e. the number of complete revolutions of the position vector when the variable s ranges from 0 to l, will, obviously, be the index of the function $G(t)$. This number is sometimes called the order of the curve Γ with respect to the origin of the coordinate system†.

If the curve Γ can in fact be constructed, the number of spires is discovered directly. It is possible to quote many examples in which the index can be determined in accordance with the form of the curve Γ. For instance, if $G(t)$ is real or imaginary non-vanishing function, then $\Gamma(t)$ is a section of straight line (traversed an even number of times) and the index of $G(t)$ is zero. If the real part $\xi(s)$ or the imaginary part $\eta(s)$ does not change the sign, it is evident that the index is also zero, etc. If the function $G(t)$ is representable in form of a product or quotient of functions constituting the limiting values of functions analytic inside or outside the contour, then the index is computed on the basis of the properties 2, 3, 4.

In the general case the index can be computed by means of the formula (12.3). Setting in it, in accordance with the formula (12.5)

$$d \arg G(t) = d \arctan \frac{\eta(s)}{\xi(s)}.$$

we obtain

$$\varkappa = \frac{1}{2\pi} \int_{\Gamma} \frac{\xi \, d\eta - \eta \, d\xi}{\xi^2 + \eta^2} = \frac{1}{2\pi} \int_0^l \frac{\xi(s) \, \eta'(s) - \eta(s) \, \xi'(s)}{\xi^2(s) + \eta^2(s)} \, ds. \tag{12.6}$$

† See, for instance, M. Morse, *Topological Methods in the Theory of Functions of a Complex Variable*, Princeton (1947).

4 a*

In some simple cases it is possible to derive an algebraic algorithm for computing the index (see Problems 1, 2, 3 at the end of the chapter).

The concept of index can be extended to the case of an open contour L, and also to the cases when $G(t)$ has discontinuities or vanishes (see § 45).

Remark. If $G(t)$ is the boundary value of an analytic function, then the contour L may be varied in a wide range, without altering the index. Evidently, the index will be altered always when the deformation of the contour is such that the latter passes through a zero or a pole of $G(z)$.

EXAMPLE. Compute the index of $G(t) = t^n$ with respect to an arbitrary contour L enclosing the origin of the coordinate system.

1st method. The function t^n is the boundary value of the function z^n which has one zero of order n inside the contour. Hence

$$\varkappa = \operatorname{Ind} t^n = n.$$

2nd method. If the argument of t is φ, the argument of t^n is $n\varphi$. When the point t, after having traversed the contour L, returns to its original value, φ acquires the increment 2π. Consequently,

$$\operatorname{Ind} t^n = n.$$

The index can be calculated by numerical methods on high-speed computers. Obviously, owing to the integer nature of the index, any approximation to within $1/2$ rounded to the nearest whole number gives the exact value.

One of the numerical methods of calculating the index will now be described. Divide the interval $[0, l]$ of variation of the parameter s which determines the equation of the contour L, into n equal parts at the points $s_0 = 0, s_1, s_2, \ldots, s_n = l$. As a result the contour L itself and the curve Γ which is the image of L is divided into n arcs by the function $G(t)$. Let γ_i be one such arc of the curve Γ, where (ξ_{i-1}, n_{i-1}), (ξ_i, η_i) are the coordinates of its end-points. Here $\xi_i = \xi_i(s)$ and $\eta_i = \eta_i(s)$. We put $\Delta\varphi_i$ for the angle between the position vectors of the end-points of the arc γ_i. Then from formula (12.1) we have:

$$\varkappa = \operatorname{Ind} G(t) = \frac{1}{2\pi} \sum_{i=1}^{n} \Delta\varphi_i.$$

For sufficiently large n

$$\Delta\varphi_i \approx \sin \Delta\varphi_i = \frac{\xi_i \eta_{i-1} - \xi_{i-1}\eta_i}{\sqrt{(\xi_{i-1}^2 + \eta_{i-1}^2)}\sqrt{(\xi_i^2 + \eta_i^2)}},$$

and therefore

$$\varkappa = \frac{1}{2\pi} \sum_{i=1}^{n} \frac{\xi_i \eta_{i-1} - \xi_{i-1} \eta_i}{\sqrt{(\xi_{i-1}^2 + \eta_{i-1}^2)} \sqrt{(\xi_i^2 + \eta_i^2)}} + R_n,$$

where R_n is the error.

It could easily be shown that if the functions $t_1(s)$ and $t_2(s)$, contained in the equation of contour L, satisfy Hölder's condition with the index α, while $G(t)$ satisfies it with the index β, then

$$|R_n| < \frac{C}{n^{3\alpha\beta - 1}},$$

where C is a constant.

Hence, if $\alpha\beta > 1/3$ and $n \to \infty$, then $R_n \to 0$. We thus obtain a convenient formula for programming:

$$\varkappa = \frac{1}{2\pi} \sum_{i=1}^{n} \frac{\xi_i \eta_{i-1} - \xi_{i-1} \eta_i}{\sqrt{(\xi_{i-1}^2 + \eta_{i-1}^2)} \sqrt{(\xi_i^2 + \eta_i^2)}}. \tag{12.7}$$

The use of this formula will now be illustrated by an example of the computation on the U.S.S.R. "Strela" computer. For the index of the function.

$$w = u + iv = \sin\frac{\pi}{2}(x + y) - \frac{9}{8(x^2 + y^2 + x) + 1} + i(x - y)$$

on a unit circle, after rounding to the third place, we have

$$\text{if } n = 32 \quad \varkappa = -0\cdot949,$$
$$\text{if } n = 64 \quad \varkappa = -0\cdot983.$$

Consequently, $\varkappa = -1$.

As shown by this example, it would have been sufficient to have taken a much smaller n to achieve this result.

The real and imaginary parts of the function under consideration are defined by analytic functions and they may be continued analytically inside the circle. The continued function $w(x, y)$ has zero at the point $(\frac{1}{2}, \frac{1}{2})$ and tends to infinity on the circle $(x + \frac{1}{2})^2 + y^2 = 1/8$. It will be seen from the example that for non-analytic complex functions there is no simple dependence of the index on the zeros or infinities of the function in the domain.

This illustration has been taken from an unpublished work of L. A. Chikin.

§ 13. Some auxiliary theorems

In subsequent considerations we shall have frequently to use certain theorems of the theory of functions of complex variable, the statements of which we now quote. The proofs may be found in all text-books of the theory.

1. THEOREM (on analytic continuation). *Suppose that the two domains D_1 and D_2 have a common smooth boundary L; in the domains D_1 and D_2 two analytic functions $f_1(z)$ and $f_2(z)$ are given. Assume that when a point z tends to the curve L both functions tend to limiting values which are continuous on the curve L and equal to each other. Under these conditions the functions $f_1(z)$, $f_2(z)$ constitute the analytic continuation of one another.*

2. LIOUVILLE'S THEOREM (generalized). *Let the function $f(z)$ be analytic in the entire plane of the complex variable, except at the points $a_0 = \infty$, a_k $(k = 1, 2, \ldots, n)$, where it has poles, and suppose that the principal parts of the expansions of the function $f(z)$ in the vicinities of the poles have the form:*

at the point a_0: $G_0(z) = c_1^0 z + c_2^0 z^2 + \cdots + c_{n_0}^0 z^{n_0}$;

at the points a_k:

$$G_k\left(\frac{1}{z - a_k}\right) = \frac{c_1^k}{z - a_k} + \frac{c_2^k}{(z - a_k)^2} + \cdots + \frac{c_{m_k}^k}{(z - a_k)^{m_k}}.$$

Then the function $f(z)$ is a rational function and is representable by the relation

$$f(z) = C + G_0(z) + \sum_{k=1}^{n} G_k\left(\frac{1}{z - a_k}\right).$$

In particular, if the only singularity of the function $f(z)$ is a pole of order m at infinity, then $f(z)$ is a polynomial of degree m:

$$f(z) = c_0 + c_1 z + \cdots + c_m z^m.$$

§ 14. The Riemann problem for a simply-connected domain

14.1. Formulation of the problem

Suppose that we are given a simple smooth closed contour L dividing the plane of the complex variable into the interior domain D^+ and the exterior domain D^-, and two functions of position on the contour, $G(t)$ and $g(t)$ which satisfy the Hölder condition, and $G(t)$ does not vanish. It is required *to find two functions*†: $\Phi^+(z)$, *analytic in the domain* D^+, *and* $\Phi^-(z)$, *analytic in the domain* D^-, *including* $z = \infty$, *which satisfy on the contour* L *either the linear relation*

$$\Phi^+(t) = G(t)\,\Phi^-(t) \quad \text{(homogeneous problem)} \tag{14.1}$$

or

$$\Phi^+(t) = G(t)\,\Phi^-(t) + g(t) \quad \text{(non-homogeneous problem)} \tag{14.2}$$

The function $G(t)$ will be called *the coefficient of the Riemann problem*, and the function $g(t)$ its *free term*.

14.2. Determination of sectionally analytic function in accordance with given jump

Consider first a Riemann problem of particular type. Suppose that on a closed contour L a function $G(t)$ is given which satisfies the Hölder condition. It is required to find a sectionally analytic function $\Phi(z)$ ($\Phi(z) = \Phi^+(z)$ for $z \in D^+$, $\Phi(z) = \Phi^-(z)$ for $z \in D^-$), vanishing at infinity and undergoing in passing through the contour L a jump $\varphi(t)$, i.e. satisfying the condition

$$\Phi^+(t) - \Phi^-(t) = \varphi(t). \tag{14.3}$$

From formula (4.9) it is evident that the function

$$\Phi(z) = \frac{1}{2\pi i} \int\limits_L \frac{\varphi(\tau)}{\tau - z}\,d\tau \tag{4.4}$$

is the solution of the problem. It is easy to prove the uniqueness of this solution. In fact, assuming existence of two solutions and considering their difference we find that the jump of this difference on the line L is zero; consequently, this function is analytic in the entire plane and vanishes at infinity. Therefore, Liouville's theorem implies that it vanishes identically.

The solution of the above problem can also be formulated as follows.

An arbitrary function $\varphi(t)$ *given on a closed contour and satisfying the Hölder condition, can be uniquely represented in the form of the difference of the functions* $\Phi^+(t)$, $\Phi^-(t)$ *constituting the boundary values of the analytic functions* $\Phi^+(z)$, $\Phi^-(z)$, *under the additional condition* $\Phi^-(\infty) = 0$.

---

† Using the terminology of § 1, we can speak of one sectionally analytic function.

If the additional condition $\Phi^-(\infty) = 0$ be discarded, it is readily observed that the solution of the problem is given by the formula

$$\Phi(z) = \frac{1}{2\pi i} \int\limits_L \frac{\varphi(\tau)}{\tau - z} d\tau + \text{const.} \tag{4.4'}$$

14.3. Solution of the homogeneous problem

Assume that the homogeneous boundary value problem (14.1) is soluble, and let the functions $\Phi^+(z)$ and $\Phi^-(z)$ be its solution†. Denote the number of zeros of the functions $\Phi^+(z)$, $\Phi^-(z)$ in the domains of definition D^+, D^-, by N^+, N^-, respectively. Computing the index of both sides of the relation (14.1) and making use of the properties 2 and 3 of § 12.1 we obtain

$$N^+ + N^- = \text{Ind } G(t) = \varkappa. \tag{14.4}$$

The index $\varkappa$ of the coefficient of the Riemann problem will be called the *index of the problem*.

Evidently, the left-hand side of the last relation is a non-negative number. Hence we have:

1. For the Riemann homogeneous boundary value problem to be soluble it is necessary that the index $\varkappa$ of the problem is non-negative (by assumption the functions $\Phi^+(z)$ and $\Phi^-(z)$ have no poles).

2. If $\varkappa > 0$ the functions $\Phi^+(z)$, $\Phi^-(z)$ constituting the solution of the problem have altogether $\varkappa$ zeros.

3. If $\varkappa = 0$ the functions $\Phi^\pm(z)$ have no zeros.

1. *The case* $\varkappa = 0$. Here $\ln G(t)$ is a single-valued function, and $\ln \Phi^+(z)$ and $\ln \Phi^-(z)$ are analytic. Taking the logarithms in the boundary condition (14.1) we have

$$\ln \Phi^+(t) - \ln \Phi^-(t) = \ln G(t). \tag{14.3'}$$

For $\ln G(t)$ any branch may be taken. It is easy to verify that the final result is independent of the choice of branch.

Thus, we have arrived at the problem of determining a sectionally analytic function $\ln \Phi(z)$ in accordance with the given jump on L. Its solution under the additional condition $\ln \Phi^-(\infty) = 0$ is given by the formula

$$\ln \Phi(z) = \frac{1}{2\pi i} \int\limits_L \frac{\ln G(\tau)}{\tau - z} d\tau. \tag{14.5}$$

† In all that follows, for brevity, we consider as insoluble the homogeneous problem having only the zero solution.

Denote for brevity

$$\ln \Phi(z) = \Gamma(z). \tag{14.6}$$

It follows directly from the Sokhotski formulae that the solution of the boundary value problem (14.1) which satisfies the condition $\Phi^-(\infty) = 1$, is given by the functions

$$\Phi^+(z) = e^{\Gamma^+(z)}, \quad \Phi^-(z) = e^{\Gamma^-(z)}. \tag{14.7}$$

If the additional condition $\Phi^-(\infty) = 1$ be discarded, the formula (14.5) should be completed by an arbitrary constant term, and the solution of the problem has the form

$$\Phi^+(z) = Ae^{\Gamma^+(z)}, \quad \Phi^-(z) = Ae^{\Gamma^-(z)}, \tag{14.8}$$

where A is an arbitrary constant. Since $\Gamma^-(\infty) = 0$, A is the value of $\Phi^-(z)$ at infinity.

Thus, in the case $\varkappa = 0$ and under the condition $\Phi^-(\infty) \neq 0$ the solution contains one arbitrary constant, i.e. there is one linearly independent solution. If $\Phi^-(\infty) = 0$, then $A = 0$ and the problem has only the trivial solution—the identical zero.

The above implies an important corollary. *An arbitrary function $G(t)$ given on the contour L, satisfying the Hölder condition and having zero index, is representable as the ratio of the functions $\Phi^+(t)$ and $\Phi^-(t)$ constituting the boundary values of functions analytic in the domains D^+, D^- and having in these domains no zeros. These functions are determined to within an arbitrary constant factor and are given by the formulae* (14.8).

2. *The case $\varkappa > 0$.* Assume specifically that the origin of the coordinate system lies in the domain D^+. The function $t^\varkappa$ has the index $\varkappa$ (example in § 12.2). Let us write the boundary condition in the form

$$\Phi^+(t) = t^\varkappa [t^{-\varkappa} G(t)] \Phi^-(t).$$

Obviously, the function $G_1(t) = t^{-\varkappa} G(t)$ has zero index. Representing it in form of the ratio $G_1(t) = \dfrac{e^{\Gamma^+(t)}}{e^{\Gamma^-(t)}}$, where

$$\Gamma(z) = \frac{1}{2\pi i} \int\limits_L \frac{\ln [\tau^{-\varkappa} G(\tau)]}{\tau - z} d\tau, \tag{14.9}$$

the boundary condition can also be written as follows:

$$\frac{\Phi^+(t)}{e^{\Gamma^+(t)}} = t^\varkappa \frac{\Phi^-(t)}{e^{\Gamma^-(t)}}.$$

The last relation indicates that the function $\Phi^+(z)/e^{\Gamma^+(z)}$, analytic in the domain D^+, and the function $z^\varkappa \Phi^-(z)/e^{\Gamma^-(z)}$, analytic in D^-, except at infinity where it can have a pole of order not higher than $\varkappa$, constitute the analytic continuation of each other through the contour L (§ 13). Consequently, they are branches of a unique analytic function which can have, in the entire plane, only one singularity—a pole of order not higher than $\varkappa$ at infinity. According to the generalized Liouville theorem (§ 13) this function is a polynomial of degree not higher that $\varkappa$ with arbitrary complex coefficients.

Hence, we obtain the general solution of the problem

$$\Phi^+(z) = e^{\Gamma^+(z)} P_\varkappa(z), \quad \Phi^-(z) = e^{\Gamma^-(z)} z^{-\varkappa} P_\varkappa(z). \qquad (14.10)$$

We now state the final result.

THEOREM. *If the index $\varkappa$ of the Riemann boundary value problem is positive, the problem has $\varkappa + 1$ linearly independent solutions*

$$\Phi_k^+(z) = z^k e^{\Gamma^+(z)}, \quad \Phi_k^-(z) = z^{k-\varkappa} e^{\Gamma^-(z)} \quad (k = 0, 1, \ldots, \varkappa).$$

The general solution contains $\varkappa + 1$ arbitrary constants and is given by the formula (14.10).

The case of zero index examined before is a particular case of the last theorem. It has already been observed that if $\varkappa < 0$ the homogeneous problem is insoluble.

The solution of the problem will be completely determined if $\varkappa + 1$ independent conditions be imposed upon the functions $\Phi^+(z)$, $\Phi^-(z)$. The additional conditions can be described in various ways. For instance, similarly to the case of the Cauchy problem for a linear differential equation of order $\varkappa + 1$, we may prescribe the values of the functions (Φ^+ or Φ^-) and its first $\varkappa$ derivatives at some point (say, at the origin of the coordinate system). Then the values of all arbitrary constants are determined from a system of $\varkappa + 1$ consistent linear equations.

In the following, in applications of the Riemann problem to the solution of singular integral equations, we shall have to seek the solution of the problem under the additional condition $\Phi^-(\infty) = 0$.

It follows from the formula (14.10) that $\Phi^-(\infty)$ is equal to the coefficient of $z^\varkappa$ in the polynomial $P_\varkappa(z)$.

Under the condition of vanishing of the solution at infinity, it is representable in the form

$$\Phi^+(z) = e^{\Gamma^+(z)} P_{\varkappa-1}(z); \quad \Phi^-(z) = e^{\Gamma^-(z)} z^{-\varkappa} P_{\varkappa-1}(z), \qquad (14.11)$$

where $P_{\varkappa-1}$ is a polynomial of $(\varkappa-1)$th degree with arbitrary coefficients. Thus, in this case the problem has $\varkappa$ linearly independent solutions.

14.4. The canonical function of the homogeneous problem

To obtain the solution of the non-homogeneous problem it is convenient to use a special particular solution of the homogeneous problem, to an investigation of which we now proceed.

DEFINITION. *By the order of the analytic function $\Phi(z)$ at a point z_0 we understand the exponent of the lowest power in the expansion of $\Phi(z)$ into the power series of $(z - z_0)$.*

This definition implies that if $\Phi(z)$ has at a point z_0 a zero of certain order m, then this number m is the order of the function. To a pole of order m there corresponds the negative order $(-m)$. If a function is analytic at z_0 and is not zero here, then its order is zero.

In the vicinity of infinity we form the expansion in powers of $1/z$; accordingly, by the order of a function at infinity we mean the exponent of the lowest power of $1/z$ in the expansion of the function, i.e. again a positive order corresponds to a zero of the function and negative to a pole†.

We shall call the *total order* of an analytic function the algebraic sum of its orders for all points of the domain. Consequently, the total order is equal to the difference between the number of zeros and the number of poles of the function.

If, in contrast to §14.3, we admit as solutions of the Riemann problem functions having poles, then a reasoning similar to that used in deriving relations (14.4) proves that the total order of the solution of the homogeneous Riemann problem is equal to the index of the problem.

We now seek a solution which has zero order in the entire plane except for one exceptional point at which the order is equal to the index. This exceptional point will, in all that follows, be the point at infinity.

DEFINITION. *We shall call the canonical function of the homogeneous Riemann problem, a sectionally analytic function satisfying the*

† Observe that this definition of the order of a function at infinity is not generally used. Sometimes, conversely, the order of a function at infinity is defined as the order of its pole under the condition that the principal part of the Laurent series contains a finite number of terms. To the zero of a function there corresponds a negative order.

boundary condition (14.1) *and having zero order everywhere in the finit* *part of the plane. At the point at infinity its order therefore is equal to $\varkappa$.* For $\varkappa \geqq 0$ the canonical function has no poles and constitutes a solution of the boundary value problem. We shall also call it the *canonical solution.*

For $\varkappa < 0$ the canonical function has a pole at infinity and, consequently, it is not a solution of the homogeneous Riemann problem (in the sense of § 14.1). Nevertheless, it will be used as an auxiliary function in solving the non-homogeneous problem.

If we write the boundary condition of the Riemann problem in the form

$$\Phi^+(t) = t^\varkappa [t^{-\varkappa} G(t)] \Phi^-(t),$$

we easily find that for an arbitrary $\varkappa$ the canonical function of the problem $X(z)$ is given by the relation

$$X^+(z) = e^{\Gamma^+(z)}, \quad X^-(z) = z^{-\varkappa} e^{\Gamma^-(z)}, \tag{14.12}$$

where $\Gamma(z)$ is given by the formula (14.9).

For $\varkappa \geqq 0$ the general solution of the homogeneous problem is expressed by the canonical function as follows:

$$\Phi(z) = X(z) P_\varkappa(z). \tag{14.13}$$

14.5. Solution of the non-homogeneous problem

If we replace the coefficient $G(t)$ of the boundary condition

$$\Phi^+(t) = G(t) \Phi^-(t) + g(t) \tag{14.2}$$

by the ratio of the boundary values of the canonical function of the homogeneous problem, $G(t) = \dfrac{X^+(t)}{X^-(t)}$, we can reduce it to the form

$$\frac{\Phi^+(t)}{X^+(t)} = \frac{\Phi^-(t)}{X^-(t)} + \frac{g(t)}{X^+(t)}.$$

In accordance with § 5.1, the function $\dfrac{g(t)}{X^+(t)}$ satisfies the Hölder condition. Let us replace it, in accordance with § 14.2, by the difference of the boundary values of the analytic functions

$$\frac{g(t)}{X^+(t)} = \Psi^+(t) - \Psi^-(t),$$

where

$$\Psi(z) = \frac{1}{2\pi i} \int\limits_L \frac{g(\tau)}{X^+(\tau)} \frac{d\tau}{\tau - z}. \tag{14.14}$$

Then the boundary condition can be written in the form

$$\frac{\Phi^+(t)}{X^+(t)} - \Psi^+(t) = \frac{\Phi^-(t)}{X^-(t)} - \Psi^-(t).$$

Let us observe that for $\varkappa \geqq 0$ the function $\dfrac{\Phi^-(z)}{X^-(z)}$ has at infinity a pole, and for $\varkappa < 0$ a zero of order $\varkappa$.

Reasoning analogous to that in § 14.3 (in solving the homogeneous problem) leads to the following results:

1. $\varkappa \geqq 0$. Then

$$\frac{\Phi^+(z)}{X^+(z)} - \Psi^+(z) = \frac{\Phi^-(z)}{X^-(z)} - \Psi^-(z) = P_\varkappa(z).$$

Hence we have the solution

$$\Phi(z) = X(z)\left[\Psi(z) + P_\varkappa(z)\right], \qquad (14.15)$$

$X(z)$, $\Psi(z)$ being given by the formulae (14.12), (14.14); $P_\varkappa$ is a polynomial of degree $\varkappa$ with arbitrary coefficients.

It is readily observed that the formula (14.15) yields the general solution of the non-homogeneous problem, since it contains as one term $X(z) P_\varkappa(z)$, the solution of the homogeneous problem.

2. $\varkappa < 0$. In this case $\dfrac{\Phi^-(z)}{X^+(z)}$ vanishes at infinity and

$$\frac{\Phi^+(z)}{X^+(z)} - \Psi^+(z) = \frac{\Phi^-(z)}{X^-(z)} - \Psi^-(z) = 0,$$

whence

$$\Phi(z) = X(z)\,\Psi(z). \qquad (14.16)$$

In the expression for the function $\Phi^-(z)$ the first factor has in view of the formula (14.12) a pole of order $-\varkappa$ at infinity, whereas the second term, as an integral of the Cauchy type (14.14), has in the general case a zero of order one at infinity. Consequently, $\Phi^-(z)$ has at infinity a pole of order not exceeding $-\varkappa - 1$. Thus, if $\varkappa < -1$, the non-homogeneous problem in general is insoluble. It is soluble only when the free term satisfies certain additional conditions. To derive the latter, let us expand the Cauchy type integral (14.14) into a series in the vicinity of infinity (see § 1)

$$\Psi^-(z) = \sum_{k=1}^{\infty} c_k z^{-k},$$

where

$$c_k = -\frac{1}{2\pi i} \int_L \frac{g(\tau)}{X^+(\tau)} \tau^{k-1} d\tau.$$

For analyticity of $\Phi^-(z)$ at infinity it is necessary that the first $-\varkappa - 1$ coefficients of the expansion of $\Psi^-(z)$ vanish. This implies that for the solubility of the non-homogeneous problem in the case of a negative index ($\varkappa < -1$) it is necessary and sufficient that the following $-\varkappa - 1$ conditions are satisfied:

$$\int_L \frac{g(\tau)}{X^+(\tau)} \tau^{k-1} d\tau = 0 \quad (k = 1, 2, \ldots, -\varkappa - 1). \quad (14.17)$$

The above investigation makes it possible to state the following conclusion.

THEOREM. *In the case $\varkappa \geqq 0$ the non-homogeneous Riemann problem is soluble for an arbitrary free term and its general solution is given by the relation*

$$\Phi(z) = \frac{X(z)}{2\pi i} \int_L \frac{g(\tau)}{X^+(\tau)} \frac{d\tau}{\tau - z} + X(z) P_\varkappa(z), \quad (14.18)$$

where the canonical function $X(z)$ is given by (14.12) and $P_\varkappa(z)$ is a polynomial of degree $\varkappa$ with arbitrary complex coefficients. If $\varkappa = -1$ the non-homogeneous problem is also soluble and has a unique solution.

In the case $\varkappa < -1$ the non-homogeneous problem is in general insoluble. In order that it be soluble it is necessary and sufficient that the free term of the problem satisfies $-\varkappa - 1$ conditions (14.17). If these latter are satisfied the unique solution of the problem is given by the formula (14.18) where we should set $P(z) \equiv 0$.

In the case when we impose on the unknown solution the additional condition of *vanishing at infinity*, then as was already pointed out in § 14.4, the polynomial of degree $\varkappa$ is to be replaced by a polynomial of degree $\varkappa - 1$. For the problem to be soluble in the case of a negative index it is also necessary that the coefficient $c_{-\varkappa}$ vanishes.

Consequently, under the condition $\Phi^-(\infty) = 0$ the solution is given for $\varkappa \geqq 0$ by the formula

$$\Phi(z) = X(z) [\Psi(z) + P_{\varkappa-1}(z)] \quad (14.18')$$

(if $\varkappa = 0$ we have to set $P(z) \equiv 0$).

If $\varkappa < 0$ the solution as before is given by the formula (14.18) where $P(z) \equiv 0$, and it is necessary that the following $-\varkappa$ conditions of solubility are satisfied:

$$\int_L \frac{g(\tau)}{X^+(\tau)} \tau^{k-1} d\tau = 0 \quad (k = 1, 2, \ldots, -\varkappa). \tag{14.19}$$

Thus the theorem on solubility of the non-homogeneous problem assumes the more symmetric form:

for $\varkappa \geqq 0$ the general solution of the non-homogeneous problem depends linearly on $\varkappa$ arbitrary constants;

for $\varkappa < 0$ the number of conditions of solubility is $-\varkappa$.

Let us note that here for $\varkappa = 0$ the problem is absolutely soluble, and the solution is unique.

14.6. Examples

On the basis of the foregoing, the solution of the Riemann problem reduces in essence to two operations:

1. Representation of the arbitrary function prescribed on the contour in the form of the difference of the boundary values of functions analytic in the domains D^+ and D^-.

2. Representation of this function in the form of the ratio of the boundary values of analytic functions.

The second operation can be reduced to the first by taking logarithms. Some complications appearing when the index does not vanish, are due only to many-valuedness of the logarithm. The first operation for the arbitrary function is equivalent to computing an integral of the Cauchy type. Accordingly, the solution of the problem given by the formulae (14.9), (14.14)–(14.16) is expressed explicitly (or as is customary to say—in a closed form) by Cauchy type integrals.

In the general case there is no simple algorithm for computing such integrals. It is necessary to apply one of the approximation methods, which leads to very cumbersome calculations. We shall here confine ourselves to the case when the coefficient and the free term of the problem are analytically continuable from the contour on the plane of the complex variable. In this case the operations 1 and 2 are elementary. If then the contour of integration L be varied, the solution alters only when there is a change of position of the zeros and singular points of the coefficient $G(z)$, with respect to the contour.

Note that the importance of the proposed method of solving the problem with rational coefficients goes beyond the example since an arbitrary continuous function (thereby satisfying Hölder's condition) can be approximated with any degree of accuracy by rational functions and the solution of the problem with rational coefficients may be the basis for an approximate solution in the general case (see § 16.3, 8).

We will solve the Riemann boundary value problem for a contour consisting of a finite number of simple curves, the coefficient of the problem being a rational function without zeros or poles on the contour:

$$\Phi^+(t) = \frac{p(t)}{q(t)} \Phi^-(t) + g(t). \tag{14.20}$$

We expand the polynomials $p(z)$, $q(z)$ as the product

$$p(z) = p_+(z)\, p_-(z), \quad q(z) = q_+(z)\, q_-(z), \tag{14.21}$$

where $p_+(z)$, $q_+(z)$ are the polynomials with roots in D^+, whilst $p_-(z)$, $q_-(z)$ have their roots in D^-. From the property 4 of an index (§ 12.1), it is easily found that $\varkappa = m_+ - n_+$, where m_+, n_+ represent the numbers of zeros of the polynomials $p_+(z)$ and $q_+(z)$.

Since the coefficient is a function which is analytically continuable in the domain $D^\pm$, it is inadvisable to avail ourselves of the general formulae†, but to obtain the latter directly by analytical continuation. Representing the boundary condition as

$$\frac{q_-(t)}{p_-(t)} \Phi^+(t) - \frac{p_+(t)}{q_+(t)} \Phi^-(t) = \frac{q_-(t)}{p_-(t)} g(t),$$

on the same grounds as in § 14.5, the solution is

$$\left.\begin{array}{l}
\Phi^+(z) = \dfrac{p_-(z)}{q_-(z)} [\Psi(z) + X^-_1(z)], \\[2mm]
\Phi^-(z) = \dfrac{q_+(z)}{p_-(z)} [\Psi(z) + X^-_1(z)], \\[2mm]
\Psi(z) = \dfrac{1}{2\pi i} \displaystyle\int_L \dfrac{q_-(\tau)}{p_-(\tau)} g(\tau) \dfrac{d\tau}{\tau - z} \\[2mm]
(\Phi^-(\infty) = 0).
\end{array}\right\} \tag{14.22}$$

---

† The derivation of the general formulae was based on reduction to the case of a zero index by introducing the multiplier $t^{-\varkappa}$. Use of this method here would lead to superfluous multipliers which later admit cancellation.

If the index is negative, it is necessary to put $P_{x-1} = 0$ and add the solubility condition:

$$\int_L \frac{q_-(\tau)}{p_-(\tau)} g(\tau)\, \tau^{k-1}\, d\tau = 0, \quad (k = 1, 2, \ldots, -\varkappa). \quad (14.23)$$

This expression agrees with the general formulae (14.14) and (14.15) if it is taken into account that the canonical function here is $X^+ = p_-/q_-$, $X^- = q_+/p_+$. The latter may be obtained from consideration of the canonical function defined in § 14.4.

Note also that in the general case for a practical solution of the Riemann problem it is worthwhile representing the coefficient as $G = p_+ p_-/q_+ q_-$, where G_1 is a function with zero index, whilst the polynomials P_+, q_+ are chosen with regard to the type of coefficient. The solution is obtained most simply if these coefficients are appropriately selected.

As an example let us solve the Riemann problem

$$\Phi^+(t) = \frac{t}{t^2 - 1}\, \Phi^-(t) + \frac{t^3 - t^2 + 1}{t^2 - t}$$

under the condition that $\Phi(\infty) = 0$ and L is an arbitrary smooth closed contour of the form:

(a) The contour L contains the point $z_1 = 0$ and does not contain the points $z_2 = 1$, $z_3 = -1$.

(b) The contour L contains the points $z_1 = 0$, $z_2 = 1$ and does not contain the point $z_3 = -1$.

(c) The contour L contains the points $z_1 = 0$, $z_2 = 1$, $z_3 = 1$.

(d) The contour L contains the points $z_2 = 1$, $z_3 = -1$ and does not contain the point $z_1 = 0$.

To solve the problem it suffices to find the canonical function $X(z)$ of the homogeneous problem (the formulae (14.12), (14.13)) and the particular solution $\Psi(z)$ of the non-homogeneous problem (the formula (14.14)). The function $X(z)$ can be found directly without computing Cauchy type integrals by making use of the preceding results.

For the solution we employ the foregoing method.

(a) In the case in question (Fig. 12), we have

$$p_+(t) = t, \quad p_-(t) = 1, \qquad m_+ = 1,$$
$$q_+(t) = 1, \quad q_-(t) = t^2 - 1, \quad n_+ = 0, \quad \varkappa = m_+ - n_+ = 1.$$

We write the boundary value problem as

$$(t^2 - 1)\,\Phi^+(t) - t\Phi^-(t) = \frac{1}{t}(t^3 - t^2 + 1)(t + 1)$$

Hence

$$\Psi(z) = \frac{1}{2\pi i} \int_L \frac{q_-(\tau)}{p_-(\tau)}\, g(\tau)\, \frac{d\tau}{\tau - z}$$

$$= \frac{1}{2\pi i} \int_L \frac{\tau^3 - \tau + 1}{\tau - z}\, d\tau + \frac{1}{2\pi i} \int_L \frac{1/\tau}{\tau - z}\, d\tau,$$

and it is implied from formulae (1.1) and (1.2) that

$$\Psi^+(z) = z^3 - z + 1, \quad \Psi^-(z) = -\frac{1}{z}.$$

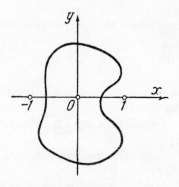

FIG. 12.

The general solution of the problem contains one arbitrary constant. From (14.22), we find that

$$\Phi^+(z) = \frac{1}{z^2 - 1}\left[(z^3 - z + 1) + c\right] = \frac{z^3 - z + 1}{z^2 - 1} + \frac{c}{z^2 - 1},$$

$$\Phi^-(z) = \frac{1}{z}\left[-\frac{1}{z} + c\right] = -\frac{1}{z^2} + \frac{c}{z},$$

where c is the arbitrary constant. Substituting $c - 1$ for c, the solution may also be written as

$$\Phi^+(z) = z + \frac{c}{z^2 - 1}, \quad \Phi^-(z) = -\frac{z + 1}{z^2} + \frac{c}{z}.$$

(b) Here
$$p_+(t) = t, \quad p_-(t) = 1; \quad q_+(t) = t - 1, \quad q_-(t) = t + 1;$$
$$m_+ = n_+ = 1, \quad \varkappa = 0;$$
$$(t + 1)\,\Phi^+(t) - \frac{t}{t - 1}\,\Phi^-(t) = \frac{(t + 1)(t^3 - t^2 + 1)}{t(t + 1)};$$

$$\Psi(z) = \frac{1}{2\pi i}\int_L \frac{\tau^2 + \tau}{\tau - z}\,d\tau + \frac{1}{2\pi i}\int_L \frac{\dfrac{\tau + 1}{\tau(\tau - 1)}}{\tau - z}\,d\tau$$

$$= \begin{cases} z^2 + z & \text{if } z \in D^+, \\ -\dfrac{z + 1}{z(z - 1)} & \text{if } z \in D^-. \end{cases}$$

The problem has a unique solution

$$\Phi^+(z) = \frac{p_-(z)}{q_-(z)}\,\Psi^+(z) = \frac{1}{z + 1}(z^2 + z) = z,$$

$$\Phi^-(z) = \frac{q_+(z)}{p_+(z)}\,\Psi^-(z) = \frac{z - 1}{z}\left[-\frac{z + 1}{z(z - 1)}\right] = -\frac{z + 1}{z^2}.$$

(c) In this case
$$p_+(t) = t, \quad p_-(t) = 1; \quad q_+(t) = t^2 - 1, \quad q_-(t) = 1; \quad m_+ = 1,$$
$$n_+ = 2, \quad \varkappa = -1;$$

$$\Psi(z) = \frac{1}{2\pi i}\int_L \frac{\tau}{\tau - z}\,d\tau + \frac{1}{2\pi i}\int_L \frac{\dfrac{1}{\tau(\tau - 1)}}{\tau - z}\,d\tau$$

$$= \begin{cases} z & \text{if } z \in D^+, \\ -\dfrac{1}{z(z - 1)} & \text{if } z \in D^-. \end{cases}$$

A solution of the problem only exists if the solubility conditions (14.23) are satisfied, or, in this case, if the one condition is fulfilled:

$$\int_L \frac{q_-(\tau)}{p_-(\tau)}\,g(\tau)\,d\tau = 0.$$

Evaluating this integral, we find that

$$\int_L \frac{\tau^3 - \tau^2 + 1}{\tau^2 - \tau}\,d\tau = \int_L \tau\,d\tau + \int_L \frac{d\tau}{\tau - 1} - \frac{d\tau}{\tau} = 0 + 2\pi i - 2\pi i = 0.$$

The solubility condition is thus fulfilled and the unique solution of the problem is

$$\Phi^+(z) = z, \quad \Phi^-(z) = -\frac{z+1}{z^2}.$$

(d) In this case

$$p_+(t) = 1, \quad p_-(t) = t; \quad q_+(t) = t^2 - 1, \quad q_-(t) = 1;$$
$$\varkappa = m_+ - n_+ = -2 < 0.$$

For solubility of the problem, two conditions must be satisfied:

$$\int_L \frac{q_-(\tau)}{p_-(\tau)} q(\tau) \tau^{k-1} d\tau = 0 \quad (k = 1, 2).$$

Evaluating this integral for $k = 1$, we find that

$$\int_L \frac{\tau^3 - \tau^2 + 1}{\tau(\tau^2 - \tau)} d\tau = \int_L \left(1 - \frac{1}{\tau} - \frac{1}{\tau^2} + \frac{1}{\tau - 1}\right) d\tau = 2\pi i \neq 0.$$

Thus, the condition of solubility is not satisfied and therefore the problem has no solution.

Observe that if we formally calculate the function $\Phi(z)$, it has a pole at infinity and, consequently, it cannot be a solution of the problem.

14.7. The Riemann problem for the semi-plane

Suppose that the contour L is the real axis. As before the Riemann problem consists in finding two functions analytic in the upper and the lower semi-planes, respectively, $\Phi^+(z)$ and $\Phi^-(z)$ (a sectionally analytic function $\Phi(z)$), the limiting values of which satisfy on the contour the boundary condition

$$\Phi^+(t) = G(t)\Phi^-(t) + g(t) \quad (-\infty < t < \infty). \tag{14.24}$$

The given functions $G(t)$ and $g(t)$ satisfy the Hölder condition both in the finite points and in the vicinity of the infinite point of the contour (see conditions (4.20), (4.27)). We also assume that $G(t) \neq 0$.

The solution is derived according to the same plan as for the finite contour. The basic difference from the case with a finite curve considered before, consists in the fact that here the point at infinity and the origin of the coordinate system lie on the contour itself, and hence cannot be taken as the exceptional point at which the canonical function is allowed to have a non-zero order. Instead of the previously used auxiliary function t having index equal to unity along L, we shall introduce here the linear fractional function

$$\frac{t - i}{t + i},$$

which has the same property on the real axis. The argument of this function

$$\arg \frac{t - i}{t + i} = \arg \frac{(t - i)^2}{t^2 + 1} = 2 \arg (t - i)$$

changes by 2π when t traverses the real axis in the positive direction. Thus

$$\operatorname{Ind} \frac{t - i}{t + i} = 1.$$

If $\operatorname{Ind} G(t) = \varkappa$ the function

$$G(t) \left(\frac{t - i}{t + i} \right)^{-\varkappa}$$

has zero index. Its logarithm is a one-valued function on the real axis.

We construct the canonical function for which the exceptional point is $-i$:

$$X^+(z) = e^{\Gamma^+(z)}, \quad X^-(z) = \left(\frac{z - i}{z + i} \right)^{-\varkappa} e^{\Gamma^-(z)}, \tag{14.25}$$

where

$$\Gamma(z) = \frac{1}{2\pi i} \int_{-\infty}^{\infty} \ln \left[\left(\frac{\tau - i}{\tau + i} \right)^{-\varkappa} G(\tau) \right] \frac{d\tau}{\tau - z}.$$

Making use of the limiting values of this function we transform the boundary condition (14.24)

$$\frac{\Phi^+(t)}{X^+(t)} = \frac{\Phi^-(t)}{X^-(t)} + \frac{g(t)}{X^+(t)}.$$

Further, we introduce the analytic function

$$\Psi(z) = \frac{1}{2\pi i} \int_{-\infty}^{\infty} \frac{g(\tau)}{X^+(\tau)} \frac{d\tau}{\tau - z}, \tag{14.26}$$

which makes it possible to transform the boundary condition to the form

$$\frac{\Phi^+(t)}{X^+(t)} - \Psi^+(t) = \frac{\Phi^-(t)}{X^-(t)} - \Psi^-(t).$$

Let us observe that in contrast to the case of a finite contour, here, in general, $\Psi^-(\infty) \neq 0$. Applying the theorem on analytic continuation and taking into account that the only singularity of the function under consideration may be a pole at the point $z = -i$ of order not exceeding $-\varkappa$ (for $\varkappa > 0$), we have, in view of the generalized Liouville theorem

$$\frac{\Phi^+(z)}{X^+(z)} - \Psi^+(z) = \frac{\Phi^-(z)}{X^-(z)} - \Psi^-(z) = \frac{P_\varkappa(z)}{(z+i)^\varkappa} \quad (\varkappa \geq 0),$$

where $P_\varkappa(z)$ is a polynomial of degree not higher than $\varkappa$, with arbitrary coefficients. This implies the general solution of the problem

$$\Phi(z) = X(z)\left[\Psi(z) + \frac{P_\varkappa(z)}{(z+i)^\varkappa}\right] \quad \text{for} \quad \varkappa \geq 0, \qquad (14.27)$$

$$\Phi(z) = X(z)\left[\Psi(z) + C\right] \qquad \text{for} \quad \varkappa < 0. \qquad (14.28)$$

For $\varkappa < 0$ the function $X(z)$ has at the point $z = -i$ a pole of order $-\varkappa$; therefore for the problem to be soluble we have to set $C = -\Psi^-(-i)$. For $\varkappa < -1$ the following additional conditions should be satisfied:

$$\int_{-\infty}^{\infty} \frac{g(\tau)}{X^+(\tau)} \frac{d\tau}{(\tau+i)^k} = 0 \quad (k = 2, \ldots, -\varkappa). \qquad (14.29)$$

Thus, we have obtained results analogous to those which we deduced in the case of a finite contour.

THEOREM. *For $\varkappa \geq 0$ the homogeneous and non-homogeneous Riemann boundary value problems for the semi-plane are absolutely soluble and the solution depends linearly on $\varkappa + 1$ arbitrary constants. For $\varkappa < 0$ the homogeneous problem is insoluble. The non-homogeneous problem for $\varkappa < 0$ is uniquely soluble, in the case $\varkappa = -1$ absolutely, and for $\varkappa < -1$ only if the $-\varkappa -1$ additional conditions (14.26) are satisfied.*

So far we have been seeking solutions of the Riemann problem which are simply bounded at infinity. Now we shall make a special investigation of solutions which vanish at infinity.

Substituting into the boundary condition $\Phi^+(\infty) = \Phi^-(\infty) = 0$ we obtain $g(\infty) = 0$. Consequently, in order that the Riemann problem has a solution vanishing at infinity, the free term of the boundary condition should also vanish at infinity. We shall hereafter assume this to be the case. To derive solutions we have to replace $P_\varkappa$ in the formula (14.27) by $P_{\varkappa-1}$, and set $C = 0$ in (14.28). Thus

$$\Phi(z) = X(z)\left[\Psi(z) + \frac{P_{\varkappa-1}(z)}{(z+i)^\varkappa}\right]. \tag{14.30}$$

For $\varkappa \leq 0$ we set $P_{\varkappa-1} \equiv 0$ in this formula.

The conditions of solubility (14.29) should be completed by one more, namely $\Psi(-i) = 0$. Thus, these conditions take the form

$$\int\limits_{-\infty}^{\infty} \frac{g(\tau)}{X^+(\tau)} \frac{d\tau}{(\tau+i)^k} = 0 \quad (k = 1, \ldots, -\varkappa). \tag{14.31}$$

Now, for $\varkappa > 0$ there exists a solution which depends on $\varkappa$ arbitrary constants; for $\varkappa \leq 0$ the solution is unique, and for $\varkappa < 0$ for its existence it is necessary and sufficient that $-\varkappa$ conditions be satisfied.

§ 15. Exceptional cases of the Riemann problem

In the formulation of the Riemann boundary value problem in § 14.1, it was required that the coefficient $G(t)$ satisfies the Hölder condition which excluded the possibility of infinite values, and that it does not vanish anywhere. It was clear from the course of the solution (using $\ln G(t)$) that these restrictions are essential. Now we shall investigate the problem relaxing these restrictions, i.e. assuming that the function $G(t)$ may have at isolated points of the contour zero or infinity, of order equal to an integer.

We assume for simplicity that the contour L consists of one closed curve.

15.1. The homogeneous problem

Let us write the boundary condition of the homogeneous Riemann problem in the form

$$\Phi^+(t) = \frac{\prod\limits_{k=1}^{\mu} (t - \alpha_k)^{m_k}}{\prod\limits_{j=1}^{\nu} (t - \beta_j)^{p_j}} G_1(t)\, \Phi^-(t). \tag{15.1}$$

Here α_k $(k = 1, 2, \ldots, \mu)$, β_j $(j = 1, 2, \ldots, \nu)$ are some points of the contour; m_k, p_j are positive integers, $G_1(t)$ is a non-vanishing function which satisfies the Hölder condition. The points α_k are the zeros of the function $G(t)$, and the points β_j its poles†. We denote

$$\text{Ind } G_1(t) = \varkappa; \quad \sum_{j=1}^{\nu} p_j = p; \quad \sum_{k=1}^{\mu} m_k = m.$$

The solution will be sought in the class of functions bounded on the contour‡.

Let $X(z)$ be the canonical function of the Riemann problem with the coefficient $G_1(t)$. Set in (15.1) $G_1(t) = \dfrac{X^+(t)}{X^-(t)}$ and write the boundary condition in the form

$$\frac{\Phi^+(t)}{X^+(t)\prod\limits_{k=1}^{\mu}(t-\alpha_k)^{m_k}} = \frac{\Phi^-(t)}{X^-(t)\prod\limits_{j=1}^{\nu}(t-\beta_j)^{p_j}}. \qquad (15.2)$$

Let us apply to this relation the theorem of analy ic continuation and the generalized Liouville theorem (cf. § 14.3, the case $\varkappa > 0$). The points α_k, β_j cannot be singular points of the unified analytic function, since this would contradict the assumption of boundedness of $\Phi^+(t)$ or $\Phi^-(t)$. Consequently, the only possible singularity is the point at infinity. The order of $X^-(z)$ at infinity is $\varkappa$, and the order of $\prod\limits_{j=1}^{\nu}(z-\beta_j)^{p_j}$ is equal to $-p$. It follows that the order at infinity of the function

$$\frac{\Phi^-(z)}{X^-(z)\prod\limits_{j=1}^{\nu}(z-\beta_j)^{p_j}}$$

is $-\varkappa + p$. For $\varkappa - p \geqq 0$, in accordance with the generalized Liouville theorem we have

$$\frac{\Phi^+(z)}{X^+(z)\prod\limits_{k=1}^{\mu}(z-\alpha_k)^{m_k}} = \frac{\Phi^-(z)}{X^-(z)\prod\limits_{j=1}^{\nu}(z-\beta_j)^{p_j}} = P_{\varkappa \cdot p}(z),$$

† Since the function $G(t)$ is not analytic the application of the term "pole" is not quite legitimate. We use this term for brevity, meaning a point at which the (non-analytic) function has an infinity of order equal to an integer.

‡ It is easy to prove that only the integrability conditions is essential, i.e. the restriction that the infinity must be of order lower than one. This implies also the boundedness (see Problem 12 at the end of this chapter).

hence

$$\left. \begin{aligned} \varPhi^+(z) &= X^+(z) \prod_{k=1}^{\mu} (z - \alpha_k)^{m_k} P_{\varkappa-p}(z), \\ \varPhi^-(z) &= X^-(z) \prod_{j=1}^{\nu} (z - \beta_j)^{p_j} P_{\varkappa-p}(z). \end{aligned} \right\} \tag{15.3}$$

If $\varkappa - p < 0$ we have to set $P_{\varkappa-p} \equiv 0$ and, consequently, the problem has no solution.

Let us call the boundary value problem with the coefficient $G_1(t)$ the reduced problem. The index of the reduced problem $\varkappa$ will also be called the index of the problem under consideration. The formulae (15.3) indicate that the degree of the polynomial $P(z)$ is by p units smaller than the index of the problem $\varkappa$.

It follows that *the number of solutions of the problem* (15.1) *in the class of functions which are bounded on the contour, is not influenced by the presence of zeros of the coefficient of the problem, and is diminished by the total order of all poles. In particular, if the index turns out to be smaller than the total order of poles, then the problem is insoluble.*

If the problem is soluble, its solution is given by the formulae (15.3), the canonical function of the reduced problem $X(z)$ being found in accordance with the formulae (14.9), (14.12) where $G(t)$ is to be replaced by $G_1(t)$. If the additional restriction $\varPhi^-(\infty) = 0$ is imposed, the number of solutions is diminished by one and the polynomial $P(z)$ should have the degree $\varkappa - p - 1$.

15.2. The non-homogeneous problem

Let us write the boundary condition in the form

$$\varPhi^+(t) = \frac{\prod\limits_{k=1}^{\mu} (t - \alpha_k)^{m_k}}{\prod\limits_{j=1}^{\nu} (t - \beta_j)^{p_j}} G_1(t) \varPhi^-(t) + g(t). \tag{15.4}$$

It is readily observed that the boundary condition cannot be satisfied by finite $\varPhi^+(t)$, $\varPhi^-(t)$, if it is assumed that $g(t)$ has poles at points distinct from β_j, or if at the latter points the order of the pole of $g(t)$ exceeds p_j. Hence we adopt the condition that $g(t)$ may have poles only at the points β_j and that their orders do not exceed p_j.

For the applicability of the further theory it is necessary to assume that the functions $G_1(t)$ and $\prod\limits_{j=1}^{\nu}(t - \beta_j)^{p_j} g(t)$ be differentiable at the exceptional points a sufficient number of times. The order of the required highest derivative will be determined below.

As in the case of the homogeneous problem, let us replace $G_1(t)$ by the ratio of the canonical functions $X^+(t)/X^-(t)$; then the boundary condition can be written in the form†

$$\prod_{j=1}^{\nu}(t - \beta_j)^{p_j} \frac{\Phi^+(t)}{X^+(t)}$$
$$= \prod_{k=1}^{\mu}(t - \alpha_k)^{m_k} \frac{\Phi^-(t)}{X^-(t)} + \prod_{j=1}^{\nu}(t - \beta_j)^{p_j} \frac{g(t)}{X^+(t)}. \quad (15.4')$$

The function $\prod\limits_{j=1}^{\nu}(t - \beta_j)^{p_j} g(t)/X^+(t)$ is integrable. Replacing it by the difference of the boundary values of the analytic functions

$$\prod_{j=1}^{\nu}(t - \beta_j)^{p_j} \frac{g(t)}{X^+(t)} = \Psi^+(t) - \Psi^-(t),$$

where

$$\Psi(z) = \frac{1}{2\pi i} \int_L \prod_{j=1}^{\nu}(\tau - \beta_j)^{p_j} \frac{g(\tau)}{X^+(\tau)} \frac{d\tau}{\tau - z}, \quad (15.5)$$

we reduce the boundary condition to the form

$$\prod_{j=1}^{\nu}(t - \beta_j)^{p_j}\frac{\Phi^+(t)}{X^+(t)} - \Psi^+(t) = \prod_{k=1}^{\mu}(t - \alpha_k)^{m_k} \frac{\Phi^-(t)}{X^-(t)} - \Psi^-(t).$$

---

† It is impossible to transform the boundary condition as in solving the homogeneous problem (formula (15.2)) leaving $\prod\limits_{j=1}^{\nu} (t - \beta_j)^{p_j}$ as the coefficient and dividing by $\prod\limits_{j=1}^{\nu} (t - \alpha_k)^{m_k}$, since the free term would then become non-integrable. Conversely, writing the condition (15.2) in the form (15.4') we arrive at the solution

$$\Phi^+(z) = \prod_{j=1}^{\nu} (z - \beta_j)^{-p_j} X^+(z) P_{\varkappa+m}(z);$$
$$\Phi^-(z) = \prod_{k=1}^{\mu} (z - \alpha_k)^{-m_k} X^-(z) P_{\varkappa+m}(z).$$

The condition of boundedness of $\Phi^+(t)$, $\Phi^-(t)$ would cause the necessity to take the polynomial in the form

$$P_{\varkappa+m}(z) = \prod_{k=1}^{\mu} (z - \alpha_k)^{m_k} \prod_{j=1}^{\nu} (z - \beta_j)^{p_j} P_{\varkappa-p}(z),$$

which would lead us to the solution (15.3) derived before.

Applying as before (§ 14.3) the theorem of analytic continuation and the generalized Liouville theorem we obtain

$$\left.\begin{aligned}
\Phi^+(z) &= \frac{X^+(z)}{\prod\limits_{j=1}^{\nu}(z-\beta_j)^{p_j}}\,[\Psi^+(z)+P_{\varkappa+m}(z)], \\[2em]
\Phi^-(z) &= \frac{X^-(z)}{\prod\limits_{k=1}^{\mu}(z-\alpha_k)^{m_k}}\,[\Psi^-(z)+P_{\varkappa+m}(z)].
\end{aligned}\right\} \qquad (15.6)$$

The last formulae yield solutions which in general have infinities at the points α_k, β_j. In order that the solution be bounded it is necessary that the function $\Psi^+(z) + P_{\varkappa+m}(z)$ has zeros of orders p_j at the points β_j, and the function $\Psi^-(z) + P_{\varkappa+m}(z)$ zeros of orders m_k at the points α_k, respectively. These requirements impose $m + p$ conditions on the coefficients of the polynomial $P_{\varkappa+m}(z)$. If these coefficients be chosen in accordance with the imposed conditions, the formulae (15.6) yield the solution of the non-homogeneous problem (15.4) in the class of bounded functions.

The above method of solution is inconvenient, both from the practical point of view (difficulties in solving systems of equations) and from the standpoint of theory; there is the necessity of proving the consistency and independence of equations following from the condition of boundedness of the solution at the points α_k, β_j, to which there can also be added conditions of boundedness of the solution at infinity.

To avoid these complications L. A. Chikin (1) suggested the construction of a special particular solution which he called the canonical function of the non-homogeneous problem.

DEFINITION. *The canonical function $Y(z)$ of the non-homogeneous problem (15.4) is the sectionally analytic function which satisfies the boundary condition (15.4), has zero order everywhere in the finite part of the plane (including the points α_k, β_j), and has at infinity the highest possible order.*

In constructing the canonical function we make use of the solution given by the formulae (15.6). Let us construct a polynomial $Q(z)$, such that it satisfies the following conditions:

$$Q^{(i)}(\beta_j) = \Psi^{+(i)}(\beta_j) \ (i = 0, 1, \ldots, p_j - 1; \ j = 1, 2, \ldots, \nu),$$

$$Q^{(l)}(\alpha_p) = \Psi^{-(l)}(\alpha_k) \ (l = 0, 1, \ldots, m_k - 1; \ k = 1, 2, \ldots, \mu).$$

Here $\Psi^{+(i)}(\beta_j)$, $\Psi^{-(l)}(\alpha_k)$ are the values of the ith and lth derivatives at the indicated points†. Thus $Q(z)$ is the interpolation polynomial for the function

$$\Psi(z) = \begin{cases} \Psi^+(z) & \text{at the points } \beta_j, \\ \Psi^-(z) & \text{at the points } \alpha_k \end{cases}$$

with the interpolation nodes β_j, α_k, the multiplicities being p_j, m_k, respectively.

It is known† that such a polynomial is uniquely determined and its degree is ϱ,

$$\varrho = m + p - 1.$$

The canonical function of the non-homogeneous problem is expressed in terms of the interpolation polynomial as follows:

$$Y^+(z) = \mathrm{X}^-(z) \frac{\Psi^+(z) - Q_\varrho(z)}{\displaystyle\prod_{j=1}^{\nu}(z - \beta_j)^{p_j}} \, ; \quad Y^-(z) = \mathrm{X}^-(z) \frac{\Psi^-(z) - Q_\varrho(z)}{\displaystyle\prod_{k=1}^{\mu}(z - \alpha_k)^{m_k}} \, .$$

$$(15.7)$$

To construct the general solution of the non-homogeneous problem (15.4) we shall make use of the fact that the general solution is composed of a particular solution of the non-homogeneous problem and the general solution of the homogeneous problem. In view of the formulae (15.3) and (15.7) we obtain

$$\left. \begin{aligned} \Phi^+(z) &= Y^+(z) + \mathrm{X}^+(z) \prod_{k=1}^{\mu}(z - \alpha_k)^{m_k} P_{\varkappa-p}(z); \\ \Phi^-(z) &= Y^-(z) + \mathrm{X}^-(z) \prod_{j=1}^{\nu}(z - \beta_j)^{p_j} P_{\varkappa-p}(z). \end{aligned} \right\}$$

$$(15.8)$$

If $\varkappa - p < 0$ we have to set $P_{\varkappa-p} \equiv 0$. Making use of the formula (15.7) it is easy to find that the order of $Y^-(z)$ at infinity is $\varkappa - p + 1$. If $\varkappa < p - 1$, then $Y^-(z)$ has at infinity a pole and the canonical function is no more a solution of the non-homogeneous problem.

If, however, the free term $g(t)$ be subject to $p - \varkappa - 1$ conditions it is possible to increase the order at infinity of the function $Y(z)$ by $p - \varkappa - 1$ units, and thus to convert the canonical function $Y(z)$

† The existence of these derivatives is ensured if the functions $\prod\limits_{j=1}^{\nu}(t - \beta_j)^p g(t)$, $G_1(t)$ be subject to the condition that at the points β_j, α_k they have derivatives of p_jth and m_kth order, respectively, which satisfy the Hölder condition.

nto a solution of the non-homogeneous problem. To this end, evidently, it is necessary and sufficient that in the expansion of the function $\Psi(z) - Q_\varrho(z)$ in the vicinity of infinity the first $p - \varkappa - 1$ coefficients vanish. This condition yields $p - \varkappa - 1$ conditions of solubility of the problem in the considered case. Let us elucidate the nature of these conditions. The expansion of $\Psi(z) - Q_\varrho(z)$ may be represented in the form

$$\Psi(z) - Q_\varrho(z) = -c_\varrho z^\varrho - c_{\varrho-1} z^{\varrho-1} - \cdots$$
$$\cdots - c_0 - c_{-1} z^{-1} + c_{-2} z^{-2} + \cdots + c_{-k} z^{-k} + \cdots,$$

where $c_0, c_1, \ldots, c_\varrho$ are the coefficients of the polynomial $Q_\varrho(z)$ and c_{-k} are the coefficients of expansion of the function $\Psi(z)$; it is easy to show that they are to be calculated by means of the formula

$$c_{-k} = -\frac{1}{2\pi i} \int_L \frac{\prod\limits_{j=1}^{\nu} (\tau - \beta_j)^{p_j} g(\tau) \tau^{k-1}}{X^+(\tau)} d\tau.$$

The above conditions of solubility have the form

$$c_\varrho = c_{\varrho-1} = \cdots = c_{\varrho-p+\varkappa+2} = 0.$$

If the solution is subject to the additional condition $\Phi^-(\infty) = 0$, then for $\varkappa - p > 0$ the polynomial $P_{\varkappa-p-1}(z)$ should replace $P_{\varkappa-p}(z)$ in the formulae (15.8); for $\varkappa - p < 0$ it is necessary to satisfy $p - \varkappa$ additional conditions.

Remark. The method of construction of the canonical function of the non-homogeneous problem can also be applied in other cases, when it is required to derive a solution which satisfies some prescribed conditions (see Problem 13 at the end of this chapter).

§16. Riemann problem for multiply-connected domain. Some new results

16.1. Formulation of the problem

Let $L = L_0 + L_1 + \cdots + L_m$ be a set of $m + 1$ non-intersecting contours, the contour L_0 enclosing the remaining ones (Fig. 13). D^+ is the $(m + 1)$-connected domain situated inside the contour L_0 and outside $L_1, \ldots, L_m$. By D^- we denote the supplement of $D^+ + L$ to the whole plane. For definiteness we suppose that the origin of

the coordinate system is located in the domain D^+. For the positive
direction on the contour L we take that which leaves the domain D^+
on the left, i.e. the contour L_0 is to be traversed anticlockwise and the
contours $L_1, \ldots, L_m$ clockwise.

Firstly we show that the jump problem

$$\Phi^+(t) - \Phi^-(t) = g(t)$$

is solved by the same formula

$$\Phi(z) = \frac{1}{2\pi i} \frac{g(\tau)d\tau}{\tau - z},$$

as in the case of a singly-connected domain. This follows from Sokhotski's formulae which, as shown in § 8.7, are the same for multiply-
connected as for simply-connected domains.

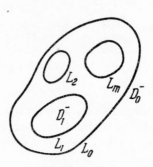

FIG. 13.

The Riemann problem (homogeneous and non-homogeneous) is
formulated in exactly the same manner as it was done in § 14.1
for the simply-connected domain.

Denote $\varkappa_k = (1/2\pi)[\arg G(t)]_{L_k}$ (all contours L_k are traversed in
the above established positive direction).

The quantity

$$\varkappa = \sum_{k=0}^{m} \varkappa_k \tag{16.1}$$

is called the index of the problem.

If all $\varkappa_k$ are zero it is readily observed that the solution of the
problem has exactly the same form as for the simply-connected
domain (cf. § 8.7).

To reduce the general case to the simplest one we introduce the function

$$\prod_{k=1}^{m}(t - z_k)^{\varkappa_k},$$

where z_k are some points lying inside the contours $L_k (k = 1, 2, \ldots, m)$, respectively.

Taking into account that $[\arg(t - z_k)]_{L_j} = 0$ if $k \neq j$, and $[\arg(t - z_j)]_{L_j} = -2\pi$ we easily obtain

$$\frac{1}{2\pi}\left[\arg \prod_{k=1}^{m}(t - z)_k)^{\varkappa_k}\right]_{L_j} = \frac{1}{2\pi}[\arg(t - z_j)^{\varkappa_j}]_{L_j} = -\varkappa_j$$

$$(j = 1, 2, \ldots, m).$$

Hence

$$\left\{\arg\left[G(t)\prod_{k=1}^{m}(t - z_k)^{\varkappa_k}\right]\right\}_{L_j} = 0 \quad (j = 1, 2, \ldots, m).$$

Let us calculate the change of the argument of the function $G(t)\prod_{k=1}^{m}(t - z_k)^{\varkappa_k}$ along the contour L_0:

$$\frac{1}{2\pi}\left\{\arg\left[G(t)\prod_{k=1}^{m}(t - z_k)^{\varkappa_k}\right]\right\}_{L_0}$$

$$= \frac{1}{2\pi}[\arg G(t)]_{L_0} + \frac{1}{2\pi}\sum_{k=1}^{m}[\varkappa_k \arg(t - z_k)]_{L_0} = \varkappa_0 + \sum_{k=1}^{m}\varkappa_k = \varkappa.$$

Since the origin is located in the domain D^+,

$$[\arg t]_{L_k} = 0 \quad (k = 1, 2, \ldots, m), \quad [\arg t]_{L_0} = 2\pi.$$

Consequently

$$\left\{\arg\left[t^{-\varkappa}\prod_{k=1}^{m}(t - z_k)^{\varkappa_k}G(t)\right]\right\}_L = 0. \tag{16.2}$$

16.2. Solution of the problem

1. *Homogeneous problem.*

Let us write the boundary condition

$$\Phi^+(t) = G(t)\Phi^-(t) \tag{14.1}$$

in the form

$$\Phi^+(t) = \frac{t^{\varkappa}}{\prod_{k=1}^{m}(t - z_k)^{\varkappa_k}}\left[t^{-\varkappa}\prod_{k=1}^{m}(t - z_k)^{\varkappa_k}G(t)\right]\Phi^-(t). \tag{16.3}$$

Since the function $t^{-\varkappa}\prod\limits_{k=1}^{m}(t-z_k)^{\varkappa_k}G(t)$ has zero index on every contour L_k $(k=0, 1, \ldots, m)$, it can be represented in the form of the ratio

$$t^{-\varkappa}\prod_{k=1}^{m}(t-z_k)^{\varkappa_k}G(t) = \frac{e^{\Gamma^+(t)}}{e^{\Gamma^-(t)}}, \qquad (16.4)$$

where

$$\Gamma(z) = \frac{1}{2\pi i}\int_L \frac{\ln\left[\tau^{-\varkappa}\prod\limits_{k=1}^{m}(\tau-z_k)^{\varkappa_k}G(\tau)\right]}{\tau-z}\,d\tau. \qquad (16.5)$$

Now the boundary condition takes the form

$$\prod_{k=1}^{m}(t-z_k)^{\varkappa_k}\frac{\Phi^+(t)}{e^{\Gamma^+(t)}} = t^{\varkappa}\frac{\Phi^-(t)}{e^{\Gamma^-(t)}}.$$

Applying as before the theorem of analytic continuation and the generalized Liouville theorem we obtain

$$\left.\begin{array}{l}\Phi^+(z) = \prod\limits_{k=1}^{m}(z-z_k)^{-\varkappa_k}e^{\Gamma^+(z)}P_{\varkappa}(z),\\[2mm] \Phi^-(z) = z^{-\varkappa}e^{\Gamma^-(z)}P_{\varkappa}(z).\end{array}\right\} \qquad (16.6)$$

We see that the solution differs from the solution of the problem derived before in the case of a simply-connected domain only by the presence of the factor $\prod\limits_{k=1}^{m}(z-z_k)^{-\varkappa_k}$ in the expression for the function $\Phi^+(z)$.

If there is the additional condition $\Phi^-(\infty) = 0$ the polynomial $P_{\varkappa}(z)$ in (16.6) should be replaced by $P_{\varkappa-1}(z)$.

The canonical function of the problem (§ 14.4) is given by the formulae

$$X^+(z) = \prod_{k=1}^{m}(z-z_k)^{-\varkappa_k}e^{\Gamma^+(z)}, \quad X^-(z) = z^{-\varkappa}e^{\Gamma^-(z)}. \qquad (16.7)$$

For subsequent considerations (§ 21.2) we shall require formulae for the limiting values on the contour of the canonical function $X(z)$. Taking limiting values in the formula (16.5), we find in view of the Sokhotski formulae

$$\Gamma^{\pm}(t) = \pm\tfrac{1}{2}\ln[t^{-\varkappa}\Pi(t)G(t)] + \Gamma(t),$$

where $\Gamma(t)$ is the principal value of the integral (16.5), and

$$\Pi(t) = \prod_{k=1}^{m}(t-z_k)^{\varkappa_k}.$$

Now, passing to the limit $z \to t$ in the relations (16.7), we obtain

$$X^+(t) = \sqrt{\left(\frac{G(t)}{t^\varkappa \Pi(t)}\right)}\, e^{\Gamma(t)}, \quad X^-(t) = \frac{1}{\sqrt{[t^\varkappa \Pi(t)\, G(t)]}}\, e^{\Gamma(t)}. \quad (16.8)$$

The sign of the root is determined by the choice of the branch of the function $\ln[t^{-\varkappa}\Pi(t)\, G(t)]$ which may be arbitrary (see § 14.2).

2. Non-homogeneous problem.

Reasoning entirely analogous to that in § 14.5 leads to the following representation of the boundary condition

$$\Phi^+(t) = G(t)\, \Phi^-(t) + g(t) \quad (14.2)$$

of the Riemann problem:

$$\frac{\Phi^+(t)}{X^+(t)} - \Psi^+(t) = \frac{\Phi^-(t)}{X^-(t)} - \Psi^-(t).$$

Here $\Psi(z)$ is given by the formula (14.14).

Hence we obtain the general solution

$$\Phi(z) = X(z)\,[\Psi(z) + P_\varkappa(z)] \quad (16.9)$$

or

$$\Phi(z) = X(z)\,[\Psi(z) + P_{\varkappa-1}(z)], \quad (16.10)$$

if the solution be subject to the condition $\Phi^-(\infty) = 0$.

For $\varkappa < 0$ the non-homogeneous problem is soluble if and only if the conditions

$$\int\limits_L \frac{g(t)}{X^+(t)}\, t^{k-1}\, dt = 0, \quad (16.11)$$

are satisfied; here k takes values from 1 to $-\varkappa - 1$ if the solutions are to be bounded at infinity, and from 1 to $-\varkappa$ if $\Phi(\infty) = 0$.

If the conditions (16.11) are satisfied the solution is obtained from the formulae (16.9) or (16.10), where we set $P(z) \equiv 0$.

We shall now show the necessary alteration in the form of the solution when the exterior contour L_0 is absent and the domain D^+ constitutes the plane with holes. Observe that now the point at infinity belongs to D^+, not to D^- as before. Therefore, now the behaviour of the function $\Phi^+(z)$ at infinity is relevant, in contrast to the previous case when we had to consider $\Phi^-(z)$ at infinity.

The principal difference as compared with the preceding case consists in the fact that now the function $\prod\limits_{k=1}^{m}(t - z_k)^{\varkappa_k} G(t)$ which does not contain the factor $t^{-\varkappa}$. Has zero index with respect to all contours L_k ($k = 1, 2, \ldots, m$). Consequently, to deduce the solution it is sufficient to repeat all previous considerations, omitting the above factor.

16.3. Some new results

Recent papers on the Riemann boundary value problem are mainly directed towards broadening the class of given and required functions. Here we shall give the main results obtained in this field.

1. B. V. Khvedelidze, in many papers (they are summarized in the article [11]) investigated the Riemann problem on the assumption that $G(t)$ satisfies Hölder's condition†, whilst $g(t) \in L_p$ (Lebesgue integrable in power p). Using the theory of Cauchy integrals in their appropriate class (see § 5.3), a complete solution of the boundary value problem has been produced. The results (both in qualitative respects—calculation of the number of solutions and conditions of solubility in relation to the index, and also algorithmically—the type of formulae which yield solutions in closed form) are the same as in the case with $g(t)$ satisfying Hölder's condition. The only difference is that here, and everywhere else that it is a question of Lebesgue integration, the corresponding limit values on the contour only exist almost everywhere (except for a set of measure zero) and only along paths which are not tangential to the contour.

2. Only very recently has it been possible to relax the restriction requiring the coefficient to satisfy Hölder's condition, retaining only the natural condition of continuity. The problem was so ripe that its solution was produced simultaneously and independently by several writers. First we shall give Simonenko's solution (1).

It is based on the following simple idea. If $G(t) = 1$ we have the jump problem which is unconditionally and soluble in class L_p. It appears that small deviations of the coefficient do not affect the problem. The transition to the general case is effected by approximating the continuous function by functions which satisfy Hölder's condition.

In the boundary condition of the Riemann problem

$$\Phi^+(t) = G(t)\,\Phi^-(t) + g(t)$$

$G(t)$ is assumed to be a continuous function and $g(t) \in L_p$. Schematically the solution may be divided into four stages.

---

† On the possibility of discontinuities at isolated points, see § 44.5.

(i) An estimate is made using Khvedelidze's results for a Cachy integral in class L_p (see § 5.3):

$$||\Phi^{\pm}(t)|| = \left|\left| \pm \frac{1}{2}\varphi + \frac{1}{2}S\varphi \right|\right| \leq \frac{1 + M_p}{2} ||\varphi||,$$

where M_p is the norm of the operator $S(S\varphi = \frac{1}{\pi i}\int_L \frac{\varphi(\tau)}{\tau - i} d\tau)$.

(ii) Using the principle of contraction mappings (of method of successive approximations) the author proves that a Riemann problem, in which the coefficient is a measurable function satisfying the condition

$$|G(t) - 1| < \frac{2}{1 + M_p},$$

has a unique solution.

(iii) Using the approximation of $\ln G(t)$ by functions satisfying Hölder's condition, the author proves the single-valued solubility of the Riemann problem with zero index.

(iv) The case of an arbitrary index is investigated by reducing it to that for the zero index. The method or reduction is very similar to that discussed in §§ 16.1 and 16.2.

These results are the same as those of Khvedelidze for G satisfying Hölder's condition. It is thus shown that the broadening of the class of the coefficient G to include just continuous functions leads to no narrowing of the class of solutions; it still coincides with the class L_p of the free term.

3. The solution proposed by Ivanov (4) and Khvedelidze (2) is based on Smirnov's theorem which provides an estimate of the order of growth of a singular Cauchy integral with continuous density. This integral is generally unbounded, though in growing infinitely it may only do so very slowly; thus, not only $S\varphi$, but also $e^{+S\varphi}$ belong to L_r for any $r > 0$†.

Let $\varkappa = \text{Ind } G(t) = 0$, $g(t) \in L_p$. From foregoing considerations the canonical function of the Riemann problem $X^{\pm}(t) = e^{\pm\frac{1}{2}\ln G\pm\frac{1}{2}S\ln G}\in L_r$. As usual we reduce the boundary condition to the jump problem

† According to the results in § 8.2 for a first order discontinuity $S\varphi$ has a logarithmic singularity and $e^{S\varphi}$ therefore has a power singularity; the jump may be selected such that the order of growth is arbitrarily large.

form:

$$\frac{\Phi^+(t)}{X^+(t)} - \frac{\Phi^-(t)}{X^-(t)} = \frac{g(t)}{X^+(t)}.$$

Using Hölder's inequality† the right-hand side belongs to the class $L_{p-\varepsilon}$, where ε is an arbitrarily small positive number. The solution of the problem with its boundary values defined by the formula

$$\Phi^\pm(t) = X^\pm \left[\pm \frac{1}{2} \frac{g}{X^+} + \frac{1}{2} S\left(\frac{f}{X^+}\right) \right],$$

belongs to the same class $L_{p-\varepsilon}$. This result is less accurate than that obtained by Simonenko's method; he established that one may put $\varepsilon = 0$.

4. Cherskii (1) has studied the Riemann boundary value problem in abstract space, assuming that in the boundary condition

$$\varphi^+ = A\varphi^- + g$$

A is a linear operator, whilst φ^+, φ^-, g are elements of a Banach space. The singular operator S is defined axiomatically. Its basic definitive property is $S^2 = I$ [I is a unit operator], provided $S \neq \pm I$. The elements $\varphi^\pm$ are formed from an arbitrary element φ by the formulae $\varphi^\pm = \frac{1}{2}(\pm \varphi + S\varphi)$. The index is defined in terms of some operator. A complete solution is given with this formulation of the problem.

The main feature of the paper is the combination of an investigation of problems in the existence of a solution with the construction of the algorithm for actually producing it. This unification arises since the abstract theory is constructed on the effective solution of the Riemann problem (§§ 14–17) by replacing the concrete operators and functions by abstract operators and elements of abstract space. The boundary value problem is considered in association with the solution of singular integral equations (see § 21.1).

5. One consequence of the abstract theory by Yu. I. Cherskii (2) is his solution of the Riemann problem in generalized functions (functionals over some space of fundamental functions).

6. Yu. L. Rodin (1) studied the Riemann boundary value problem for domains on Riemann surfaces. As the apparatus of the solution two special integrals were constructed $\Phi^\pm(z) = \dfrac{1}{2\pi i} \displaystyle\int_L \varphi(\tau) A^\pm(\tau, z) \, d\tau$, which are counterparts of the Cauchy integral. Their limit values on

† See, for example, Hardy, Littlewood *et al.*, Inequalities, 1946, p. 169.

the contour are given by Sokhotski's formulae:

$$\Phi^{\pm}(t) = \pm \frac{1}{2}\varphi(t) + \frac{1}{2\pi i}\int_L \varphi(\tau)\,A^{\pm}(t, \tau)\,d\tau.$$

In this paper the boundary value problem is solved by reducing it to a singular integral equation with a Cauchy kernel (see § 21.1). In another paper, as yet unpublished, use is made of the usual method of constructing a canonical function and reducing it to the jump problem. For surfaces of the zero order the results completely coincide in qualitative respects with what we had for a plane domain. For a surface of the order $p > 0$ the number of solutions or solubility conditions depends not only on the index $\varkappa$, but also on the order p. Let us formulate some of the results. If $\varkappa \geqq p$ the homogeneous and inhomogeneous problems are absolutely soluble. If $0 \leqq \varkappa < p$ the homogeneous and inhomogeneous problems are only soluble if certain conditions are fulfilled.

7. Chikin (2), (3) studied aspects of the stability of the solution of the Riemann problem for small deformations of the contour satisfying various limitations. The main result was: if $G(t)$ and $g(t)$ satisfy Hölder's condition and $G(t) \neq 0$, then the homogeneous and inhomogeneous problems are stable.

8. Questions of approximate solutions of the Riemann boundary value problem were considered by Batyrev (1). Two methods are indicated. Both are based on an approximation of the coefficient of the boundary condition by a rational function. In the first method the coefficient $G(t)$ is approximated by the function $r_n(t) = \sum\limits_{k=-n}^{n} a_k t^k$. The homogeneous problem with the coefficient $r_n(t)$ is solved by the method of § 14.6. In the second method, by taking logarithms, the homogeneous problem is reduced to the jump problem; $\ln G(t)$ is then approximated by the same function $r_n(t)$. An estimate of the error was made.

Other papers on this topic deal with singular integral equations. They are considered in § 21.5.

§ 17. Riemann boundary value problem with shift

17.1. Formulation of the problem and general remarks

Suppose that we are given a simple closed contour L dividing the plane into the interior domain D^+ and exterior domain D^-. There are given on this contour the functions $G(t)$, $g(t)$ which satisfy the Hölder condition, and a function $\alpha(t)$ which establishes a one to one mapping of the contour L onto itself, the direction being preserved, and possesses the derivative $\alpha'(t)$ which satisfies the Hölder condition and does not vanish anywhere. The function $\alpha(t)$ will be called the map function.

It is required to determine a function $\Phi^+(z)$ analytic in the domain D^+ and a function $\Phi^-(z)$ analytic in D^-, the limiting values of which on the contour are continuous and satisfy the linear relation

$$\Phi^+[\alpha(t)] = G(t)\Phi^-(t) \qquad \text{(homogeneous problem)} \qquad (17.1)$$

or

$$\Phi^+[\alpha(t)] = G(t)\Phi^-(t) + g(t) \quad \text{(non-homogeneous problem)} \quad (17.2)$$

The above formulated problem is a generalization of the Riemann boundary value problem. In the particular case when $\alpha(t)$ is the boundary value of a function analytic in the domain D^+ the problem can be reduced to the Riemann problem. To this end it is sufficient to introduce a new unknown sectionally analytic function

$$\Phi_1^+(z) - \Phi^+[\alpha(z)], \quad \Phi_1^-(z) = \Phi^-(z).$$

In the general case this shift problem may be reduced to the ordinary Riemann boundary value problem. But first it is necessary to give the solution of the jump problem:

$$\Phi^+[\alpha(t)] - \Phi^-(t) = g(t). \qquad (17.3)$$

This is based on an investigation of the simplest problem of a zero jump:

$$\Phi^+[\alpha(t)] - \Phi^-(t) = 0. \qquad (17.3')$$

A similar problem underlies the solution of the ordinary Riemann problem

$$\Phi^+(t) - \Phi^-(t) = 0,$$

but the latter is solved directly by using the theorem of analytic continuability and Liouville's theorem. In contrast (17.3') cannot be solved so simply and the theory of Fredholm integral equations has to be employed.

17.2. Problem with zero jump

It is required to determine a sectionally analytic function $\Phi(z)$ which satisfies on the contour the boundary condition

$$\Phi^+[\alpha(t)] - \Phi^-(t) = 0 \qquad (17.4)$$

and the additional condition $\Phi^-(\infty) = 0$.

We form the integral equation of the problem by using the conditions (4.11) and (4.12) for $\Phi^+(t)$ to be the boundary values of the functions in the respective domains. Bearing in mind the study of the solubility of the resultant equation, we introduce two new analytic functions, taking the required functions for the density of Cauchy integrals:

$$\Psi^+(z) = \frac{1}{2\pi i} \int_L \frac{\Phi^-(\tau)}{\tau - z} d\tau, \quad \Psi^-(z) = \frac{1}{2\pi i} \int_L \frac{\Phi^+(\tau)}{\tau - z} d\tau. \quad (17.5)$$

By virtue of the Sokhotski formulae and conditions (4.11), (4.12), we have on the contour

$$\Psi^+(t) = \frac{1}{2}\Phi^-(t) + \frac{1}{2\pi i} \int_L \frac{\Phi^-(\tau)}{\tau - t} d\tau = 0, \qquad (17.6)$$

$$\Psi^-(t) = -\frac{1}{2}\Phi^+(t) + \frac{1}{2\pi i} \int_L \frac{\Phi^+(\tau)}{\tau - t} d\tau = 0. \qquad (17.7)$$

The right-hand side of the equalities only holds when $\Phi^\pm(t)$ are boundary values of functions analytic in $D^\pm$.

Let us construct the equation for $\Phi^-(t)$; to this end consider the expression

$$\Psi^+(t) - \Psi^-[\alpha(t)] = \tfrac{1}{2}\{\Phi^-(t) + \Phi^+[\alpha(t)]\} +$$

$$+ \frac{1}{2\pi i} \int_L \frac{\Phi^-(\tau)}{\tau - t} d\tau - \frac{1}{2\pi i} \int_L \frac{\Phi^+(\tau)}{\tau - \alpha(t)} d\tau = 0. \quad (17.8)$$

To eliminate from this relation the function $\Phi^+(t)$ by means of the boundary condition (17.4) we make the substitution $\tau = \alpha(\tau_1)$ in the second integral and denote again the variable τ_1 by τ. Then we obtain for $\Phi^-(t)$ the integral equation †

$$\Psi^+(t) - \Psi^-[\alpha(t)] \equiv \Phi^-(t) - \frac{1}{2\pi i} \int\limits_L \left[\frac{\alpha'(\tau)}{\alpha(\tau) - \alpha(t)} - \frac{1}{\tau - t} \right] \Phi^-(\tau)\,d\tau = 0.$$

$$(17.9)$$

We shall prove that the kernel of the equation

$$K(t, \tau) = \frac{\alpha'(\tau)}{\alpha(\tau) - \alpha(t)} - \frac{1}{\tau - t}$$

has at the point $\tau = t$ a singularity of order lower than one.
Subtracting the obvious relations

$$\alpha(t) - \alpha(\tau) = \int\limits_\tau^t \alpha'(u)\,du,$$

$$\alpha'(\tau)(t - \tau) = \int\limits_\tau^t \alpha'(\tau)\,du,$$

we obtain

$$\alpha(t) - \alpha(\tau) - \alpha'(\tau)(t - \tau) = \int\limits_\tau^t [\alpha'(u) - \alpha'(\tau)]\,du.$$

To estimate the latter integral, write $|u - \tau| = r$ and make use of the inequality (4.3) of §4.1, $|du| = |ds| < m\,dr$. Taking into account the Hölder condition $|\alpha'(u) - \alpha'(\tau)| < A r^\lambda$ we obtain

$$|\alpha(t) - \alpha(\tau) - \alpha'(\tau)(t - \tau)| < Am \int\limits_0^\varrho r^\lambda dr = \frac{Am}{\lambda + 1} \varrho^{1+\lambda}$$

$$= \frac{Am}{1 + \lambda} |t - \tau|^{1+\lambda},$$

where ϱ is a sufficiently small number. Hence

$$|K(t, \tau)| < M |t - \tau|^{\lambda - 1}$$

where M is a constant.

---

† Since we have not required that the functions $\Phi^\pm(t)$ satisfy the Hölder condition we cannot state that the Sokhotski formulae for the limiting values hold everywhere on the contour; they hold only almost everywhere. In the equation (17.9), however, the first term is a continuous function and the second, in view of a property of the kernel proved below, is also a continuous function. Now, if a relation in which all terms are continuous is satisfied almost everywhere, it is satisfied everywhere. Thus, the equation (17.9) holds on the whole contour.

Thus, the integral equation (17.9) is a Fredholm equation. It is readily observed that a function $\Phi^-(t)$ which satisfies the boundary condition (17.4) and the analyticity condition (17.6), satisfies the integral equation (17.9). This fact follows from the very manner of constructing this equation. The question as to whether every solution of the integral equation is a solution of the boundary value problem is for the time being left aside.

We now proceed to establish a result which is fundamental for the problem under consideration.

LEMMA I. *The generalized problem with zero jump and the additional condition $\Phi^-(\infty) = 0$ is insoluble*†.

Assume the converse, i.e. that this problem is soluble. Let the sectionally analytic function $\Phi(z)$ be its solution. We infer directly from the form of the boundary condition (17.4) that the functions

$$\Phi_k(z) = [\Phi(z)]^k$$

are also solutions of the problem, for k being an arbitrary positive integer (it is easy to carry out the proof for arbitrary k as well). It is evident that all these functions are linearly independent. As was pointed out above, the functions $\Phi_k^-(t)$ satisfy the integral equation (17.9) which therefore has an infinite number of solutions. Since a Fredholm integral equation can have only a finite number of solutions, the resulting contradiction proves the lemma.

We cannot as yet conclude on the basis of the above proved lemma, that the equation (17.9) is also insoluble, since we have not yet proved that every solution of this equation is a solution of the boundary value problem. We now consider this problem.

LEMMA II. *The homogeneous Fredholm integral equation* (17.9) *is insoluble.*

Assume that the homogeneous Fredholm integral equation (17.9) has a solution $\Phi^-(t)$ different from zero. By means of this solution we construct a new function $\Phi^+(t) = \Phi^-[\beta(t)]$ where $\beta(t)$ is a function inverse to $\alpha(t)$ $(\beta[\alpha(t)] \equiv t)$. Next, in accordance with the formulae (17.5) we determine the two analytic functions $\Psi^+(z)$, $\Psi^-(z)$.

From integral equation (17.9)

$$\Psi^+(t) = \Psi^-[\alpha(t)]$$

† Here, as before, a homogeneous boundary value problem or a homogeneous integral equation whose only solution is identically zero will be called insoluble.

or

$$\Psi^+[\beta(t)] = \Psi^-(t) \qquad (17.10)$$

with the additional condition $\Psi^-(\infty) = 0$.

The function $\beta(t)$, as well as $\alpha(t)$, maps the contour onto itself. Hence, the problem (17.10) is of the same type as the problem (17.4); consequently, on the basis of Lemma I, $\Psi^+(z) = \Psi^-(z) \equiv 0$.

In view of the formulae (17.6), (17.7) the last result implies that $\Phi^+(t)$, $\Phi^-(t)$ constitute the boundary values of functions analytic in the domains D^+, D^-, respectively. The functions $\Phi^+(z)$, $\Phi^-(z)$ therefore are the solution of the boundary value problem (17.4) and again, in accordance with Lemma I they vanish identically.

Consequently, $\Phi^-(t) = 0$ as required.

17.3. Problem with given jump

Consider now the boundary value problem

$$\Phi^+[\alpha(t)] - \Phi^-(t) = g(t). \qquad (17.3)$$

Having in mind further applications let us somewhat generalize the problem, seeking the solution in the class of functions which may have at infinity a pole of given order n.

In view of the formulae (4.11)–(4.13) the conditions of analyticity in this case have the form

$$\left. \begin{array}{l} \dfrac{1}{2}\,\Phi^+(t) + \dfrac{1}{2\pi i} \displaystyle\int_L \dfrac{\Phi^+(\tau)}{\tau - t}\, d\tau = 0, \\[4mm] \dfrac{1}{2}\,\Phi^-(t) + \dfrac{1}{2\pi i} \displaystyle\int_L \dfrac{\Phi^-(\tau)}{\tau - t}\, d\tau - P_n(t) = 0, \end{array} \right\} \qquad (17.11)$$

where $P_n(z) = c_0 + c_1 z + \cdots + c_n z^n$ is the polynomial representing the principal part of the function $\Phi^-(z)$ at infinity.

Introducing as before the sectionally analytic function

$$\left. \begin{array}{l} \Psi^+(z) = \dfrac{1}{2\pi i} \displaystyle\int_L \dfrac{\Phi^-(\tau)}{\tau - z}\, d\tau - P_n(z), \\[4mm] \Psi^-(z) = \dfrac{1}{2\pi i} \displaystyle\int_L \dfrac{\Phi^+(z)}{\tau - z}\, d\tau, \end{array} \right\} \qquad (17.12)$$

we write the conditions (17.11) in the form of the two relations:

$$
\left.
\begin{aligned}
\Psi^+(t) &= \frac{1}{2}\,\Phi^-(t) + \frac{1}{2\pi i}\int_L \frac{\Phi^-(\tau)}{\tau - t}\,d\tau - P_n(t) = 0; \\[2mm]
\Psi^-(t) &= -\frac{1}{2}\,\Phi^+(t) + \frac{1}{2\pi i}\int_L \frac{\Phi^+(\tau)}{\tau - t}\,d\tau = 0.
\end{aligned}
\right\}
\quad (17.13)
$$

Proceeding as in the preceding section we arrive at a non-homogeneous Fredholm equation for $\Phi^-(t)$:

$$
\Psi^+(t) - \Psi^-[\alpha(t)]
$$
$$
\equiv \Phi^-(t) - \frac{1}{2\pi i}\int_L \left[\frac{\alpha'(\tau)}{\alpha(\tau) - \alpha(t)} - \frac{1}{\tau - t}\right]\Phi^-(\tau)\,d\tau - P_n(t) +
$$
$$
+ \frac{1}{2}g(t) - \frac{1}{2\pi i}\int_L \frac{g(\tau)\,\alpha'(\tau)}{\alpha(\tau) - \alpha(t)}\,d\tau = 0. \quad (17.14)
$$

The corresponding homogeneous equation is identical with the equation (17.9) which, in accordance with Lemma II is insoluble. Consequently, the equation (17.14) is soluble for an arbitrary right-hand side (see § 20.2).

Let $R(t,\tau)$ be the resolvent of the equation (17.14). Then it is known that its solution is representable in the form

$$
\Phi^-(t) = g_1(t) + P_n(t) + \int_L R(t,\tau)\,[g_1(\tau) + P_n(\tau)]\,d\tau, \quad (17.15)
$$

where

$$
g_1(t) = -\frac{1}{2}g(t) + \frac{1}{2\pi i}\int_L \frac{g(\tau)\,\alpha'(\tau)}{\alpha(\tau) - \alpha(t)}\,d\tau. \quad (17.16)
$$

It may be proved without difficulty that if $g(t)$ satisfies Hölder's condition, the solution of equation (17.14) also satisfies this condition. In fact, the resolvent $R(t,\tau)$ of the equation is known to be representable as a sum of iterative kernels. Iteration of non-singular kernels improves their properties; consequently, the functional properties of the resolvent are the same as those of the kernel. In § 5 it was proved that a singular integral with a Cauchy kernel transforms functions which satisfy Hölder's condition onto itself. This holds good more so for an integral with a weaker singularity†.

† Actually, kernels of this type improve the properties of a function.

Reasoning similar to that in the proof of Lemma II leads to the result that every solution of the equation (17.14) is a solution of the boundary value problem (17.3). From the boundary condition (17.3) we determine the boundary value $\Phi^+(t)$:

$$\Phi^+(t) = \Phi^-[\beta(t)] + g[\beta(t)] \quad (\beta[\alpha(t)] \equiv t). \qquad (17.17)$$

Now the Cauchy formula makes it possible to construct the functions $\Phi^+(z)$, $\Phi^-(z)$.

Thus, we have proved the following theorem.

THEOREM. *The jump problem* (17.3) *in the class of functions which have a given principal part at infinity, is soluble unconditionally and single-valuedly.*

It could easily be proved that if the coefficients of the polynomial $P_n(z)$ were deemed to be arbitrary, the equation would have $(n + 1)$ linearly independent solutions. But we do not need this and we shall not dwell on it further†.

17.4. The homogeneous problem with zero index

Suppose that

$$\Phi^+[\alpha(t)] = G(t)\,\Phi^-(t) \quad (G(t) \neq 0). \qquad (17.1)$$

Since $\alpha(t)$ maps the contour L in a one-to-one way onto itself preserving the direction, we have

$$\text{Ind } \Phi^+[\alpha(t)] = \text{Ind } \Phi^+(t) = N^+.$$

Hence

$$\varkappa = \text{Ind } G(t) = N^+ + N^-.$$

Thus, as in § 14.3 we obtain the following assertions.

1. For $\varkappa < 0$ the homogeneous problem is insoluble.
2. For $\varkappa = 0$ the solution of the homogeneous problem has no zeros.

We seek the solution which satisfies the condition $\Phi^-(\infty) = 1$. Taking logarithms in the relation (17.1) we obtain

$$\ln \Phi^+[\alpha(t)] - \ln \Phi^-(t) = \ln G(t).$$

---

† If, as in the first edition of the book, the shift problem is solved directly without reducing it to the Riemann problem, this fact is the basis for calculating the number of linearly independent solutions of the problem.

For the sectionally analytic function $\Gamma(z) = \ln \Phi(z)$ we obtain the boundary condition of the type (17.3) with the additional condition $\Gamma(\infty) = 0$.

An explicit expression for the function $\Gamma(z)$ is given by the formulae (17.15), (17.17) in which we set $g(t) = \ln G(t)$ and $P_n(z) \equiv 0$.

Consequently, the problem has a unique solution given by the relations

$$\Phi(z) = e^{\Gamma(z)}; \quad \Gamma^{\pm}(z) = \frac{1}{2\pi i} \int_L \frac{\Gamma^{\pm}(\tau)}{\tau - z} d\tau, \qquad (17.18)$$

$$\left.\begin{array}{l} \Gamma^-(t) = g_1(t) + \int R(t, \tau)\, g_1(\tau)\, d\tau, \\[2mm] \Gamma^+(t) = \Gamma^- [\beta(t)] + \ln G [\beta(t)], \end{array}\right\} \qquad (17.19)$$

$g_1(t)$ being determined by the formula (17.16) where $g(t) = \ln G(t)$, and $R(t,\tau)$ is the resolvent of the integral equation (17.14).

We solve a special problem that will be required later. Find the sectionally analytic function $\gamma^+(z)$ $(\gamma^-(\infty) = 0)$ from the boundary condition

$$\gamma^+ [a(t)] = \frac{1}{\alpha'(t)} \gamma^-(t). \qquad (17.20)$$

We prove that $\alpha'(t)$ has a zero index. Consider the ratio $[\alpha(\tau) - \alpha(t)]/(\tau - t)$. On traversing the point τ of the contour L the numerator and denominator obtain the same increment π.

The increment of the argument of the fraction is zero for any nonequal t and τ. Passing to the limit as $\tau \to t$, we obtain what is required. It is also known that $\alpha'(t) \neq 0$. Consequently, the problem (17.20) is unconditionally and single-valuedly soluble.

17.5. Reducing the shift problem to the ordinary Riemann problem

The close analogy between the shift problem and the ordinary Riemann boundary value problem suggests the presence of some direct relationship. It appears that such a relationship can in fact be established. To this end it is sufficient to carry out a conformal mapping of the domains $D^{\pm}$ on two new domains $D_1^{\pm}$ separated by a common contour and mutually complementing each other for a total plane so that the points t and $\alpha(t)$ of the original contour L change to the same points of the new contour L_1. After this the shift problem may be reduced by a simple change of variables to the ordinary Riemann problem. The justification rests on our solution of the jump problem and the homogeneous problem with zero index.

We begin by finding the conformal mapping.

We solve the auxiliary problem with zero jump

$$\omega^+[\alpha(t)] - \omega^-(t) = 0 \qquad (17.4')$$

on condition that $\omega^-(z)$ has at infinity a simple pole having the expansion

$$\omega^-(z) = z + \frac{c_1}{z} + \dots. \qquad (17.21)$$

We form the integral equation of the problem. Putting in (17.14) $g(t) = 0$, $P_n(t) = t$ and $\Phi^-(t) = \omega^-(t)$, we have the unconditionally and single-valuedly soluble equation

$$\omega^-(t) - \frac{1}{2\pi i} \int_L \left[\frac{\alpha'(\tau)}{\alpha(\tau) - \alpha(t)} - \frac{1}{\tau - t} \right] \omega^-(\tau)\, d\tau = t. \qquad (17.22)$$

Finding $\omega^-(t)$ from here, and $\omega^+(t) = \omega^-[\beta(t)]$ from the boundary condition (17.4'), the required functions $\omega^+(z)$ are obtained from the Cauchy formula.

Let $\sigma(t)$ be the inverse function of $\omega^-(t)$ $(\omega^-[\sigma(t)] \equiv t)$. Replacing t in (17.4') by $\sigma(t)$, we have the identity

$$\omega^+ \{\alpha[\sigma(t)]\} \equiv t. \qquad (17.23)$$

The sectionally analytic functin $\Phi_1(z)$ is introduced by the equalities

$$\Phi^+(z) = \Phi_1^+[\omega^+(z)], \quad \Phi^-(z) = \Phi_1^-[\omega^-(z)]. \qquad (17.24)$$

In the new functions the boundary condition (17.2) of the shift problem is written as

$$\Phi_1^+ \{\omega^+[\alpha(t)]\} = G(t)\, \Phi_1^-[\omega^-(t)] + g(t).$$

Substituting $t = \sigma(t_1)$, we get

$$\Phi_1^+\{\omega^+[\alpha(\sigma(t_1))]\} = G[\sigma(t_1)]\, \Phi_1^-(t_1) + g[\sigma(t_1)].$$

Having regard to the identity (17.23) and putting $G_1 = G$, $g_1 = g$, we obtain Riemann boundary value problem

$$\Phi_1^+(t) = G_1(t)\, \Phi_1^-(t) + g_1(t). \qquad (17.25)$$

The shift boundary value problem is thus reduced to the ordinary Riemann boundary problem. For the total solution, however, it still remains to prove the "equivalence" of the two problems, i.e. that the relations (17.24) set a corresponding definite solution of the

Riemann problem (17.25) to each solution of the shift problem (17.2) and vice versa.

Suppose that the analytic functions $\omega^{\pm}(z)$ in the domains $D^{\pm}$ map them on the domains $D_1^{\mp}$ bounded by their respective contours $L_1^{\mp}$. We establish the following:

1. The domains $D_1^{\mp}$ have no interior points in common and they are separated by the common boundary L_1, forming a complement of each other up to the total plane.

2. The domains $D^{\pm}$, $D_1^{\mp}$ stand in one to one correspondence.

Proof. The contours $L_1^{\mp}$ are respectively defined by the equations

$$\zeta = \omega^+(t), \quad \zeta = \omega^-(t) \quad (t \in L).$$

But, by virtue of the boundary condition,

$$\omega^-(t) = \omega^+[\alpha(t)] = \omega^+(t_1).$$

When t traverses the contour L once, $t_1 = \alpha(t)$ also goes round once in the same direction. The contour L^- therefore coincides with $L_1^{\mp}$. We shall denote it simply as L_1.

To establish the one-to-one correspondence of the domains, it is sufficient to establish this property for the contours; by virtue of the properties of a conformal mapping† the same also holds for the domains.

Firstly we establish that the contour values of the solution of the boundary problem (17.4′) have a derivative which satisfies Hölder's condition. We consider the equality obtained by formal differentiation of the boundary condition‡ (17.4′)

$$\alpha'(t)\gamma^+[\alpha(t)] = \gamma^-(t) \quad \left(\gamma(z) = \frac{d\omega}{dz}\right). \tag{17.26}$$

By virtue of (17.21) the boundary problem (17.26) has to be solved on condition that $\gamma^-(\infty) = 1$. The latter problem in essence coincides with the problem (17.20) solved in § 17.4. Its solution therefore exists and $\gamma^{\pm}(t)$ therefore satisfies Hölder's condition.

We go over to the solution of the original problem (17.4′) by integration. It only remains to see that integration introduces no logarithmic term. It is therefore necessary to prove that the residue of an infinitely remote point is zero.

† See, for example, V. I. Smirnov, 3, pt. 2, p. 90.
‡ There is no need to justify the legitimacy of differentiation since the latter is only inductive reasoning. All the reasoning may be conducted by regarding (17.26) as the original.

In fact

$$\int_L \gamma^-(t)\, dt = \int_L \alpha'(t)\, \gamma^+[\alpha(t)]\, dt = \int_L \gamma^+(t_1)\, dt_1 = 0.$$

Integrating $\gamma(z)$ and appropriately selecting an arbitrary constant, we arrive at a solution of the problem (17.4′) which, as shown, therefore has a derivative on the contour which satisfies Hölder's condition.

We may now pass on to the main part. Suppose that the converse is true, i.e. that for $t_1 \neq t_2$

$$\omega^-(t_1) = \omega^-(t_2) = \omega_0.$$

Consider the differences $\omega_1^\pm(z) = \omega^\pm(z) - \omega_0$. The functions $\omega_1^\pm(z)$ satisfy the boundary condition (17.4′) and the following equalities hold:

$$\omega_1^-(t_1) = \omega_1^-(t_2) = \omega_1^+[\alpha(t_1)] = \omega_1^+[\alpha(t_2)] = 0.$$

We now turn to functions which do not vanish on the contour and introduce:

$$\omega_2^-(z) = \frac{\omega_1^-(z)}{(z - t_1)(z - t_2)}, \quad \omega_2^+(z) = \frac{\omega_1^+(z)}{[z - \alpha(t_1)]\,[z - \alpha(t_2)]}. \quad (17.27)$$

Expressing the functions $\omega_1^\pm(t)$ in the boundary condition (17.4′) in terms of $\omega_2^+(t)$, we arrive at the following boundary condition

$$\omega_2^+[\alpha(t)] = \frac{(t - t_1)(t - t_2)}{[\alpha(t) - \alpha(t_1)]\,[\alpha(t) - \alpha(t_2)]}\, \omega_2^-(t).$$

Since the index of the coefficient is zero† and according to (17.27) $\omega^-(\infty) = 0$, then according to § 17.4 $\omega_2^\pm(z) \equiv 0$. Hence $\omega^\pm(z) = \omega_0 = $ const., which is absurd since $\omega^-(z)$ must have a pole at infinity. The one to one correspondence of the mapping is proved as required.

This reducibility of the shift problem to the Riemann problem yields the following theorem.

THEOREM. *The homogeneous shift problem with $\varkappa \geqq 0$ has $(\varkappa + 1)$ linearly independent solutions. The inhomogeneous problem is unconditionally soluble and its solution depends linearly on $(\varkappa + 1)$ arbitrary constants. If $\varkappa \leqq -1$ the homogeneous problem is insoluble, whilst for the solubility of the inhomogeneous problem fulfilment of $(-\varkappa - 1)$ conditions is required.*

† See the discussion at the end of § 17.4.

Note the fundamental importance of the proven fact that the boundary values of the solution of the problem (17.4′) have derivatives which satisfy Hölder's condition. It follows that if the contour L in the initial shift problem is a Lyapounov curve, then on reduction to the ordinary Riemann problem the contour L_1 belongs to the same class. From these considerations all the results related to the broadening of the class of coefficients of the Riemann problem (§ 16.3) can be transferred without further validation to the shift problem.

This method of reducing the shift problem to the Riemann boundary value problem was indicated in a short note in the paper by Mandzhavidze and Khvedelidze (1). The condition of existence of a second derivative in Hölder's class was imposed on $\alpha(t)$. The justification of the method was taken from an unpublished paper by I. B. Simonenko.

18. Other generalized problems

18.1. Formulation of the problems and notation

The problem considered in the preceding section can be generalized in a simple way. Instead of requiring that the function $\alpha(t)$ maps the contour L onto itself it may be assumed that it maps the contour L onto another contour L_1. The theory presented in the preceding section requires little change to be made to this case. There is however no need to do so, since it will be proved below that the new problem can be reduced to the one already investigated by means of a conformal mapping.

Besides the limiting values of the analytic functions we can introduce into the boundary condition the corresponding complex conjugate values. We at once observe that problems whose boundary conditions contain functions complex conjugate to analytic functions arise frequently in problems of a practical nature. Further, there appeared an essential condition in solving the problem of § 17, that the function $\alpha(t)$ maps the contour L onto itself preserving the direction (this fact constitutes the basis of the relation $\operatorname{Ind}\Phi^+[\alpha(t)] = \operatorname{Ind}\Phi(t)$). If we assume that in mapping the direction is changed, essentially new problems arise.

We now formulate four boundary value problems of the above indicated type, assuming that L is a simple smooth closed contour, the tangent of which makes with a constant direction an angle satisfying the Hölder condition with respect to the arc s of the contour (such curves will be called the *Lyapounov curves*†); we moreover assume that $G(t)$ and $g(t)$ are functions which satisfy the same conditions as in § 17.

The function $\alpha(t)$ is supposed to have a derivative which satisfies the Hölder condition and does not vanish; it maps the contour L in a one-to-one way onto another contour L_1 having the same properties as L. The domains bounded by the contour L_1 will be denoted by D_1^+ and D_1^-.

PROBLEM I. It is required to find a function $\Phi_1^+(z)$ analytic in the domain D^+ and a function $\Phi^-(z)$ analytic in the domain D^-, which satisfy on L the boundary condition

$$\Phi_1^+[\alpha(t)] = G(t)\,\Phi^-(t) + g(t),$$

$\alpha(t)$ being such that the direction of traversing the contour is preserved.

PROBLEM II. It is required to find a function $\Phi_1^+(z)$ analytic in D_1^+, and a function $\Phi^+(z)$ analytic in D^+, which satisfy on L the boundary condition

$$\Phi_1^+[\alpha(t)] = G(t)\,\Phi^+(t) + g(t),$$

$\alpha(t)$ being such that the direction of traversing the contour is changed.

PROBLEM III. It is required to find a function $\Phi_1^+(t)$ analytic in D_1^+ and a function $\Phi^-(z)$ analytic in D^-, which satisfy on L the boundary condition

$$\Phi_1^+[\alpha(t)] = G(t)\,\overline{\Phi^-(t)} + g(t),$$

$\alpha(t)$ being such that the direction of traversing the contour is changed.

PROBLEM IV. It is required to find a function $\Phi_1^+(z)$ analytic in D_1^+ and a function $\Phi^+(z)$ analytic in D^+, which satisfy on L the boundary condition

$$\Phi_1^+[\alpha(t)] = G(t)\,\overline{\Phi^+(t)} + g(t),$$

$\alpha(t)$ being such that the direction of traversing the contour is preserved.

---

† This name is analogous to the notion of "Lyapounov surface" employed in the theory of potential, the normal to the surface satisfying an identical condition. The contour will certainly be a Lyapounov curve if it has a continuous curvature.

For subsequent considerations we shall need some new notation.

(a) The symbol $\overline{f(z)}$—a bar over the whole function—denotes as usual the function complex conjugate to that under consideration.

(b) The symbol $f(\bar{z})$ denotes the function obtained from $f(z)$ by replacing in it z by $\bar{z}$, i.e. y by $-y$.

(c) The symbol $\bar{f}(z)$—the bar only over f—denotes the function defined by the condition

$$\bar{f}(z) = \overline{f(\bar{z})}.$$

Replacing in the latter relation z by $\bar{z}$ we obtain

$$\bar{f}(\bar{z}) = \overline{f(z)}.$$

For instance, if $f(z)$ is given by the series $f(z) = \sum c_n z^n$, then

$$\overline{f(z)} = \sum \bar{c}_n \bar{z}^n, \quad f(\bar{z}) = \sum c_n \bar{z}^n, \quad \bar{f}(z) = \sum \bar{c}_n z^n.$$

If $f(z) = u(x, y) + iv(x, y)$, then

$$\overline{f(z)} = u(x, y) - iv(x, y), \quad j(\bar{z}) = u(x, -y) + iv(x, -y),$$
$$\bar{f}(z) = u(x, -y) - iv(x, -y).$$

For a function represented by the Cauchy-type integral

$$f(z) = \frac{1}{2\pi i} \int\limits_L \frac{\varphi(\tau)\,d\tau}{\tau - z},$$

we have

$$\overline{f(z)} = -\frac{1}{2\pi i} \int\limits_L \frac{\overline{\varphi(\tau)\,d\tau}}{\bar{\tau} - \bar{z}}, \quad f(\bar{z}) = \frac{1}{2\pi i} \int\limits_L \frac{\varphi(\tau)\,d\tau}{\tau - \bar{z}},$$

$$\bar{f}(z) = -\frac{1}{2\pi i} \int\limits_L \frac{\overline{\varphi(\tau)\,d\tau}}{\bar{\tau} - z}.$$

18.2. Reduction to the simplest case

We shall show that by means of a conformal mapping the Problems I–IV can be reduced to analogous problems in which the contour L is a circle and $\alpha(t)$ maps L onto itself.

Let the functions $w = w^+(z)$, $w = w^-(z)$ map conformally the domains D_1^+, D^- onto the interior of a unit circle C of the plane w, respectively. By $z^+(w)$, $z^-(w)$ we denote the inverse functions. It is known from the theory of conformal mappings that under the conditions imposed upon L, L_1 not only the functions $w^+(z)$, $w^-(z)$, $z^+(w)$, $z^-(w)$ but also their first derivatives are continuously continuable onto L_1, L, C, respectively, and satisfy there the Hölder condition.

Introduce the new functions

$$\Psi^+(w) = \Phi_1^+[z^+(w)], \quad \Psi^-(w) = \Phi^-[z^-(w)].$$

Then the boundary condition of Problem I takes the form

$$\Psi^+[\alpha_1(\zeta)] = G_1(\zeta)\,\Psi^-(\zeta) + g_1(\zeta) \quad \text{(on } C\text{)}, \tag{I'}$$

where

$$G_1(\zeta) = G[z^-(\zeta)]; \quad g_1(\zeta) = g[z^-(\zeta)]; \quad \alpha_1(\zeta) = w^+\{\alpha[z^-(\zeta)]\}.$$

The functions $G_1(\zeta)$, $g_1(\zeta)$ and $\alpha_1(\zeta)$ defined on C have the same properties as the functions $G(t)$, $g(t)$ and $\alpha(t)$ on L. Thus, Problem I has been reduced to Problem I', where the contour is a unit circle and the function $\alpha(t)$ maps L onto itself. Problem II is reduced in an analogous way to the corresponding particular case.

In case of Problem II and Problem IV we map the domains D_1^+, D^+ by means of the functions $w = w_1^+(z)$, $w = w_2^+(z)$ onto the interior of a unit circle. Then the function $\alpha_1(\zeta) = w_1^+\{\alpha[z_2^+(\zeta)]\}$ maps this unit circle onto itself in a one to one way, the direction of traversing being preserved or changed, depending on the corresponding property of the function $\alpha(t)$.

Thus, we may assume that the boundary conditions of Problems I–IV are given on a unit circle and $\alpha(t)$ establishes a one to one mapping of this circle onto itself. We now prove that under these conditions Problems II, III, IV are reduced to Problem I.

Consider Problem II. We introduce the new functions

$$\Psi^+(w) = \Phi_1^+(z), \quad \Psi^-(w) = \Phi^+\left(\frac{1}{z}\right).$$

It is evident that the function $\Psi^-(w)$ is analytic outside the unit circle. We first arrive at the boundary condition

$$\Psi^+[\alpha(t)] = G(t)\,\Psi^-\overline{(t)} + g(t),$$

and next, introducing the new variable ζ which varies on the circle, we make the substitution $\bar{t} = \zeta$ and obtain the boundary condition

$$\Psi^+[\alpha_1(\zeta)] = G_1(\zeta)\,\Psi^-(\zeta) + g_1(\zeta),$$

where

$$\alpha_1(\zeta) = \alpha(\overline{\zeta}); \quad G_1(\zeta) = G(\overline{\zeta}); \quad g_1(\zeta) = g(\overline{\zeta}).$$

Thus, we arrive at the boundary value Problem I in the particular case when L is a circle.

Problem III is reduced to Problem II by means of the substitution

$$\Phi_1^+(z) = \overline{\Phi}^-\left(\frac{1}{\bar{z}}\right) = \overline{\Phi^-\left(\frac{1}{z}\right)}.$$

Obviously, $\Phi_1^+(z)$ is analytic inside the circle. The boundary condition of Problem III is then written in the form

$$\Phi^+[\alpha(t)] = G(t)\,\Phi_1(\ ) + g(t).$$

This is already the boundary condition of Problem II.

Problem IV is reduced to Problem I by means of the substitution

$$\Phi^-(z) = \overline{\Phi^+}\left(\frac{1}{\bar{z}}\right) = \overline{\Phi^+\left(\frac{1}{z}\right)};$$

$\Phi^-(z)$ is a function analytic outside C. The boundary condition of Problem IV is reduced to the form

$$\Phi_1^+[\alpha(t)] = G(t)\,\Phi^-(t) + g(t) \quad \text{(on } C\text{)},$$

which again is the boundary condition of Problem I.

On the basis of the above results, it is easy to transfer all facts concerning the solubility of Problem I and dependence of the number of its solution on the index of the coefficient $G(t)$, to the case of Problems II–IV.

§ 19. Historical notes

The boundary value problem (14.1) called here the Riemann problem† was first encountered in a paper of Riemann on differential equations with algebraic coefficients [22], p. 177. The homogeneous problem was formulated by him for the case of n pairs of unknown functions, in connection with the problem of determination of a differential equation the solutions of which undergo a prescribed linear substitution on encircling a singular point (equation with a given monodromy group).

This topic is outside the scope of this book and we shall not deal with it; readers interested in this problem are referred to the paper [4].

Riemann made no attempt to solve the problem formulated by himself. The first solution of the homogeneous boundary value problem (14.1) was given by Hilbert. Using the conditions ensuring that an arbitrary complex function constitutes the boundary value of an analytic function (see Problem 24, Chapter I). Hilbert derived a Fredholm integral equation which should be satisfied by the solution of the problem. Investigating this equation he proved the alternative: one of the two problems with the coefficient $G(t)$ or $\overline{G(t)}$ is soluble. Later authors considering the general case of the boundary value problem (Picard [19], Privalov (1)) proceeded along the same lines of reducing the problem to an integral equation, making use of Cauchy type integrals. Instead of the Hilbert alternative an alternative for the coefficients $G(t)$ and $\frac{1}{G(t)}$ was then obtained. This method has up to now been applied in solving the Riemann problem with many unknown functions.

There are two different papers in which solutions in closed form were examined. Plemelj (1) incidentally proved that in the case of single-valued $\ln G(t)$ the solution of the homogeneous problem can be obtained explicitly in terms of Cauchy type integrals. Carleman (1) solving a singular integral equation with Cauchy kernel (see Chapter III) solves at the same time the non-homogeneous Riemann problem

† Various authors employ for this problem also the following names: "Hilbert problem", "problem of linear relationship", "Hilbert–Privalov problem", "Riemann–Privalov problem". Up to 1944 only the above adopted name was used.

with constant coefficient $G(t)$ for the case when the contour is a section of the real axis [0, 1]. The methods of solution of Plemelj and Carleman are essentially the same as those given in this chapter, but they may only be applied to very special cases in which the peculiar properties of the problem are not revealed. First of all there is no notion of the index which, as is known to the reader who studied this chapter, is a fundamental characteristic of both the Riemann problem itself and its generalizations.

We should also mention the paper of Wiener and Hopf (1), devoted to the solution of singular integral equations of convolution type. This problem has not so far been mentioned in connection with the Riemann problem; they are, however, closely related. The solution of these integral equations is reduced to a problem of the same type as the Riemann boundary value problem, the difference consisting in the fact that the relation between analytic functions is given not on a contour but at points of a domain (strip bounded by two straight lines parallel to the x-axis). The method of solution is distinct from that used in the case of the Riemann boundary value problem only in secondary details.

A full solution of the Riemann problem for a simply-connected domain in approximately the same form as this exposition†, was presented by the author in 1936, (1). In 1941 B. V. Khvedelidze (1) generalized this solution to the case of a multiply-connected domain. Exceptional cases of the Riemann problem (§ 15) were first examined in the papers of the author (2), [3] and then in a more general form by L. A. Chikin (1).

The Riemann map problem (§ 17) is first encountered in a paper of Haseman (1). By a method analogous to that which Hilbert applied to the solution of the Riemann problem, he reduces it to a Fredholm integral equation and deduces the same alternative as Hilbert for the Riemann problem.

A full solution of the map problem and its other modifications (§ 18) was obtained by D. A. Kveselava whose first paper (1) was published in 1946; a review of the subject is contained in the paper (2). Our treatment (§ 17) follows in essence the exposition of this author, although the integral equation constituting the basis of the solution of the problem, is more alike to the equation of Haseman than to that of D. A. Kveselava. Reduction of other generalized problems to the fundamental generalized problem is given in the paper of L. I. Chibrikova and V. S. Rogozhin (1).

Other information on the Riemann problem will be presented in Chapter V in connection with its various generalizations.

Problems on Chapter II

1*. Prove that the index of the function

$$\xi(s) + i\eta(s) \quad (0 \leq s \leq l) \quad (\xi(s + l) = \xi(s)), \quad \eta(s + l) = \eta(s)$$

is equal to half the difference between the number of the cases when the function $\eta(s)/\xi(s)$ is infinite in passing from positive to negative values, and the number of the cases when it is infinite in passing from negative to positive values (Hermite [8], p. 233).

---

† The notion of a canonical function was introduced by N. I. Muskhelishvili.

Remark. Cauchy, who introduced the notion of the index, defined it as the difference between the number of infinities of the function in passing from positive to negative values and in passing from negative to positive values. Consequently, the index in the Cauchy sense is equal to twice the index in the sense of our definition. The notion of the Cauchy index is also applicable to non-periodic functions. Then, expression

$$\arctan \frac{\eta(l)}{\xi(l)} - \arctan \frac{\eta(0)}{\xi(0)} \quad \left(-\frac{\pi}{2} < \arctan \varkappa < \frac{\pi}{2} \right)$$

is omitted.

2*. Let $\xi(t), \eta(t)$ be polynomials with respect to t $(a \le t \le b)$. We construct the sequence

$$\eta, \ \xi, \ -r_1, \ r_2, \ \ldots, \ -r_{n-1}, \ -r_n, \tag{1}$$

where r_1 is the remainder of the quotient η/ξ, r_2 the remainder of the quotient ξ/r_1, etc. Prove that the index of the function $\xi(s) + i\eta(s)$ is equal to half the difference between the numbers of changes of the signs in the sequence (1) (Goursat [7], p. 169).

3*. Let $\Phi(x, y), \varphi(x, y), \psi(x, y)$ be three functions of two independent variables x, y, having continuous first partial derivatives. The following is assumed: (1) the functions Φ, φ, ψ do not vanish simultaneously; (2) their Jacobian also does not vanish; (3) the curves $\Phi = 0, \varphi = 0, \psi = 0$ are closed. The points of the plane for which $\varphi, \psi < 0$ will be called interior, and those for which $\varphi, \psi > 0$, exterior. Passing from the exterior domain into the interior will be called the entrance, and the reverse the exit.

By the characteristic $\varkappa(\Phi, \varphi, \psi)$ of the functions Φ, φ, ψ Kronecker meant half the difference between the number of points of entrance and points of exit of the curve $\Phi(x, y) = 0$ through the curve $\varphi(x, y) = 0$, or what is equivalent through the curve $\psi(x, y) = 0$. Prove that:

(1) the characteristic of $\varkappa(\Phi, \varphi, \psi)$ coincides with the index of the function $G(x + iy) = \varphi(x, y) + i\psi(x, y)$ with respect to the curve $\Phi(x, y) = 0$;

(2) the characteristic is unaltered by a cyclic permutation of the functions Φ, φ, ψ, and its sign is changed under a permutation of two of them (N. G. Chetayev, *Kronecker characteristics*, Uchenye zapiski Kazanskovo Universiteta, v. 98, ch. 9, 1938).

Remark. Kronecker defined the notion of a characteristic for the general case of $n + 1$ functions of n variables.

In all that follows the contour L is a simple closed curve dividing the plane into two domains: the interior D^+ and the exterior D^-; c_i are arbitrary constants.

4. Solve the Riemann boundary value problem

$$\Phi^+(t) = \frac{(t-i)(t-2i)}{(t+i)(t+2i)} \Phi^-(t) + \frac{2t}{(t+i)^2(t+2i)(t-i)},$$

assuming that the points $i, 2i$ belong to the domain D^+ and the points $-i, -2i$ to D^-.

Answer.

$$\Phi^+(z) = \frac{1}{(z+i)^2(z+2i)} + \frac{c_0 + c_1 z + c_2 z^2}{(z+i)(z+2i)},$$

$$\Phi^-(z) = -\frac{1}{(z-i)^2(z-2i)} + \frac{c_0 + c_1 z + c_2 z^2}{(z-i)(z-2i)}.$$

5. For what value of the parameter a is the following Riemann boundary value problem soluble:

$$\Phi^+(t) = \frac{(t-i)(t-2i)}{(t+i)(t+2i)} \Phi^-(t) + \frac{2t(t-2i)(at+1)}{(t+i)(t+2i)}, \quad \Phi^-(\infty) = 0?$$

The contour L is such that the point $-i$ lies in D^+ and the points $i, 2i, -2i$ in D^-. Solve the problem for the determined value of a.

Answer.
$$a = -i, \quad \Phi^+(z) = \frac{-2iz(z-2i)}{z+2i}, \quad \Phi^-(z) = 0.$$

6. Solve the Riemann boundary value problem

$$\Phi^+(t) = \frac{t}{\ln \dfrac{t}{t-1}} \Phi^-(t) + \sin t + e^{\frac{1}{t}}, \quad \Phi^-(\infty) = 0$$

for a contour such that the points $0, 1$ lie in the domain D^+.

Answer.
$$\Phi^-(z) = \frac{1}{z} e^{\frac{1}{z}} \ln\left(1 - \frac{1}{z}\right) - \left(\frac{c_0}{z} + c_1\right) \ln\left(1 - \frac{1}{z}\right).$$

7. Solve the Riemann boundary value problem

$$\Phi^+(t) = (t-1)\,\Phi^-(t) + t^2 - 1$$

making the following assumptions:
  (a) the point 1 belongs to D^+;
  (b) the point 1 belongs to D^-;
  (c) the point 1 belongs to L.

Answer.

  (a) $\Phi^+(z) = z^2 - 1 + c_0 + c_1 z, \quad \Phi^-(z) = (z-1)^{-1}(c_0 + c_1 z);$

  (b), (c) $\Phi^+(z) = z^2 - 1 + c_0(z-1), \quad \Phi^-(z) = c_0.$

8. Solve the Riemann boundary value problem

$$\Phi^+(t) = (t-1)^{-1}\,\Phi^-(t) + t + (t-1)^{-1}$$

making the same assumptions as in the preceding problem.

Answer.

(a), (b) $\Phi^+(z) = z, \quad \Phi^-(z) = 1;$

  (c) $\Phi^+(z) = (z-1)^{-1}(z^2 - z + 1 + c_0), \quad \Phi^-(z) = c_0.$

9. Prove that the Riemann problem

$$\Phi^+(t) = \frac{1}{(t-1)(t-2)} \Phi^-(t) + t + 1 + \frac{1}{t} - \frac{a}{t-3}$$

in the case when the points $0, 3$ belong to D^+ and the points $1, 2$ lie on the contour L, is soluble only for $a = 1$. Solve the problem for $a = 1$.

Answer.

$$\Phi^+(z) = z + 1, \quad \Phi^-(z) = 3\frac{(z-1)(z-2)}{z(z-3)}.$$

10. Solve the Riemann boundary value problem

$$\Phi^+(t) = \frac{2t - ih}{t + i}\,\Phi^-(t) + \frac{t + ib}{t^2 + 1} \quad (h \neq 0, \ -\infty < t < \infty):$$

(a) for the class of functions satisfying the condition $|\Phi^\pm(t)| < A|t|^{-\alpha}$ $(\alpha > 0)$ for large $|t|$;

(b) for the class of functions satisfying the condition

$$|\Phi_1(t) - \Phi_1^\pm(\infty)| < A|t|^{-\alpha}$$

for large $|t|$.

Answer. (a) For $h > 0$

$$\Phi^+(z) = \frac{1 - b}{2(z + i)} + \frac{c_1}{z + i}, \quad \Phi^-(z) = \frac{z + i}{2z - ih}\left[-\frac{1 + b}{2(z - i)} + \frac{c_1}{z + i}\right];$$

for $h < 0$ $\quad \Phi^+(z) = \frac{h + 2b}{(h - 2)(z + i)}, \quad \Phi^-(z) = \frac{1 + b}{(h - 2)(z - i)};$

(b) for $h > 0$ $\quad \Phi_1^+(z) = \Phi^+(z) + \frac{cz}{z + i}, \quad \Phi_1^-(z) = \Phi^-(z) + \frac{cz}{2z - ih},$

for $h < 0$ $\quad \Phi_1^+(z) = \Phi^+(z) + c\frac{2z - ih}{z + i}, \quad \Phi_1^-(z) = \Phi^-(z) + c.$

11. Solve the Riemann boundary value problem

$$\Phi^+(t) = \frac{t - ih}{t - i}\,\Phi^-(t) + \frac{t + ib}{t^2 + 1} \quad (h \neq 0, \ -\infty < t < \infty)$$

for the classes of functions of the preceding problem.

Answer. (a) For $h > 0$

$$\Phi^+(z) = \frac{1 - b}{2(z + i)}, \quad \Phi^-(z) = -\frac{1 + b}{2(z - ih)},$$

for $h < 0$

$$\Phi^+(z) = \frac{1}{z + i}, \quad \Phi^-(z) = 0,$$

the condition of solvability $b = -1$ being fulfilled.

(b) for $h > 0$ $\quad \Phi_1^+(z) = \Phi^+(z) + c, \quad \Phi_1^-(z) = \Phi^-(z) + c\frac{z - i}{z - ih};$

for $h < 0$ $\quad \Phi_1^+(z) = \frac{h + b}{(h - 1)(z + i)} + \frac{(b + 1)(z - ih)}{2i(h - 1)(z + i)}, \quad \Phi_1^-(z) = \frac{b + 1}{2i(h - 1)}.$

12*. Prove that if we admit as solutions functions which take infinite values of an integrable order on the contour, the class of solutions is not extended. The class of solutions in which the above singularities on the contour are admissible coincides with the solution found in § 14 which is continuous including the contour (Gakhov (3)).

13. Construct the solution of the non-homogeneous Riemann boundary value problem (14.2) which satisfies the following conditions:

$$\frac{d^k \Phi(z_\nu)}{dz_k'} = a_k^\nu \left(c = 1, 2, \ldots, m_\nu; \quad \nu = 1, 2, \ldots, n; \quad \sum_{\nu=1}^{n} m_\nu = m \right),$$

assuming that $\varkappa \geq m$; z_ν are given points of the domains D^+, D^-, and a_k^ν are known numbers.

Hint. Make use of the method of deducing the canonical function of the non-homogeneous problem (§ 15.2).

14*. Solve the Riemann map problem in the exceptional case when the coefficient of the problem may vanish or have an infinity of order which is an integer,

$$\Phi^+[\alpha(t)] = \frac{\prod\limits_{k=1}^{\mu} (t - a_k)^{m_k}}{\prod\limits_{j=1}^{\nu} (t - b_j)^{p_j}} \, G_1(t) \, \Phi^-(t).$$

Prove that the number of solutions is equal to Ind $G_1(t) - \sum\limits_{j=1}^{\nu} p_j$ (L. I. Chibrikova (1)).

Hint. Represent the boundary condition in the form

$$\frac{\Phi^+[\alpha(t)]}{\prod\limits_{k=1}^{\mu} [\alpha(t) - a_k]^{m_k}} = \prod\limits_{k=1}^{\mu} \left[\frac{t - a_k}{\alpha(t) - a_k} \right]^{m_k} G_1(t) \, \frac{\Phi^-(t)}{\prod\limits_{j=1}^{\nu} (t - b_j)^{p_j}}.$$

Consider the analogous case of the non-homogeneous problem.

15*. Prove that the necessary condition of solvability of the boundary value problem

$$\Phi^+[\alpha(t)] = G(t) \, \Phi^+(t) + g(t) \quad \text{on} \quad L,$$

where $\Phi^+(z)$ is a function analytic in D^+, $\alpha(t)$ maps the contour L onto itself changing the direction and satisfies the identity $\alpha[\alpha(t)] \equiv t$, is constituted by on of the following conditions:

(1) the expression

$$\frac{g[\alpha(t)] + g(t) \, G[\alpha(t)]}{1 - G(t) \, G[\alpha(t)]}$$

is a continuous function which is the boundary value of a function analytic in D^+;

(2) $G(t) \, G[\alpha(t)] = 1$; $g(t) + g[\alpha(t)] \, G(t) = 0$ (D. A. Kveselava (2)).

CHAPTER III

SINGULAR INTEGRAL EQUATIONS WITH CAUCHY KERNEL

In this chapter we shall present one of the most important theoretical applications of the Riemann boundary value problem—the investigation of singular integral equations with Cauchy kernel. Since the Riemann boundary value problem has so far been considered for closed contours only, the integral in the integral equations under investigation will be taken along contours of the same kind. Later, in connection with generalizations in the formulation of the Riemann problem we shall accordingly also investigate other types of integral equations.

§ 20. Basic concepts and notation

20.1. Singular integral equation

If the kernel of the linear integral equation

$$\varphi(t) + \int_L K(t, \tau)\, \varphi(\tau)\, d\tau = f(t)$$

has the form

$$K(t, \tau) = \frac{M(t, \tau)}{(\tau - t)^\alpha} \quad (0 \leqq \alpha < 1),$$

where $M(t, \tau)$ is a continuous function, then it is known that by means of iterations it can be reduced to an integral equation with a continuous kernel†. Such an equation possesses all the properties of a Fredholm equation and is called a *quasi-Fredholm* or simply *a Fredholm equation*.

† See for instance I. I. Privalov, *Integral Equations*, ONTI, Moscow, 1935, p. 81 (in Russian).

If $\alpha = 1$ the integral becomes singular and the above indicated method of reduction to a Fredholm equation is not valid. Equations of this kind require a special theory.

We shall presently consider equations with Cauchy kernel of the form

$$K\varphi \equiv a(t)\varphi(t) + \frac{1}{\pi i} \int_L \frac{M(t, \tau)}{\tau - t} \varphi(\tau)\, d\tau = f(t). \qquad (20.1)$$

The integral in the sense of its principal value is taken over the contour L which in the general case is composed of $m+1$ closed smooth curves, $L = L_0 + L_1 + \cdots + L_m$, situated as in Fig. 13 (§ 16.1). The functions $a(t)$, $f(t)$, $M(t, \tau)$ prescribed on L are assumed to satisfy the Hölder condition, the last function with respect to both variables.

The letter K denotes the operation carried out on the function $\varphi(t)$ in the left-hand side of the equation.

Making the transformation

$$\frac{M(t, \tau)}{\tau - t} = \frac{M(t, \tau) - M(t, t)}{\tau - t} + \frac{M(t, t)}{\tau - t}$$

and introducing the notation

$$M(t, t) = b(t), \quad \frac{1}{\pi i} \frac{M(t, \tau) - M(t, t)}{\tau - \tau} = k(t, \tau), \qquad (20.2)$$

we can write equation (20.1) in the form

$$K\varphi \equiv a(t)\varphi(t) + \frac{b(t)}{\pi i} \int_L \frac{\varphi(\tau)}{\tau - t}\, d\tau + \int_L k(t, \tau)\, \varphi(\tau)\, d\tau = f(t). \qquad (20.3)$$

It follows from formulae (20.2) that the function $b(t)$ satisfies the Hölder condition on the entire contour L and $k(t, \tau)$ everywhere except for the points $\tau = t$ where the estimate

$$|k(t, \tau)| < \frac{A}{|\tau - t|^{1-\lambda}} \quad (0 < \lambda \leqq 1).$$

is valid. Equation (20.3) will be called *the complete singular integral equation*. If $f(t)$ does not vanish we have a non-homogeneous equation, while in the opposite case the equation is homogeneous.

The expression

$$K^\circ \varphi \equiv a(t)\, \varphi(t) + \frac{b(t)}{\pi i} \int_L \frac{\varphi(\tau)}{\tau - t}\, d\tau \qquad (20.4)$$

is referred to as the *dominant part* of the equation, and the term $\int_L K(t,\tau)\varphi(\tau)\,d\tau$ is called its *regular part*. The equation

$$K^\circ\varphi \equiv a(t)\,\varphi(t) + \frac{b(t)}{\pi i}\int_L \frac{\varphi(t)}{\tau - t}\,dt = f(t) \qquad (20.5)$$

is called the *dominant equation* corresponding to the complete equation (20.3), and the operator K° the *dominant operator*.

Introducing for the regular part of the equation the notation

$$k\varphi \equiv \int_L k(t,\tau)\,\varphi(\tau)\,d\tau,$$

we can write the complete equation in the form

$$K\varphi \equiv K^\circ\varphi + k\varphi = f(t).$$

The equation

$$K'\psi \equiv a(t)\,\psi(t) - \frac{1}{\pi i}\int_L \frac{b(\tau)\,\psi(\tau)}{\tau - t}\,d\tau + \int_L k(\tau,t)\,\psi(\tau)\,d\tau = 0, \qquad (20.3')$$

obtained from the homogeneous equation $K\varphi = 0$ by the transposition of the variables in the kernel, is called the *adjoint equation*. The operator K' is referred to as the *operator adjoint* to K.

In particular, the equation

$$K^{\circ\prime}\psi \equiv a(t)\,\psi(t) - \frac{1}{\pi i}\int_L \frac{b(\tau)}{\tau - t}\,\psi(\tau)\,d\tau = 0 \qquad (20.5')$$

is adjoint to the dominant equation (20.5). Let us observe here that the operator $K^{\circ\prime}$ adjoint to the dominant operator K° is not identical with the operator K'° dominant for the adjoint operator. The latter is given by the formula†

$$K'^\circ\psi \equiv a(t)\,\psi(t) - \frac{b(t)}{\pi i}\int_L \frac{\psi(\tau)}{\tau - t}\,d\tau. \qquad (20.6)$$

† In formula (20.3) the transformation

$$\int_L \frac{b(\tau)\,\psi(\tau)}{\tau - t}\,d\tau = \int_L \frac{b(\tau) - b(t)}{\tau - t}\,\psi(\tau)\,d\tau + b(t)\int_L \frac{\psi(\tau)}{\tau - t}\,d\tau$$

should be performed, and the first integral referred to the regular part,

In all that follows we shall seek solutions of singular equations, which satisfy the Hölder condition.

20.2. Fundamental results of the theory of Fredholm integral equations

For the reader's convenience we present for the sake of reference the fundamental results of the theory of linear Fredholm integral equations of second kind

$$a(t)\,\varphi(t) + \lambda \int_L K(t, \tau)\,\varphi(\tau)\,d\tau = f(t);$$

some of these results have already been employed in § 17.3.

Assuming that $a(t)$ does not vanish we can, without any loss of generality, take $a(t) \equiv 1$. We shall therefore consider hereafter Fredholm equations of the form

$$\varphi(t) + \lambda \int_L K(t, \tau)\,\varphi(\tau)\,d\tau = f(t) \quad \text{(non-homogeneous)} \quad (20.7)$$

and

$$\varphi(t) + \lambda \int_L K(t, \tau)\,\varphi(\tau)\,d\tau = 0 \quad \text{(homogeneous)} \quad (20.8)$$

DEFINITION. If for some value of the parameter $\lambda = \lambda_0$ the homogeneous Fredholm equation has a non-trivial solution, then λ_0 is called an *eigenvalue* and the solutions (linearly independent) $\varphi_1(t), \ldots, \varphi_n(t)$, *eigenfunctions* of the kernel $K(t, \tau)$, or which is the same, of the homogeneous equation (20.8).

1. Fredholm theorems.

THEOREM I. *If $\lambda = \lambda_0$ is not an eigenvalue of the kernel, i.e. if the homogeneous equation* (20.8) *is insoluble†, then the non-homogeneous equation* (20.7) *is soluble for an arbitrary right-hand side $f(t)$.*

The general solution is given by the formula

$$\varphi(t) = f(t) - \int_L R(t, \tau) f(\tau)\,d\tau, \quad (20.9)$$

where the function $R(t, \tau)$—the resolvent of the equation—is in a certain way determined by the kernel $K(t, \tau)$.

---

† Here and in all that follows we call a homogeneous equation having only trivial solutions, insoluble.

THEOREM II. *If $\lambda = \lambda_0$ is an eigenvalue of the homogeneous equation* (20.8), *then it is also an eigenvalue of the adjoint equation*†

$$\psi(t) + \lambda \int_L K(\tau, t)\,\psi(\tau)\,d\tau = 0, \qquad (20.7')$$

both equations having the same number of linearly independent solutions (eigenfunctions belonging to the eigenvalue λ_0).

The general solution of the homogeneous equation has the form

$$\varphi(t) = \sum_{k=1}^{n} c_k\,\varphi_k(t),$$

where $\varphi_1(t), \ldots, \varphi_n(t)$ is a complete system of linearly independent eigenfunctions belonging to the eigenvalue λ_0 and c_k are arbitrary constants.

THEOREM III. *If the homogeneous equation is soluble, then in general the non-homogeneous equation is insoluble. It is soluble if and only if the following conditions are satisfied:*

$$\int_L f(t)\,\psi_k(t)\,dt = 0. \qquad (20.10)$$

Here $\psi_k(t)$ $(k = 1, 2, \ldots, n)$ constitute a complete system of eigenfunctions of the adjoint equation (20.7), *belonging to the given eigenvalue λ_0.*

If the conditions (20.10) are satisfied, the general solution of the non-homogeneous equation is given by the formula

$$\varphi(t) = f(t) - \int_L H(t, \tau)f(\tau)\,d\tau + \sum_{k=1}^{n} c_k\varphi_k(t), \qquad (20.11)$$

where $H(t,\tau)$ is *the generalized resolvent* and $\sum_{k=1}^{n} c_k\varphi_k(t)$ is the general solution of the corresponding homogeneous equation.

2. *The spectrum of the equation.*

DEFINITION. The set of eigenvalues of the parameter λ of the integral equation is called its *spectrum.*

Let us quote two theorems characterizing the spectrum of the Fredholm equation.

† Instead of equation (20.7′), by the adjoint equation is frequently meant the equation $\quad \psi(t) + \bar{\lambda} \int_L \overline{K(\tau, t)}\,\psi(\tau)\,\overline{\tau'(\sigma)}\,d\sigma = 0 \quad (\tau = \tau(\sigma)).$

THEOREM IV. *The set of eigenvalues of a Fredholm integral equation has no limit points within a finite distance. If the set of the eigenvalues is infinite, then its limit point is the point at infinity.*

In other words the spectrum of a Fredholm integral equation is a discrete set.

THEOREM V. *To every eigenvalue there belongs a finite number of eigenfunctions.*

The latter property has already been employed in Theorem III.

§ 21. The dominant equation

21.1. Reduction to the Riemann boundary value problem

Let us begin the exposition of the theory of singular integral equation by examining its simplest type—the dominant equation (20.5)

$$K^\circ \varphi \equiv a(t)\,\varphi(t) + \frac{b(t)}{\pi i} \int_L \frac{\varphi(\tau)}{\tau - t}\,d\tau = f(t). \tag{21.1}$$

In this simplest case the solution of the equation can be reduced to the solution of the Riemann boundary value problem (Chapter II) and the results can be obtained in a closed form.

Introduce a sectionally analytic function represented by an integral of Cauchy type, the density of which is the required solution of the dominant equation

$$\Phi(z) = \frac{1}{2\pi i} \int_L \frac{\varphi(\tau)}{\tau - z}\,d\tau. \tag{21.2}$$

In accordance with the Sokhotski formulae (4.9) and (4.10)

$$\left. \begin{aligned} \varphi(t) &= \Phi^+(t) - \Phi^-(t), \\ \frac{1}{\pi i} \int_L \frac{\varphi(\tau)}{\tau - t}\,dt &= \Phi^+(t) + \Phi^-(t). \end{aligned} \right\} \tag{21.3}$$

Introducing the values of

$$\varphi(t) \quad \text{and} \quad \frac{1}{\pi i} \int_L \frac{\varphi(\tau)}{\tau - t}$$

into equation (21.1) and solving it with respect to $\Phi^+(t)$ we find that the sectionally analytic function $\Phi(z)$ is a solution of the Riemann boundary value problem

$$\Phi^+(t) = G(t)\,\Phi^-(t) + g(t), \tag{21.4}$$

where

$$G(t) = \frac{a(t) - b(t)}{a(t) + b(t)}, \quad g(t) = \frac{f(t)}{a(t) + b(t)}. \tag{21.5}$$

Since the required function $\Phi(z)$ is represented by an integral of Cauchy type it should satisfy the additional condition

$$\Phi^-(\infty) = 0. \tag{21.6}$$

The index of the coefficient

$$\frac{a(t) - b(t)}{a(t) + b(t)}$$

of the Riemann problem (21.4) will be called *the index of the integral equation* (21.1).

Having solved the boundary value problem (21.4) by means of formula (21.3) we determine the solution of equation (21.1).

Thus the integral equation (21.1) is reduced to the Riemann boundary problem (21.4). The solution of the initial equation is obtained from the first formula of (21.3). To establish the equivalence of the equation and boundary value problem, it is necessary to prove that, conversely, $\varphi(t)$, as found in the indicated manner from the solution of the boundary value problem, satisfies the equation (21.1). To this end it must be established that the second formula of (21.3) holds good. We will prove this.

If the solution of problem (21.4) is represented by the Cauchy integral (21.2), both formulae in (21.3) hold good and one arrives unambiguously at the initial equation from the boundary problem. Suppose now that there is another function $\Phi_1(z)$ which satisfies the same conditions. By definition, the following equality holds for the difference $\Phi_2 = \Phi - \Phi_1$

$$\Phi_2^+(t) - \Phi_2^-(t) = 0.$$

In accordance with Liouville's theorem and the theorem of analytic continuability (having regard to (21.6)), $\Phi_2(z) \equiv 0$. Consequently $\Phi_1(z) = \Phi(z)$ as required.

We shall first consider the normal (non-exceptional) case, when the coefficient $G(t)$ of the Riemann problem (21.4) does not take the value zero at infinity, which, for equation (21.1), corresponds to the condition
$$a(t) \pm b(t) \neq 0. \tag{21.7}$$

For simplicity in the following formulae we first divide the whole equation (21.1) by $\sqrt{[a^2(t) - b^2(t)]}$, i.e. we assume that the coefficients of this equation satisfy the condition
$$a^2(t) - b^2(t) = 1. \tag{21.7'}$$

The exceptional cases will be examined at the end of this chapter (§ 25).

It should be observed that it is also easy to construct the singular integral equation corresponding to the given boundary value problem (21.4). Introducing into the boundary condition (21.4) the limiting values of the Cauchy type integral, in accordance with formulae (4.8) we arrive at the dominant singular integral equation
$$\frac{1}{2}[1 + G(t)]\,\varphi(t) + \frac{1 - G(t)}{2\pi i} \int_L \frac{\varphi(\tau)}{\tau - t}\, d\tau = g(t). \tag{21.8}$$

In view of formula (21.2) the solutions of the latter equation yield the solution of the Riemann boundary value problem. Since there exists no method of solving the dominant equation independent of the solution of the Riemann problem, equation (21.8) should be regarded not as a new method of solving the Riemann problem, but simply as a formula which enables us to quickly construct the dominant singular integral equation corresponding to the given Riemann boundary value problem.

21.2. Solution of the dominant equation

Let us write down in accordance with formulae (16.10), (16.8) the solution of the Riemann boundary value problem (21.4), assuming that $\varkappa \geq 0$, and let us calculate by means of the Sokhotski formulae the limiting values of the corresponding functions
$$\Phi^+(t) = X^+(t)\left[\frac{1}{2}\,\frac{g(t)}{X^+(t)} + \Psi(t) - \frac{1}{2}\,P_{\varkappa-1}(t)\right],$$
$$\Phi^-(t) = X^-(t)\left[-\frac{1}{2}\,\frac{g(t)}{X^+(t)} + \Psi(t) - \frac{1}{2}\,P_{\varkappa-1}(t)\right],$$

where $\Psi(t)$ is the singular integral obtained by replacing z by t in formula (14.14). The arbitrary polynomial is taken in the form $-\frac{1}{2}P_{\varkappa-1}(t)$, for convenience of further writing.

Hence, in view of formula (21.3)

$$\varphi(t) = \frac{1}{2}\left[1 + \frac{X^-(t)}{X^+(t)}\right]g(t) + X^+(t)\left[1 - \frac{X^-(t)}{X^+(t)}\right]\left[\Psi(t) - \frac{1}{2}P_{\varkappa-1}(t)\right].$$

In accordance with the boundary condition we replace $[X^-(t)]/[X^+(t)]$ by $1/G(t)$ and the function $\Psi(t)$ by its expression given by formula (14.14). Then

$$\varphi(t) = \frac{1}{2}\left[1 + \frac{1}{G(t)}\right]g(t) +$$
$$+ X^+(t)\left[1 - \frac{1}{G(t)}\right]\left[\frac{1}{2\pi i}\int_L \frac{g(\tau)}{X^+(\tau)}\frac{d\tau}{\tau - t} - \frac{1}{2}P_{\varkappa-1}(t)\right].$$

Finally, substituting for $X^+(t)$ its expression from formula (16.8) and the values of $G(t)$ and $g(t)$ from (21.5) we obtain

$$\varphi(t) = a(t)f(t) - \frac{b(t)Z(t)}{\pi i}\int_L \frac{f(\tau)}{Z(\tau)}\frac{d\tau}{\tau - t} + b(t)Z(t)P_{\varkappa-1}(t), \quad (21.9)$$

where

$$Z(t) = [a(t) + b(t)]X^+(t) = [a(t) - b(t)]X^-(t) = \frac{e^{\Gamma(t)}}{\sqrt{[t^\varkappa\Pi(t)]}}. \quad (21.10)$$

$$\Gamma(t) = \frac{1}{2\pi i}\int_L \frac{\ln\left[\tau^{-\varkappa}\Pi(\tau)\dfrac{a(\tau) - b(\tau)}{a(\tau) + b(\tau)}\right]}{\tau - t}d\tau,$$

$$\Pi(t) = \prod_{k=1}^m (t - z_k)^{\varkappa_k} \quad (21.11)$$

and the coefficients $a(t)$, $b(t)$ satisfy the condition (21.7′). Since the functions $a(t)$, $b(t)$, $f(t)$ satisfy the Hölder condition, on the basis of the properties of the limiting values of a Cauchy type integral the function $\varphi(t)$ also satisfies the Hölder condition.

The last term in formula (21.9) represents the general solution of the homogeneous equation $(f(t) \equiv 0)$ and the first two terms a particular solution of the non-homogeneous equation.

Comparing formula (21.9) with formula (20.11) we observe that the function

$$\frac{b(t)Z(t)}{\pi i Z(\tau)(\tau - t)}$$

plays the part of the resolvent for the dominant equation, and its eigenfunctions are $\varphi_k(t) = b(t) Z(t) t^{k-1}$ $(k = 1, 2, \ldots, \varkappa)$. The particular solution of equation (21.1) can be written in the form Rf where R is the operator defined by the relation†

$$Rf = a(t)f(t) - \frac{b(t) Z(t)}{\pi i} \int_L \frac{f(\tau)}{Z(\tau)} \frac{d\tau}{\tau - t}.$$

Then the general solution of equation (21.1) takes the form

$$\varphi(t) = Rf + \sum_{k=1}^{\varkappa} c_k \varphi_k(t). \tag{21.12}$$

Let us note that in the cases when the Riemann problem (21.4) can be solved directly by analytic continuation (see § 14.6), it is more convenient to seek the solution of the dominant equation not in the form (21.9) but in accordance with formula (21.3).

If $\varkappa < 0$ we know that the Riemann problem (21.4) is in general insoluble. Its conditions of solubility

$$\int_L \frac{g(\tau)}{X^+(\tau)} \tau^{k-2} d\tau = 0 \quad (k = 1, 2, \ldots, -\varkappa) \tag{16.11}$$

are also the conditions of solubility of equation (21.1).

Substituting for $g(\tau)$ and $X^+(\tau)$ their expressions (21.5) and (21.10) we can write the conditions of solubility in the form

$$\int_L \frac{f(\tau)}{Z(\tau)} \tau^{k-1} d\tau = 0 \quad (k = 1, 2, \ldots, -\varkappa). \tag{21.13}$$

If the conditions of solubility are satisfied the solution of the non-homogeneous equation (21.4) is given by formula (21.9) where $P_{\varkappa-1}(t) \equiv 0$.

We now state the results of our investigation.

1. *If $\varkappa > 0$ the homogeneous equation $K°\varphi = 0$ has $\varkappa$ linearly independent solutions*

$$\varphi_k(t) = b(t) Z(t) t^{k-1} \quad (k = 1, 2, \ldots, \varkappa).$$

---

† The operator R is inverse to the operator $K°$ in the sense that the operator $K°$ converts the function φ into f, whereas R converts the function f back into φ.

2. *If $\varkappa \leqq 0$ the homogeneous equation is insoluble* (the zero solution is not taken into account).

3. *If $\varkappa \geqq 0$ the non-homogeneous equation is soluble for an arbitrary right-hand side $f(t)$, and its general solution depends linearly on $\varkappa$ arbitrary constants.*

4. *If $\varkappa < 0$ the non-homogeneous equation is soluble if and only if its right-hand side $f(t)$ satisfies $-\varkappa$ conditions*

$$\int_L \psi_k(t) f(t) \, dt = 0, \tag{21.14}$$

where $\psi_k(t) = [1/Z(t)] \, t^{k-1}$.

Comparing the above properties of the dominant singular integral equation with the properties of the Fredholm integral equation (§ 20.2) we observe significant differences. For the Fredholm equation, when the homogeneous equation is soluble the non-homogeneous equation is in general insoluble, and conversely, when the first is insoluble, the latter is absolutely soluble. For a singular equation, however, when the homogeneous equation is soluble the non-homogeneous is also absolutely soluble; if now the first is insoluble, the latter is also in general insoluble.

As in the case of a Fredholm equation let us introduce in the kernel of the dominant equation a parameter λ and consider the equation

$$K_\lambda^\circ \varphi \equiv a(t) \, \varphi(t) + \frac{\lambda b(t)}{\pi i} \int_L \frac{\varphi(\tau)}{\tau - t} \, d\tau = 0.$$

It has been proved that the latter equation is soluble if

$$\varkappa = \operatorname{Ind} \frac{a(t) - \lambda b(t)}{a(t) + \lambda b(t)} > 0.$$

The index of a continuous function (§ 12.1) varies discontinuously, but only for values of λ for which $a(t) \mp \lambda b(t) = 0$. If we draw on the complex plane $\lambda = \lambda_1 + i\lambda_2$ the curves $\lambda = \pm \dfrac{a(t)}{b(t)}$, they divide the plane into regions of constant index. Thus, the eigenvalues of the dominant singular integral equation fill entire domains; consequently, *the spectrum of such an equation, as distinct from the spectrum of a Fredholm equation, is not discrete but continuous.*

21.3. The solution of the equation adjoint to the dominant equation

The equation

$$K^{\circ\prime}\psi = a(t)\,\psi(t) - \frac{1}{\pi i} \int_L \frac{b(\tau)\,\psi(\tau)}{\tau - t}\,d\tau = h(t), \qquad (21.15)$$

adjoint to the dominant equation $K^\circ \varphi = f$ is not a dominant equation. However, by means of the substitution

$$b(t)\,\psi(t) = \omega(t) \qquad (21.16)$$

it is converted into the dominant equation with respect to the function $\omega(t)$:

$$a(t)\,\omega(t) - \frac{b(t)}{\pi i} \int_L \frac{\omega(\tau)}{\tau - t}\,d\tau = b(t)\,h(t). \qquad (21.17)$$

Having determined from the latter equation $\omega(t)$, in accordance with formula (21.16) we determine the required function $\psi(t)$. The case when $b(t)$ vanishes at certain isolated points does not lead to any complications, since in view of (21.17)

$$\omega(t) = \frac{b(t)}{a(t)}\left[\frac{1}{\pi i} \int_L \frac{\omega(\tau)}{\tau - t}\,d\tau + h(t) \right],$$

i.e. $\omega(t)$ has zeros at the same points as $b(t)$, their order being not less than those of $b(t)$.

Introducing the sectionally analytic function

$$\Omega(z) = \frac{1}{2\pi i} \int_L \frac{\omega(\tau)}{\tau - z}\,d\tau, \qquad (21.18)$$

in the same way as in the preceding section, we arrive at the Riemann boundary value problem

$$\Omega^+(t) = \frac{a(t) + b(t)}{a(t) - b(t)}\,\Omega^-(t) + \frac{b(t)\,h(t)}{a(t) - b(t)}. \qquad (21.19)$$

The coefficient of the latter problem is a quantity inverse to the coefficient of the Riemann problem (21.5) which corresponds to the equation $K^\circ \varphi = f$. Consequently

$$\varkappa' = \operatorname{Ind} \frac{a(t) + b(t)}{a(t) - b(t)} = -\operatorname{Ind} \frac{a(t) - b(t)}{a(t) + b(t)} = -\varkappa. \qquad (21.20)$$

Bearing in mind formula (14.12) determining the canonical function of the homogeneous Riemann problem, we observe that the canonical functions corresponding to equation (21.19), $X'(z)$, and to equation (21.4), $X(z)$, are inverse to each other, i.e.

$$X'(z) = \frac{1}{X(z)}.$$

Proceeding as in the preceding section we obtain the solution of the singular integral equation (21.15) for $\varkappa' = -\varkappa \geqq 0$, in the form

$$\psi(t) = a(t)h(t) + \frac{1}{\pi i Z(t)} \int_L \frac{b(\tau) Z(\tau) h(\tau)}{\tau - t} d\tau + \frac{1}{Z(t)} Q_{\varkappa'-1}(t),$$
$$(21.21)$$

where $Z(t)$ is given by formula (21.10), and $Q_{\varkappa'-1}(t)$ is a polynomial of degree $\varkappa' - 1$ with arbitrary coefficients (if $\varkappa' = 0$ then $Q(t) \equiv 0$).

The reader is advised, as an exercise, to derive formula (21.21) independently.

If $\varkappa' = -\varkappa < 0$, for the solubility of equation (21.15) it is necessary and sufficient that the following conditions are satisfied:

$$\int_L b(t) Z(t) h(t) t^{k-1} dt = 0 \quad (k = 1, 2, \ldots, -\varkappa'). \quad (21.22)$$

If they are satisfied the solution is given by formula (21.21) where we set $Q_{\varkappa'-1}(t) \equiv 0$.

We observe that the conditions of solubility (21.13) and (21.22) have the form required by the third Fredholm theorem. We shall further prove that the same occurs also in the general case, for the complete singular integral equation.

The results of a simultaneous investigation of the dominant equation and its adjoint equation, indicate a new significant distinction from the properties of the Fredholm equation. In spite of the second Fredholm theorem the adjoint homogeneous singular dominant equations are never simultaneously soluble. They are either both insoluble ($\varkappa = 0$), or, when the index is not zero, that which has a positive index is soluble.

The difference between the numbers of their solutions is equal to the index $\varkappa$. This property which, as will be shown below, is valid also for complete equations, is most typical for singular integral equations.

21.4. Examples

We have already noted that the main difficulty in practical solution of the Riemann problem and consequently of the dominant singular integral equation, consists in computation of singular integrals appearing in formula (21.9). This computation in the general case may be performed by an application of approximate methods of integration; so far, however, methods of approximate computation of singular integrals have not been elaborated, either from the theoretical or practical point of view.

We shall confine ourselves to an example, when the solution of the Riemann boundary value problem (21.4) corresponding to an integral equation, was found by elementary methods—by means of analytic continuation (§ 14.6). The solution of the integral equation will directly be obtained from the Sokhotski formulae (21.3).

Consider the integral equation

$$K°\varphi \equiv (t^2 + t - 1)\,\varphi(t) +$$

$$+ \frac{t^2 - t - 1}{\pi i} \int_L \frac{\varphi(\tau)}{\tau - t}\,d\tau = 2\left(t^3 - t + 1 + \frac{1}{t}\right).$$

Here

$$G(t) = \frac{a(t) - b(t)}{a(t) + b(t)} = \frac{t}{t^2 - 1}, \quad g(t) = \frac{f(t)}{a(t) + b(t)} = \frac{t^3 - t^2 + 1}{t^2 - t},$$

and, consequently, the boundary condition (21.4) takes the form

$$\Phi^+(t) = \frac{t}{t^2 - 1}\,\Phi^-(t) + \frac{t^3 - t^2 + 1}{t^2 - t}.$$

The solution of this boundary value problem satisfying the condition $\Phi^-(\infty) = 0$ was derived in § 14.6, for various contours L.

1°. If L is the contour in Fig. 12 (the case (a)), then

$$\Phi^+(z) = z + \frac{c}{z^2 - 1}, \quad \Phi^-(z) = -\frac{z + 1}{z^2} + \frac{c}{z}$$

and the solution of the integral equation is provided by the function

$$\varphi(t) = \Phi^+(t) - \Phi^-(t) = \frac{t^3 + t + 1}{t^2} + C\frac{1 + t - t^2}{t^2 - t}.$$

$2°$. If L is the contour of example (c) the index is -1; the condition of solubility of the Riemann problem is satisfied and

$$\Phi^+(z) = z \quad \Phi^-(z) = -\frac{z+1}{z^2}.$$

The solution of the integral equation is given by the formula

$$\varphi(t) = \Phi^+(t) - \Phi^-(t) = \frac{t^3 + t + 1}{t^2}.$$

An examination of the remaining cases of position of the contour is left to the reader.

21.5. Aproximate solution

The solution of the dominant singular integral equation is equivalent to finding the contour values of the solution of the Riemann boundary value problem. Any method of solving the boundary value problem (see, for example, § 16.3, No. 8) is therefore a method of solving the integral equation too. But it may be simpler to find some contour values than to find the whole solution of the boundary value problem. Special methods of solving the dominant equation are based on this. Two methods will now be discussed (see Ivanov (2)).

Let

$$a(t)\,\varphi(t) + \frac{b(t)}{\pi i} \int_\gamma \frac{\varphi(\tau)}{\tau - 1}\, d\tau = f(t) \tag{21.1}$$

be the defined equation and

$$\Phi^+(t) - G(t)\,\Phi^-(t) = g(t) \quad \left(G = \frac{a-b}{a+b},\; g = \frac{f}{a+b}\right) \tag{21.4}$$

the corresponding Riemann problem. It is assumed that the contour γ is circular and that

$$\varkappa = \text{Ind}\, \frac{a-b}{a+b} \geqq 0.$$

1. *Method of moments.* The formula (14.18′) shows that there are $\varkappa$ linearly independent solutions of the Riemann problem of order $1, 2, \ldots, \varkappa$ respectively at infinity. To be specific, it is required to find the solution of highest possible order at infinity. The approximate solution is required in the form

$$\Phi^+(t) = \sum_{k=1}^n c_k t^k; \quad \Phi^-(t) = \sum_{k=-1}^{-n} c_k t^{k-\varkappa} \tag{21.23}$$

Substituting into (21.4), and then multiplying successively by t^{-j+1}, $j = 0, \pm 1, \ldots, \pm n$, and integrating with respect to γ, we obtain $(2n + 1)$ equations† for the constants c :

$$\int_{\gamma} \left[\sum_{k=0}^{n} c_k t^k - G(t) \, t^{-\varkappa} \sum_{k=-1}^{-n} c_k t^k \right] t^{-j+1} \, dt = \int_{\gamma} g(t) \, t^{-j+1} \, dt. \quad (21.24)$$

Using general approximate methods [L. V. Kantorovich and G. P. Akilov, *Functional analysis in normed spaces*, Chapter XIV, English translation published by Pergamon Press, Oxford, 1964], it can be proved that the set (21.24) is unconditionally and single-valuedly soluble if n is sufficiently large.

2. *Mixed method.* Here the solution is also required in the form (21.23), but the equations for the constants c_k are formed, not by the equalization of moments, but from consideration of the interpolating requirement of coincidence of the left- and right-hand sides of (21.24) at $(2n + 1)$ points of the circle. For simplicity they can be equally spaced $t_j = \left(e^{j \frac{\theta}{2n+1} t} \right)$. The set is written as

$$\sum_{k=1}^{n} \alpha_k t_j^k - G(t_j) t_j^{-\varkappa} \sum_{k=-1}^{-n} \alpha_k t_j^k = f(t_j). \quad 21.25)$$

It may be proved that the set is unconditionally and single-valuedly soluble for sufficiently large n.

The solution of (21.1) is found by the formula

$$\varphi(t) = \Phi^+(t) - \Phi^-(t) = \sum_{k=0}^{\varkappa} c_k t^k - \sum_{k=-1}^{-n} c_k t^k. \quad (21.26)$$

For $\varkappa < 0$, if the solubility conditions (16.11) are fulfilled, the solution may be sought as

$$\varphi(t) = \sum_{k=1}^{n} c_k t^k - \sum_{k=-\varkappa-1}^{-n} c_k t^{k-\varkappa}. \quad (21.27)$$

The same methods, with slight modifications, are also applicable to the solution of the complete equation (20.3). In the cited work, using the theory of approximation of functions, V. V. Ivanov indicated the order of error. But the estimates contain constants which are hard to determine so that their practical value is possibly in doubt.

---

† This is equalization of the "moments" on the left- and right-hand sides of (21.4). Hence the name of the method.

21.6. The behaviour of the solution at corner points

From Sokhotski's formulae for corner points (4.18), (4.19) we obtain the pair of equivalent formulae:

$$\varphi(t) = \Phi^+(t) - \Phi^-(t), \tag{21.28}$$

$$\frac{1}{\pi i} \int_L \frac{\varphi(\tau)}{\tau - t} d\tau = \frac{\alpha}{\pi} \Phi^+(t) + \left(2 - \frac{\alpha}{\pi}\right) \Phi^-(t). \tag{21.29}$$

In analysing the solution of the Riemann boundary value problem it was seen that only one of the Sokhotski formulae was used, namely, the jump formula (21.28). But this formula is just the same in form for a smooth point of the contour as for a corner point. Hence it follows that the solution of the Riemann problem does not change because of the presence of contour corner points and all that we obtained for this problem on a smooth contour applies equally to contours with corner points†.

We will now see how matters stand with the solution of the dominant equation

$$a(t) \varphi(t) + \frac{b(t)}{\pi i} \int_L \frac{\varphi(\tau)}{\tau - t} d\tau = f(t). \tag{21.1}$$

At a smooth point on the contour ($\alpha = \pi$) the following formula holds for a singular integral:

$$\frac{1}{\pi i} \int \frac{\varphi(\tau)}{\tau - t} d\tau = \Phi^+(t) + \Phi^-(t).$$

For an isolated corner point its limit values to the left and right are expressed by the same formula. But the value of the singular integral at the corner point must be calculated from the formula (21.29). Hence the corner point for a singular integral will be a point of removable discontinuity. Hence it is easy to deduce that equation (21.1) cannot, generally speaking, have a continuous solution at the corner point t_0. In actual fact, if we only assume that $b(t_0) \neq 0$ and that $\varphi(t_0)$ is continuous, we come across a contradiction since one term in the equation is discontinuous whilst the rest are continuous.

We require the solution of the equation in the form

$$\varphi(t) = \varphi_1(t) + \xi(t), \tag{21.30}$$

where $\varphi_1(t)$ is a function that satisfies Hölder's condition, and $\xi(t)$ is everywhere zero except at corner points. Substituting (21.30)

---

† For the present only a finite number of corner points is considered.

into the equation and noting that $\int\limits_L \dfrac{\xi(\tau)}{\tau-t}\,d\tau \equiv 0$, we obtain

$$a(t)\,\varphi_1(t) + \frac{b(t)}{\pi i} \int\limits_L \frac{\varphi_1(\tau)}{\tau - t}\,d\tau = f(t) - a(t)\,\xi(t). \qquad (21.31)$$

Here φ_1 and ξ have to be such that the equation is satisfied.

Introducing the Cauchy integral

$$\Phi(z) = \frac{1}{2\pi i} \int \frac{\varphi_1(\tau)}{\tau - z}\,d\tau$$

and remembering that at all non-corner points the usual Sokhotski formulae must be satisfied for its limit values and $\xi(t) = 0$, it is found that for the smooth points equation (21.1) reduces to the usual Riemann boundary problem written as

$$[a(t) + b(t)]\,\Phi^+(t) - [a(t) - b(t)]\,\Phi^-(t) = f(t). \qquad (21.32)$$

In this relation all the terms are continuous functions† and therefore, being satisfied almost everywhere, it is necessarily fulfilled everywhere. Consequently, the boundary condition (21.32) holds good for the whole contour including the corner points. But, on the other hand, the Sokhotski formulae (21.28), (21.29) (in which φ has to be replaced by φ_1) hold for the corner points. We therefore find from (21.31) that the boundary condition for the contour corner points is

$$\left[a(t) + \frac{\alpha}{\pi} b(t)\right] \Phi^+(t) - \left[a(t) - \left(2 - \frac{\alpha}{\pi}\right) b(t)\right] \Phi^-(t)$$
$$= f(t) - a(t)\,\xi(t). \qquad (21.33)$$

Eliminating $f(t)$ from (21.33) by (21.32), the expression for $\xi(t)$ is

$$\xi(t) = \frac{b(t)}{a(t)}\left(1 - \frac{\alpha}{\pi}\right)[\Phi^+(t) - \Phi^-(t)]. \qquad (21.34)$$

Here it is presupposed that $a(t) \neq 0$ at the corner point. The case $a(t) = 0$ is exceptional.

Thus for all non-corner points the solution of the singular equation is obtained from the solution of the Riemann problem (21.32) by the usual formula

$$\varphi(t) = \varphi_1(t) = \Phi^+(t) - \Phi^-(t),$$

but for corner points it is necessary to add (21.34) to the solution.

---

† The discourse on the nature of the limit values of a Cauchy integral in § 5 also remains valid for contours with corner points.

Regarding the angle between the left and right tangential vectors of the contour as a function of the point $\alpha = \alpha(t)$ (for smooth points $\alpha(t) = \pi$), the solution of the equation for the whole contour can be combined into one formula:

$$\varphi(t) = \varphi_1(t) + \xi(t) = \left\{1 + \frac{b(t)}{a(t)}\left[1 - \frac{\alpha(t)}{\pi}\right]\right\} [\Phi^+(t) - \Phi^-(t)], \quad (21.35)$$

where $\Phi^\pm(t)$ is the solution of the boundary problem (21.32).

This result has also been obtained in a different way by Kveselava (3).

So far it has been assumed that there is only a finite number of corner points on the contour. In this case, as we have seen, the problem is completely soluble from elementary considerations. It is natural to pose the question: what is the widest class of contours for which the foregoing arguments are valid? To answer this question, it is necessary to use the theory of functions of a real variable. The result obtained by Alekseyev (1) (2) will now be given.

If the contour L is a closed curve with left- and right-hand tangents at each point which are continuous to the left and right respectively, then formula (21.35) still holds for the solution of equation (21.1). We show that such curves (Alekseyev called them class-R curves) may have a denumerable set of corner points. But here the number of corner points for which the jump in the angle of the tangent to the axis of the abscissae exceeds a given number $\varepsilon > 0$, is always only finite. Class-R contours include Radon's class of curves with limited rotation (I. Radon, Über die Randwertaufgaben beim logarithmischen Potential, *Sitzungsber. Österr. Akad. Wiss.*, Wien, 128, 1919).

§ 22. Regularization of the complete equation

22.1. Product of singular operators

It was shown in the preceding section that the simplest singular integral equations—the dominant equations—can be solved in a closed form.

Below (§ 51) we shall distinguish some particular types of complete singular integral equations which can also be solved in a closed form. In the general case however the solution of a singular integral equation is carried out by reduction to a Fredholm integral equation.

The procedure of reduction of a singular integral equation to a Fredholm (regular) equation is called the *regularization*. Below we shall give various methods of regularization, the more important of which consist in an application to the singular operator of another especially chosen singular operator.

Let K_1 and K_2 be singular operators

$$K_1\varphi \equiv a_1(t)\,\varphi(t) + \frac{1}{\pi i}\int_L \frac{M_1(t,\tau)}{\tau-t}\,\varphi(\tau)\,d\tau, \qquad (22.1)$$

$$K_2\omega \equiv a_2(t)\,\omega(t) + \frac{1}{\pi i}\int_L \frac{M_2(t,\tau)}{\tau-t}\,\omega(\tau)\,d\tau. \qquad (22.2)$$

The operator K defined by the formula $K\varphi = K_2(K_1\varphi)$ is called the product of the operators K_1 and K_2 in the indicated order (the product of operators in general is not commutative).

Construct the expression for the operator K:

$$K\varphi = K_2 K_1 \varphi \equiv a_2(t)\left[a_1(t)\,\varphi(t) + \frac{1}{\pi i}\int_L \frac{M_1(t,\tau)}{\tau-t}\,\varphi(\tau)\,d\tau \right] +$$

$$+ \frac{1}{\pi i}\int_L \frac{M_2(t,\tau)}{\tau-t}\left[a_1(\tau)\,\varphi(\tau) + \frac{1}{\pi i}\int_L \frac{M_1(\tau,\tau_1)}{\tau_1-\tau}\,\varphi(\tau_1)\,d\tau_1 \right] d\tau \qquad (22.3)$$

and separate its dominant part. To this end let us perform the following transformations:

$$\left.\begin{aligned}
\int_L \frac{M_1(t,\tau)}{\tau-t}\,\varphi(\tau)\,d\tau &= M_1(t,t)\int_L \frac{\varphi(\tau)}{\tau-t}\,d\tau + \\
&\qquad + \int_L \frac{M_1(t,\tau)-M_1(t,t)}{\tau-t}\,\varphi(\tau)\,d\tau, \\[2ex]
\int_L \frac{a_1(\tau)\,M_2(t,\tau)}{\tau-t}\,\varphi(\tau)\,d\tau &= a_1(t)\,M_2(t,t)\int_L \frac{\varphi(\tau)}{\tau-t}\,d\tau + \\
&\qquad + \int_L \frac{a_1(\tau)\,M_2(t,\tau)-a_1(t)\,M_2(t,t)}{\tau-t}\,\varphi(\tau)\,d\tau, \\[2ex]
\frac{1}{\pi i}\int_L \frac{M_2(t,\tau)}{\tau-t}\,d\tau \, \frac{1}{\pi i}\int_L & \frac{M_1(\tau,\tau_1)}{\tau_1-\tau}\,\varphi(\tau_1)\,d\tau_1 \\
&= M_2(t,t)\,M_1(t,t)\,\varphi(t) + \\
&\quad + \frac{1}{\pi i}\int_L \varphi(\tau_1)\,d\tau_1\,\frac{1}{\pi i}\int_L \frac{M_2(t,\tau)\,M_1(\tau,\tau_1)}{(\tau_1-\tau)(\tau-t)}\,d\tau.
\end{aligned}\right\} \quad (22.4)$$

in the last line we have made use of the formula for the change of integration order (7.7)).

It is readily observed that all kernels of the integrals in the last terms of the right-hand sides of formulae (22.4) have at the point $= t$ a singularity of order not exceeding $|\tau - t|^{\lambda-1}$ ($\lambda < 1$).

This is evident for the first two kernels, if we take into account that he functions $a_1(t)$, $M_1(t,\tau)$, $M_2(t,\tau)$ satisfy the Hölder condition. For the last kernel, introducing the notation $M_2(t,\tau) \times M_1(\tau,\tau_1) = M(t,\tau,\tau_1)$ we perform the transformation

$$\int \frac{M(t, \tau, \tau_1)}{(\tau_1 - \tau)(\tau - t)}\, d\tau = \frac{1}{\tau_1 - t}\left[\int_L \frac{M(t, \tau, \tau_1)}{\tau - t}\, d\tau - \right.$$
$$\left. - \int_L \frac{M(t, \tau, \tau_1)}{\tau - \tau_1}\, d\tau \right] = \frac{\omega_1(t, \tau_1) - \omega_2(t, \tau_1)}{\tau_1 - t},$$

where

$$\omega_1(t, \tau_1) = \int_L \frac{M(t, \tau, \tau_1)}{\tau - t}\, d\tau, \quad \omega_2(t, \tau_1) = \int_L \frac{M(t, \tau, \tau_1)}{\tau - \tau_1}\, d\tau.$$

On the basis of § 5.1 the functions $\omega_1(t,\tau_1)$, $\omega_2(t,\tau_1)$ satisfy the Hölder condition and, moreover, $\omega_1(t, t) = \omega_2(t, t)$; hence the last kernel as also at the point $\tau = t$ a singularity not stronger than $|\tau - t|^{\lambda-1}$.

Denoting as in § 20.1

$$M_1(t, t) = b_1(t), \quad M_2(t, t) = b_2(t), \tag{22.5}$$

we find that the dominant operator K° of the product K of two singular operators K_1 and K_2 is given by the formula

$$K^\circ\varphi = (K_2 K_1)^\circ\varphi = [a_2(t)\, a_1(t) + b_2(t)\, b_1(t)]\, \varphi(t) +$$
$$+ \frac{a_2(t)\, b_1(t) + b_2(t)\, a_1(t)}{\pi i}\int_L \frac{\varphi(\tau)}{\tau - t}\, d\tau. \tag{22.6}$$

We now write the operators K_1, K_2 in the form (20.3) with the dominant part separated out

$$K_1\varphi \equiv a_1(t)\, \varphi(t) + \frac{b_1(t)}{\pi i}\int_L \frac{\varphi(\tau)}{\tau - t}\, d\tau + \int_L k_1(t, \tau)\, \varphi(\tau)\, d\tau, \tag{22.7}$$

$$K_2\omega \equiv a_2(t)\, \omega(t) + \frac{b_2(t)}{\pi i}\int_L \frac{\omega(\tau)}{\tau - t}\, d\tau + \int_L k_2(t, \tau)\, \omega(\tau)\, d\tau. \tag{22.8}$$

Thus, the coefficients $a(t)$, $b(t)$ of the dominant part of the product of operators are the following:

$$a(t) = a_2(t) a_1(t) + b_2(t) b_1(t); \quad b(t) = a_2(t) b_1(t) + b_2(t) a_1(t). \quad (22.9)$$

These formulae do not contain the regular kernels k_1 and k_2, and they are symmetric with respect to the indices 1 and 2. We therefore infer the following:

1. *The dominant part of the product of singular operators is independent of regular parts of these operators and, moreover, it is independent of the order of the operators in the product.*

Thus, the change of the order of the operators and a change of their regular parts influence only the regular part of the product of the operators, and do not alter its dominant part.

We now compute the coefficient of the Riemann problem corresponding to the dominant operator $(K_2 K_1)°$. In view of formulae (21.5), (21.9) we have

$$G(t) = \frac{a(t) - b(t)}{a(t) + b()t} = \frac{[a_2(t) - b_2(t)] [a_1(t) - b_1(t)]}{[a_2(t) + b_2(t)] [a_1(t) + b_1(t)]} = G_2(t) G_1(t),$$
$$(22.10)$$

where
$$G_1(t) = \frac{a_1(t) - b_1(t)}{a_1(t) + b_1(t)}, \quad G_2(t) = \frac{a_2(t) - b_2(t)}{a_2(t) + b_2(t)} \quad (22.11)$$

denote the coefficients of the Riemann problems corresponding to the operators $K_1°$, $K_2°$. Thus, we have the following important conclusion.

2. *The coefficient of the Riemann problem for the operator $(K_2 K_1)°$ is equal to the product of the coefficients of the Riemann problems for the operators $K_1°$ and $K_2°$ and, consequently, the index of the product of singular operators is equal to the sum of the indices of the multiplied operators,* $\varkappa = \varkappa_1 + \varkappa_2.$ $\quad (22.12)$

The full expression for the operator $K_2 K_1$ is the following:

$$K_2 K_1 \varphi \equiv a(t) \varphi(t) + \frac{b(t)}{\pi i} \int_L \frac{\varphi(\tau)}{\tau - t} d\tau + \int_L k(t, \tau) \varphi(\tau) d\tau.$$

Here $a(t)$, $b(t)$ are given by formulae (22.9). For the regular kernel $k(t, \tau)$ it is easy on the basis of formulae (22.4) to write down the explicit expression; since however we shall not use this expression anywhere, we do not quote it here.

22.2. Regularizing operator

If the singular operator K_2 is such that the operator K_2K_1 is a regular (Fredholm) operator, i.e. it does not contain a singular integral $(b(t) \equiv 0)$, then K_2 is called the *regularizing operator* with respect to the singular operator K_1, or, briefly, its *regularizator*.

Observe that if K_2 is the regularizator, then in view of the Property 1° of the product of operators, the operator K_1K_2 is also regular.

We now proceed to find the general form of the regularizing operator.

According to the definition, we have the relation

$$b(t) = a_2(t)\,b_1(t) + b_2(t)\,a_1(t) = 0, \tag{22.13}$$

which implies that

$$a_2(t) = u(t)\,a_1(t), \quad b_2(t) = -\,u(t)\,b_1(t), \tag{22.14}$$

where $u(t)$ is an arbitrary function satisfying the Hölder condition.

Consequently, if K is a singular operator

$$K\varphi \equiv a(t)\,\varphi(t) + \frac{b(t)}{\pi i} \int_L \frac{\varphi(\tau)}{\tau - t}\,d\tau + \int_L k(t,\,\tau)\,\varphi(\tau)\,d\tau, \tag{22.15}$$

then the general form of its regularizator, which we denote by $\tilde{K}$, is the following:

$$\tilde{K}\omega \equiv u(t)\,a(t)\,\omega(t) - \frac{u(t)\,b(t)}{\pi i} \int_L \frac{\omega(\tau)}{\tau - t}\,d\tau + \int_L \tilde{k}(t,\,\tau)\,\omega(\tau)\,d\tau. \tag{22.16}$$

Here $\tilde{k}(t,\tau)$ is an arbitrary Fredholm kernel and $u(t)$ is an arbitrary function satisfying the Hölder condition.

Since the index of the regular operator $(b(t) \equiv 0)$ is obviously zero, the Property 2 of the product of operators implies that *the absolute value of the index of the regularizing operator is equal to that of the regularized operator, the signs being opposite*. This result can be inferred directly from the form of the regularizing operator (22.16), since

$$\tilde{G}(t) = \frac{\tilde{a}(t) - b(t)}{\tilde{a}(t) + \tilde{b}(t)} = \frac{a(t) + b(t)}{a(t) - b(t)} = \frac{1}{G(t)}.$$

Thus, for any singular operator with Cauchy kernel (22.15) of normal type $(a(t) \pm b(t) \neq 0)$ there exists an infinite set of regularizing operators (22.16) the dominant part of which depends on an arbitrary function $u(t)$ and contains an arbitrary regular kernel $\tilde{k}(t,\tau)$.

The arbitrary quantities $u(t)$ and $\tilde{k}(t,\tau)$ can in case of need, be so chosen that the regularizing operator satisfies certain additional conditions. If no conditions are imposed we obtain the simplest regularizators if we set in formula (22.16):

1°. $u(t) \equiv 1$, $\tilde{k}(t,\tau) \equiv 0$, whence the regularizator takes the form

$$\tilde{K}\omega = K'^{\circ}\omega \equiv a(t)\,\omega(t) - \frac{b(t)}{\pi i} \int_L \frac{\omega(t)}{\tau - t}\,d\tau; \qquad (22.17)$$

2°. $u(t) \equiv 1$; $\tilde{k}(t,\tau) = -\frac{1}{\pi i}\frac{b(\tau) - b(t)}{\tau - t}$;

then

$$\tilde{K}\omega = K^{\circ\prime}\omega \equiv a(t)\,\omega(t) - \frac{1}{\pi i} \int_L \frac{b(\tau)\,\omega(\tau)}{\tau - t}\,d\tau. \qquad (22.18)$$

The simplest operators K'° and $K^{\circ\prime}$ are most frequently employed as regularizators.

Since the multiplication of operators is not commutative, two types of regularization are to be distinguished: regularization from left when the result is the operator $\tilde{K}K$, and regularization from right which leads to the operator $\tilde{K}K$. In view of a remark made above it follows that a regularizator from the left is also a regularizator from the right, and conversely. Thus, *the operation of regularization is commutative.*

It follows from the Property 1 of §22.1 that *this operation is mutual*, i.e. if the operator $\tilde{K}$ is regularizing for the operator K, then also the converse is true—the operator K is regularizing for the operator $\tilde{K}$.

The operators K_1K_2 and K_2K_1 may differ only in the regular part. Denoting the regular (Fredholm) operator by T, the symbolic connection between the products of singular operators differing in the order of multiplied operators, has the form

$$K_1K_2 - K_2K_1 = T.$$

22.3. Methods of regularization

Consider the complete singular integral equation

$$K\varphi \equiv a(t)\,\varphi(t) + \frac{b(t)}{\pi i} \int_L \frac{\varphi(\tau)}{\tau - t}\, d\tau + \int_L k(t, \tau)\,\varphi(\tau)\, d\tau = f(t). \quad (22.19)$$

It was already indicated at the beginning of this section that the solution (and also the theoretical investigation) of such an equation is carried out by means of regularization†. Three methods of regularization are in use. The first two are based on the product of the singular operator under consideration with its regularizator (regularization from left and from right). The third method is essentially different from the first two; here the elimination of the singular integral is performed by solving the corresponding dominant equation.

In constructing the theory of singular equations (Noether's Theorems) we shall employ only the first two methods of regularization; hence, we first treat these methods and postpone the exposition of the third method to the end of § 24, in order not to interrupt the logical chain.

1. *Regularization from the left.* Consider the regularizing operator (22.16)

$$\tilde{K}\omega \equiv u(t)\,a(t)\,\omega(t) - \frac{u(t)\,b(t)}{\pi i} \int_L \frac{\omega(\tau)}{\tau - t}\, d\tau + \int_L \tilde{k}(t, \tau)\,\omega(\tau)\, d\tau. \quad (22.20)$$

Introducing into $\tilde{K}\omega$ instead of the function ω the expression $\tilde{K}\varphi - f$ we arrive at the integral equation

$$\tilde{K}K\varphi = \tilde{K}f. \quad (22.21)$$

By definition ($\tilde{K}$ is the regularizator) the operator $\tilde{K}K$ is a Fredholm operator; consequently equation (22.21) is a Fredholm equation. Thus we have transformed the singular integral equation (22.19) into a Fredholm integral equation (22.21) for the same unknown function $\varphi(t)$.

This is the essence of the first method of regularization—regularization from the left. Observe that this method was already employed

---

† The methods of direct solution (in general approximate) of a complete singular equation have so far been little elaborated.

by the originators of the theory of singular integral equations, Hilbert and Poincaré (see §26, Historical notes).

2. *Regularization from the right.* Substituting in equation (22.19) for the unknown function expression (22.20)

$$\varphi(t) = \tilde{K}\omega, \qquad (22.22)$$

where $\omega(t)$ is a new unknown function, we are led to the integral equation

$$K\tilde{K}\omega = f, \qquad (22.23)$$

which also is a Fredholm equation. Thus, we have passed from the singular integral equation (22.19) for the unknown function $\varphi(t)$ to a Fredholm integral equation for the new unknown function $\omega(t)$.

Having solved the Fredholm equation (22.23) we find the solution of the original equation (22.19) by means of formula (22.22). Application of the latter formula requires only computation of integrals (one ordinary and one singular integral).

This is the essence of the second method of regularization—regularization from right.

22.4. Relation between solutions of singular and regularized equations

In reducing the singular integral equation to a regular one we performed a functional transformation on the equation. In general this transformation can either introduce superfluous solutions which do not satisfy the original equation, or cause a loss of some of them. Therefore, in general, the equation obtained is not equivalent to the original one. We proceed to establish the relation between the solutions of these two equations.

1°. *Regularization from the left.*

Let

$$K\varphi = f \qquad (22.24)$$

be the given singular equations, and

$$\tilde{K}K\varphi = \tilde{K}f \qquad (22.25)$$

the corresponding regular equation.

Represent the latter in the form

$$\tilde{K}(K\varphi - f) = 0. \qquad (22.25')$$

Since the operator $\tilde{K}$ is homogeneous every solution of the original equation (22.24) (a function making the expression $K\varphi - f$ zero) also satisfies equation (22.25). Consequently, regularization from the left does not lead to any loss of solutions.

Consider now the converse question, are all solutions of the regularized equation also solutions of the original equation? It is easy to show that this is not always the case.

Consider the singular integral equation corresponding to the regularized operator

$$\tilde{K}\omega = 0. \tag{22.26}$$

Let $\omega_1(t), \ldots, \omega_p(t)$ be a complete set of solutions, i.e. the set of all linearly independent eigenfunctions of the regularizing operator $\tilde{K}$. Regarding equation (22.25′) as a singular equation of the form (22.26), with unknown function $\omega = K\varphi - f$ we have

$$K\varphi - f = \sum_{j=1}^{p} \alpha_j \omega_j \tag{22.27}$$

where α_j are constants.

Thus, the regularized equation is equivalent not to the original equation (22.24), but to equation (22.27).

If we apply to equation (22.27) the operator $\tilde{K}$ we arrive at equation (22.25) the constants α_j being arbitrary. It seems that we may now infer that equation (22.25) is equivalent to equation (22.27), for arbitrary α_j (a beginner invariably draws such an erroneous conclusion). *In fact this is true only when equation* (22.27) *is soluble for arbitrary* α_j. However it can turn out that it is soluble only for a special choice of the considered constants. This means that if in the expression $K\varphi - f$ we substitute for φ the solutions of equation (22.25). it is converted only into such a solution of the equation $\tilde{K}\omega = 0$ for which the equation

$$K\varphi - f = \sum_{j=1}^{p} a_j \omega_j$$

is soluble. Consequently, *equation* (22.25) *is equivalent to equation* (22.27) *in which* α_j *are arbitrary or definite constants*, depending on the circumstances. It may turn out that equation (22.27) is soluble only under the condition that all $\alpha_j = 0$. In this case equation (22.25) is equivalent to the original equation (22.24) and the regularizator is equivalent. In particular, if the regularizator has no eigenfunctions the right-hand side of equation (22.27) vanishes identically and it certainly is equivalent. Such an operator necessarily exists if $\varkappa \geqq 0$; for instance it may be the regularizator K'° which in this case has no eigenfunctions†.

Problems of equivalent regularization in the general case will be examined later.

2°. *Regularization from the right.*

Let

$$K\varphi = f \tag{22.24}$$

be the original singular equation, and

$$K\tilde{K}\omega = f \tag{22.28}$$

the regularized equation obtained by the substitution

$$\tilde{K}\omega = \varphi. \tag{22.29}$$

If ω_j is a solution of equation (22.28), then formula (22.29) yields the corresponding solution of the original equation

$$\varphi_j = \tilde{K}\omega_j.$$

Consequently, the regularization from the right cannot lead to superfluous solutions.

Conversely, let φ_k be a solution of the original equation. Then the solution of the regularized equation (22.28) can be obtained as the solution of the non-homogeneous singular equation

$$\tilde{K}\omega = \varphi_k,$$

which however can turn out to be insoluble.

Thus, regularization from the right may lead to a loss of solutions. This loss will not occur if equation (22.29) is soluble for an arbitrary right-hand side. In this case the operator $\tilde{K}$ is the equivalent regularizator from right. Below (§ 23.3) we shall find that this is the case when $\varkappa \leqq 0$, because then the index of the operator $\tilde{K}$ is non-negative and equation (22.29) is absolutely soluble.

† This follows from the argument in § 21.2, since the index of the regularizator K'° (dominant) is equal to $-\varkappa \leqq 0$.

§ 23. Fundamental properties of singular equations

23.1. Some properties of adjoint operators

In what follows we shall use two properties of adjoint operators. They are not typical for singular operators, but occur for all linear operators. For the reader's sake we shall present their derivation. Since the presence of a singular integral does not complicate the proofs we shall for simplicity denote the singular kernel by $K(t, \tau)$.

Let K be the singular operator

$$K\varphi \equiv a(t)\,\varphi(t) + \int_L K(t, \tau)\,\varphi(\tau)\,d\tau;$$

and K' its adjoint operator

$$K'\psi \equiv a(t)\,\psi(t) + \int_L K(\tau, t)\,\psi(\tau)\,d\tau.$$

1ST PROPERTY. *For arbitrary functions $\varphi(t)$ and $\psi(t)$ which satisfy the Hölder condition, the following identity is valid:*

$$\int_L \psi K\varphi\,dt = \int_L \varphi K'\psi\,dt. \tag{23.1}$$

Proof.† We have

$$\int_L \psi K\varphi\,dt = \int_L \psi(t)\left[a(t)\,\varphi(t) + \int_L K(t, \tau)\,\varphi(\tau)\,d\tau\right]dt$$

$$= \int_L \varphi(t)\,a(t)\,\psi(t)\,dt + \int_L \varphi(\tau)\left[\int_L K(t, \tau)\,\psi(t)\,dt\right]d\tau.$$

Denoting the variable τ in the double integral by t and t by τ and rearranging appropriately the terms we obtain the required relation (23.1).

Observe that the identity (23.1) completely defines the adjoint operator and that it is sometimes taken as its definition.

2ND PROPERTY. *The following identity holds:*

$$(K_2 K_1)' = K_1' K_2'. \tag{23.2}$$

---

† Let us recall that according to Lemma of § 7.1 the change of the order of integration in the double integral in which only one integral is singular, is permissible.

Proof. Constructing as in § 22.1 the product of the singular operator (22.3) and performing some simple transformations including an application of the formula (7.7) for the change of integration order we obtain

$$K_2 K_1 \varphi = [a_1(t)\, a_2(t) + b_1(t)\, b_2(t)]\, \varphi(t) +$$
$$+ \frac{1}{\pi i} \int\limits_L \frac{a_1(\tau)\, M_2(t,\, \tau) + a_2(t)\, M_1(t,\, \tau)}{\tau - t}\, \varphi(\tau)\, d\tau +$$
$$+ \frac{1}{\pi i} \int\limits_L \left[\frac{1}{\pi i} \int\limits_L \frac{M_1(\tau_1,\, \tau)\, M_2(t,\, \tau_1)}{(\tau_1 - t)\,(\tau - \tau_1)}\, d\tau_1 \right] \varphi(\tau)\, d\tau$$

Hence, in accordance with the definition of the adjoint operator (§ 20.1)

$$(K_2 K_1)'\, \varphi = [a_1(t)\, a_2(t) + b_1(t)\, b_2(t)]\, \varphi(t) -$$
$$- \frac{1}{\pi i} \int\limits_L \frac{a_1(t)\, M_2(\tau,\, t) + a_2(\tau)\, M_1(\tau,\, t)}{\tau - t}\, \varphi(\tau)\, d\tau +$$
$$+ \frac{1}{\pi i} \int\limits_L \left[\frac{1}{\pi i} \int\limits_L \frac{M_1(\tau_1,\, t)\, M_2(\tau,\, \tau_1)}{(\tau_1 - t)\,(\tau - \tau_1)}\, d\tau_1 \right] \varphi(\tau)\, d\tau$$

Constructing now the product of the operators

$$K_2'\varphi \equiv a_2(t)\, \varphi(t) - \frac{1}{\pi i} \int\limits_L \frac{M_2(\tau,\, t)\, \varphi(\tau)}{\tau - t}\, d\tau$$

and

$$K_1'\psi \equiv a_1(t)\, \psi(t) - \frac{1}{\pi i} \int\limits_L \frac{M_1(\tau,\, t)\, \psi(\tau)}{\tau - t}\, d\tau$$

and comparing the results we establish the validity of (23.2).

23.2. Fundamental theorems on singular integral equations (Noether's theorems)

Basic properties of Fredholm integral equations are described by three theorems called the Fredholm theorems (§ 20.2). Considering in § 21 the simplest singular integral equations—the dominant equations—we observed that some of these properties (the conditions of solubility of the non-homogeneous equation) are identical with the properties of Fredholm equations, but others (for instance the relation between the numbers of solutions of adjoint equations) are significantly different. It was then observed that the fundamental properties of dominant equations are also valid for the general case of a complete equation. We shall prove this assertion here and hence we shall establish theorems characterizing fundamental properties of singular equations of the type under consideration.

Consider the complete singular integral equation

$$K\varphi = f. \tag{23.3}$$

It is known that the number of solutions of a Fredholm integral equation (the number of eigenfunctions belonging to a given eigenvalue) is finite. It is easy to prove the validity of the same property for singular equations.

THEOREM I. *The number of solutions of the singular integral equations* (23.3) *is finite.*

The proof follows directly from the possibility of regularization of a singular equation. It was established (§ 22.4) that the regularization from the left does not lead to any loss of solutions. Consequently, the number of solutions of the singular equation (23.3) does not exceed the number of solutions of the Fredholm equation $\tilde{K}K\varphi = \tilde{K}f$. This implies the assertion of the theorem, which is in complete agreement with Theorem V (§ 20.2).

We now prove that the conditions of solubility of a singular equation have the same form as for the Fredholm equations (third Fredholm theorem).

THEOREM II. *A necessary and sufficient condition of solubility of the singular equation* (23.3) *is given by the relations*†

$$\int_L f(t)\,\psi_j(t)\,dt = 0 \quad (j = 1, 2, 3, \ldots, n'), \tag{23.4}$$

where $\psi_1(t), \ldots, \psi_{n'}(t)$ is a complete system of linearly independent solutions of the adjoint homogeneous equation $K'\psi = 0$.

The necessity of conditions (23.4) is a simple consequence of the identity (23.1). In fact, setting in integral (23.4) $f(t) = K\varphi$ and making

---

† Since the functions under consideration are complex, condition (23.4) cannot be regarded as the condition of orthogonality of the functions $f(t)$, $\psi(t)$ (see § 24.2).

use of the identity (23.1) and the fact that $K'\psi_j = 0$ we arrive at the relations (23.4):

$$\int_L f(t)\,\psi_j(t)\,dt = \int_L \psi_j \, K\varphi \, dt = \int_L \varphi K'\psi_j \, dt \equiv 0$$

$$(j = 1, 2, \ldots, n').$$

The proof of sufficiency is different for two cases.

1. $\varkappa \geqq 0$. The regularizing operators have the index $-\varkappa \leqq 0$; hence from the set of these operators it is always possible to choose an operator $\tilde{K}$ which has no eigenfunctions (for instance K'°). Therefore the Fredholm equation

$$\tilde{K}K\varphi = \tilde{K}f \qquad (23.5)$$

is equivalent to the original equation (23.3). Consequently the equations (23.5) and (23.3) are simultaneously soluble or insoluble.

We write the conditions of solubility of equation (23.5)

$$\int_L \chi_j \tilde{K}f\,dt = 0. \qquad (23.6)$$

where $\chi_j(t)$ are the solutions of the equation

$$K'\tilde{K}'\chi = 0, \qquad (23.5')$$

which in view of identity (23.2) is adjoint to equation (23.5).

Applying identity (23.1) we reduce condition (23.6) to the form

$$\int_L f\tilde{K}'_j\chi\,dt = 0. \qquad (23.6')$$

Regarding equation (23.5) as a singular integral equation with operator K' and unknown function $\tilde{K}'\chi$ we observe that the function $\tilde{K}'\chi$ is an eigenfunction of the operator K'. Denoting it by $\psi_j(t)$ we arrive at condition (23.4).

2. $\varkappa < 0$. Let us apply regularization from the right†. Making the substitution

$$\varphi = K^{\circ'}\omega, \qquad (23.7)$$

we obtain the Fredholm equation

$$KK^{\circ'}\omega = f, \qquad (23.8)$$

---

† In this case the regularizator from the left with no eigenfunctions does not exist see § 23.3, Corollary 2°).

which is equivalent to the original equation (23.3) in the sense that they are both either soluble or insoluble, and formula (23.7) associates with every solution of equation (23.8) a definite solution of equation (23.3), and conversely, formula (23.7) associates with every solution of equation (23.3) some solutions of equation (23.8) (see § 22.4).

The conditions of solubility of equation (23.8) have the form

$$\int_L f(t)\,\chi_j(t)\,dt = 0,$$

where $\chi_j(t)$ is a complete system of solutions of the equation

$$K^\circ K' \chi = 0, \tag{23.8'}$$

adjoint to equation (23.8).

Regarding equation (23.8) as singular with operator K° and unknown function $K'\chi$ let us observe that the operator K°, as a dominant operator with a negative index, has no eigenfunctions; hence

$$K'\chi = 0.$$

This means that $\chi_j(t)$ is an eigenfunction of the operator K'. Denoting it by $\psi_j(t)$ we again arrive at conditions (23.4).

This completes the proof of the theorem.

We now proceed to the proof of a theorem which constitutes the principal item in the theory of singular equations with Cauchy kernels. According to the second Fredholm theorem, adjoint Fredholm equations have the same number of solutions. In contrast to this the numbers of solutions of adjoint singular integral equations are in general distinct. The following theorem establishes an exact relation between them.

THEOREM III. *The difference between the number n of linearly independent solutions of the singular equation $K\varphi = 0$ and the number n' of linearly independent solutions of the adjoint equation $K'\psi = 0$ depends only on the dominant part of the operator K and is equal to its index, i.e.*

$$n - n' = \varkappa. \tag{23.9}$$

Proof. Assume that $\varkappa \geqq 0$. Take as the regularizing operator $K^{\circ\prime}$. Then the Fredholm equation $K^{\circ\prime}K\varphi = 0$ is equivalent to the original equation $K\varphi = 0$ and consequently also has n solutions. In accordance with the second Fredholm theorem the adjoint equation

$$K'K^\circ \psi = 0 \tag{23.10}$$

7*

also has n solutions. The latter equation is equivalent (see § 22.4, 1°) to the dominant equation

$$K°\psi = \alpha_1\psi_1 + \cdots + \alpha_{n'}\psi_{n'}. \qquad (23.11)$$

Since $\varkappa \geqq 0$, according to the results of § 21.2 the latter equation is soluble for an arbitrary right-hand side, whence it follows that all constants α_i in it are arbitrary. Equation (23.11) has the same number of solutions (i.e. n) as (23.10). But it constitutes a non-homogeneous dominant equation with index $\varkappa \geqq 0$ and according to formula (21.12) its solution has the form

$$\psi(t) = \sum_{j=1}^{n'} \alpha_j R\psi_j + \sum_{j=1}^{\varkappa} c_j \varphi_j(t).$$

It is easy to prove that the $n' + \varkappa$ functions entering the right-hand side of the last relation are linearly independent. In fact, assume that a relation

$$\sum_{j=1}^{n'} \alpha_j R\psi_j + \sum_{j=1}^{\varkappa} c_j \varphi_j(t) \equiv 0$$

exists, in which at least one α_j is different from zero. This leads to a contradiction, since the left-hand side of this relation is the solution of the non-homogeneous equation $K°\varphi = \Sigma\, a_j\psi_j$, and consequently cannot vanish. The identity $\Sigma\, c_j\varphi_j(t) \equiv 0$ also can be satisfied only for all c_j equal to zero, since the functions $\varphi_j(t)$ are by definition linearly independent. †

Thus, equation (23.10) has $n' + \varkappa$ linearly independent solutions. Hence $n = n' + \varkappa$ which is identical with (23.9).

No special proof is required for the case $\varkappa < 0$. Since the property of adjointness of operators is mutual, for the original let us take the adjoint operator K' which has the index $\varkappa' = -\varkappa > 0$.

On the basis of the property proved above we have

$$n' - n = -\varkappa,$$

which again leads to relation (23.9). This completes the proof of the theorem.

The three theorems proved in this section are usually referred to as *Noether's theorems* (see § 26, Historical notes).

---

† The result may be formulated as follows: if the right-hand side of the equation contains a linear combination n' of linearly independent functions with arbitrary coefficients, whilst the corresponding homogeneous equation has $\varkappa$ linearly independent solutions, the inhomogeneous equation has $n' + \varkappa$ linearly independent solutions. One has frequently to use this result in various reasoning.

23.3. Some corollaries

Theorem III (Noether's theorem), which represents, as has already been emphasized, a most typical property of the singular integral equation with Cauchy kernel, enables us to deduce some important corollaries which are frequently used in applications of these equations.

1. All Fredholm theorems in the usual formulation (§ 20.2) do not hold for a singular equation and should be replaced by the above established Noether's theorems. However, the difference is not the same for all theorems. The second Fredholm theorem asserting the equality of the numbers of solutions of adjoint equations decidedly contradicts the third (Noether) theorem and the latter should be used. Now, the first and third Fredholm theorems cannot be extended to singular equations only, in view of the particular formulation they possess. This formulation can so be altered that the theorems hold also for singular equations. To this end the given homogeneous equation should be replaced by the adjoint equation in the formulation.

FREDHOLM THEOREMS I and III (combined and modified).

If the homogeneous equation adjoint to the given equation is insoluble, the non-homogeneous equation is absolutely soluble. If the adjoint homogeneous equation is soluble, the non-homogeneous equation is soluble only under the conditions

$$\int\limits_L f(t)\,\psi_k(t)\,dt = 0,$$

where $\psi_k(t)$ is a complete system of eigenfunctions of the adjoint operator.

If the index of the operator $\varkappa = 0$ the Noether Theorem III is identical with Fredholm theorem II. In this case the first and the third Fredholm theorems are suitable for the singular equation in the same formulation as for the Fredholm equation. Thus, *all Fredholm theorems hold for a singular equation with zero index*. This fact enabled N. I. Muskhelishvili ([17*], p. 236) to call singular integral equations *with zero index, quasi-Fredholm equations†*.

2. The Noether Theorem III leads to the following important corollary. *Among all singular equations having a given index $\varkappa$, the dominant equations have the smallest number of solutions.*

---

† In the general theory of singular operators constructed in functional analysis, the singular operator with zero index is reckoned among Fredholm operators.

In fact, the number of solutions of the dominant equation is exactly $\varkappa$ for $\varkappa > 0$ and zero for $\varkappa \leqq 0$. Since the number of solutions n' of the adjoint equation is non-negative, then

$$n = \varkappa + n' \geqq \varkappa,$$

where n is the number of solutions of the given singular equation.

Thus, it is impossible by a choice of the regular part of the kernel in the complete equation to diminish the number of its solutions in comparison with its dominant equation. Conversely, it is easy to prove that by a suitable choice of the regular part it is possible to ascertain that the complete equation has a given number of solutions greater than $\varkappa$ (see Example 15 at the end of this chapter).

It follows from the results proved above that *an operator K with a negative index has no regularizing operator K without eigenfunctions.*

This follows directly from the fact that the index of the regularizator has a different sign from the index of the original operator, and from the corollary established above.

§ 24. Equivalent regularization. The third method of regularization

24.1. Statement of the problem. Various interpretations of the concept of equivalent regularization

In § 22.4 it was indicated that the operation of regularization leads in general to an equation which is not equivalent to the original one. Both an appearance of superfluous solutions (for regularization from left) and a loss of solutions (for regularization from right) may occur. It is of great importance (both theoretical and practical) to establish *under what conditions and in what way a singular equation can be reduced to an equivalent Fredholm equation,* i.e. to an equation which contains all solutions of the original equation and also all solutions of which satisfy the original equation. This problem can be tackled from various points of view, and its statement requires a detailed specification.

We may demand the equivalence of the given and the regularized equations for an arbitrary right-hand side. In this case we are faced with an equivalent regularization of not a single equation with a definite right-hand side, but of a whole class of equations with a prescribed singular operator K. Thus in essence we solve here the problem of equivalent regularization of the operator K. This problem can also be stated for a definite equation with a given right-hand side. Here we already account for the properties of the right-hand side and we may find that an equation with the same operator admits an equivalent regularization for one right-hand side but not for another.

The solution of the stated problem essentially depends also on whether we demand that the regularized equation contains the same unknown function as the original one, or we allow construction of a regularized equation for a new function. In other words it depends on whether we necessarily demand a regularization from the left or allow also a regularization from the right.

Firstly we shall solve the simpler problem of an equivalent regularization of an operator.

24.2. Equivalent regularization of an operator

1. We allow only a regularization from the left.

LEMMA. *In order that an operator is an equivalent regularizator for the singular equation $\tilde{K}K\varphi = f$ for an arbitrary right-hand side, it is necessary and sufficient that it has no eigenfunctions.*

Proof. If the condition is satisfied, it follows at once from $\tilde{K}(K\varphi - f) = 0$ that $K\varphi - f = 0$ which establishes the sufficiency of the condition.

We now prove the necessity.

By assumption, there exists an operator $\tilde{K}$ which equivalently regularizes the equation $K\varphi = f$ for an arbitrary f. Let us assume contrary to the condition of the lemma that the operator $\tilde{K}$ has an eigenfunction ω and set $f = \omega$. Then, by assumption, the equations $K\varphi = \omega$ and $\tilde{K}K\varphi = \tilde{K}\omega = 0$ are equivalent, but the latter, and hence also the first equations, have non-zero solutions. Thus, it is necessary that $\omega \equiv 0$. This completes the proof.

It was proved in § 23 that an operator with a non-negative index has a regularizator having no eigenfunctions, and an operator with a negative index has no regularizator of this kind. Comparison of this result with the above proved lemma yields the following conclusion.

THEOREM. *In order that the singular integral equation* $K\varphi = f$ *has an equivalent regularization for an arbitrary right-hand side f, it is necessary and sufficient that the index of the operator* K *is non-negative.*

This theorem completely solves the problem of equivalent regularization of an operator from the left.

The above result indicates that the demand to apply only regularization from the left makes impossible an equivalent regularization of the whole class of operators having a negative index.

2. We now admit both regularizations, from the left and from the right. The regularization from the right will be regarded as equivalent, when to every solution φ of the original equation $K\varphi = f$ there corresponds in accordance with the substitution formula $\tilde{K}\omega = \varphi$ a definite solution ω of the regularized equation $K\tilde{K}\omega = f$, and conversely. It is easy to show that in this case all restrictions drop out and an equivalent regularization of a whole class of singular operators becomes possible.

All results necessary to justify this assertion were given in § 22.4; here we shall present only the final conclusions.

The operator $\tilde{K} = K'^{\circ}$ *is an equivalent regularizator for an arbitrary index*, and

when $\varkappa \geqq 0$ regularization from the left is to be applied,
when $\varkappa < 0$ regularization from the right should be used.

In the latter case we arrive at an equation for a new function ω, but if this function is known we can find all solutions of the original equation by quadrature, and in view of the properties of regularization from the right no superfluous solutions can arise. Thus, in the absence of limitations on the method of regularization and generalizing the concept of equivalence, any singular operator can be equivalently regularized. This fact has already been employed in proving the Noether theorems, without stating it as a special property.

24.3. Equivalent regularization from the left of the singular equation. The necessary condition

Let us examine the following practically important problem. A singular equation

$$K\varphi = f \qquad (24.1)$$

is given, the right-hand side being a prescribed function. It is necessary to find under what conditions this equation has a regularizator from the left $\tilde{K}$ which leads to an equivalent Fredholm equation

$$\tilde{K}K\varphi = \tilde{K}f. \tag{24.2}$$

It is also necessary to construct this operator, when it exists.

If $\varkappa \geq 0$, then we know that there exists an operator which has no eigenfunctions (for instance K'°). Evidently it is an equivalent regularizator from the left for any f. It is therefore of interest to examine the case $\varkappa < 0$, in which there exists no regularizator without eigenfunctions.

It is easy to find the necessary condition of existence of an equivalent regularizator for a given equation. It was established in §22.4, 1, that equation (24.2) is equivalent to the equation

$$K\varphi = f + \sum_{j=1}^{p} \alpha_j \omega_j, \tag{24.3}$$

where $\omega_1, \omega_2, \ldots, \omega_p$ is a complete system of eigenfunctions of the operator $\tilde{K}, \alpha_1, \ldots, \alpha_p$ are constants which can be either arbitrary or definite.

By definition, equation (24.2) is equivalent to the original equation (24.1) if and only if all α_j turn out to be zeros. Let us see how this can happen.

Let $\psi_1, \psi_2, \ldots, \psi_q$ be a complete system of eigenfunctions of the adjoint operator K'. Multiplying equation (24.3) by $\psi_1, \psi_2, \ldots, \psi_q$ and integrating over the contour L we obtain, taking into account the identities

$$\int_L K\varphi\psi_j \, dt = 0,$$

the system of equations

$$\sum_{j=1}^{p} a_{ij}\alpha_j = -f_i \quad (i = 1, 2, \ldots, q), \tag{24.4}$$

where

$$a_{ij} = \int_L \psi_i\omega_j \, dt, \quad f_i = \int_L f\psi_i \, dt.$$

Obviously, the system (24.4) may be satisfied by zero values of α_j only when all $f_i = 0$. But the latter relations are the conditions of solubility of equation (24.1). Hence we have the following result.

If the regularizing operator has non-vanishing eigenfunctions, then in order that the given singular integral equation (24.1) *could be reduced to the equivalent Fredholm equation* (24.2), *it is necessary that it is soluble.*

Is this condition sufficient? The answer is yes, but it is not easy to prove it. Auxiliary information is required and we shall now present it.

24.4. Conjugate equation. Another form of the conditions of solubility of the non-homogeneous equation

The concept of an adjoint operator introduced in § 20.1 is distinct from the analogous concept of a conjugate operator used in the theory of linear complex operators. In this connection the condition of solubility also took a different form. For instance, we could nowhere formulate it as a condition of orthogonality of the free term to the solution of the conjugate equation. Now we shall introduce the concept of a conjugate operator and we shall find a new form of the conditions of solubility of the non-homogeneous equation.

For the same reason as in § 23.1 we shall not separate out the singularity of the kernel.

Let K be a singular operator defined by the formula

$$K\varphi \equiv a(t)\,\varphi(t) + \int_L K(t,\,\tau)\,\varphi(\tau)\,d\tau \qquad (24.5)$$

and let

$$\tau = \tau(\sigma) \quad (0 \leqq \sigma \leqq l) \qquad (24.6)$$

be the equation of the contour L, σ being the arc coordinate. When σ takes the particular value s the corresponding complex coordinate t will be denoted by $t(s)$.

We shall denote functions of the real variables s, σ obtained by the substitution (24.6), by the same letters as the original functions of the complex variables t, τ. For instance

$$\varphi(t) = \varphi\,[t(s)] \sim \varphi(s), \quad K(t,\,\tau) = K[t(s),\,\tau(\sigma)] \sim K(s,\,\sigma), \quad \text{etc.}$$

Now the operator (24.5) can be written in the form

$$K\varphi \equiv a(s)\,\varphi(s) + \int_0^l K(s,\,\sigma)\,\tau'(\sigma)\,\varphi(\sigma)\,d\sigma.$$

The operator K* defined by the formula

$$K^*\psi^* \equiv \overline{a(s)}\,\psi^*(s) + \int_0^l \overline{K(\sigma,\,s)\,t'(s)}\,\psi^*(\sigma)\,d\sigma, \qquad (24.7)$$

will be called the operator *conjugate* to K.

The equations $K\varphi = 0$ and $K^*\varphi^* = 0$ will be called *conjugate* equations. It is evident that the conjugate operator satisfies the condition
$$(K^*)^* = K.$$

The expression
$$(\varphi, \psi) = \int_0^l \varphi(s)\,\overline{\psi(s)}\,ds \tag{24.8}$$

will be called the *scalar product* of the two functions φ and ψ. Obviously
$$(\psi, \varphi) = \int_0^l \psi(s)\,\overline{\varphi(s)}\,ds = \overline{(\varphi, \psi)}.$$

If $(\varphi, \psi) = 0$ the functions φ and ψ are said to be *orthogonal*.

As in § 23.1 we can prove the identity
$$(K\varphi, \psi) = (\varphi, K^*\psi). \tag{24.9}$$

We proceed to establish a connection between the solutions of the adjoint equation
$$K'\psi \equiv a(t)\,\psi(t) + \int_L K(\tau, t)\,\psi(\tau)\,d\tau = 0 \tag{24.10}$$

and the conjugate equation
$$K^*\psi^* \equiv \overline{a(s)}\,\psi^*(s) + \int_0^l \overline{K(\sigma, s)\,t'(s)}\,\psi^*(\sigma)\,d\sigma = 0. \tag{24.11}$$

Passing to the complex conjugate quantities and making use of notation of § 18.1 we obtain
$$\overline{K}^*\,\overline{\psi}^* \equiv t'(s)\left[a(t)\,\frac{\overline{\psi^*(t)}}{t'(s)} + \int_L K(\tau, t)\,\frac{\overline{\psi^*(\tau)}}{\tau'(s)}\,d\tau\right] = 0. \tag{24.11'}$$

A comparison of equations (24.10) and (24.11) indicates that the solutions of the adjoint and conjugate equations are related as follows:
$$\frac{\overline{\psi^*(t)}}{t'(s)} = \psi(t) \quad \text{or} \quad \overline{\psi^*(t)} = t'(s)\,\varphi(t). \tag{24.12}$$

Incidentally, this implies that the equations conjugate and adjoint to a given equation have the same number of solutions.

Making use of formulae (24.12) we can write the conditions of solubility of the non-homogeneous singular equation (24.1)
$$K\varphi = f$$
in a different form.

7 a*

We have (formula (23.4))

$$\int_L f(t)\,\psi_j(t)\,dt = 0, \quad \text{or} \quad \int_0^l f(t)\,\psi_j(t)\,t'(s)\,ds = 0.$$

Hence, taking into account (24.12) we obtain

$$(f,\psi_j^*) = \int_0^l f(t)\,\overline{\psi_j^*(t)}\,ds = 0 \quad (j = 1, 2, \ldots, k'). \quad (24.13)$$

The conditions of solubility (24.13) can be stated in the following form.

In order that the singular non-homogeneous equation (24.1) *be soluble, it is necessary and sufficient that the free term of the equation is orthogonal to all solutions of the conjugate homogeneous equation* (24.11).

We now derive one more auxiliary identity

$$(K^*K\omega, \omega) = (K\omega, K\omega) = \int_0^l |K\omega|^2\,ds. \quad (24.14)$$

Regarding the function $K^*K\omega$ as the result of application of the operator K^* to the function $K\omega$ we obtain the first part of the relation from identity (24.9), the second part following from the definition of the scalar product (24.8).

Let us now return to the problem of equivalent regularization of an equation.

24.5. Theorem on equivalent regularization of an equation

We shall need the following auxiliary result.

LEMMA. *Suppose that the non-homogeneous singular integral equation*

$$K\varphi = f \quad (24.1)$$

is soluble. Then it is equivalent to the equation

$$K^*K\varphi = K^*f. \quad (24.15)$$

By the assumption of the theorem, there exists a function $\varphi_1(t)$ which satisfies equation (24.1). In view of the homogeneity of the operator K^* it is also a solution of equation (24.15). Let us prove that any other function $\varphi_2(t)$ which satisfies equation (24.15), satisfies also equation (24.1).

We have

$$K^*K\varphi_1 = K^*f, \quad K^*K\varphi_2 = K^*f.$$

Subtracting the first relation from the second we find that the function $\omega = \varphi_2 - \varphi_1$ is a solution of the homogeneous equation

$$K^*K\omega = 0. \tag{24.16}$$

We now construct the scalar product of the functions $K^*K\omega$ and ω. Making use of relation (24.16) and identity (24.14) we obtain

$$0 = (K^*K\omega, \omega) = \int_0^l |K\omega|^2 ds.$$

Hence $K\omega = 0$ or

$$K\varphi_2 = K\varphi_1.$$

But $K\varphi_1 = f$; consequently

$$K\varphi_2 = f.$$

This completes the proof.

We are now in a position to deduce the principal assertion.

By a direct calculation it is easy to prove that in general the operator K^* is not a regularizing operator for the operators K (it is only so when the operator K is real). Consequently, equation (24.15) is not a Fredholm equation. The index of the conjugate operator K^* is identical with the index of the adjoint operator K' and, consequently, it is equal to the index of the given operator K, taken with negative sign. Since the index of the product of two operators is equal to the sum of the indices of the multiplied operators (§ 22.1), equation (24.15) is a singular integral equation with zero index. Now we know (§ 22.4) that such an equation has a regularizator without eigenfunctions, for instance the operator $(K^*K)'^\circ$.

Thus, we have proved the following principal theorem on equivalent regularization of an equation.

THEOREM. *If the right-hand side of the equation*

$$K\varphi = f \tag{24.1}$$

is such that this equation is soluble, then there exists a regularizing operator from the left, which reduces it to an equivalent Fredholm equation.

According to the proof given above, for such an equivalent regularizator we may take the operator

$$\tilde{K} = (K^*K)'^\circ K^*, \tag{24.17}$$

where K* is the conjugate operator defined by formula (24.4) and $(K^*K)'^\circ$ is the dominant operator defined in terms of the operator K*K by formula (20.6).

Consequently, although in the case of the absence of a regularizator without eigenfunctions ($\varkappa < 0$) there does not exist (as it was shown in § 24.2) an operator accomplishing the regularization for an arbitrary right-hand side, if the free term f satisfies the condition of solubility of the singular integral equation $K\varphi = f$ there exists an equivalent regularizing operator which can effectively be constructed. It is remarkable that the form of this operator is independent of the right-hand side, so that for all soluble equations with a given operator K the same equivalent regularizator can be taken.

24.6. Regularization by solving the dominant equation (the method of Carleman–Vekua)

In the above theory of singular equations we have only made use of regularization based on a product of singular operators; the third method of regularization mentioned in § 22.3, consisting in using the explicit solution of the dominant equation, has so far remained unused.

We now proceed to expound this method, which is important from the standpoint of theory and applications.

Taking the regular term of the singular equation to the right-hand side we have

$$a(t)\varphi(t) + \frac{b(t)}{\pi i} \int\limits_L \frac{\varphi(\tau)}{\tau - t}\, d\tau = f(t) - \int\limits_L k(t, \tau)\, \varphi(\tau)\, d\tau \quad (24.18)$$

or, symbolically,

$$K^\circ \varphi = f - k\varphi. \quad (24.18')$$

Let us solve the latter equation as dominant, regarding for the time being the right-hand side as a known function†. Applying

---

† The concept appearing in the last method of regularization is frequently employed in the theory of equations of mathematical physics, for reducing a boundary value problem to an integral equation (see for instance the reduction to an integral equation of the Sturm–Liouville problem, or the reduction to an integral or integro-differential equation of the boundary value problem for an equation of elliptic type).

formula (21.9) we obtain

$$\varphi(t) = \left[a(t)f(t) - \frac{b(t)Z(t)}{\pi i} \int_L \frac{f(\tau)}{Z(\tau)} \frac{d\tau}{\tau - t} + a(t)Z(t)P_{\varkappa-1}(t) \right] -$$

$$- \left[a(t) \int_L k(t, \tau)\varphi(\tau)\,d\tau - \frac{b(t)Z(t)}{\pi i} \int_L \frac{d\tau_1}{Z(\tau_1)(\tau_1 - t)} \times \right.$$

$$\left. \times \int_L k(\tau_1, \tau)\varphi(\tau)\,d\tau \right]; \quad (24.19)$$

for $\varkappa \leqq 0$ we set $P_{\varkappa-1}(t) \equiv 0$.

Changing the order of integration in the double integral we can write the term in the last brackets in the form

$$\int_L \left[a(t)k(t, \tau) - \frac{b(t)Z(t)}{\pi i} \int_L \frac{k(\tau_1, \tau)}{Z(\tau_1)(\tau_1 - t)}\,d\tau_1 \right] \varphi(\tau)\,d\tau.$$

Since $Z(t)$ satisfies the Hölder condition (and consequently it is bounded) and $k(\tau_1, \tau)$ has near $\tau_1 = \tau$ the estimate $|k(\tau_1, \tau)| < c/|\tau_1 - \tau|$, it is readily observed that also the whole integral

$$\int_L \frac{k(\tau_1, \tau)}{Z(\tau_1)(\tau_1 - t)}\,d\tau_1$$

obeys the same estimate as $k(\tau_1, \tau)$. Consequently, the kernel

$$K(t, \tau) = R_t k(t, \tau) = a(t)k(t, \tau) - \frac{b(t)Z(t)}{\pi i} \int_L \frac{k(\tau_1, \tau)}{Z(\tau_1)(\tau - t)}\,d\tau_1$$

$$(24.20)$$

is a Fredholm kernel. Taking all terms containing $\varphi(t)$ to the left-hand side we obtain

$$\varphi(t) + \int_L K(t, \tau)\varphi(\tau)\,d\tau = f_1(t), \quad (24.21)$$

where $K(t, \tau)$ is a Fredholm kernel defined by formula (24.20), and $f_1(t)$ is the free term which has the form

$$f_1(t) = a(t)f(t) - \frac{b(t)Z(t)}{\pi i} \int_L \frac{f(\tau)}{Z(\tau)} \frac{d\tau}{\tau - t} + b(t)P_{\varkappa-1}(t). \quad (24.22)$$

If the index of equation (24.19) $\varkappa < 0$, then according to § 21.2, formula (21.9), in which we have to set $P_{\varkappa-1}(t) \equiv 0$, holds only if the conditions (21.13) are satisfied. In our case in the latter condition $f(t)$ should be replaced by

$$f(t) - \int_L k(t, \tau)\, \varphi(\tau)\, d\tau.$$

Consequently, for $\varkappa < 0$ besides the Fredholm equation (24.21), the function has to satisfy the relations

$$\int_L \left[\int_L \frac{k(t, \tau)}{Z(\tau)}\, t^{k-1}\, dt \right] \varphi(\tau)\, d\tau = \int_L \frac{f(t)}{Z(t)}\, t^{k-1}\, dt$$

$$(k = 1, 2, \ldots, -\varkappa). \qquad (24.23)$$

Let us state the derived result.

If $\varkappa \geq 0$ the solution of the complete singular integral equation (24.18) can be reduced to solving the Fredholm integral equation (24.21). If $\varkappa < 0$ equation (24.19) can be reduced to equation (24.21) (in which we have to set $P_{\varkappa-1}(t) \equiv 0$) completed by the functional conditions (24.23). The latter conditions can be written in the form

$$\int_L \varrho_k(\tau)\, \varphi(\tau)\, d\tau = f_k \quad (k = 1, 2, \ldots, -\varkappa), \qquad (24.24)$$

where

$$\varrho_\varkappa(t) = \int_L \frac{k(t, \tau)}{Z(t)}\, t^{k-1}\, dt$$

are known functions and

$$f_k = \int_L \frac{f(t)}{Z(t)}\, t^{k-1}\, dt$$

are known numbers.

Conditions (24.24) are fulfilled if the original singular equation is soluble. The original singular equation (24.18) is satisfied only by these solutions of the regular equation for which relations (24.24) hold (see the example in § 24.7).

Problems of equivalence of the considered method of regularization can be solved comparatively easily. Omitting the details let us state these results.

For $\varkappa \geq 0$ the regularized equation (24.21) is equivalent to the original singular equation. For $\varkappa < 0$ the singular equation is equivalent to the regularized equation completed by the additional conditions (24.24). In the latter case, having solved the regularized equation, it is necessary to substitute the resulting solutions into relations (24.24). Only those solutions of the regularized equation are solutions of the original one, which satisfy the above indicated conditions (see the example in § 24.7). If conditions (24.24) are not satisfied for any of the solutions, this means that the given singular equation is insoluble.

Observe that on the basis of the above method of regularization it is possible to construct the entire theory of singular integral equations. For the details the reader is referred to a paper of I. N. Vekua (2).

24.7. Example

As an example, to illustrate the above theory, we shall perform by various methods the regularization of the singular integral equation

$$K_\varphi \equiv (t + t^{-1})\,\varphi(t) + \frac{t - t^{-1}}{\pi i} \int_L \frac{\varphi(\tau)}{\tau - t}\,d\tau -$$

$$- \frac{1}{2\pi i} \int_L (t + t^{-1})\,(\tau + \tau^{-1})\,\varphi(\tau)\,d\tau = 2t^2, \qquad (24.25)$$

where L is the unit circle.

The regular part of the kernel is degenerate, whence it follows that we can reduce the equation to the union of the dominant equation and a linear algebraic equation, by the same method which is used in solving Fredholm equations with degenerate kernels. Consequently, our equation can be solved in a closed form. Thus, there is no necessity for regularization. However the considered equation is convenient for an illustration of the general methods, since all computations can be completed.

For convenience of the following we first solve the given equation. Denoting

$$\frac{1}{2\pi i} \int_L (\tau + \tau^{-1})\,\varphi(\tau)\,d\tau = A, \qquad (24.26)$$

we write it in the form of a dominant equation

$$(t + t^{-1})\,\varphi(t) + \frac{t - t^{-1}}{\pi i} \int_L \frac{\varphi(\tau)}{\tau - t}\,d\tau = 2t^2 + A(t + t^{-1}).$$

The index of the corresponding Riemann boundary value problem

$$\Phi^+(t) = t^{-2}\,\Phi^-(t) + t + \frac{A}{2}\,(1 + t^{-2}),\qquad(24.27)$$

is $\varkappa = -2$ and the conditions of solubility (see § 21.2) are satisfied only when $A = 0$. We also have then

$$\Phi^+(z) = z,\quad \Phi^-(z) = 0.$$

This yields the solution of equation (24.25).

$$\varphi(t) = \Phi^+(t) - \Phi^-(t) = t.$$

Substituting the latter expression into relation (24.26) we find that it is satisfied when $A = 0$. Consequently, the equation under consideration is soluble and it has the unique solution

$$\varphi(t) = t.$$

1. *Regularization from the left.* It follows from the boundary condition (24.27) that the index of the equation $\varkappa = -2 < 0$. Consequently, any regularizing operator has eigenfunctions (not less than two) and in general a regularization from the left leads to an equation which is not equivalent to the original one.

Consider first regularization from the left by means of the simplest operator K'°. Let us find the eigenfunctions of the equation.

$$K'^\circ\,\omega \equiv (t + t^{-1})\,\omega(t) - \frac{t - t^{-1}}{\pi i}\int\limits_L \frac{\omega(\tau)}{\tau - t}\,d\tau = 0.$$

The corresponding Riemann boundary value problem

$$\Phi^+(t) = t^2\,\Phi^-(t),$$

has the index $\varkappa = 2$. Finding in accordance with the formulae of § 21.2 the eigenfunctions of the operator K'° we have

$$\omega_1(t) = 1 - t^{-2},\quad \omega_2(t) = t - t^{-1}.$$

It follows from the general theory (§ 22.4, 1°) that the regular equation $K'^\circ K\varphi = K'^\circ f$ is equivalent to the singular equation

$$K\varphi = f + \alpha_1\,\omega_1 + \alpha_2\,\omega_1,\qquad(24.28)$$

where α_1, α_2 are constants which may, or may not, be arbitrary. Taking into account (24.26) we write equation (24.28) in the form of a dominant equation

$$(t + t^{-1})\,\varphi(t) + \frac{t - t^{-1}}{\pi i} \int_L \frac{\varphi(\tau)}{\tau - t}\, d\tau = 2t^2 + A(t + t^{-1}) +$$
$$+ \alpha_1(1 - t^{-2}) + \alpha_2(t - t^{-1}).$$

The corresponding Riemann boundary value problem has the form

$$\Phi^+(t) = t^{-2}\,\Phi^-(t) + t + \frac{A}{2}(1 + t^{-2}) +$$
$$+ \frac{\alpha_1}{2}(t^{-1} - t^{-3}) + \frac{\alpha_2}{2}(1 - t^{-2}).$$

Its solution can formally be represented as follows:

$$\Phi^+(z) = z + \tfrac{1}{2}A + \tfrac{1}{2}\alpha_2,$$
$$\Phi^-(z) = \tfrac{1}{2}z^2[\alpha_1 z^{-3} + (\alpha_2 - A)z^{-2} - \alpha_1 z^{-1}].$$

The conditions of solubility yield the relations

$$\alpha_1 = 0, \quad \alpha_2 = A.$$

Then the solution of equation (24.28) is given by the relation

$$\varphi(t) = \Phi^+(t) - \Phi^-(t) = t + A.$$

Substituting the obtained value of φ into relation (24.26) we arrive at the identity $A = A$. Consequently, the constant $\alpha_2 = A$ is arbitrary and the regularized equation is equivalent not to the original equation but to the equation $\quad K\varphi = f + \alpha_2\,\omega_2,$

which has the solution $\varphi(t) = t + A$ where A is an arbitrary constant. The latter solution satisfies the original equation only when $A = 0$.

We now proceed to find the equivalent regularizator from the left. Since the considered equation is soluble, it has such a regularizator, which can effectively be constructed in accordance with the theorem of § 24.5.

First we construct the conjugate operator K^*. By definition

$$K^*\,\psi^* \equiv (\bar{t} + \bar{t}^{-1})\,\psi^*(t) - \frac{1}{\pi i}\int_L (\bar{\tau} - \bar{\tau}^{-1})\frac{\varphi^*(\tau)}{\bar{\tau} - \bar{t}}\,\bar{t}'\,d\sigma +$$
$$+ \frac{1}{2\pi i}\int_L (\bar{\tau} - \bar{\tau}^{-1})(\bar{t} + \bar{t}^{-1})\,\psi^*(\tau)\,\bar{t}'\,d\sigma.$$

Since

$$\bar{t} = t^{-1}, \quad \bar{\tau} = \tau^{-1}, \quad \bar{t}' = -\,it^{-1}, \quad d\sigma = (i\tau)^{-1}\,d\tau,$$

we obtain

$$K^* \psi^* \equiv (t + t^{-1}) \psi^*(t) - \frac{1}{\pi i} \int_L (\tau - \tau^{-1}) \frac{\psi^*(\tau)}{\tau - t} d\tau -$$

$$- \frac{1}{2\pi i} \int_L (1 + t^{-2})(1 + \tau^{-2}) \psi^*(\tau) d\tau. \quad (24.29)$$

By virtue of the lemma of § 24.5 the equation $K^*K\varphi = K^*f$ is equivalent to the original one, but in general it is not regular. The significance of the transformation by means of the operator K^* consists in the general case in reducing the given soluble singular equation to an equivalent singular equation with zero index; the latter is then regularized. In the case under consideration, however, the dominant part of the operator K^* is identical with the operator K'° which, as is known, is a regularizator for the operator K. Hence the equation $K^*K\varphi = K^*f$ is a Fredholm equation.

Performing all the calculations (the reader is advised to actually perform them, making use of the general scheme of § 24.3), we obtain the Fredholm equation

$$\varphi(t) + \frac{1}{4\pi i} \int_L (\tau t^{-2} - \tau^{-1}) \varphi(\tau) d\tau = t. \quad (24.30)$$

The equivalence of the latter equation to the original one follows from the lemma of § 24.5 and therefore does not require a special justification. However, let us indicate, in view of other analogous cases, in which such a justification may be required, the method of carrying it out, without using the above mentioned lemma.

Since equation (24.30) has a degenerate kernel, the simplest way is to solve it. It turns out that similarly to the original singular equation, this equation has the unique solution $\varphi(t) = t$. A more general way consists in making use of the method by which we established before the non-equivalence of the equation $K'^\circ K\varphi = K'^\circ f$ to the original one. Let us outline the investigation.

Solving the equation $K^*\psi^* = 0$ we find the eigenfunctions of the operator K^*

$$\psi_1^* = t + t^{-1} + t^{-3}, \quad \psi_2^* = 1.$$

Next, solving the singular equation with degenerate regular kernel

$$K\varphi = f + \alpha_1 \omega_1^* + \alpha_2 \omega_2^*,$$

which is equivalent to the Fredholm equation (24.30) we find that it is soluble only when $\alpha_1 = \alpha_2 = 0$. This completes the proof of equivalence.

2. *Regularization from the right.* For the regularizator from the right we take the simplest operator K'°. Setting

$$\varphi(t) = K'^\circ \omega \equiv (t + t^{-1}) \omega(t) - \frac{t - t^{-1}}{\pi i} \int_L \frac{\omega(\tau)}{\tau - t} d\tau, \quad (24.31)$$

we obtain (the reader is advised to perform the required transformations for himself) the Fredholm equation for the function $\omega(t)$:

$$K K'^{\circ} \omega \equiv \omega(t) - \frac{1}{4\pi i} \int_L [t(\tau^2 - 1 + \tau^{-2}) +$$

$$+ 2\tau^{-1} + t^{-1}(\tau^2 + 3 + \tau^{-2}) - 2t^{-2}\tau^{-1}] \omega(\tau) d\tau = \frac{t^2}{2}. \qquad (24.32)$$

Solving this equation as degenerate, we have

$$\omega(t) = \frac{t^2}{2} + \alpha(t - t^{-1}) + \beta(1 - t^{-2}),$$

where α, β are arbitrary constants.

Thus, the regularized equation has with respect to $\omega(t)$ two linearly independent solutions, since the original equation (24.25) had a unique solution. Substituting the obtained value of $\omega(t)$ into formula (24.31) we obtain

$$\varphi(t) = K'^{\circ}\left[\frac{t^2}{2} - \alpha(t + t^{-1}) + \beta(1 - t^{-2})\right] = t$$

which constitutes the solution of the original singular equation. This result is in accordance with the general theory, since for a negative index, the regularization from the right is equivalent.

3. *Regularization by solving the dominant equation.* This method of regularization is carried out in accordance with formulae (24.19)–(24.22). It should be borne in mind that these formulae are valid for equations for which the condition $a^2 - b^2 = 1$ is satisfied; hence, it is necessary to first divide equation (24.25) by two.

Then we have

$$a = \tfrac{1}{2}(t + t^{-1}), \quad b = \tfrac{1}{2}(t - t^{-1}), \quad f(t) = t^2,$$

$$K(t, \tau) = -\frac{1}{4\pi i}(t + t^{-1})(\tau + \tau^{-1}),$$

$$Z(t) = (a + b)X^+ = t,$$

$$Rf = \frac{1}{2}(t + t^{-1})t^2 - \frac{(t - t^{-1})t}{2\pi i}\int_L \frac{\tau^2}{\tau}\frac{d\tau}{\tau - t} = t,$$

$$R_t K(t, \tau) = -\frac{1}{2}(t + t^{-1})\frac{1}{4\pi i}(t + t^{-1})(\tau + \tau^{-1}) +$$

$$+ \frac{(t - t^{-1})t(\tau + \tau^{-1})}{2\pi i \cdot 4\pi i}\int_L \frac{\tau_1 + \tau_1^{-1}}{\tau_1}\frac{d\tau_1}{\tau_1 - t} = -\frac{1}{2\pi i}(\tau + \tau^{-1}).$$

The regularized equation has the form

$$\varphi(t) - \frac{1}{2\pi i} \int_L (\tau + \tau^{-1})\, \varphi(\tau)\, d\tau = t. \tag{24.33}$$

It should be completed by the conditions of the form (24.24) for $k = 1, 2$.

Solving equation (24.33) as degenerate we find $\varphi(t) = t + A$ where A is an arbitrary constant. Thus, the regularized equation is not equivalent to the original one. Conditions (24.24) enable us to separate out from the entire set of solutions of the regularized equation those which are solutions of the original equation. In our case these conditions have the form

$$-\frac{1}{4\pi i} \int_L \left[\int_L \frac{(t + t^{-1})(\tau + \tau^{-1})}{t}\, dt \right] (\tau + A)\, d\tau = \int_L \frac{t^2}{t}\, dt,$$

$$-\frac{1}{4\pi i} \int_L \left[\int_L \frac{(t + t^{+1})(\tau + \tau^{-1})}{t}\, t\, dt \right] (\tau + A)\, d\tau = \int_L \frac{t^2}{t}\, t\, dt.$$

The first equation is satisfied identically, while the second is satisfied for $A = 0$. Hence we arrive at the solution of the original equation $\varphi(t) = t$.

The above equation was used by I. N. Vekua (2), for illustration of the method of regularization by solving the dominant equation.

§ 25. Exceptional cases of singular integral equations

In investigating singular integral equations we excluded the cases in which the functions $a(t) \pm b(t)$ vanish on the contour L. This was due to the fact that the coefficient $G(t) = \dfrac{a(t) - b(t)}{a(t) + b(t)}$ of the Riemann problem to which the dominant singular equation is reduced, has in these cases zeros and poles on the contour, and, consequently, a problem of this kind cannot be tackled by the general theory. Relying on the results of § 15 we shall carry out a special investigation of these exceptional cases.

We assume here and in the next section that the coefficients of the investigated singular equations are such that the additional requirement of differentiability, which was stated in § 15, is satisfied in solving the exceptional cases of the Riemann problem.

25.1. Solution of the dominant equation

Consider the dominant equation (21.1)

$$K^\circ \varphi \equiv a(t)\,\varphi(t) + \frac{b(t)}{\pi i} \int_L \frac{\varphi(\tau)\,d\tau}{\pi i} = f(t) \qquad (25.1)$$

assuming that the functions $a(t) - b(t)$ and $a(t) + b(t)$ have zeros on the contour at the points $\alpha_1, \alpha_2, \ldots, \alpha_\mu$ and $\beta_1, \beta_2, \ldots, \beta_\nu$ respectively, of integral orders, and, consequently, are representable in the form

$$a(t) - b(t) = \prod_{k=1}^{\mu} (t - \alpha_k)^{m_k}\, r(t),$$

$$a(t) + b(t) = \prod_{j=1}^{\nu} (t - \beta_j)^{p_j}\, s(t),$$

where $r(t)$ and $s(t)$ do not vanish anywhere.

It is convenient do divide equation (25.1) throughout by $\sqrt{[s(t)\,r(t)]}$ and assume hereafter that its coefficients satisfy the condition

$$a^2(t) - b^2(t) = \prod_{k=1}^{\mu} (t - \alpha_k)^{m_k} \prod_{j=1}^{\nu} (t - \beta_j)^{p_j} = \Pi_0. \qquad (25.2)$$

In the same way as in § 21.1 equation (25.1) is reduced to the Riemann problem

$$\Phi^+(t) = \frac{\prod\limits_{k=1}^{\mu} (t - \alpha_k)^{m_k}}{\prod\limits_{j=1}^{\nu} (t - \beta_j)^{p_j}}\, G_1(t)\, \Phi^-(t) + \frac{f(t)}{\prod\limits_{j=1}^{\nu} (t - \beta_j)^{p_j}\, s(t)}, \qquad (25.3)$$

where $G_1(t) = \dfrac{r(t)}{s(t)}$. The solution of this problem, in the class of functions which satisfy the condition $\Phi(\infty) = 0$, is given by formulae (15.7) and (15.8). Applying them we obtain

$$\left.\begin{aligned}
\Phi^+(z) &= \frac{X^+(z)}{\prod\limits_{j=1}^{\mu} (z - \beta_j)^{p_j}} [\Psi^+(z) - Q_\varrho(z) + \Pi_0 P_{\varkappa - p - 1}(z)], \\[2ex]
\Phi^-(z) &= \frac{X^-(z)}{\prod\limits_{k=1}^{\mu} (z - \alpha_k)^{m_k}} [\Psi^-(z) - Q_\varrho(z) + \Pi_0 P_{\varkappa - p - 1}(z)],
\end{aligned}\right\} \qquad (25.4)$$

where
$$\Psi(z) = \frac{1}{2\pi i} \int\limits_L \frac{f(\tau)}{s(\tau)\, X^+(\tau)}\, \frac{d\tau}{\tau - z}, \tag{25.5}$$

and $Q_\varrho(z)$ is the interpolation polynomial for the function $\Psi(z)$ of degree $\varrho = m + p - 1$ with nodes at the points $\varkappa_k$ and β_j respectively of multiplicities m_k and p_j, where $m = \Sigma m_k$ and $p = \Sigma p_j$.

Let us regard the polynomial $Q_\varrho(z)$ as an operator associating with the free term $f(t)$ of equation (25.1) a polynomial interpolating in the indicated sense the integral of Cauchy type (25.5). Denote this operator by
$$\tfrac{1}{2} Tf = Q_\varrho(z). \tag{25.6}$$

The coefficient $\tfrac{1}{2}$ is taken for convenience in further transformations.

Proceeding as in § 21.2 we obtain from (25.4)

$$\Phi^+(t) = \frac{X^+(z)}{\prod\limits_{k=1}^{\mu} (t - \beta_k)^{p_k}} \left[\frac{1}{2}\, \frac{f(t)}{s(t)\, X^+(t)} + \frac{1}{2\pi i} \int\limits_L \frac{f(\tau)}{s(\tau)\, X^+(\tau)}\, \frac{d(\tau)}{\tau - t} - \right.$$
$$\left. - \frac{1}{2} Tf(t) - \frac{1}{2}\, \Pi_0 P_{\varkappa - p - 1}(t) \right],$$

$$\Phi^-(t) = \frac{X^-(t)}{\prod\limits_{j=1}^{\nu} (t - \alpha_j)^{m_j}} \left[-\frac{1}{2}\, \frac{f(t)}{s(t)\, X^+(t)} + \frac{1}{2\pi i} \int\limits_L \frac{f(\tau)}{s(\tau)\, X^+(\tau)}\, \frac{d\overset{\circ}{\tau}}{\tau - t} - \right.$$
$$\left. - \frac{1}{2} Tf(t) - \frac{1}{2}\, \Pi_0 P_{\varkappa - p - 1}(t) \right].$$

The coefficient $-\tfrac{1}{2}$ in the last terms of these formulae can be introduced in view of the arbitrariness of the coefficients of the polynomial $P_{\varkappa - p - 1}(t)$.

It follows from these formulae that
$$\varphi(t) = \Phi^+(t) - \Phi^-(t)$$
$$= \frac{1}{2} \left(\frac{X^+(t)}{\prod\limits_{j=1}^{\nu} (t - \beta_j)^{p_j}} + \frac{X^-(t)}{\prod\limits_{k=1}^{\mu} (t - \alpha_k)^{m_k}} \right) \frac{f(t)}{s(t)\, X^+(t)} +$$
$$+ \frac{1}{2} \left(\frac{X^+(t)}{\prod\limits_{j=1}^{\nu} (t - \beta_j)^{p_j}} - \frac{X^-(t)}{\prod\limits_{k=1}^{\mu} (t - \alpha_k)^{m_k}} \right) \times$$
$$\times \left(\frac{1}{\pi i} \int\limits_L \frac{f(\tau)}{s(\tau)\, X^+(\tau)}\, \frac{d\tau}{\tau - t} - Tf - \Pi_0 P_{\varkappa - p - 1}(t) \right). \tag{25.7}$$

Introducing the notation

$$Z(t) = s(t) X^+(t) = r(t) X^-(t). \tag{25.8}$$

and making use of relation (25.2) we can represent formula (25.7) in the form

$$\varphi(t) = \frac{1}{\Pi_0}\left[a(t)f(t) - \frac{b(t) Z(t)}{\pi i} \int_L \frac{f(\tau)}{Z(\tau)} \frac{d\tau}{\tau - t} + b(t) Z(t) Tf\right] +$$
$$+ b(t) Z(t) P_{\varkappa - p - 1}(t).$$

Introducing the operator

$$Rf \equiv \frac{1}{\Pi_0}\left[a(t)f(t) - \frac{b(t) Z(t)}{\pi i} \int \frac{f(\tau)}{Z(\tau)} \frac{d\tau}{\tau - t} + b(t) Z(t) Tf\right], \tag{25.9}$$

we finally obtain

$$\varphi(t) = Rf + b(t) Z(t) P_{\varkappa - p - 1}(t). \tag{25.10}$$

Formula (25.10) represents the solution of equation (25.1) when $\varkappa - p > 0$, which depends linearly on $\varkappa - p - 1$ arbitrary constants. When $\varkappa - p < 0$ the solution exists only if $p - \varkappa + 1$ special conditions of solubility are satisfied, which constitute a restriction of the free term $f(t)$, and follow from the conditions of solubility of the Riemann problem (25.3) corresponding to the considered case. We shall not state these conditions explicitly.

25.2. Regularization of the complete equation

Consider the complete singular equation

$$K \varphi \equiv a(t)\varphi(t) + \frac{b(t)}{\pi i} \int_L \frac{\varphi(\tau)}{\tau - t} d\tau + \int_L k(t, \tau) \varphi(\tau) d\tau = f(t) \tag{25.11}$$

under the same assumptions with respect to the functions $a(t) \pm b(t)$ as in the preceding section.

Let us represent it in the form

$$K^\circ \varphi = f(t) - \int_L k(t, \tau)\varphi(\tau) d\tau,$$

and let us apply the method of regularization by solving the dominant equation. According to formula (25.10) we have

$$\varphi(t) = R\left[f(t) - \int_L k(t, \tau) \varphi(\tau) d\tau\right] + b(t) Z(t) P_{\varkappa - p - 1}(t),$$

or

$$\varphi(t) + R\left[\int_L k(t, \tau)\,\varphi(\tau)\,d\tau\right] = Rf + b(t)\,Z(t)\,P_{\varkappa - p - 1}(t). \quad (25.12)$$

We shall prove that in the expression $R\left[\int_L k(t,\tau)\,\varphi(\tau)\,d\tau\right]$ the operation R with respect to the variable t and the operation of integration with respect to τ, are commutative.

Transform this equation by means of formula (25.9)

$$R\left[\int_L k(t, \tau)\,\varphi(\tau)\,d\tau\right] = \frac{1}{\Pi_0}\left\{a(t)\int_L k(t, \tau)\,\varphi(\tau)\,d\tau - \right.$$

$$- \frac{b(t)\,Z(t)}{\pi i}\int_L \frac{d\tau}{Z(\tau)(\tau - t)}\int_L k(\tau, \sigma)\,\varphi(\sigma)\,d\sigma +$$

$$\left. + b(t)\,Z(t)\,T\left[\int_L k(t, \tau)\,\varphi(\tau)\,d\tau\right]\right\}. \quad (25.13)$$

In the second term the change of the integration order is permissible (§ 7.1). Performing it and changing the notations of the integration variables, τ into σ and σ into τ, we obtain

$$\frac{b(t)\,Z(t)}{\pi i}\int_L \frac{d\tau}{Z(\tau)(\tau - t)}\int_L k(\tau, \sigma)\,\varphi(\sigma)\,d\sigma$$

$$= \int_L \varphi(\tau)\,d\tau\,\frac{b(t)\,Z(t)}{\pi i}\int_L \frac{k(\sigma, \tau)}{Z(\sigma)(\sigma - t)}\,d\sigma.$$

In the third term, the operator $T\left[\int_L k(t,\tau)\,\varphi(\tau)\,d\tau\right]$, in accordance with its definition in § 25.1 (formulae (25.5), (25.6)), yields the polynomial interpolating the function

$$\frac{1}{\pi i}\int_L \frac{d\tau}{Z(\tau)(\tau - z)}\int_L k(\tau, \sigma)\,\varphi(\sigma)\,d\sigma = \int_L \varphi(\tau)\,d\tau\,\frac{1}{\pi i}\int_L \frac{k(\sigma, \tau)\,d\sigma}{Z(\sigma)(\sigma - z)}.$$

It follows from the possibility of this change of the integration order that

$$T\left[\int_L k(t, \tau)\,\varphi(\tau)\,d\tau\right] = \int_L \varphi(\tau)\,T_t[k(t, \tau)]\,d\tau.$$

The index t on the symbol of the operator T_t indicates that the operation (25.6) is carried out with respect to the variable t.

Now relation (25.13) can be transformed as follows:

$$R\left[\int_L k(t,\tau)\,\varphi(\tau)\,d\tau\right]$$

$$= \int_L \varphi(\tau)\,\frac{1}{\Pi_0}\left\{a(t)\,k(t,\tau) - \frac{b(t)\,Z(t)}{\pi i}\int_L \frac{k(\sigma,\tau)\,d\sigma}{Z(\sigma)\,(\sigma-t)} + \right.$$

$$\left. + b(t)\,Z(t)\,T_t[k(t,\tau)]\right\}d\tau = \int_L R_t[k(t,\tau)]\,\varphi(\tau)\,d\tau.$$

The commutativity of the operations has thus been proved.

Returning to equation (25.12) we represent it in the form

$$\varphi(t) + \int_L R_t[k(t,\tau)]\,\varphi(\tau)\,d\tau = Rf + b(t)\,Z(t)\,P_{\varkappa-p-1}(t). \qquad (25.14)$$

Since the operator R is bounded the derived integral equation (25.14) is a Fredholm equation. Consequently, the problem of regularization of the singular equation (25.11) is solved.

It follows from the general theory of regularization that when $\varkappa - p \geqq 0$ equation (25.14) is equivalent to equation (25.11), and when $\varkappa - p < 0$ to equation (25.14) and a system of certain functional equations.

In conclusion we observe that for the above cases of singular integral equations the Noether theorems are not in general valid.

§ 26. Historical notes

The theory of singular integral equations with the integral in the sense of its principal value was originated almost simultaenously with the theory of Fredholm equations. In the first papers (Poincaré [20], Hilbert (1), in 1903–1904) integral equations with the kernel $\cot\dfrac{\sigma-s}{2}$ were examined (see § 31). In the above paper of Hilbert, errors were made, which were considered in a paper by the present author [2]. The general theory of singular equations was first elaborated in Noether's paper (1), in 1921 $\left(\text{also with the kernel } \cot\dfrac{\sigma-s}{2}\right)$. She established the general properties of singular equations, which are now called the Noether theorems. It should be observed that the methods used by Noether (solution of the Hilbert problem, see Chapter IV) do not make it possible to deduce these theorems for complex equations with the Cauchy kernel. Noether's paper is written in a very heavy style and is difficult to understand.

Soon after this a paper of Carleman (1) appeared, which has already bee mentioned in the preceding chapter. He presented a solution of the dominar equation with the Cauchy kernel, in the particular case when the coefficients a constant and the contour is a segment of the x-axis [0, 1]. Moreover, the metho of regularization by solving the dominant equation was given. A general theor of singular equations was not dealt with in Carleman's paper.

After the appearance of the author's paper in 1937, (1), the solution of the Rie mann boundary value problem was applied to investigating singular equations The first steps were taken by S. G. Mikhlin in 1939, (2).

Later, in 1940 I. N. Vekua in a paper (1) presented a complete solution of th dominant equation and the foundations of the general theory of singular equations This theory was developed by him in a number of works following shortly after wards (1941–1943), the most important being (2) and (3). He elaborated particularl thoroughly the method of regularization by solving the dominant equation (2 (N. I. Muskhelishvili named it the method of Carleman–Vekua). Vekua was als the first to introduce the method of regularization from the right, which also wa made use of in developing the general theory. A simple proof of the seconc Noether theorem given in § 23.2 is due to I. N. Vekua.

At the same time (from 1941) a number of papers of V. D. Kupradze were published, the most important being (1) and (2), in which a general theory o singular equations was worked out. Making use only of the classical method o regularization—regularization from the left—he presented a few variants of proofs of the basic properties of the singular equations, i.e. of the Noether theorems and the theorems on equivalent regularization. The proof of Theorem III of Noether exposed in § 23.2 is due to V. D. Kupradze (2).

The present author in the thesis (2), in 1941 (see also [3]) gave an outline of an investigation of exceptional cases of singular integral equations, by making use of the corresponding results for the Riemann boundary value problem. In 1952 the same problem by the same method was investigated in detail by L. A. Chikin (1). In 1948 D. I. Sherman (3), independently of the present author's work which had not then been published, examined by different methods the exceptional cases of systems of singular integral equations; in particular, his work contains the solution for the case considered here.

The problems of equivalent regularization were first considered by S. G. Mikh-lin in 1938 (1). This paper, however, as well as all following papers dealt only with the statement of the problem of equivalent regularization for an arbitrary right-hand side (see § 24.1). Although in the first papers he employed cumber-some methods (for instance the method of successive approximations (1)) the importance of these papers should be emphasized, since they were the first on this problem and attracted the attention of scientists to problems of equivalent regulari-zation. The theorem on equivalent regularization of equations in the form of § 24.5 was first announced in the author's thesis (2) in 1941.

A proof of this theorem was before long derived, first by V. D. Kupradze (2) and then in a different way by I. N. Vekua (3). The proof given in § 24.5 in essence is identical with Vekua's proof, although the form is very different. It also differs little from the proof of S. G. Mikhlin [16].

It should be observed that the development of the theory of equivalent regulariza-tion turned out to be most difficult and a great number of errors were made here.

A complete and comparatively simple solution of the problem of equivalent regularization based on the above mentioned papers is presented in this book, and belongs to the most significant achievements of the theory of singular integral equations.

We cannot fail to mention the papers of D. I. Sherman (1)–(4); although they are not directly concerned with regularization, they are closely connected with it. He worked out a method which makes it possible to reduce, with the help of special integral representations of analytic functions, boundary value problems directly to Fredholm integral equations, thus avoiding singular equations regarded as an intermediate stage (for the details see § 34).

The elaboration of the fundamental problems of the theory of singular integral equations with Cauchy kernel (including also cases of discontinuous coefficients and open contours, see Chapter VI) based on the solution of the Riemann boundarn value problem, was carried out in 1940–1944, mainly in the works of Georgian mathematicians. To fully appreciate the progress in this field the reader should compare the exposition of this book with that given in the paper of Noether (1) or even S. G. Mikhlin's paper (1) written in 1938, when the work on this theory was first begun.

We have to mention one more method of constructing the theory of singular equations, which has not so far been dealt with. The outline of this method was indicated by Carleman (1) and then worked out by V. D. Kupradze. The latter author proved [12], that the whole theory of singular equations can be deduced by this method. However, even he notes that this method has no advantage (from the standpoint of theory) over the method given in this book which is based on the solution of the Riemann boundary value problem; thus the theoretical vahu of the new method consists so far in the fact that every new method of tackling a problem makes it possible to better understand the latter. V. D. Kupradze also notes that the problem of a practical use of the method can only be solved after an actual application to solving singular equations.

Further development of the theory consists in various generalizations. First we mention the papers of S. G. Mikhlin which are important because they connect the theory of singular equations with concepts of functional analysis. In this connection we note also the paper of Z. I. Khalilov [10]. There is a large number of works of Georgian mathematicians (L. G. Magnardze, B. V. Khvedelidze, I. N. Kartsivadze) in which the theory of singular equations is generalized by making use of the ideas of the theory of functions of real variable (integrability in the Lebesgue sense, etc.). We do not touch upon these questions, since they are outside the scope of our book.

Problems on Chapter III

In all problems the contour L is a simple closed curve, dividing the plane into two domains: the interior D^+ and the exterior D^-; the letters $c_1, c_2, \ldots$ denote arbitrary constants.

1. Solve the dominant singular integral equation

$$t(t-2)\,\varphi(t) + \frac{t^2 - 6t + 8}{\pi i} \int_L \frac{\varphi(\tau)}{\tau - t}\, d\tau = \frac{1}{t},$$

assuming that the point $z = 0$ belongs to the domain D^+ and the point $z =$
to the domain D^-.

Answer. $\varphi(t) = \dfrac{-t^2 + 6t - 4}{8t(t-2)^2}$.

Hint. In solving Problems 2–8 the appropriate results of Problems 4–8, 10, 1
of Chapter II should be used.

2. Solve the dominant singular integral equation

$$(t^2 - 2)\,\varphi(t) + \frac{3t}{\pi} \int\limits_L \frac{\varphi(t)}{\tau - t}\,d\tau = \frac{2t}{t^2+1},$$

assuming that the points i, $2i$ belong to the domain D^+ and the points $-i$, -2
to the domain D^-

Answer. $\varphi(t) = \dfrac{2t(t^2 - 5)}{(t^2+1)^2(t^2+4)} + \dfrac{t(c_0 + c_1 t)}{(t^2+1)(t^2+4)}$.

3. Prove that the equation

$$(t^2 - 2)\,\varphi(t) + \frac{3t}{\pi} \int\limits_L \frac{\varphi(\tau)}{\tau - t}\,d\tau = 2t(t - 2i)(1 + at)$$

is soluble only when $a = -i$, assuming that the point $-i$ belongs to D^+ and the
points i, $2i$, $-2i$ to D^-. Derive the solution of this equation for the above valu
of the parameter.

Answer. $\varphi(t) = \dfrac{-2it(t-2i)}{t+2i}$.

4. Solve the equation

$$\left[t - \ln\left(1 - \frac{1}{t}\right)\right]\varphi(t) - \frac{t + \ln\left(1 - \dfrac{1}{t}\right)}{\pi i} \int\limits_L \frac{\varphi(\tau)}{\tau - t}\,d\tau = 2\ln\frac{t}{t-1}\left(\sin t + e^{\frac{1}{t}}\right)$$

for a contour L such that the points 0, 1 remain in the domain D^+.

Answer.

$$\varphi(t) = \sin t - \frac{1}{t}\,e^{\frac{1}{t}}\ln\left(1 - \frac{1}{t}\right) + (c_0 + c_1 t)\left[1 + \frac{1}{t} \times \ln\left(1 - \frac{1}{t}\right)\right].$$

5. Solve the equation

$$t\varphi(t) - \frac{t-2}{\pi i} \int\limits_L \frac{\varphi(t)}{\tau - t}\,d\tau = 2(t^2 - 1)$$

under the following assumptions:

(a) the point 1 belongs to D^+;
(b) the point 1 belongs to D^-;
(c) the point 1 belongs to L.

Answer. (a) $\varphi(t) = t^2 - 1 + c_0 \dfrac{t-2}{t-1}$;

(b) and (c) $\varphi(t) = t^2 - 1$.

6. Solve the equation

$$t\varphi(t) + \frac{t-2}{\pi i} \int_L \frac{\varphi(\tau)}{\tau - t}\, d\tau = 2(t^2 - t + 1)$$

under the same assumptions as in the preceding problem.

Answer. (a) and (c) The equation is insoluble.

(b) $\varphi(t) = t + \dfrac{1}{t-1}$.

7. Solve the equation

$$[3t - i(h-1)]\varphi(t) - \frac{t - i(h+1)}{\pi i} \int_{-\infty}^{\infty} \frac{\varphi(\tau)\, d\tau}{\tau - t} = \frac{2(t+ib)}{t-i} \quad (h \neq 0,\ -\infty < t < \infty):$$

(a) in the class of functions which satisfy the condition $|\varphi(t)| < A\,|t|^{-\alpha}\ (\alpha > 0)$ for large $|t|$;

(b) in the class of functions which satisfy the condition $|\varphi_1(t) - \varphi_1(\infty)| < A\,|t|^{-\alpha}\ (\alpha > 0)$ for large $|t|$ (see § 4.6).

Answer. (a) for $h > 0$

$$\varphi(t) = \frac{1-b}{2(t+i)} + \frac{(t+i)(1+b)}{2(2t-ih)(t-i)} + c_1\left[\frac{1}{t+i} - \frac{1}{2t-ih}\right],$$

for $h < 0$

$$\varphi(t) = \frac{h+2b}{(h-2)(t+i)} - \frac{1+b}{(h-2)(t-i)};$$

(b) for $h > 0$

$$\varphi_1(t) = \varphi(t) + c\left[\frac{t}{t+i} - \frac{t}{2t-ih}\right],$$

for $h < 0$

$$\varphi_1(t) = \varphi(t) + c\left[\frac{2t-ih}{t+i} - 1\right].$$

8. Solve the equation

$$[2t - i(h+1)]\varphi(t) + \frac{h-1}{\pi} \int_{-\infty}^{\infty} \frac{\varphi(\tau)\, d\tau}{\tau - t} = \frac{2(t+ib)}{t+i}$$

$$(h \neq 0,\ -\infty < t < \infty)$$

in the classes of Problem 7.

Answer. (a) for $h > 0$

$$\varphi(t) = \frac{1-b}{2(t+i)} + \frac{1-b}{2(t-hi)},$$

for $h < 0$
$$\varphi(t) = \frac{1}{t+i}$$

assuming that the condition of solubility $b = -1$ is satisfied;

(b) $\varphi_1(t) = \varphi(t)$.

Remark. An increase of the number of solutions in the case (b) cannot occur for the coefficient $G(t)$ of the corresponding Riemann problem has the property $G(\infty) = 1$; hence the quantity $\varphi_1(\infty)$ cannot be different from zero.

9. Prove that for the equation

$$K\varphi \equiv (1+t)\,\varphi(t) + \frac{1-t}{\pi i} \int_L \frac{\varphi(\tau)}{\tau-t}\,d\tau + \frac{1}{2\pi i} \int_L e^{t\tau} \varphi(\tau)\,d\tau = 2t$$

the operator K'° is an equivalent regularizator from the left, assuming that the origin of the coordinate system belongs to the domain D^+. Construct the regularized solution.

Answer. $\varphi(t) + \dfrac{1}{4\pi i} \int \left(e^{t\tau} - \dfrac{2}{t} + 2 \right) \varphi(\tau)\,d\tau = t.$

The equivalence follows from the fact that the operator K'° is dominant with negative index and hence has no eigenfunctions.

10. For the equation

$$K\varphi \equiv (1+t^{-2})\,\varphi(t) + \frac{1-t^{-2}}{\pi i} \int_{|\tau|=1} \frac{\varphi(\tau)}{\tau-t}\,d\tau - \frac{1}{\pi i} \int_{|\tau|=1} \frac{\varphi(\tau)}{t^2\tau}\,d\tau = 2t$$

prove that:

(a) the simplest regularizator K'° is not an equivalent regularizator from the left

(b) the conjugate operator K^* is an equivalent regularizator from the left. Construct the regularized equation $K^*K\varphi = K^*f$.

Answer. (a) The Fredholm equation $K'^\circ K\varphi = K'^\circ f$ is equivalent to the equation

$$K\varphi = 2 + \alpha \left(1 - \frac{1}{t^2} \right)$$

for an arbitrary α;

(b) $\varphi + \dfrac{3}{4\pi i} \int_{|\tau|=1} \dfrac{1}{\tau}\,\varphi(\tau)\,d\tau = t.$

11. Prove that for the equation

$$K\varphi \equiv (t^2 + t^{-2})\,\varphi(t) + \frac{(t^2 - t^{-2})}{\pi i} \int_L \frac{\varphi(\tau)}{\tau-t}\,d\tau - \frac{(t+t^{-2})}{2\pi i} \int_L (\tau + \tau^{-2})\,\varphi(\tau)\,d\tau = 2t$$

the operator K'° is an equivalent regularizator from the left, if the origin of the coordinate system lies in D^+. Construct the Fredholm equation $K'^\circ K\varphi = K'^\circ f$.

Answer. $\varphi(t) - \dfrac{1+t^{-1}}{4\pi i} \int_L (\tau + \tau^{-2})\,\varphi(\tau)\,d\tau = t^2.$

12. For the equation

$$K\varphi \equiv \frac{t+2}{t-2}\,\varphi(t) + \frac{1}{\pi i} \int\limits_{|\tau|=1} \frac{\varphi(\tau)}{\tau - t}\,d\tau + \frac{1}{2\pi i} \int\limits_{|\tau|=1} \left(1 + \frac{1}{t}\right) \tau^2 \varphi(\tau)\,d\tau = 2t:$$

(a) construct the equivalent regularizator from the left

$$\tilde{K} = (K^*K)'^{\circ} K^*;$$

(b) prove that the dominant operator K'° is also an equivalent regularizator from the left. Construct the regularized equation $K'^\circ K\varphi = K'^\circ f$.

Answer.

(a) $(K^*K)'^\circ K^* \psi \equiv \dfrac{8t\,(t+2)}{(1-2t)^2\,(t-2)}\,\psi(t) - \dfrac{8t}{(1-2t)^2}\,\dfrac{1}{\pi i} \displaystyle\int\limits_{|\tau|=1} \dfrac{\psi(\tau)}{\tau-t}\,d\tau +$

$$+ \frac{1}{\pi i} \int\limits_{|\tau|=1} \left[\frac{2}{t^2(t-2)(1-2t)}\left(1 + \frac{1}{\tau}\right) + \frac{24t}{(t-2)(1-2t)^2(1-2\tau)} \right] \psi(\tau)\,d\tau;$$

(b) $\varphi(t) + \dfrac{1}{8\pi i} \displaystyle\int\limits_{|\tau|=1} \left(1 - \dfrac{2}{t}\right)\left(3\tau^2 + \dfrac{4}{\tau-2}\right) \varphi(\tau)\,d\tau = t - 2.$

13. Derive the inversion formula, for the singular integral of Cauchy type

$$\psi(t) = \frac{1}{\pi i} \int\limits_L \frac{\varphi(\tau)}{\tau - t}\,d\tau$$

(§ 7.3) from the formula for the solution of the dominant equation (21.9).

14. Deduce the basic properties of singular integral equations (Noether theorems, § 23.2), on the basis of regularization by solving the dominant equation (I. N. Vekua (2)).

15. Consider a singular integral equation

$$K\varphi \equiv a(t)\,\varphi(t) + \frac{b(t)}{\pi i} \int\limits_L \frac{\varphi(\tau)}{\tau - t}\,d\tau + \int\limits_L k\,(t,\tau)\,\varphi(\tau)\,d\tau = f(t)$$

(soluble or insoluble). Prove that it is possible without altering the dominant part, to change the regular part of the kernel $K(t,\tau)$ in such a way that the equation

$$K_1\varphi \equiv a(t)\,\varphi(t) + \frac{b(t)}{\pi i} \int\limits_L \frac{\varphi(\tau)}{\tau - t}\,d\tau + \int\limits_L k_1(t,\tau)\,\varphi(\tau)\,d\tau = f(t)$$

has not less than a certain prescribed number of solutions.

16. Prove that the generalized singular integral equation

$$a(t)\,\varphi[x(t)] + b(t)\varphi(t) + \frac{a(t)}{\pi i} \int\limits_L \frac{\varphi(\tau)}{\tau - \alpha(t)}\,d\tau - \frac{b(t)}{\pi i} \int\limits_L \frac{\varphi(\tau)}{\tau - t}\,d\tau = f(t)$$

can be reduced to the generalized Riemann boundary value problem

$$\Phi^+[\alpha(t)] = G(t)\,\Phi^-(t) + g(t),$$

where

$$\Phi(z) = \frac{1}{2\pi i} \int_L \frac{\varphi(\tau)}{\tau - z}\, d\tau; \quad G(t) = \frac{b(t)}{a(t)}; \quad g(t) = \frac{1}{2}\frac{f(t)}{a(t)}$$

(D. A. Kveselava, (2)).

17. Prove that the solution of the singular integral equation

$$a(t)\varphi(t) + \frac{b(t)}{\pi i} \int_L \frac{\varphi(\tau)}{\tau - t}\, d\tau + \int_L k(t,\tau)\,\varphi(\tau)\, d\tau = f(t),$$

the regular kernel of which $k(t,\tau) = \sum_{j=1}^{n} \alpha_j(t)\beta_j(\tau)$ is degenerate, can be reduced to solving a system of linear algebraic equations. Carry out an investigation of the equation, on the basis of the theorems of solubility of systems of linear algebraic equations.

18*. Suppose that the regular part of the kernel of a singular integral equation contains a parameter λ, i.e.

$$K\varphi = a(t)\,\varphi(t) + \frac{b(t)}{\pi i} \int_L \frac{\varphi(\tau)}{\tau - t}\, d\tau + \lambda \int_L k(t,\tau)\,\varphi(\tau)\, d\tau = f(t).$$

Prove that:

(a) For $\varkappa \geq 0$ the general solution constitutes a meromorphic function of the parameter λ, which contains linearly $\varkappa$ arbitrary constants. Except for, perhaps a set of discrete eigenvalues $\lambda_1, \lambda_2, \ldots$ the homogeneous equation has exactly $\varkappa$ linearly independent solutions which also are meromorphic functions in λ. For the eigenvalues of λ the number of solutions of the equation is greater than $\varkappa$.

(b) For $\varkappa = -\varkappa' < 0$, for all values of λ which are not eigenvalues, conditions 23.4) of solubility of the equation have the form of $\varkappa'$ relations

$$\int_L \omega_j(t, \lambda)\,f(t)\, dt = 0,$$

where $\omega_j(t, \lambda)$ are meromorphic functions in the parameter λ, the poles of which are the eigenvalues of λ (I. N. Vekua, (3)).

19*. Prove without solving the Riemann problem that the difference between the numbers of solutions of adjoint equations is independent of the regular part of the kernel.

(Noether (1), N. I. Muskhelishvili [17*], § 53.)

CHAPTER IV

HILBERT BOUNDARY VALUE PROBLEM AND SINGULAR INTEGRAL EQUATIONS WITH HILBERT KERNEL

In this chapter we shall consider the second fundamental boundary value problem of analytic functions, the so-called Hilbert problem, and singular integral equations with a Hilbert kernel which are closely connected with it. Moreover, we shall examine a number of topics which are auxiliary to solving the Hilbert problem (the Schwarz symbol, the regularizing factor). Finally, an application of the Hilbert problem will be given, in solving boundary value problems for polyharmonic and polyanalytic functions.

In all that follows only simply-connected domains are considered. In contrast to the Riemann problem in which the cases of simply-connected and multiply-connected domains are not essentially different, in the case of the Hilbert problem the transition from a simply-connected to a multiply-connected domain involves great difficulties. The Hilbert problem for multiply-connected domains will be examined in the next chapter.

§ 27. Formulation of the Hilbert problem and some auxiliary formulae

27.1. Formulation of the Hilbert problem

Consider a simple smooth closed contour L and real functions of its arc s, $a(s)$, $b(s)$, $c(s)$ which obey the Hölder condition.

By the Hilbert boundary value problem we understand the following problem. *It is required to find in the domain D^+ a function*

$$f(z) = u(x, y) + iv(x, y),$$

which is continuous on the contour and the limiting values of the real and imaginary parts of which satisfy on the contour the linear relation†

$$a(s)\,u(s) + b(s)\,v(s) = c(s). \qquad (27.1)$$

When $c(s) \equiv 0$ we are faced with the homogeneous problem, while when $c \not\equiv 0$ we have the non-homogeneous problem (the analogous "exterior" problem will be considered in § 29.4).

In order to solve the above problem we shall need some auxiliary formulae which we now proceed to consider.

27.2. The Schwarz operator for simply-connected domain

Frequently we meet with the problem of determining an analytic function $F(z) = u + iv$ inside a domain from its defined real part $u(s)$ on the boundary. In the following we shall call this the Schwarz problem. In essence it reduces to the Dirichlet problem‡. The harmonic function $u(x, y)$ inside the domain is found from $u(s)$ and then the conjugate harmonic function $v(x, y)$ is obtained therefrom by integration of the total differential (Cauchy–Riemann equation) to within an arbitrary term. The Schwarz problem is thus solved to within an arbitrary constant. The solution is fully determined if the value of the imaginary part of the required analytic function is defined at some point inside the domain. To solve the Schwarz problem, we introduce the concept of a Schwarz operator. Its use simplifies the formulation of many results.

By the Schwarz operator we understand an operator which establishes an analytic function in terms of the limiting values of its real part. More rigorously we can state it as follows.

Suppose that a real function $u(s)$ satisfying the Hölder condition is given on a smooth contour. The Schwarz operator S is an operator which determines the analytic function $F(z)$, the limiting value of the real part of which coincides on the contour with the function $u(s)$, and the imaginary part of which vanishes at a given point z_0.

We can write the above statement as follows:

$$F(z) = u(x, y) + iv(x, y) = Su.$$

If L is the unit circle the Schwarz operator is identical with the familiar Schwarz integral (see § 6.1); if L is the real axis the Schwarz

---

† In all that follows, functions obtained by replacing the coordinates by the arc s will for simplicity be denoted by the same letter. For instance $u[x(s), y(s)] = u(s)$, etc.

‡ There is an essential difference for a multiply-connected domain (see § 36.2).

operator is simply the Cauchy type integral. For an arbitrary contour, an explicit expression for the Schwarz operator can be given in terms of the Green function.

Let
$$G(x, y; \xi, \eta) = \ln \frac{1}{r} + g(x, y; \xi, \eta)$$

be the Green's function of the Laplace operator for the domain D^+, where $r = \sqrt{[(x - \xi)^2 + (y - \eta)^2]}$ and $g(x, y; \xi, \eta)$ is a function harmonic in the two pairs of variables x, y and ξ, η, which takes the value $\ln r$ when one of the points (x, y), (ξ, η) is located on the contour.

We shall regard $G(x, y; \xi, \eta)$ as a function of two complex variables $z = x + iy$, $\zeta = \xi + i\eta$, both of which vary in the domain D^+; it will hereafter be denoted by $G(z, \zeta)$.

It is known from the theory of harmonic functions that the solution of the first boundary value problem for harmonic functions—the Dirichlet problem—is given by the formula

$$u(x, y) = \frac{1}{2\pi} \int_L \frac{\partial G(z, \tau)}{\partial n} u(\sigma) \, d\sigma, \qquad (27.2)$$

where $\tau = \tau(\sigma)$ is the complex coordinate of a point of the contour and n is the interior normal.

Let $H(z, \zeta)$ be the harmonic function complex conjugate to $G(z, \zeta)$ with respect to the variable z. It is determined on the basis of the Cauchy–Riemann equations by the relation

$$H(z, \zeta) = \int_{z_0}^{z} \left(-\frac{\partial G}{\partial y} \partial x + \frac{\partial G}{\partial x} \partial y \right), \qquad (27.3)$$

where z_0 is a fixed point of the domain D^+.

Since D^+ is simply-connected the function H is determined uniquely and in view of the last relation it obeys the condition

$$H(z_0, \zeta) \equiv 0.$$

The function

$$M(z, \zeta) = G(z, \zeta) + i H(z, \zeta)$$

is called the *complex Green's function for the domain* D^+†. It is analytic in z everywhere except at the point $z = \zeta$ where it has a logarithmic singularity.

† This term was introduced by S. G. Mikhlin [15].

By virtue of relations (27.2) and (27.3) the formula

$$v(x, y) = \frac{1}{2\pi} \int_L \frac{\partial H(z, \tau)}{\partial n} u(\sigma) \, d\sigma$$

determines the harmonic function $v(x, y)$ complex conjugate to the function $u(x, y)$.

Hence, the relation

$$F(z) = u(x, y) + iv(x, y) = \frac{1}{2\pi} \int_0^l \frac{\partial M(z, \tau)}{\partial n} u(\sigma) \, d\sigma$$

yields the analytic function the real part of which is equal to the given function $u(\sigma)$ on the contour. This function satisfies also the additional condition $v(z_0) = 0$.

Consequently, the Schwarz operator is given by the formula

$$Su \equiv \frac{1}{2\pi} \int_0^l \frac{\partial M(z, \tau)}{\partial n} u(\sigma) \, d\sigma. \tag{27.4}$$

If the condition $v(z_0) = 0$ be discarded, then

$$F(z) = Su + i\beta_0, \tag{27.5}$$

where β_0 is an arbitrary constant equal to $v(z_0)$.

The function $T(z, \tau) = \partial M(z, \tau)/\partial n$ is called the Schwarz kernel for the contour L. In a conformal mapping of the domain D^+ onto the unit circle it becomes the Schwarz kernel for the circle $T(z, \tau) = (\tau + z)/(\tau - z)$ (see § 6.1).

27.3. Determination of an analytic function possessing a pole, in terms of the value of its real part on the contour (Problem A)

In the preceding section we determined a function analytic in a domain, in terms of the value of its real part on the contour. In subsequent considerations we shall require a solution of the more general problem consisting in allowing for the presence of a pole of the unknown analytic function. This problem will hereafter be called *Problem A*. We now present a rigorous formulation.

Consider a simple smooth closed contour L bounding an interior domain D^+. It is required to determine a function $F(z)$ which is analytic in the domain D^+ except at the point z_0 where it may possess a pole of order not exceeding n, and the real part of which is equal on the contour L to a prescribed function $u(s)$.

The corresponding homogeneous problem $(u(s) \equiv 0)$ will be called *Problem A_0* and its solution will be denoted by $Q(z)$.

We first present a solution of Problem A_0 for the case when the contour L is the *unit circle* and $z_0 = 0$.

Let us write down the expansion of the unknown function in the vicinity of the origin of the coordinate system

$$Q(z) = \sum_{k=-n}^{\infty} c_k z^k.$$

According to the stated condition,

$$\operatorname{Re} \sum_{k=-n}^{\infty} c_k e^{iks} = 0.$$

Denoting $c_k = \alpha_k + i\beta_k$ we obtain

$$\sum_{k=-n}^{\infty} (\alpha_k \cos ks - \beta_k \sin ks) = 0.$$

In view of uniqueness of expansion into a Fourier series we have

$$\alpha_0 = 0, \quad \alpha_{-k} = -\alpha_k, \quad \beta_{-k} = \beta_k, \quad k = 1, 2, \ldots, n,$$
$$\alpha_k = \beta_k = 0, \quad k = n+1, n+2, \ldots$$

whence

$$c_0 = i\beta_0, \quad c_{-k} = -\bar{c}_k \quad (k = 1, 2, \ldots, n), \quad c_k = 0 \quad (k > n).$$

Consequently the required function $Q(z)$ has the form

$$Q(z) = i\beta_0 + \sum_{k=1}^{n} (c_k z^k - \bar{c}_k z^{-k}). \tag{27.6}$$

Remark. If in the formulation of Problem A_0 for the unit circle the domain D^+ interior for the unit circle be replaced by the exterior domain D^-, then both the reasoning and the final result remain unaltered and, consequently, formula (27.6) yields the solution of Problem A_0 also for the domain D^- exterior with respect to the unit circle. The point at which the function has a pole is then the point at infinity.

The solution of Problem A_0 for an arbitrary domain and also for the unit circle but with a pole at z_0, can be deduced from the above found solution by means of conformal mapping. Let

$$w = \omega(z) \tag{27.7}$$

be the function mapping conformally the domain D^+ of the plane z onto the unit circle of the plane w in such a way that a point z_0 is transformed into the origin of the coordinate system $\omega(z_0) = 0$, and $\omega'(z_0) > 0$.

Then the formula

$$Q(z) = i\beta_0 + \sum_{k=1}^{n} \left\{ c_k [\omega(z)]^k - \bar{c}_k [\omega(z)]^{-k} \right\} \tag{27.8}$$

can easily be verified to yield the *solution of Problem A_0 for an arbitrary domain D^+*.

If by $\omega(z)$ we understand the function mapping conformally the domain D^- onto the domain exterior with respect to the unit circle and transforming the point at infinity into the point at infinity, and moreover $\omega'(\infty) > 0$, then formula (27.8) gives the solution of Problem A_0 for the domain D^-.

We are now in a position to solve Problem A.

In accordance with the definition of the Schwarz operator, Su is a function analytic in D^+ the real part of which is equal to the function $u(s)$ on the contour. Evidently, the difference $F(z) - Su$ is a function which satisfies the conditions of Problem A_0. Consequently

$$F(z) - Su = Q(z),$$

whence

$$F(z) = Su + Q(z), \tag{27.9}$$

where the Schwarz operator Su is determined by formula (27.4) and the solution of the homogeneous Problem A_0, $Q(z)$, by formula (27.8).

Formulae (27.8), (27.9) indicate that the solutions of Problem A_0 and Problem A contain $2n + 1$ arbitrary real constants

$$\beta_0, \ \alpha_k, \ \beta_k, \quad k = 1, 2, \ldots, n \quad (c_k = \alpha_k + i\beta_k).$$

In particular when the pole is absent ($n = 0$) formula (27.9) becomes formula (27.5).

To construct the function $F(z)$ possessing the above enumerated properties in terms of the limiting values of its imaginary part $v(s)$ it is sufficient to note that

$$\text{Im } F(t) = \text{Re}[-iF(t)] = v(s),$$

whence

$$-iF(z) = Sv + Q(z),$$

i.e.

$$F(z) = iSv + iQ(z). \tag{27.10}$$

§ 28. Regularizing factor

28.1. Definition of the regularizing factor

Let L be a simple smooth closed contour bounding the interior domain D^+ and

$$G(t) = G[t(s)] = a(s) + ib(s)$$

be a complex function prescribed on this contour and satisfying the Hölder condition; moreover, this function does not vanish on L. It was already indicated in § 4.3 that in general an arbitrary function $G(t)$ is not a boundary value of a function analytic in the domain D^+.

Let us find out whether it is possible *to determine a function of position on the contour, $R(t)$, which on multiplication by the function $G(t)$ becomes the boundary value of a function $\Phi^+(z)$ analytic in D^+*, i.e.

$$G(t) R(t) = \Phi^+(t). \tag{28.1}$$

The function $R(t)$ possessing this property will be called the *regularizing factor*.

The formulation of the problem can be generalized by allowing for certain singularities for the function $\Phi^+(z)$, at some points of the domain D^+.

The problem formulated above is not definite, since if it has a solution, then, obviously, it has an infinite number of other solutions derived from the first by multiplying it by a boundary value of an arbitrary function analytic in D^+. Thus, the formulation of the problem needs a better specification. To make the problem definite the regularized factor should be subjected to some additional conditions which may be of various types.

We have already met one case of finding the regularizing factor, in solving the homogeneous Riemann boundary value problem. In fact, writing the boundary condition of this problem in the form

$$G(t) \Phi^-(t) = \Phi^+(t),$$

we observe that it is a relation of type (28.1) where $R(t) = \Phi^-(t)$. Thus, the Riemann boundary value problem can be regarded as the problem of determining of a regularizing factor which is subjected to the additional condition that it is the boundary value of a function analytic in the exterior domain D^-. The solution of the Riemann problem establishes the existence of such a factor when $\varkappa \geqq 0$.

There are two more types of regularizing factors, namely those:

(1) with constant argument;

(2) with constant modulus.

Without any loss of generality we may assume in the first case that the constant argument is zero and hence we may regard the factor as real and positive; in the second case we may assume that the modulus is unity. In the next two sections we shall deal with the determination of regularizing factors of the above types.

28.2. Real regularizing factor

We require that the regularizing factor $R(t)$ is a real function; denoting it in this case by $p(s)$ we have

$$p(s)[a(s) + ib(s)] = \Phi^+(t). \tag{28.2}$$

Computing the index of both sides and taking into account that the index of $p(s)$ is zero we obtain

$$\varkappa = \text{Ind}\,[a(s) + ib(s)] = \text{Ind}\,\Phi^+(t).$$

Bearing in mind the properties of the index (§ 12.1) we have the following results:

1. For $\varkappa = 0$ the function $\Phi^+(z)$ cannot have any zeros in the domain D^+.

2. For $\varkappa > 0$ the function $\Phi^+(z)$ has in D^+ $\varkappa$ zeros.

3. For $\varkappa < 0$ the function $\Phi^+(z)$ cannot be analytic in D^+; it should have in this domain not less than $-\varkappa$ poles.

In order to ensure the solubility and, moreover, uniqueness of the problem of determining the regularizing factor of the considered type, for an arbitrary index, we shall define this concept somewhat differently, so that the required analytic function may have a pole in the domain D^+.

For simplicity we suppose that the origin of the coordinate system lies in the domain D^+.

DEFINITION. *By the regularizing factor of the complex function* $a(s) + ib(s)$ *prescribed on the contour L we shall understand a real positive function of position on the contour, $p(s)$, such that the product $p(s)[a(s) + ib(s)]$ is the boundary value of a function $\chi^+(z)$ analytic in D^+, which is of zero order everywhere in this domain, except perhaps at the origin of the coordinate system where its order is equal to the* *index $\varkappa$ of the function $a(s) + ib(s)$* (definition of the order is given to § 14.4).

We now proceed to prove the existence of the regularizing factor and to derive formulae for its calculation.

1. $\varkappa = 0$. Since the function $\chi^+(z)$ has no zeros in the domain D^+ it can be represented by means of the exponential function

$$\chi^+(z) = e^{i\gamma(z)}, \quad \gamma(z) = \omega(x, y) + i\omega_1(x, y). \tag{28.3}$$

(the factor i in the exponent is introduced for convenience of further computations).

According to the definition of the regularizing factor we have

$$p(s)[a(s) + ib(s)] = e^{i\gamma(t)} = e^{-\omega_1(s)}e^{i\omega(s)}. \tag{28.4}†$$

Equating the moduli and arguments of both sides of the latter relation we obtain

$$p(s)\sqrt{[a^2(s) + b^2(s)]} = e^{-\omega_1(s)}, \quad \omega(s) = \arctan\frac{b(s)}{a(s)}. \tag{28.5}$$

This formula gives the boundary value of the harmonic function $\omega(x, y)$. The function $\omega(x, y)$ itself is to be found by solving the Dirichlet problem. Then the Cauchy–Riemann equations serve to determine the complex conjugate harmonic function $\omega_1(x, y)$; the required regularizing factor $p(s)$ is expressed in terms of the boundary value of the latter function.

The results acquire the simplest form by introducing the Schwarz operator.

In accordance with (27.5) and (28.5) we have

$$\gamma(z) = \omega(x, y) + i\omega_1(x, y) = S\arctan\frac{b}{a}; \tag{28.6}$$

for definiteness we have imposed the following condition on $\gamma(z)$:

$$\operatorname{Im}\gamma(z_0) = \omega_1(z_0) = 0. \tag{28.7}$$

The regularizing factor is given by the formula

$$p(s) = \frac{e^{-\omega_1(s)}}{\sqrt{[a^2(s) + b^2(s)]}}. \tag{28.8}$$

Let us consider the problem of the uniqueness of the regularizing factor. Assume that there exist two distinct regularizing factors $p(s)$

---

† We adopt the simplified notation (see footnote on p. 208).

and $p_1(s)$. Then we have the relations

$$p(s)[a(s) + i b(s)] = \chi^+(s), \quad p_1(s)[a(s) + i b(s)] = \chi_1^+(s).$$

Dividing side by side these relations and denoting $\chi_1^+(z)/\chi^+(z) = \psi(z)$ we obtain

$$\frac{p_1(s)}{p(s)} = \psi(s).$$

The imaginary part of the function $\psi(z)$ vanishes on the contour. Consequently, in view of the uniqueness of the solution of the Dirichlet problem $\text{Im } \psi(z) = 0$ everywhere in the domain D^+. Hence, $\psi(z) = \text{const.}$

Thus, the regularizing factor is determined to within a constant multiplier. Condition (28.7) determines this constant multiplier.

Thus we have proved the existence and uniqueness of the regularizing factor in the case $\varkappa = 0$, under the additional condition (28.7).

2. $\varkappa \gtreqless 0$. By definition we have

$$p(s)[a(s) + i b(s)] = t^\varkappa e^{i\gamma(t)} = t^\varkappa e^{-\omega_1(s)} e^{i\omega(s)}. \tag{28.9}$$

Taking again the moduli and arguments of both sides we obtain

$$\left.\begin{array}{l} p(s) = \dfrac{|t|^\varkappa e^{-\omega_1(s)}}{\sqrt{[a^2(s) + b^2(s)]}}, \\[3mm] \omega(s) = \arg\{t^{-\varkappa}[a(s) + i b(s)]\} = \arctan \dfrac{b}{a} - \varkappa \arg t. \end{array}\right\} \tag{28.10}$$

Hence

$$\gamma(z) = S\left[\arctan \frac{b}{a} - \varkappa \arg t\right]. \tag{28.11}$$

The regularizing factor is given by formula (28.10). The uniqueness of the solution follows from the foregoing result, since according to relation (28.9) the problem consists in finding the regularizing factor for the function $t^{-\varkappa}[a(s) + ib(s)]$, the index of which is zero.

Thus, we have proved the following assertion.

THEOREM. *Any function of position on the contour has a real regularizing factor which, when subjected to the additional condition* (28.7), *is given by formulae* (28.6), (28.8), (28.10), (28.11) *in a unique way.*

The investigation shows that the determination of the regularizing factor in the form of a real function is equivalent to the determination of an analytic function, the values of its argument being prescribed on the contour; this leads to the solution of the Dirichlet problem.

28.3. Regularizing factor with constant modulus

We now seek the regularizing factor in the form

$$R(t) = e^{i\theta(s)}. \qquad (28.12)$$

Let us represent the given complex function also in polar coordinates, $a + ib = re^{i\alpha}$. According to the definition we have

$$re^{i(\alpha+\theta)} = \Phi^+(t). \qquad (28.13)$$

We seek the function $\theta(s)$ under the condition that in traversing the contour L it acquires an increment $2\pi v$ where v is a given integer. Taking the index of the both sides of the last relation we obtain

$$\varkappa + v = n, \qquad (28.14)$$

where n is the number of the zeros of the function $\Phi^+(z)$ in the domain D^+; thus, it depends not only on the index of the given function $a + ib$ but also on the index of the regularizing factor itself which, in contrast to the former, does not necessarily vanish.

Let us subject the unknown function $\Phi^+(z)$ to the condition that its order is n at the origin which we assume to belong to the domain D^+. Representing the required function $\Phi^+(z)$ in the form $z^n e^{\gamma(z)}$ we have

$$re^{i(\theta+\alpha)} = t^n e^{\gamma(t)} = t^n e^{\omega(s)} e^{i\omega_1(s)}. \qquad (28.15)$$

Equating the moduli and the arguments in the last relation we obtain

$$\omega = \ln(r|t|^{-n}), \qquad (28.16)$$

$$\omega_1 = \theta + \alpha - n \arg t. \qquad (28.17)$$

Consequently, we know the boundary value of the real part ω of the required analytic function $\gamma(z)$. Solving the Dirichlet problem we find the harmonic function $\omega(x, y)$. Next, finding the complex conjugate function ω_1 we determine in accordance with formula (28.17) the function θ in terms of which, by means of formula (28.13), the regularizing factor is expressed.

The result can easily be written in terms of the Schwarz operator

$$\gamma(z) = \omega(x, y) + i \omega_1(x, y) = S \ln(r|t|^{-n}), \qquad (28.18)$$

where for definiteness we have taken

$$\operatorname{Im} \gamma(z_0) = 0. \qquad (28.19)$$

The regularizing factor is given by the formula

$$R(t) = e^{i(\omega_1 - \alpha + n \arg t)}. \tag{28.20}$$

Formulae (28.18)–(28.20) uniquely determine the regularizing factor of the required form.

The solution shows that the determination of the regularizing factor with constant modulus is equivalent to the determination of an analytic function the boundary values of its modulus being given; this finally reduces to the solution of the Dirichlet problem.

28.4. Other forms of the regularizing factor

The general definition of the regularizing factor given in § 28.1 leaves great possibilities of choice of the form of this factor. We can observe a far reaching similarity between the definition of the regularizing factor and that of the familiar integrating factor in the theory of differential equations. For both of them great arbitrariness is allowed and the problem of their determination becomes definite only when we give up the general formulation and seek a factor of a definite form.

In § 28.2 and § 28.3 we examined the simplest types of the regularizing factor, when either the modulus or the argument are constant. These cases admit a geometric interpretation.

Let us consider the specification of a complex function $a(s) + ib(s) = re^{i\alpha}$ as the specification of a curve Γ the parametric equation of which has in the polar coordinates the form

$$r = r(s), \quad \alpha = \alpha(s).$$

Multiplying the function $re^{i\alpha}$ by a real positive regularizing factor $p(s)$ we perform over the points of the curve Γ a transformation consisting in changing the magnitudes of its position vectors with respect to $p(s)$, their directions remaining unaltered; in other words we carry out the transformation of stretching (with a variable modulus of stretching). The considered problem consists in such a choice of the continuously varying modulus of stretching that the transformed curve Γ_1, the equation of which is $r_1 = p(s) r(s)$, $\alpha_1 = \alpha(s)$, yields the geometric transformation (in the sense indicated above) of the complex function $\Phi^+(t)$ which is the boundary value of a function analytic in the domain D^+.

The multiplication of the complex function $re^{i\alpha}$ by a factor of the form $e^{i\theta(s)}$ can be interpreted geometrically as a rotation of the position vectors of the curve Γ through a continuously varying angle $\theta(s)$, this angle being so chosen that the transformed curve, having the equation $r_1 = r(s)$, $\alpha_1 = \alpha(s) + \theta(s)$, represents the geometric image of the boundary value of a function analytic in the domain D^+ (except perhaps the origin where it can have a pole).

If, according to the preceding example, we construct the geometric image of the regularizing factor itself, then in the first case it is given by the equation $\varrho = p(s)$, $\varphi = 0$, and is represented geometrically by a segment of the positive semi-axis traversed twice, or, in general, an even number of times. In the second case it has the equation $\varrho = 1$, $\varphi = \theta(s)$ and, consequently, it is represented geometrically by a unit circle traversed $n - \varkappa$ times (see (28.14)).

Generalizing the formulation of the problem in a natural way we seek the regularizing factor the geometric image of which is a curve prescribed beforehand.

Let there be given on the contour L a complex function $a(s) + ib(s) = re^{i\alpha}$, the geometric image of which is the curve Γ. Moreover, consider a curve C, the polar equation of which is $\varrho = f(\theta)$. It is required to find for the function $a(s) + ib(s)$ the regularizing factor $R(t) = \varrho(s) e^{i\theta(s)}$, the modulus and argument of which are related by a relation $\varrho = f(\theta)$ such that the geometric image of $R(t)$ is the prescribed curve C.

Equating the increment of $\alpha(s) + \theta(s)$ in traversing the contour L to $2\pi n$ we assume that the required analytic function $\Phi^{+}(z)$ has order n at the origin of the coordinate system.

In accordance with the definition we have

$$\varrho(s)\, r(s)\, e^{i[\alpha(s) + \theta(s)]} = t^{n} e^{\omega(s)} e^{i\omega_1(s)}. \tag{28.21}$$

Finding $\varrho(s)$ and $\theta(s)$ from the last relation and making use of the equation of the curve C we obtain

$$r(s) f[\omega_1(s) - \alpha(s) + n \arg t] = |t|^{n} e^{\omega(s)}. \tag{28.22}$$

This relation is the boundary value connecting the real (ω) and imaginary (ω_1) parts of the analytic function $\gamma(z)$. Thus, the general formulation of the problem of determination of the regularizing factor leads in general to non-linear boundary value problems. For problems of this type methods of solution are known only in some of the simplest cases. Hence, as a rule, a solution of the boundary value problem (28.21) is impossible.

Nevertheless, in some cases, the problem becomes linear and admits a solution. For instance, if the curve C is a logarithmic spiral the equation of which is $\varrho = e^{\theta}$, then the boundary condition (28.22) becomes linear of the type of the Hilbert problem

$$\omega_1 - \omega = \alpha - \ln r + n \ln |t| - n \arg t.$$

Writing the latter in the form

$$- \mathrm{Re}\,[(1 + i)\,\gamma(t)] = \alpha(s) - \ln r(s) + n \ln |t(s)| - n \arg t(s),$$

we easily obtain the function $\gamma(z) = \omega + i\omega_1$ and hence the solution of the whole problem.

Observe that we could take as the basis the equation of the curve C in Cartesian coordinates (see Problem 10 at the end of this chapter).

The above considerations, based on establishing the relation between the modulus and the argument or between the real and the imaginary parts of the regularizing factor, indicate only one of the possible ways of constructing it. The factor can be determined on the basis of an entirely different reasoning; this is indicated by the example of the factor appearing in solving the Riemann problem—this fact was indicated in the beginning of the section.

It seems to be desirable to investigate the determination of regularizing factors which are subject to some conditions, as when solving problems of the theory of the integrating factor. Perhaps it would be possible to find cases when the regularizing factor would be suitable to solve a practical problem and on the other hand its determination could be accomplished by means of a practically realizable algorithm. These hopes end my brief remarks on this problem.

§ 29. The Hilbert boundary value problem for simply-connected domains

29.1. The homogeneous problem

We now at last proceed to investigate the fundamental problem of this chapter—to solve the Hilbert boundary value problem. Let us write the boundary condition of the homogeneous problem

$$a(s)\,u(s) + b(s)\,v(s) = 0. \tag{29.1}$$

We assume that the coefficients $a(s)$ and $b(s)$ do not simultaneously vanish. Dividing the boundary condition by $\sqrt{[a^2(s) + b^2(s)]}$ we reduce it to the case when the coefficients satisfy the condition

$$a^2(s) + b^2(s) = 1. \tag{29.2}$$

In all that follows we assume that this condition is satisfied. We have

$$\frac{u + iv}{a + ib} = (a - ib)\,(u + iv) = au + bv + i(av - bu).$$

Hence, the boundary condition (29.1) can be written in the two equivalent forms

$$\mathrm{Re}\left\{\frac{F(t)}{a + ib}\right\} = 0, \tag{29.3}$$

$$\mathrm{Re}\{(a - ib)\,F(t)\} = 0, \tag{29.3'}$$

where

$$F(z) = u(x, y) + iv(x, y) \tag{29.4}$$

is the unknown function.

The index $\varkappa$ of the function $a(s) + ib(s)$ will be called the *index of the Hilbert problem.*

Dividing relation (29.3) throughout by the regularizing factor of the function $a + ib$ (formulae (28.9), (28.10)) we reduce it to the form

$$\mathrm{Re}\left[\frac{F(t)}{t^{\varkappa}e^{i\gamma(t)}}\right] = 0. \tag{29.5}$$

Let us consider separately the following cases.

1. $\varkappa = 0$. In this case the boundary condition (29.5) is the condition of the Dirichlet problem. In view of the uniqueness of the solution of this problem we have in the domain D^+

$$\mathrm{Re}\left[\frac{F(z)}{e^{i\gamma(z)}}\right] = 0,$$

whence

$$F(z) = i\beta_0 e^{i\gamma(z)}, \tag{29.6}$$

where β_0 is an arbitrary constant.

2. $\varkappa > 0$. The boundary condition (29.5) is the condition of Problem A_0. On the basis of the results of § 27.3 we obtain

$$F(z) = z^{\varkappa} e^{i\gamma(z)} Q(z), \tag{29.7}$$

where $Q(z)$ is given by formulae (27.7), (27.8). Consequently, the problem has $2\varkappa + 1$ linearly independent solutions.

3. $\varkappa < 0$. *There does not exist* an analytic function constituting a solution of the problem. The problem would have a solution if we admitted functions having in the domain D^+ not less than $-\varkappa$ poles.

29.2. The non-homogeneous problem

Let us write the boundary condition (27.1) in the form

$$\mathrm{Re}\left[\frac{F(t)}{a(s) + ib(s)}\right] = c(s). \tag{29.8}$$

Similarly to the case of the homogeneous problem let us divide both sides of the boundary condition by the regularizing factor of the function $a(s) + ib(s)$

$$\mathrm{Re}\left[\frac{F(t)}{t^{\varkappa} e^{i\gamma(t)}}\right] = |t|^{-\varkappa} e^{\omega_1(s)} c(s). \tag{29.9}$$

Let us again consider the cases of various indices.

1. $\varkappa = 0$. In this case the boundary condition (29.9) is the condition of the Dirichlet problem. Applying the Schwarz operator to both sides we find for $F(z)$:

$$F(z) = e^{i\gamma(z)}[S(e^{\omega_1(s)} c(s)) + i\beta_0]. \tag{29.10}$$

2. $\varkappa > 0$. The boundary condition (29.9) is the condition of Problem A, and in view of the results of § 27.3 we have

$$F(z) = z^z e^{i\gamma(z)}[S(|t|^{-\varkappa} e^{\omega_1(s)} c(s)) + Q(z)]. \tag{29.11}$$

Since formulae (29.10) and (29.11) contain general solutions of the corresponding homogeneous problems, clearly, they yield the general solution of the non-homogeneous problem.

3. $\varkappa < 0$. Proceeding as before we obtain

$$F(z) = z^{\varkappa} e^{i\gamma(z)} [S(|t|^{-\varkappa} e^{\omega_1(s)} c(s)) + iC]. \tag{29.12}$$

In view of the presence of the factor $z^{\varkappa}$ the function $F(z)$ defined by the last formulae can have a pole of order $-\varkappa$. In order to obtain an analytic solution we have to set in formula (29.12) $C = 0$ and to demand that the function

$$S(|t|^{-\varkappa} e^{\omega_1(s)} c(s))$$

has a zero of order $-\varkappa$ at the origin of the coordinate system. It can be proved by a direct calculation that this leads to $-2\varkappa - 1$ conditions which should be satisfied by the free term, in order that the solution be possible. Since we have no explicit expression for the Schwarz operator for an arbitrary contour, it is not possible to write down these conditions for solubility in the general case. When such an expression is known these conditions can explicitly be written. In the next section we shall do so for a unit circle.

We now state the results obtained above.

THEOREM. *If the index of the complex function* $a(s) + ib(s)$, $\varkappa \geqq 0$, *then the homogeneous Hilbert boundary value problem* (29.1) *and the non-homogeneous problem* (29.8) *are absolutely soluble. The homogeneous problem has* $2\varkappa + 1$ *linearly independent solutions, a linear combination of which enters the general solution as an additive term.*

If $\varkappa < 0$ *the homogeneous problem is insoluble. The non-homogeneous problem is soluble if and only if the free term satisfies* $-2\varkappa - 1$ *conditions of solubility.*

The theory presented here proves that essentially the solution of the Hilbert problem reduces to the solution of three Dirichlet problems. The first problem to be solved is the determination of the regularizing factor, the second the finding of the particular solution of the non-homogeneous problem, and, finally, the determination of the function $Q(z)$ for an arbitrary domain is reduced to finding the function which maps conformally the considered domain onto a circle, which in turn is equivalent to solving the Dirichlet problem.

29.3. Problem for the unit circle

For the unit circle the *Schwarz operator* is identical with the *Schwarz integral*. Denoting

$$F(z) = u(x, y) + iv(x, y),$$

$$\gamma(z) = \frac{1}{2\pi} \int_0^{2\pi} \left[\arctan \frac{b(\sigma)}{a(\sigma)} - \varkappa\sigma \right] \frac{e^{i\sigma} + z}{e^{i\sigma} - z} \, d\sigma,$$

$$\gamma(z) = \omega(x, y) + i\omega_1(x, y), \tag{29.13}$$

$$Q(z) = i\beta_0 + \sum_{k=1}^{\varkappa} (c_k z^k - \bar{c}_k z^{-k}), \tag{29.14}$$

we find that the solution of the Hilbert problem

$$a(s)\, u(s) + b(s)\, v(s) = c(s)$$

is given by the following formulae.

1. For $\varkappa = 0$

$$F(z) = e^{i\gamma(z)} \left[\frac{1}{2\pi} \int_0^{2\pi} e^{\omega_1(\sigma)} c(\sigma) \frac{e^{i\sigma} + z}{e^{i\sigma} - z} \, d\sigma + i\beta_0 \right]. \tag{29.15}$$

2. For $\varkappa > 0$

$$F(z) = z^{\varkappa} e^{i\gamma(z)} \left[\frac{1}{2\pi} \int_0^{2\pi} e^{\omega_1(\sigma)} c(\sigma) \frac{e^{i\sigma} + z}{e^{i\sigma} - z} \, d\sigma + Q(z) \right]. \tag{29.16}$$

3. For $\varkappa < 0$

$$F(z) = z^{\varkappa} e^{i\gamma(z)} \frac{1}{2\pi} \int_0^{2\pi} e^{\omega_1(\sigma)} c(\sigma) \frac{e^{i\sigma} + z}{e^{i\sigma} - z} \, d\sigma. \tag{29.17}$$

In the last case, for the problem to be soluble, the function

$$\frac{1}{2\pi} \int_0^{2\pi} e^{\omega_1(\sigma)} c(\sigma) \frac{e^{i\sigma} + z}{e^{i\sigma} - z} \, d\sigma = -\frac{1}{2\pi} \int_0^{2\pi} e^{\omega_1(\sigma)} c(\sigma) \, d\sigma +$$

$$+ \frac{1}{\pi} \int_0^{2\pi} e^{\omega_1(\sigma)} c(\sigma) \frac{d\sigma}{1 - ze^{-i\sigma}} \tag{29.18}$$

should have a zero of order $-\varkappa$ at the origin. Expanding it into a power series in the vicinity of the origin and equating to zero the first $-\varkappa - 1$ coefficients we obtain

$$\int_0^{2\pi} e^{\omega_1(\sigma)} c(\sigma) e^{-ik\sigma} d\sigma = 0 \quad (k = 0, 1, \ldots, -\varkappa - 1) \tag{29.19}$$

or, in the form of real relations

$$\int_0^{2\pi} e^{\omega_1(\sigma)} c(\sigma) \cos k\sigma \, d\sigma = 0, \quad \int_0^{2\pi} e^{\omega_1(\sigma)} c(\sigma) \sin k\sigma \, d\sigma = 0 \tag{29.20}$$

$$(k = 0, 1, 2, \ldots, -\varkappa - 1).$$

Formulae (29.20) constitute $-2\varkappa - 1$ conditions of solubility to be satisfied by the free term in order that the non-homogeneous Hilbert problem with a negative index be soluble.

29.4. The Hilbert problem for exterior domain

We shall now briefly show how the above reasoning should be changed to deduce the solution of the Hilbert problem in the case of an infinite domain D^-, exterior with respect to a simple smooth contour L.

Bearing in mind future applications, as in the case of an interior domain D^+, we shall assume (in determining the index) that the positive direction of traversing the contour L is anticlockwise[†]. We shall define in the considered case the regularizing factor, assuming that the origin of the coordinate system belongs to the domain D^+.

Suppose that $\quad \text{Ind}\,[a(s) + ib(s)] = \varkappa.$

The regularizing factor for the complex function $a(s) + ib(s)$ prescribed on the contour L, with respect to the exterior domain D^- is a real positive-definite function $p(s)$, such that the product $p(s)[a(s) + ib(s)]$ is the boundary value of a function analytic in D^- and having zero order everywhere in D^-, except possibly at the point at infinity where its order is equal to the index taken with negative sign $(-\varkappa)$.

Consequently, according to the definition, we have[‡]

$$p(s)\,[a(s) + ib(s)] = t^\varkappa e^{i\gamma(t)} \tag{29.21}$$

[†] In other words, in the positive direction with respect to the domain D^+ and in the negative direction with respect to D^-.

[‡] See § 28.2.

where $\gamma(z) = \omega(x, y) + i\omega_1(x, y)$ is a function analytic everywhere in D^- including the point at infinity.

Reasoning similar to that of § 28.2 leads to the relations

$$\gamma(z) = S\left[\arctan\frac{b}{a} - \varkappa \arg t\right], \quad p(s) = \frac{|t|^\varkappa e^{-\omega_1(s)}}{\sqrt{[a^2(s) + b^2(s)]}}, \quad (29.22)$$

where S is the Schwarz operator for the domain D^-.

We now proceed to the Hilbert problem.

Assuming as before that the condition

$$a^2(s) + b^2(s) = 1$$

is satisfied, we can write the boundary condition of the Hilbert problem in the form

$$\mathrm{Re}\left[\frac{F(t)}{a(s) + ib(s)}\right] = c(s). \quad (29.23)$$

Dividing by the regularizing factor we obtain

$$\mathrm{Re}\left[\frac{F(t)}{t^\varkappa e^{i\gamma(t)}}\right] = |t|^{-\varkappa} e^{\omega_1(s)} c(s). \quad (29.24)$$

The reasoning is now analogous to that in § 29.2, the only difference being that instead of the origin, the exceptional point is the point at infinity. Thus, we obtain the following results.

1. For $\varkappa \leq 0$

$$F(z) = z^\varkappa e^{i\gamma(z)}[S(|t|^{-\varkappa} e^{\omega_1(s)} c(s)) + Q(z)], \quad (29.25)$$

where $Q(z)$ is a function analytic everywhere in D^- except at the point at infinity where it has a pole of order not exceeding $-\varkappa$, the real part of this function vanishing on the contour. It is easily observed that the function $Q(z)$ given for the unit circle by formula (27.6) can also be employed in the case of the exterior of the unit circle. It is more convenient to use the expression for $Q(z)$ in a somewhat different form, replacing z by $1/z$ in formula (27.6). Thus, the function $Q(z)$ for the domain exterior with respect to the unit circle can be written in the form

$$Q(z) = i\beta_0 + \sum_{k=1}^{-\varkappa} (c_k z^{-k} - \bar{c}_k z^k). \quad (29.26)$$

For an arbitrary domain D^- it has the form

$$Q(z) = i\beta_0 + \sum_{k=1}^{-\varkappa} \{c_k[\omega(z)]^{-k} - \bar{c}_k[\omega(z)]^k\}, \qquad (29.27)$$

where $\omega(z)$ is the function which maps conformally the domain D^- onto the domain exterior with respect to the unit circle and transforming the point at infinity into the point at infinity. In particular, for $\varkappa = 0$ the function $Q(z)$ is an arbitrary imaginary constant $i\beta_0$.

2. $\varkappa > 0$. The solution for this case can be obtained from formula (29.25), setting $Q(z) \equiv 0$. In this case the function given by formula (29.25) has a pole of order $\varkappa$ at the point at infinity. Hence, the homogeneous Hilbert problem is in this case insoluble and the non-homogeneous problem is soluble only if certain conditions are satisfied; the latter can be obtained by expanding the function $S|t|^{\varkappa}e^{\omega_1(s)} c(s)$ into series in the vicinity of infinity and equating to zero the first $\varkappa$ coefficients of expansion.

In the case of a domain exterior with respect to the unit circle, it is known† that the Schwarz operator is equal to the Schwarz integral taken with negative sign. Expanding the Schwarz integral into power series of $1/z$ and equating to zero the first $\varkappa$ coefficients we find that the $2\varkappa - 1$ conditions of solubility of the Hilbert problem for the exterior of the unit circle are given by the same formulae (29.20) as in the case of the unit circle.

We finally state the deduced results.

In the case $\varkappa \leqq 0$ the homogeneous Hilbert problem has $-2\varkappa + 1$ linearly independent solutions and the non-homogeneous problem is absolutely soluble and its solution depends linearly on $-2\varkappa + 1$ arbitrary real constants (the solution is given by formula (29.25)).

In the case $\varkappa > 0$ the homogeneous Hilbert problem is insoluble and the non-homogeneous problem is soluble only if $2\varkappa - 1$ conditions are satisfied. If the latter conditions are satisfied the non-homogeneous problem has a unique solution.

It is easy to see that the theorem stated can be deduced from the corresponding theorem of § 29.2 for the domain D^+, if in the latter $\varkappa$ be replaced by $-\varkappa$. If, in defining the index we took the direction on the contour not anticlockwise but, which is more natural for the domain D^-, clockwise, then our last result would be identical with the corresponding theorem for the domain D^+.

---

† See for instance L. V. Kantorovitch and V. I. Krylov, *Approximate Methods of Higher Analysis*, p. 594 (1962).

29.5. Examples

1. Solve the Hilbert boundary value problem

$$(\cos 2s - \cos s + \tfrac{1}{4}) u(s) + (\sin 2s - \sin s)\, v(s) = (\tfrac{5}{4} - \cos s)^2 (\cos 2s + h_1 \cos s + h_2),$$

where h_1, h_2 are some constants, in the case of the unit circle.

We write the boundary condition in the form (29.3′)

$$\mathrm{Re}\,\{[(\cos s - \tfrac{1}{2})^2 - \sin^2 s - i 2\sin s\,(\cos s - \tfrac{1}{2})]\,(u + iv)\}$$
$$= (\tfrac{5}{4} - \cos s)^2 (\cos 2s + h_1 \cos s + h_2).$$

Performing the transformation

$$a(s) - ib(s) = (\cos s - \tfrac{1}{2})^2 - \sin^2 s - 2i\sin s\,(\cos s - \tfrac{1}{2})$$
$$= [(\cos s - \tfrac{1}{2}) - i\sin s]^2 = \frac{(\tfrac{5}{4} - \cos s)^2}{[(\cos s - \tfrac{1}{2}) + i\sin s]^2} = \frac{(\tfrac{5}{4} - \cos s)^2}{(t - \tfrac{1}{2})^2}$$

we represent it as follows:

$$\mathrm{Re}\left[\frac{F(t)}{(t - \tfrac{1}{2})^2}\right] = \cos 2s + h_1 \cos s + h_2.$$

The left-hand side is the boundary value of a function analytic in the unit circle, except for the point $z = \tfrac{1}{2}$ where it may have a pole of second order. In the general considerations we applied a preliminary transformation to reduce the boundary condition to such a form that the possible pole was located at the origin of the coordinate system. However, in solving definite examples, when the regularizing factor can be found by the simple method of the analytic continuation and when it is known beforehand what is the position of the point with non-zero order of the solution, this transformation can be avoided. (In this case it would lead to complicated computations.) Reasoning analogous to that in the general case leads to the relation

$$F(z) = (z - \tfrac{1}{2})^2 [S(\cos 2s + h_1 \cos s + h_2) + Q_1(z)],$$

where $Q_1(z)$ is a function analytic in the unit circle except at the point $z = \tfrac{1}{2}$ where it has a pole of second order, the real part of which vanishes on the unit circle. This function can be computed by means of the formula

$$Q_1(z) = Q(\omega(z)),$$

where $\omega(z)$ is the function which maps conformally the unit circle onto itself in such a way that the point $z = \tfrac{1}{2}$ is transformed into the origin of the coordinate system. For $\omega(z)$ we may take the function

$$\omega(z) = \frac{z - \tfrac{1}{2}}{1 - \tfrac{1}{2}z} = \frac{2z - 1}{2 - z}.$$

Since for $z = e^{is}$

$$\cos 2s + h_1 \cos s + h_2 = \mathrm{Re}\,[z^2 + h_1 z + h_2],$$

in view of the uniqueness of the analytic function defined by the Schwarz integral, we have

$$S(\cos 2s + h_1 \cos s + h_2) = z^2 + h_1 z + h_2.$$

Taking into account formula (29.14) for the function Q we finally obtain

$$F(z) = \left(z - \frac{1}{2}\right)^2 \left[z^2 + h_1 z + h_2 + i\beta + c_1 \frac{2z-1}{2-z} + \right.$$
$$\left. + c_2 \left(\frac{2z-1}{2-z}\right)^2 - \bar{c}_1 \left(\frac{2-z}{2z-1}\right) - \bar{c}_2 \left(\frac{2-z}{2z-1}\right)^2\right].$$

In view of the absolute solubility of the problem ($\varkappa = 2 > 0$) the constants h_1 and h_2 can be arbitrary.

2. Solve the Hilbert boundary value problem

$$(\cos 2s - \cos s + \tfrac{1}{4})u - (\sin 2s - \sin s)v = \cos 2s + h_1 \cos s + h_2$$

for the unit circle.

Since the index of the problem $\varkappa = -2 < 0$ the problem, generally speaking, is insoluble. It will be soluble only for a special choice of the constants h_1 and h_2.

Transformations analogous to those of the preceding problem lead to the following form of the boundary condition:

$$\text{Re}\,[(t - \tfrac{1}{2})^2 F(t)] = \cos 2s + h_1 \cos s + h_2.$$

Hence

$$F(z)(z - \tfrac{1}{2})^{-2} (z{=}^2 + h_1 z + h_2).$$

For arbitrary h_1 and h_2 the function given by the last formula has a pole of second order at the point $z = \tfrac{1}{2}$. To eliminate it the constants should be $h_1 = 1$, $h_2 = \tfrac{1}{4}$. Then the problem has a unique solution.

§ 30. Relation between the Hilbert and Riemann problems

30.1. Comparison of the formulae representing the solutions of the boundary value problems

Comparing the above solution of the Hilbert boundary value problem with the solution of the Riemann problem (Chapter II) the reader may observe a similarity. This suggests that there is a connection between the solutions themselves as well, despite the differences in their formulations. For the simplest contours (a straight line and a circle) for which there is a simple connection between the Schwarz operator and the Cauchy type integral, a direct reduction of the Hilbert problem to the Riemann problem can be proved to occur. For other contours there is no direct connection of this kind and it can be established only by means of the conformal mapping of the domains under consideration onto a circle or the semi-plane.

We shall indicate two methods of such a reduction. First we present a method given for the first time by the author in 1941, based on a comparison of the solutions. In the next section we shall present another method given later by N. I. Muskhelishvili, which establishes a connection between the boundary conditions.

Suppose that L is the unit circle ($t = e^{is}$).

THEOREM. *The "interior" function $\Phi^+(z)$ of the Riemann boundary value problem*

$$\Phi^+(t) = G(t)\,\Phi^-(t) + g(t) \tag{30.1}$$

with the coefficients

$$G(t) = \frac{a(s) + ib(s)}{a(s) - ib(s)}, \quad g(t) = \frac{2c(s)}{a(s) - ib(s)} \tag{30.2}$$

gives the solution of the Hilbert problem with the boundary condition

$$a(s)\,u(s) + b(s)\,v(s) = c(s) \tag{30.3}$$

provided that a special choice of the arbitrary constants entering the solution is made. The proof consists of three different cases.

1. $\varkappa = \operatorname{Ind}[a(s) + ib(s)] = 0$.

The function $\Phi^+(z)$ is given by formula (14.15)†:

$$\Phi^+(z) = X^+(z)\,[\Psi^+(z) + P_{2\varkappa}(z)], \tag{30.4}$$

where

$$X^+(z) = e^{\Gamma^+(z)}, \quad \Gamma(z) = \frac{1}{2\pi i}\int_L \frac{\ln G(\tau)}{\tau - z}\,d\tau,$$

$$\Psi(z) = \frac{1}{2\pi i}\int_L \frac{g(\tau)}{X^+(\tau)}\,\frac{d\tau}{\tau - z}, \quad P_{2\varkappa}(z) = C.$$

Substituting the value of $G(t)$ from (30.2) and making use of formula (6.4) relating the Cauchy type integral to the Schwarz integral, we have

$$\Gamma(z) = \frac{1}{2\pi i}\int_L \ln \frac{a(\sigma) + ib(\sigma)}{a(\sigma) - ib(\sigma)}\,\frac{d\tau}{\tau - z} = \frac{1}{2\pi i}\int_L 2i \arctan \frac{b(\sigma)}{a(\sigma)}\,\frac{d\tau}{\tau - z}$$

$$= \frac{1}{2\pi}\int_0^{2\pi} \arctan \frac{b(\sigma)}{a(\sigma)}\,\frac{e^{i\sigma} + z}{e^{i\sigma} - z}\,d\sigma + iC = i\gamma(z) + iC$$

† It should be borne in mind that the index of the Riemann problem, Ind $G(t)$, is $2\varkappa$, since the indices of complex conjugate functions have different signs (see § 12.1).

(here we made use of the formula $\ln z = \ln |z| + i \arg z$ and of the fact that $\left| \dfrac{a+ib}{a-ib} \right| = 1$). Hence, to within a constant factor

$$e^{\Gamma^+(z)} = e^{i\gamma(z)}. \tag{30.5}$$

Consequently, the solutions of the homogeneous Riemann and Hilbert problems are identical. Let us proceed to the non-homogeneous problems. On account of (30.5) we have

$$\Psi^+(z) = \frac{1}{2\pi i} \int_L \frac{2c(\sigma)}{a(\sigma) - ib(\sigma)} e^{-i\gamma(\tau)} \frac{d\tau}{\tau - z}.$$

In view of formulae (28.4), (28.8) and (29.2)

$$(a - ib) e^{i\gamma} = p(a^2 + b^2) = e^{-\omega_1}.$$

Hence

$$\Psi^+(z) = \frac{1}{\pi i} \int_L c(\sigma) e^{\omega_1(\sigma)} \frac{d\tau}{\tau - z} = \frac{1}{2\pi} \int_0^{2\pi} c(\sigma) e^{\omega_1(\sigma)} \frac{e^{i\sigma} + z}{e^{i\sigma} - z} d\sigma + i\beta_0.$$

Substituting the latter expression into formula (30.4) and assuming that the arbitrary constant C is a purely imaginary number we arrive at formula (29.15), which completes the proof.

2. $\varkappa > 0$. As before

$$\Gamma^+(z) = \frac{1}{2\pi i} \int_L \ln \left[\tau^{-2\varkappa} \frac{a(\sigma) + ib(\sigma)}{a(\sigma) - ib(\sigma)} \right] \frac{d\tau}{\tau - z}$$

$$= \frac{1}{\pi} \int_L \left[\arctan \frac{b(\sigma)}{a(\sigma)} - \varkappa\sigma \right] \frac{d\tau}{\tau - z} = i\gamma(z) + i\beta_0$$

and moreover, we find that

$$\Psi^+(z) = \frac{1}{2\pi} \int_0^{2\pi} c(\sigma) e^{\omega_1(\sigma)} e^{\varkappa i\sigma} \frac{e^{i\sigma} + z}{e^{i\sigma} - z} d\sigma + i\beta_0.$$

We now write down also the solution of the Riemann problem obtained by the substitution into formula (30.4) of the last expression,

nd the solution of the Hilbert problem in accordance with for-
nula (29.16)

$$\Phi^+(z) = e^{i\gamma(z)} \left[\frac{1}{2\pi} \int\limits_0^{2\pi} e^{\omega_1(\sigma)} c(\sigma) \frac{e^{i\sigma} + z}{e^{i\sigma} - z} e^{\varkappa i\sigma} d\sigma + P_{2\varkappa}(z) \right],$$

$$F(z) = e^{i\gamma(z)} \left[\frac{1}{2\pi} \int\limits_0^{2\pi} e^{\omega_1(\sigma)} c(\sigma) \frac{e^{i\sigma} + z}{e^{i\sigma} - z} z^\varkappa d\sigma + z^\varkappa Q(z) \right],$$

where $P_{2\varkappa}(z)$ is a polynomial of degree $2\varkappa$ with arbitrary coefficients,
and $Q(z)$ is given by formula (29.14).

Expanding in the last formulae the functions represented by Schwarz
integrals into power series in the vicinity of the origin of the coordinate system, we easily find that these expansions differ only in the
first $\varkappa$ terms. Including these terms in $P_{2\varkappa}(z)$ and then choosing its
arbitrary constants in such a way that the relation

$$P_{2\varkappa}(z) = z^\varkappa Q(z),$$

is satisfied, we discover that these solutions are identical.

3. $\varkappa < 0$. Reasoning analogous to that used previously leads to the
formulae

$$\Phi^+(z) = e^{i\gamma(z)} \frac{1}{2\pi} \int\limits_0^{2\pi} e^{\omega_1(\sigma)} c(\sigma) \frac{e^{i\sigma} + z}{e^{i\sigma} - z} e^{\varkappa i\sigma} d\sigma,$$

$$F(z) = e^{i\gamma(z)} \frac{1}{2\pi} \int\limits_0^{2\pi} e^{\omega_1(\sigma)} c(\sigma) \frac{e^{i\sigma} + z}{e^{i\sigma} - z} z^\varkappa d\sigma.$$

The conditions of solvability of the Riemann problem are given by
formulae (14.17), which in our case have the form

$$\int\limits_0^{2\pi} e^{\omega_1(\sigma)} c(\sigma) e^{-ki\sigma} d\sigma = 0 \quad (k = 0, 1, \ldots, -\varkappa - 1).$$

Consequently, they are identical with the conditions of solvability
(29.19) of the Hilbert problem. If these conditions are satisfied the
expressions for $\Phi^+(z)$ and $F(z)$ have the same expansion into the Taylor
series

$$\tfrac{1}{2} c_{-\varkappa} + \sum_{k=1}^\infty c_{-\varkappa+k} z^k$$

where c_k are the complex coefficients of the Fourier expansions of the function $e^{\omega_1(\sigma)} c(\sigma)$. This completes the proof of the theorem.

Now suppose that the contour L is the real axis. We show the validity of the theorem also for this case.

The reasoning is the same as before. Let $\varkappa \geqq 0$. From the formulae (14.22)–(14.24) we rewrite the solution of the Riemann problem slightly altering the form of the arbitrary polynomial:

$$\Phi^+(z) = X^+(z)\left[\Psi^+(z) + P_{2\varkappa}\left(\frac{z-i}{z+i}\right)\right], \qquad (14.22')$$

where

$$X^+(z) = e^{\Gamma^+(z)},$$

$$\Gamma(z) = \frac{1}{2\pi i} \int\limits_{-\infty}^{+\infty} \ln\left[\left(\frac{t-i}{t+i}\right)^{-2\varkappa} \frac{a+ib}{a-ib}\right] \frac{dt}{t-z},$$

$$\Psi(z) = \frac{1}{\pi i} \int\limits_{-\infty}^{\infty} \frac{c(t)}{(a-ib)X^+(t)} \frac{dt}{t-z}.$$

We establish the relationship between $\Gamma^+(z)$ and the determined formula (29.13') for $\gamma(z)$. Since

$$\left|\left(\frac{t-i}{t+i}\right)^{-2\varkappa} \frac{a+ib}{a-ib}\right| = 1,$$

herefore

$$\Gamma^+(z) = \frac{1}{2\pi i} \int\limits_{-\infty}^{+\infty} i \arg\left[\left(\frac{t-i}{t+i}\right)^{-2\varkappa} \frac{a+ib}{a-ib}\right] \frac{dt}{t-z}$$

$$= \frac{1}{\pi} \int\limits_{-\infty}^{+\infty} \frac{\theta(t)}{t-z} dt = i\gamma(z).$$

Consequently

$$X^+(z) = e^{i\gamma(z)}.$$

Reasoning as in the case of the circle, we likewise have

$$X^+(t) \cdot (a-ib) = e^{i\gamma(z)}(a-ib) = \left(\frac{t-i}{t+i}\right)^{-2\varkappa} e^{-\omega_1}.$$

We put the latter in (14.22') and write out together a the solutions of the Riemann and Hilbert problems:

$$\Phi^+(z) = e^{i\gamma(z)} \left[\frac{1}{\pi i} \int\limits_{-\infty}^{+\infty} \left(\frac{t-i}{t+i}\right)^{-2\varkappa} \frac{e^{\omega_1(t)} c(t)}{t-z} dt + P_{2\varkappa}\left(\frac{z-i}{z+i}\right) \right],$$

$$F(z) = e^{i\gamma(z)} \left[\frac{1}{\pi i} \int\limits_{-\infty}^{+\infty} \left(\frac{z-i}{z+i}\right)^{2\varkappa} \frac{e^{\omega_1(t)} c(t)}{t-z} dt + \left(\frac{z-i}{z+i}\right)^{2\varkappa} Q(z) \right].$$

Expanding the integrals into power series of $(z-i)/(z+i)$ it is easy to see that these expansions differ only in $2\varkappa$ least significant terms of the expansion. Incorporating these terms in the polynomial $P_{2\varkappa}$, and then selecting its arbitrary coefficients such that the following equality is satisfied,

$$P_{2\varkappa}\left(\frac{z-i}{z+i}\right) = \left(\frac{z-i}{z+i}\right)^{2\varkappa} Q(z),$$

we establish that the solutions coincide.

If $\varkappa < 0$ the reasoning is the same; we need not dwell on this.

30.2. Connection between the boundary conditions

As before we suppose that the contour L is a straight line or a circle.

The difference in the formulations of the Riemann and Hilbert boundary value problems consists above all in the fact that in the first problem we seek a sectionally analytic function defined in the entire plane, whereas in the second problem we seek a function defined only in the domain D^+ and the complementary domain D^- is not considered at all. Hence, in order to establish a connection between the above boundary value problems on the basis of their formulations, first of all it is necessary to find an expedient method of complementary definition of the analytic function $\Phi^+(z)$, constituting the solution of the Hilbert problem in the domain D^- in such a way that a sectionally analytic function is obtained, defined on the entire plane.

Since, according to the assumption, the contour L is a straight line or a circle, this complementary definition can be performed, as the further investigation proves, by the symmetric continuation exactly in the same manner as is done in the well-known theory of analytic functions (the principle of symmetry). The difference consist only in the fact that since in our case the conditions of this principl cannot be valid (a segment of a straight line or an arc of a circl should be mapped by this function onto a segment of a straight lin or an arc of a circle), our continuation, generally speaking, is no analytic. However, we do not require analyticity, since all we want i to construct in the domain D^- a new function $\Phi^-(z)$ which woul be connected with the function $\Phi^+(z)$ determined from the Hilber boundary value problem, in such a way that the boundary conditio of the latter problem becomes equivalent to the boundary conditio of the Riemann problem which relates $\Phi^+(z)$ to $\Phi^-(z)$.

We shall consider the symmetry with respect to a straight line i the usual sense, and the symmetry with respect to a circle as inversio with respect to this circle. A point symmetric to a point z we denot by z_*. If the contour is the real axis, $z_* = \bar{z}$, whereas when it is th unit circle we have $z_* = 1/z$.

We shall hereafter employ the notation of § 18.1.

Let us continue the function $\Phi^+(z)$ given in the domain D^+ (uni circle or the upper semi-plane) into the domain D^- (exterior of th unit circle or the lower semi-plane), assuming that at points symmetri with respect to the contour the functions take the complex conjugat values, i.e.

$$\Phi^-(z) = \overline{\Phi^+(\bar{z})} = \overline{\Phi}^+(z) \tag{30.6}$$

in the case of the real axis, and

$$\Phi^-(z) = \overline{\Phi^+\left(\frac{1}{\bar{z}}\right)} = \overline{\Phi}^+\left(\frac{1}{z}\right) \tag{30.7}$$

in the case of the unit circle.

It is easy to establish that the function $\Phi^-(z)$ defined in this wa is analytic in the domain D^-. The proof is given in books o functions of a complex variable, in proving the principle of symmetry therefore we shall not dwell on this topic here.

The union of functions $\Phi^+(z)$ and $\Phi^-(z)$ can be regarded as on sectionally analytic function. Let us now find the connection betwee their limiting values on the contour.

Let the point z tend to a point t of the contour, from the domain D^- Then, taking into account that for both the real axis and the circl we have $t_* = \bar{t} = t$ and $t_* = 1/\bar{t} = 1/e^{-is} = e^{is} = t$, respectively we obtain from formulae (30.6) and (30.7)

$$\Phi^-(t) = \overline{\Phi^+(\bar{t})} = \overline{\Phi^+(t)}, \tag{30.8}$$

$$\Phi^-(t) = \Phi^+\left(\frac{1}{\bar{t}}\right) = \overline{\Phi^+(t)}. \tag{30.9}$$

Thus, the limiting values of the functions $\Phi^+(z)$ and $\Phi^-(z)$ constitute complex conjugate functions on the contour. Consequently, a function $\overline{\Phi^+(t)}$ prescribed on the contour can be regarded as the limiting value of the function $\Phi^-(t)$†.

We are now in a position to prove the possibility of reducing the Hilbert problem to the Riemann problem.

Let us write the boundary condition of the Hilbert problem

$$\text{Re}\,[(a - ib)\,(u + iv)] = c$$

in the form

$$(a - ib)\,\Phi^+(t) + (a + ib)\,\overline{\Phi^+(t)} = 2c. \tag{30.10}$$

Replacing now $\overline{\Phi^+(t)}$ by $\Phi^-(t)$ and solving for $\Phi^+(t)$ we arrive at the boundary condition of the Riemann problem

$$\Phi^+(t) = -\frac{a(s) + ib(s)}{a(s) - ib(s)}\Phi^-(t) + \frac{2c(s)}{a(s) - ib(s)}. \tag{30.11}$$

Consequently, the solution of the Hilbert boundary value problem (30.3) is equivalent to the determination of the function $\Phi^+(z)$ from the solution of the Riemann boundary value problem (30.11), under the condition that for any z not lying on the contour

$$\Phi^-(z) = \overline{\Phi^+(z_*)}. \tag{30.12}$$

In the foregoing section it was shown that the solution of the Riemann problem becomes that of the Hilbert problem for specially selected coefficients of the polynomial $P_{2\varkappa}$. This implies that the solution of the Riemann problem satisfies condition (30.12) with the appropriate coefficients.

Comparing the boundary condition (30.11) with the boundary condition (30.1) of the preceding section, we observe that their coefficients $G(t)$ differ in sign. However, taking into account that $\ln(-G) = \ln G + \pi i$ we readily observe that a change of the sign

† In the simplest case when $\Phi^+(t)$ is real, the continuation is analytic.

of $G(t)$ results in the appearance of a constant factor in $X(z)$, which has no influence on the general solution.

The method of reduction of the Hilbert problem to the Riemann problem, given in this section, has the merit over the first method that it is not based on an explicit solution of these problems. Hence, this method can also be used in the case when such explicit solutions do not exist, for instance in the corresponding problems with many unknown functions (see for instance H. P. Vekua [31], p. 172). Moreover, instead of using an independent solution of the Hilbert problem we could employ the solution of the Riemann problem, used in the book of N. I. Muskhelishvili [17*]. I considered it, however, expedient to present an independent solution of the Hilbert problem, in view of its remarkable originality. It should be observed that all considerations and operations required to adapt the solution of the Riemann problem to the solution of the Hilbert problem, having particularly in mind the use of the latter in investigating singular integral equations with Hilbert kernel, are not shorter than those required for a direct solution of the Hilbert problem.

§ 31. Singular integral equation with Hilbert kernel

31.1. Connection of the dominant equation with the Hilbert boundary value problem

It was mentioned in § 6.2 that by the Hilbert kernel we understand the function $\cot \frac{1}{2}(\sigma - s)$. The singular integral equation of the form

$$a(s)u(s) - \frac{b(s)}{2\pi} \int_0^{2\pi} u(\sigma)\cot\frac{\sigma - s}{2} d\sigma = c(s), \qquad (31.1)$$

where $a(s)$, $b(s)$, $c(s)$ are given functions satisfying the Hölder condition, is called *the dominant singular equation with Hilbert kernel*. If $c(s) = 0$ it is called homogeneous, in the opposite case non-homogeneous. The solution will be sought in the class of functions satisfying the Hölder condition.

Similarly to the dominant singular integral equation with Cauchy kernel which is connected with the Riemann boundary value problem, the dominant equation (31.1) with Hilbert kernel is even more closely related to the Hilbert problem. We shall now prove this assertion

We assume that the function $u(s)$ which constitutes the solution of equation (31.1) is the boundary value of a function $u(x, y)$ harmonic in the unit circle. In view of the absolute solubility of the Dirichlet problem in the case of an arbitrary continuous boundary value, this assumption is legitimate. Then in view of formula (6.6) we have

$$v(s) = -\frac{1}{2\pi} \int_0^{2\pi} u(\sigma) \cot \frac{\sigma - s}{2} d\sigma + \int_0^{2\pi} v(\sigma) d\sigma. \qquad (31.2)$$

Setting

$$\int_0^{2\pi} v(\sigma) d\sigma = 0, \qquad (31.3)$$

we pass from the singular equation (31.1) to the Hilbert boundary value problem

$$a(s) u(s) + b(s) v(s) = c(s) \qquad (31.4)$$

for the function $F(z) = u(x, y) + iv(x, y)$ analytic in the unit circle. Thus, we have established the following fact.

THEOREM. *The dominant singular equation with Hilbert kernel* (21.1) *is equivalent to the Hilbert problem* (31.4) *for the circle with the additional condition* (31.3), *in the sense that if* $F(z)$ *is the solution of the boundary value problem* (31.4) *satisfying condition* (31.3), *then the boundary value of its real part is the solution of equation* (31.1), *and conversely, given* $u(s)$, *the solution of equation* (31.1), *by making use of the Schwarz operator and condition* (31.3) *we obtain the solution of the Hilbert problem* (31.4).

We shall later employ the solution of the Hilbert problem deduced in § 29 to solve the dominant equation (31.1).

31.2. The homogeneous equation

In view of the theorem proved above, to obtain the solution of the homogeneous dominant equation

$$a(s) u(s) - \frac{b(s)}{2\pi} \int_0^{2\pi} u(\sigma) \cot \frac{\sigma - s}{2} d\sigma = 0 \qquad (31.5)$$

it is necessary to find the solution of the homogeneous Hilbert boundary value problem

$$a(s) u(s) + b(s) v(s) = 0, \qquad (31.6)$$

satisfying the additional condition (31.1), and to take the boundary value of its real part.

It follows from the properties of harmonic functions that condition (31.3) implies vanishing of the harmonic function $v(x, y)$ at the origin of the coordinate system.

We shall call the index of the function $a(s) + ib(s)$ *the index of the dominant equation* (31.5). Let us consider separately various cases.

1. $\varkappa = \mathrm{Ind}\,[a(s) + ib(s)] = 0$. The function $F(z) = i\beta_0 e^{i\gamma(z)}$ is the solution of the Hilbert problem, $\gamma(z)$ being given by formula (29.13); β_0 is an arbitrary real constant.

If $t = e^{is}$, then

$$F(t) = u(s) + iv(s) = i\beta_0 e^{i\gamma(t)}.$$

Setting

$$ie^{i\gamma(t)} = \xi(s) + i\eta(s) \tag{31.7}$$

we obtain

$$u(s) = \beta_0 \xi(s), \quad v(s) = \beta_0 \eta(s).$$

The solubility condition of equation (31.5) takes the form

$$\int_0^{2\pi} \eta(\sigma)\,d\sigma = 0. \tag{31.8}$$

Thus, when the index is zero two cases are possible.

(i) If condition (31.8) is satisfied, the homogeneous dominant equation (31.5) has the solution †

$$u(s) = \beta_0 \xi(s). \tag{31.9}$$

(ii) If $\int_0^{2\pi} \eta(\sigma)\,d\sigma \neq 0$ the equation has no solution.

2. $\varkappa > 0$. The solution of the Hilbert problem has the form

$$F(z) = z^{\varkappa} e^{i\gamma(z)} Q(z), \tag{31.10}$$

$Q(z)$ being given by formula (29.14).

Introducing as before the notation

$$i t^{\varkappa} e^{i\gamma(t)} = \xi(s) + i\eta(s)$$

and taking into account that $\mathrm{Re}\,Q(t) = 0$ we obtain

$$u(s) = \xi(s)\,\mathrm{Im}\,Q(t), \quad v(s) = \eta(s)\,\mathrm{Im}\,Q(t).$$

Consequently, equation (31.5) has the solution

$$u(s) = \xi(s)\,\mathrm{Im}\,Q(t) \tag{31.11}$$

---

† This constitutes the difference between equations with Hilbert and Cauchy kernels, the latter being in the corresponding case always insoluble.

provided the additional condition

$$\int_0^{2\pi} \eta(\sigma)\,\mathrm{Im}\,Q(\tau)\,d\sigma = 0. \tag{31.12}$$

is satisfied.

Setting in formula (29.14) $c_k = \alpha_k + i\beta_k$ we obtain

$$\mathrm{Im}\,Q(t) = \beta_0 + 2\sum_{k=1}^{\varkappa}(\alpha_k \sin ks + \beta_k \cos ks), \tag{31.13}$$

whence formulae (31.11), (31.12) take the form

$$u(s) = \xi(s)\left[\beta_0 + 2\sum_{k=1}^{\varkappa}(\alpha_k \sin ks + \beta_k \cos ks)\right], \tag{31.14}$$

$$\beta_0 \int_0^{2\pi} \eta(\sigma)\,d\sigma +$$

$$+ 2\sum_{k=1}^{\varkappa}\left[\alpha_k \int_0^{2\pi} \eta(\sigma)\sin k\sigma\,d\sigma + \beta_k \int_0^{2\pi} \eta(\sigma)\cos k\sigma\,d\sigma\right] = 0. \tag{31.15}$$

We now prove that the latter condition cannot identically be satisfied and, consequently, it determines one of the arbitrary constants entering the general solution (31.14).

The function $iz^\varkappa e^{i\gamma(z)}$ has a zero of order $\varkappa$ at the origin; thus, the first $\varkappa$ coefficients of its expansion into the power series in the vicinity of this point vanish, i.e.

$$c_k = a_k - ib_k = 0 \quad (k = 0, 1, 2, \ldots, \varkappa - 1).$$

The quantities a_k, b_k, however, are the Fourier coefficients of the expansion of $\eta(s)$. Hence

$$a_k = \frac{1}{\pi}\int_0^{2\pi} \eta(\sigma)\cos k\sigma\,d\sigma = 0, \quad b_k = \frac{1}{\pi}\int_0^{2\pi} \eta(\sigma)\sin k\sigma\,d\sigma = 0$$

$$(k = 0, 1, \ldots, \varkappa - 1).$$

Consequently, relation (31.15) is reduced to the following one:

$$\alpha_\varkappa \int_0^{2\pi} \eta(\sigma)\sin \varkappa\sigma\,d\sigma + \beta_\varkappa \int_0^{2\pi} \eta(\sigma)\cos \varkappa\sigma\,d\sigma = 0; \tag{31.15'}$$

the coefficients

$$a_\varkappa = \frac{1}{\pi} \int_0^{2\pi} \eta(\sigma) \cos \varkappa \sigma \, d\sigma, \quad b_\varkappa = \frac{1}{\pi} \int_0^{2\pi} \eta(\sigma) \sin \varkappa \sigma \, d\sigma$$

cannot simultaneously vanish, since $c_\varkappa = a_\varkappa - ib_\varkappa \neq 0$.

These results imply that condition (31.15) makes it possible to determine one of the coefficients $\alpha_\varkappa$, $\beta_\varkappa$ and we arrive at the following conclusion. *The homogeneous dominant singular integral equation with Hilbert kernel (31.5) the index of which $\varkappa > 0$, has exactly $2\varkappa$ linearly independent solutions.*

It follows from the preceding investigation that the case $\varkappa = 0$ cannot be obtained as a particular case of the case $\varkappa > 0$, as it occured in the case of dominant equations with Cauchy kernel.

3. $\varkappa < 0$. In this case the Hilbert problem is insoluble in the class of analytic functions and, consequently, the singular equation (31.5) is also insoluble.

31.3. The non-homogeneous equation

By virtue of the theorem proved in § 31.1 the solution of the non-homogeneous dominant equation

$$a(s) u(s) - \frac{b(s)}{2\pi} \int_0^{2\pi} u(\sigma) \cot \frac{\sigma - s}{2} \, d\sigma = c(s) \tag{31.1}$$

can be deduced by solving the Hilbert boundary value problem

$$a(s) u(s) + b(s) v(s) = c(s), \tag{31.4}$$

with the additional condition

$$\int_0^{2\pi} v(\sigma) \, d\sigma = 0, \tag{31.3}$$

and taking the boundary value of its real part.

1. $\varkappa = 0$. The solution of the Hilbert problem has in accordance with (29.15) the form

$$F(z) = e^{i\gamma(z)} \left[\frac{1}{2\pi} \int_0^{2\pi} c(\sigma) e^{\omega_1(\sigma)} \frac{e^{i\sigma} + z}{e^{i\sigma} - z} \, d\sigma + i\beta_0 \right]. \tag{31.16}$$

Setting

$$ie^{i\gamma(z)} = \xi + i\eta, \quad \frac{1}{2\pi} \int_0^{2\pi} c(\sigma) \, e^{\omega_1(\sigma)} \frac{e^{i\sigma} + z}{e^{i\sigma} - z} \, d\sigma = \Psi(z),$$

taking into account the properties of the Schwarz integral, we have on the contour,

$$\operatorname{Re}\Psi(t) = c(s) \, e^{\omega_1(s)}, \quad \operatorname{Im}\Psi(t) = -\frac{1}{2\pi} \int_0^{2\pi} c(\sigma) \, e^{\omega_1(\sigma)} \cot \frac{\sigma - s}{2} \, d\sigma.$$

$$\tag{31.17}$$

Taking the contour values in formulae (31.16) and bearing in mind formulae (31.17) we obtain

$$\left.\begin{aligned}
u(s) &= \beta_0 \xi(s) + \eta(s) \, c(s) \, e^{\omega_1(s)} - \\
&\qquad - \xi(s) \frac{1}{2\pi} \int_0^{2\pi} c(\sigma) \, e^{\omega_1(\sigma)} \cot \frac{\sigma - s}{2} \, d\sigma, \\
v(s) &= \beta_0 \eta(s) - \xi(s) \, c(s) \, e^{\omega_1(s)} - \\
&\qquad - \eta(s) \frac{1}{2\pi} \int_0^{2\pi} c(s) \, e^{\omega_1(\sigma)} \cot \frac{\sigma - s}{2} \, d\sigma.
\end{aligned}\right\} \tag{31.18}$$

In view of (31.3) the solubility condition

$$\beta_0 \int_0^{2\pi} \eta(s) \, ds - \int_0^{2\pi} \xi(s) \, c(s) \, e^{\omega_1(s)} \, ds -$$

$$- \int_0^{2\pi} \eta(s) \, ds \frac{1}{2\pi} \int_0^{2\pi} c(\sigma) \, e^{\omega_1(\sigma)} \cot \frac{\sigma - s}{2} \, d\sigma = 0$$

should be satisfied.

Changing the order of integration in the double integral and taking into account that, according to the Hilbert formula (6.7),

$$\frac{1}{2\pi} \int_0^{2\pi} \eta(s) \cot \frac{s - \sigma}{2} \, ds = \xi(\sigma) - \frac{1}{2\pi} \int_0^{2\pi} \xi(\sigma) \, d\sigma,$$

we obtain

$$\beta_0 \int_0^{2\pi} \eta(\sigma) \, d\sigma + \int_0^{2\pi} c(\sigma) \, e^{\omega_1(\sigma)} \, d\sigma \cdot \frac{1}{2\pi} \int_0^{2\pi} \xi(\sigma) \, d\sigma = 0. \tag{31.19}$$

Observe that the integrals

$$\frac{1}{2\pi}\int\limits_0^{2\pi}\xi(\sigma)\,d\sigma, \quad \frac{1}{2\pi}\int\limits_0^{2\pi}\eta(\sigma)\,d\sigma$$

represent the values of the real and imaginary parts of the function $ie^{i\gamma(z)} = \xi(x,y) + i\eta(x,y)$ at the origin and hence they cannot simultaneously vanish.

The following two cases are possible.

(i) If $\int\limits_0^{2\pi}\eta(\sigma)\,d\sigma \neq 0$, i.e. the corresponding homogeneous equation is insoluble, then condition (31.19) makes it possible to determine the arbitrary constant β_0 and, consequently, the non-homogeneous equation (31.1) has the unique solution (31.18).

(ii) If $\int\limits_0^{2\pi}\eta(\sigma)\,d\sigma = 0$, i.e. the corresponding homogeneous equation is soluble, then the non-homogeneous equation is in general insoluble.

Taking into account that $\int\limits_0^{2\pi}\xi(\sigma)\,d\sigma \neq 0$, we obtain from (31.19) its solubility condition

$$\int\limits_0^{2\pi}c(\sigma)\,e^{\omega_1(\sigma)}\,d\varphi = 0. \tag{31.20}$$

2. $\varkappa > 0$. In this case formula (29.16) yields the solution of the Hilbert problem

$$F(z) = z^\varkappa e^{i\gamma(z)}\left[\frac{1}{2\pi}\int\limits_0^{2\pi}c(\sigma)\,e^{i\omega_1(\sigma)}\frac{e^{i\sigma}+z}{e^{i\sigma}-z}\,d\sigma + Q(z)\right].$$

In the usual notation we have

$$u(s) = \eta(s)\,c(s)\,e^{\omega_1(s)} - \xi(s)\frac{1}{2\pi}\int\limits_0^{2\pi}c(\sigma)\,e^{\omega_1(\sigma)}\cot\frac{\sigma-s}{2}\,d\sigma +$$

$$+ \xi(s)\,\mathrm{Im}\,Q(t), \tag{31.21}$$

$$v(s) = -\,\xi(s)\,c(s)e^{\omega_1(s)} - \eta(s)\frac{1}{2\pi}\int\limits_0^{2\pi}c(\sigma)\,e^{\omega_1(\sigma)}\cot\frac{\sigma-s}{2}\,d\sigma +$$

$$+ \eta(s)\,\mathrm{Im}\,Q(t), \tag{31.21'}$$

assuming that condition (31.3) is satisfied.

Inserting into expression (31.21) the value of Im $Q(t)$ from formula (31.13) we obtain the solution

$$u(s) = \eta(s)\, c(s)\, e^{\omega_1(s)} - \xi(s) \frac{1}{2\pi} \int_0^{2\pi} c(\sigma)\, e^{\omega_1(\sigma)} \cot \frac{\sigma - s}{2}\, d\sigma +$$

$$+ \xi(s) \left[\beta_0 + 2 \sum_{k=1}^{\varkappa} (\alpha_k \sin ks + \beta_k \cos ks) \right]. \quad (31.22)$$

Reasoning similar to that for the corresponding case of the homogeneous equation proves that condition (31.3) determines one arbitrary constant, and consequently, the general solution given by formula (31.22) contains linearly $2\varkappa$ arbitrary real constants. This result can also be deduced from the fact that the solution of the non-homogeneous equation contains the general solution of the homogeneous equation.

3. $\varkappa < 0$. In this case, according to formula (29.17), the solution of the Hilbert problem has the form

$$F(z) = z^{\varkappa}\, e^{i\gamma(z)} \frac{1}{2\pi} \int_0^{2\pi} c(\sigma)\, e^{\omega_1(\sigma)} \frac{e^{i\sigma} + z}{e^{i\sigma} - z}\, d\sigma,$$

and for the solubility of the problem it is necessary that conditions (29.20) be satisfied. The solution of the singular equation can be derived from formula (31.21) if we set there $Q(t) \equiv 0$, i.e.

$$u(s) = \eta(s)\, c(s)\, e^{\omega_1(s)} - \xi(s) \frac{1}{2\pi} \int_0^{2\pi} c(\sigma)\, e^{\omega_1(\sigma)} \cot \frac{\sigma - s}{2}\, d\sigma, \quad (31.23)$$

assuming that condition (31.3) is satisfied.

It can be proved that conditions (29.20) and (31.3) are independent and that the latter can be written in the form

$$\int_0^{2\pi} \int_0^{2\pi} c(\sigma)\, \xi(s)\, e^{\omega_1(\sigma)} \sin \varkappa(\sigma - s)\, ds\, d\sigma = 0. \quad (31.24)$$

Completing the above by conditions (29.20)

$$\int_0^{2\pi} c(\sigma)\, e^{\omega_1(\sigma)} \cos k\sigma\, d\sigma = 0, \quad \int_0^{2\pi} c(\sigma)\, e^{\omega_1(\sigma)} \sin k\sigma\, d\sigma = 0 \quad (31.25)$$

$$(k = 0, 1, \ldots, -\varkappa - 1),$$

we obtain the following result.

In the case of a negative index $\varkappa$, the dominant non-homogeneous singular equation is soluble if and only if $-2\varkappa$ solubility conditions (31.24) and (31.25) are satisfied. If they are satisfied the solution is represented by formula (31.23).

31.4. Equation with constant coefficients

Consider the special case when the coefficients a and b of equations (31.5) and (31.1) are constant numbers. It is assumed that division by $\sqrt{(a^2 + b^2)}$ has already been carried out so that the condition $a^2 + b^2 = 1$ is satisfied.

1. The homogeneous equation is

$$au(s) - \frac{b}{2\pi} \int_0^{2\pi} u(\sigma) \cot \frac{\sigma - s}{2} d\sigma = 0. \qquad (31.5')$$

We can directly obtain the solution of the corresponding Hilbert problem (31.6):

$$F(z) = u + iv = i\beta_0 e^{i\gamma(z)} = i\beta_0(a + ib).$$

Hence

$$\xi(s) = -b, \quad \eta(s) = a, \quad e^{\omega_1(s)} = 1.$$

The solubility condition (31.8) is written as

$$\int_0^{2\pi} \eta(\sigma) d\sigma = a \cdot 2\pi = 0.$$

Consequently, if $a \neq 0$, the homogeneous equation has no non-trivial solutions.

If $a = 0$, the equation

$$\int_0^{2\pi} u(\sigma) \cot \frac{\sigma - s}{2} d\sigma = 0$$

has an arbitrary constant $u(s) = \beta_0$ as its solution.

2. The non-homogeneous equation

$$au(s) - \frac{b}{2\pi} \int_0^{2\pi} u(\sigma) \cot \frac{\sigma - s}{2} d\sigma = C(s) \qquad (31.1')$$

is unconditionally soluble by virtue of the foregoing if $a \neq 0$.

According to (31.19), we have

$$2\pi a\beta_0 - b \int_0^{2\pi} c(\sigma)\, d\sigma = 0.$$

Substituting for β_0 in (31.18), the solution of (31.1′) is

$$u(s) = ac(s) + \frac{b}{2\pi} \int_0^{2\pi} c(\sigma) \cot \frac{\sigma - s}{2}\, d\sigma - \frac{b^2}{2\pi a} \int_0^{2\pi} c(\sigma)\, d\sigma.$$

If $a = 0$, then for solubility of the equation

$$\frac{1}{2\pi} \int_0^{2\pi} u(\sigma) \cot \frac{\sigma - s}{2}\, d\sigma = c(s)$$

the following condition must be fulfilled, in accordance with (31.19),

$$\int_0^{2\pi} c(\sigma)\, d\sigma = 0. \qquad (31.26)$$

If it is fulfilled, then by putting $\eta(s) = 0$ and $\xi(s) = -b = 1$ in (31.18), we obtain the solution

$$u(s) = -\frac{1}{2\pi} \int_0^{2\pi} c(\sigma) \cot \frac{\sigma - s}{2} + C, \qquad (31.27)$$

where C is an arbitrary constant.

This latter solution could have been obtained from Hilbert's inversion formulae (§ 6.2). Setting in (6.7) $u_0 = 0$ and regarding $u(s)$ as given and $v(s)$ as the required function, it is seen that formula (6.6) gives a solution of equation (6.5) which coincides with (31.27). The condition

$$u_0 = \frac{1}{2\pi} \int_0^{2\pi} u(\sigma)\, d\sigma = 0.$$

coincides with (31.26).

31.5. The complete equation and its regularization

By a complete singular integral equation with Hilbert kernel we understand an equation of the form

$$K\varphi \equiv a(s)\,\varphi(s) - \frac{b(s)}{2\pi} \int_0^{2\pi} \varphi(\sigma) \cot \frac{\sigma - s}{2}\, d\sigma +$$

$$+ \int_0^{2\pi} K(s, \sigma)\,\varphi(\sigma)\, d\sigma = f(s), \quad (31.28)$$

where $a(s)$, $b(s)$, $f(s)$, $K(s, \sigma)$ are real functions†, $a(s)$, $b(s)$, $f(s)$ satisfying the Hölder condition and the kernel $K(s, \sigma)$ satisfying this condition everywhere except possibly at the point $\sigma = s$ where the following estimate holds:

$$|K(s, \sigma)| < \frac{A}{|\sigma - s|^\lambda} \quad (0 \leqq \lambda < 1).$$

The integral $\int_0^{2\pi} K(s, \sigma)\,\varphi(\sigma)\, d\sigma$ is called the regular part of the equation.

The dominant equation examined in the preceding section is, obviously, a particular case of the complete equation, when $K(s, \sigma) \equiv 0$.

There occur also equations in which the upper bound of integration is an arbitrary number l and the singular kernel has the form $\cot[\pi(\sigma - s)/l]$. This corresponds to an integral taken over a curve of length l, in which a change of the variable leads to real quantities. There is no need for a special investigation of such equations, since the change of the variables $s_1 = (2\pi/l)s$, $\sigma_1 = (2\pi/l)\sigma$ reduces them to those now considered.

Similarly to the analogous complete integral equation with Cauchy kernel, the solution of the complete equation (31.28) is performed by means of regularization. All three methods of regularization described in § 22 and § 24 are suitable in this case as well. In view of the complete analogy of the procedures of regularization of equation (31.28) and equation (22.19) of § 22.3, we shall omit the details and confine ourselves to fundamental facts only.

---

† We could also consider complex functions but it is not necessary to investigate such equations, since it is easier to reduce them to equations with Cauchy kernel (see § 31.6).

The general expression for the operator $\tilde{K}$ regularizing the operator K (from left and from right) can be written in the form

$$\tilde{K}\omega \equiv u(s)\,a(s)\,\omega(s) +$$

$$+ \frac{u(s)\,b(s)}{2\pi} \int_0^{2\pi} \omega(\sigma) \cot \frac{\sigma - s}{2}\, d\sigma + \int_0^{2\pi} K_1(s, \sigma)\,\omega(\sigma)\,d\sigma, \quad (31.29)$$

where $u(s)$ is an arbitrary function satisfying the Hölder condition and $K_1(s, \sigma)$ is an arbitrary kernel having the same properties as the given kernel $K(s, \sigma)$. The simplest regularizator is the operator deduced from (31.29) by setting $u(s) \equiv 1$ and $K_1(s, \sigma) \equiv 0$.

The regularization by solving the dominant equation can be carried out using formulae (31.18) and (31.22). Replacing $c(s)$ by the expression

$$f(s) - \int_0^{2\pi} K(s, \sigma)\,\varphi(\sigma)\,d\sigma$$

and performing transformations similar to those in § 24.6, we arrive at a Fredholm integral equation.

31.6. Basic properties of equation with Hilbert kernel

For equation (31.28) we could carry out all investigations of § 23 and § 24 for the equation with Cauchy kernel, the only difference being that instead of using the solution of the Riemann boundary value problem we would now have to use the solution of the Hilbert problem deduced in § 30. Since the equation considered now is real, all considerations would be simpler. For instance it can be proved that the operator K' conjugate to that under consideration is the equivalent regularizator for the former one (see Problem 19 at the end of this chapter); this property does not occur for complex operators. Observe that the basic properties of singular equations which are now called the Noether theorems, were first derived by her for equations of the form (31.28) by using the solution of the Hilbert problem, and only 20 years later these properties were deduced for complex equations with Cauchy kernel by applying the solution of the Riemann problem (see the Historical Notes for Chapter III and Chapter IV).

However, there is no necessity to derive independently the properties of the equation examined now. It is simpler to establish the reducibility of equation (31.28) to an equation with Cauchy kernel; then the entire theory of the latter equation can be applied directly to the former. Since the regular parts of the both types of equations are of the same nature, it is sufficient to find a relation between the Hilbert kernel $\cot[(\sigma - s)/2]$ and the Cauchy kernel $d\tau/(\tau - t)$; this is carried out in the same manner as in § 6.1.

Suppose that the contour of integration is the unit circle. Then

$$t = e^{is}, \quad \tau = e^{i\sigma}, \quad d\tau = ie^{i\sigma}\,d\sigma$$

and

$$\frac{d\tau}{\tau - t} = \frac{ie^{i\sigma}\,d\sigma}{e^{i\sigma} - e^{is}}.$$

Multiplying the numerator and denominator of the last expression by $e^{-i(\sigma+s)/2}$ and making use of the Euler formulae we obtain

$$\frac{d\tau}{\tau - t} = \left(\frac{1}{2}\cot\frac{\sigma - s}{2} + \frac{i}{2}\right)d\sigma,$$

or

$$\frac{1}{2}\cot\frac{\sigma - s}{2}\,d\sigma = \frac{d\tau}{\tau - t} - \frac{1}{2}\frac{d\tau}{\tau}. \qquad (31.30)$$

Replacing in equation (31.28) the Hilbert kernel by the latter expression and substituting $s = (1/i)\ln t$, $\sigma = (1/i)\ln\tau$, $d\sigma = (1/i)\,d\tau/\tau$ we reduce equation (31.28) to the equation with Cauchy kernel

$$a_1(t)\,\varphi_1(t) - \frac{ib_1(t)}{\pi i}\int_L \frac{\varphi_1(\tau)}{\tau - t}\,d\tau + \int_L k_1(t, \tau)\,d\tau = f_1(t). \qquad (31.31)$$

The coefficient of the Riemann problem corresponding to equation (31.31) has the form

$$G(t) = \frac{a_1(t) + ib_1(t)}{a_1(t) - ib_1(t)} = \frac{a(s) + ib(s)}{a(s) - ib(s)}. \qquad (31.32)$$

Hence

$$\operatorname{Ind} G(t) = 2\operatorname{Ind}[a(s) + ib(s)]. \qquad (31.33)$$

The last relation implies that if we attempt to express the properties of equation (31.28) in terms of its own index†

$$\varkappa = \operatorname{Ind}[a + ib],$$

then in all results of § 23 and § 24, $\varkappa$ has to be replaced by $2\varkappa$.

---

† N. I. Muskhelishvili [17*] by the index of equation (31.28) means the index of $G(t)$; and the same for the Hilbert problem. Hence, his index is equal to twice ours (see the footnote on p. 229).

For instance, Theorem III of Noether has the following form;

The difference of the numbers of linearly independent solutions of the equation $K\varphi = 0$ (31.26) *and its adjoint* $K'\psi = 0$ *is* $2\varkappa$, *where* $\varkappa = \mathrm{Ind}\,[a(s) + ib(s)]$.

§ 32. Boundary value problems for polyharmonic and polyanalytic functions, reducible to the Hilbert boundary value problem

32.1. Representation of polyharmonic and polyanalytic functions by analytic functions

First we quote the definition of a polyharmonic function.

DEFINITION. *By a polyharmonic function* $u(x, y)$ *of order n we understand any solution of the polyharmonic equation*

$$\Delta^n U = 0, \qquad (32.1)$$

where $\Delta = \dfrac{\partial^2}{\partial x^2} + \dfrac{\partial^2}{\partial y^2}$ *is the Laplace operator and the exponent* n *indicates that the operators is applied* n *times.*

A polyharmonic function is called *regular* in a domain D, if in this domain it is continuous up to the derivatives of order n.

We shall hereafter assume that the polyharmonic functions under consideration are regular.

For simplicity we shall consider only simply-connected domains.

Now we present a brief derivation of the representation of polyharmonic functions by analytic functions.

The simplest way to derive the required representation is the transition from the real variables x, y to the complex variables

$$z = \varkappa + iy, \quad \bar{z} = \varkappa - iy.$$

We regard z and $\bar{z}$ as independent variables†. Then, according to the rules of differentiation of an implicit function we have

$$\frac{\partial}{\partial x} = \frac{\partial}{\partial z} + \frac{\partial}{\partial \bar{z}}, \quad \frac{\partial}{\partial y} = i\frac{\partial}{\partial z} - i\frac{\partial}{\partial \bar{z}}. \qquad (32.2)$$

---

† Transition to the complex variables is legitimate by virtue of the familiar fact of analyticity of a solution of an elliptic equation with analytic coefficients. The reader will find a number of various proofs of analyticity of solutions of elliptic equations in vol. VIII (1940) of the journal *Progress in Mathematical Sciences* (Uspekhi matematitcheskikh nauk). A detailed investigation of the problem of complex representations of the solutions of a polyharmonic equation is given in the book of I. N. Vekua [30], p. 170.

Hence, we readily observe that

$$\Delta = \frac{\partial^2}{\partial x^2} + \frac{\partial^2}{\partial y^2} = 4\frac{\partial^2}{\partial z\, \partial \bar{z}}. \tag{32.3}$$

Consequently

$$\Delta^n U \equiv 4^n \frac{\partial^{2n} U}{\partial z^n\, \partial \bar{z}^n},$$

and the polyharmonic equation can be written in the form

$$\frac{\partial^{2n} U}{\partial z^n\, \partial \bar{z}^n} = 0. \tag{32.4}$$

Writing this equation in the form

$$\frac{\partial^2}{\partial z\, \partial \bar{z}} \left(\frac{\partial^{n-2} U}{\partial z^{n-1}\, \partial \bar{z}^{n-1}} \right) = 0$$

and applying the well-known method of solving the equation $\partial^2 v / \partial x\, \partial y = 0$ we obtain

$$\frac{\partial^{2n-2} U}{\partial z^{n-1}\, \partial \bar{z}^{n-1}} = \varphi(z) + \varphi_*(\bar{z}),$$

where $\varphi(z)$ and $\varphi_*(\bar{z})$ are analytic functions of their arguments. Since the left-hand side of the last equation is real, the functions $\varphi(z)$ and $\varphi_*(\bar{z})$ are complex conjugates. Hence we have

$$\frac{\partial^{2n-2} U}{\partial z^{n-1}\, \partial \bar{z}^{n-1}} = \varphi(z) + \overline{\varphi(z)} = 2\,\mathrm{Re}\,\varphi(z).$$

If the polyharmonic function be regarded as known, the last formula shows that the real part of $\varphi(z)$ is uniquely determined. The function $\varphi(z)$ itself, therefore, is determined to within an arbitrary imaginary term.

Assuming that the origin of the coordinate system belongs to the domain of regularity of the polyanalytic function and setting $\varphi(z) = \sum\limits_{k=1}^{\infty} a_k z^k$ we have

$$\frac{\partial^{2n-2} U}{\partial z^{n-1}\, \partial \bar{z}^{n-1}} = \sum_{k=0}^{\infty} a_k z^k + \sum_{k=0}^{\infty} \bar{a}_k \bar{z}^k.$$

Integrating both sides with respect to z and then with respect to $\bar{z}$ we obtain

$$\frac{\partial^{2n-4} U}{\partial z^{n-2}\, \partial \bar{z}^{n-2}}$$
$$= z\bar{z}\,[\varphi_1(z) + \bar{\varphi}_1(\bar{z})] + \psi(z) + \overline{\psi(z)} = 2\,\mathrm{Re}\,[z\bar{z}\varphi_1(z) + \psi(z)],$$

where

$$\varphi_1(z) = \sum_{k=0}^{\infty} \frac{a_k}{k+1} z^k$$

is uniquely determined by $\varphi(z)$, and $\psi(z)$ is a new analytic function which, similarly to $\varphi(z)$, is determined by U to within an imaginary term.

Repeating the above procedure $n-2$ times more we arrive at the required representation of the polyharmonic function

$$U(x, y) = \operatorname{Re} \sum_{p=0}^{n-1} z^p \bar{z}^p \varphi_p(z), \qquad (32.5)$$

where all $\varphi_p(z)$ are analytic functions determined by $U(x, y)$ to within an arbitrary imaginary term.

The function

$$V(x, y) = \operatorname{Im} \sum_{p=0}^{n-1} z^p \bar{z}^p \varphi_p(z), \qquad (32.6)$$

which, obviously, is also a solution of the polyharmonic equation, will be called the polyharmonic function *conjugate* to $U(x, y)$.

DEFINITION. *If U and V are conjugate polyharmonic functions, then the function*
$$F(z, \bar{z}) = U(x, y) + iV(x, y) \qquad (32.7)$$

is called polyanalytic.

It is evident that the polyanalytic function, similarly to the polyharmonic function, is a solution of the polyharmonic equation, and on the basis of the definition we have the representation

$$F(z, \bar{z}) = \sum_{p=0}^{n-1} z^p \bar{z}^p \varphi_p(z). \qquad (32.8)$$

There exist representations of polyharmonic and polyanalytic functions which differ from the above in form†; we shall not however need them and they are not presented here.

If $n = 2$ the equation $\Delta^2 U = 0$ is called *biharmonic* and its solutions are known as *biharmonic functions*. The solutions of the biharmonic equation are closely connected with the fundamental concepts of the theory of elasticity (stresses and displacements) and the solution of the biharmonic equation constitutes one of the basic problems

† See for instance, I. N. Vekua [30], § 32, formulae (32.18), (32.19), (32.22).

of this theory. Let us also observe that the representation of the biharmonic function by analytic functions is most frequently taken in the form

$$U(x, y) = \text{Re}\,[\bar{z}\varphi(z) + \psi(z)].$$

32.2. Formulation of the boundary value problems for polyanalytic functions

First let us introduce some notations and let us establish certain auxiliary relationships.

Suppose that

$$t = x(s) + iy(s)$$

is the equation of the contour L, in the complex form. Then

$$t' = \frac{dt}{ds} = \frac{dx}{ds} + i\,\frac{dy}{ds}.$$

Denoting as usual $\bar{t} = x(s) - iy(s)$ and taking into account that $|dt/ds| = 1$ we have

$$\bar{t}' = \frac{\overline{dt}}{ds} = \frac{1}{t'}. \tag{32.9}$$

If $W(z, \bar{z})$ is a function differentiable a sufficient number of times, then we have the following expression for the derivative with respect to the arc of the contour:

$$\frac{\partial W}{\partial s} = \frac{\partial W}{\partial t}\,t' + \frac{\partial W}{\partial \bar{t}}\,\bar{t}'. \tag{32.10}$$

Assuming that the normal is interior, i.e. the tangent and the normal are located as the x and y axes, respectively, we have

$$\frac{\partial x}{\partial s} = \frac{\partial y}{\partial n}, \quad \frac{\partial x}{\partial n} = -\frac{\partial y}{\partial s}.$$

Hence we obtain

$$\frac{\partial t}{\partial n} = it', \quad \frac{\partial \bar{t}}{\partial n} = -i\bar{t}',$$

and now the expression for $\partial^k W/\partial n^k$ takes the form

$$\frac{\partial^k W}{\partial n^k} = \left[i\left(t'\,\frac{\partial}{\partial t} - \bar{t}'\,\frac{\partial}{\partial \bar{t}}\right)\right]^k W$$

$$= i^k \sum_{m=0}^{k} (-1)^m\, C_m^k\, t'^{k-m}\bar{t}'^m\, \frac{\partial^k W}{\partial t^{k-m}\,\partial \bar{t}^m}. \tag{32.11}$$

We now proceed to formulate the basic boundary value problems for polyanalytic functions.

Similarly to any solution of an equation of elliptic type of order $2n$, a polyharmonic function is determined by prescribing n independent boundary conditions. These conditions can have a great variety of forms. We shall present four most frequently encountered boundary conditions, stating them for polyanalytic functions.

Let L be a closed simple contour the equation of which has derivatives up to the order $(2n + 2)$, which satisfy the Hölder condition. It is required to find in the domain D^+ a polyanalytic function of order n,

$$F(z, \bar{z}) = U(x, y) + iV(x, y),$$

which is continuous on the contour, in accordance with the following boundary conditions.

PROBLEM I.
$$a_k(s)\, \Delta^k U + b_k(s)\, \Delta^k V = c_k(s) \tag{32.12}$$
$$(k = 0, 1, \ldots, n - 1).$$

PROBLEM II.
$$\begin{aligned} a_k(s)\, \Delta^k U + b_k(s)\, \Delta^k V &= c_k(s), \\ a_k^1(s)\, \frac{\partial \Delta^k U}{\partial n} + b_k^1(s)\, \frac{\partial \Delta^k V}{\partial n} &= c_k^1(s) \end{aligned} \tag{32.13}$$

$(k = 0, 1, \ldots, (n - 1)/2$; if n is odd the last condition is omitted).

PROBLEM III.
$$a_k(s)\, \frac{\partial^k U}{\partial n^k} + b_k(s)\, \frac{\partial^k V}{\partial n^k} = c_k(s) \tag{32.14}$$
$$(k = 0, 1, \ldots, n - 1).$$

PROBLEM IV.
$$a_k(s)\, \frac{\partial^{n-1} U}{\partial x^{n-k}\, \partial y^{k-1}} + b_k(s)\, \frac{\partial^{n-1} V}{\partial x^{n-k}\, \partial y^{k-1}} = c_k(s) \tag{32.15}$$
$$(k = 1, 2, \ldots, n),$$

where $a_k(s)$, $b_k(s)$, $a_k^1(s)$, $b_k^1(s)$ are known real functions of the length of the arc of the contour L, which satisfy the Hölder condition up to the derivatives of order $2n - k - 2$† and $a_k^2(s) + b_k^2(s) \neq 0$,

---

† The boundary conditions contain derivatives of the unknown functions up to the order $2n - 2$. Hence, the coefficients of the boundary conditions must be subjected to restrictions sufficient for existence of the limiting values of the highest derivative on the contour, which satisfy the Hölder condition. The present restriction ensures this for all boundary value problems under consideration.

$[a_k^1(s)]^2 + [b_k^1(s)]^2 \neq 0$. We suppose also that

$$a_k^2(s) + b_k^2(s) = 1, \quad [a_k^1(s)]^2 + [b_k^1(s)]^2 = 1. \tag{32.16}$$

Computing the left-hand sides of the boundary conditions by means of the representation (32.8), we arrive at the boundary value problems for the analytic functions $\varphi_p(z)$. Then, in general, the boundary conditions all contain functions φ_p and their derivatives. The solution of such problems is very complicated. We confine ourselves here to the simplest contours (circle and straight line); in this case the Problems I–IV can be reduced to a union of Hilbert boundary value problems each of them containing only one unknown function. On the basis of the theory of § 29 the solution of all of the above problems can be given in a closed form.

Let us note that all results presented in the following subsections are due to V. S. Rogozhin (1), and to M. P. Ganin (1) and (2); we shall particularly make use of the unpublished paper of the latter "Boundary value problems of the theory of polyanalytic and polyharmonic functions".

When $b_k(s) = 0$, $a_k(s) = 0$ the considered boundary value problems for polyanalytic functions reduce to problems for polyharmonic functions.

32.3. Boundary value problems for a circle

The boundary conditions (32.12) of Problem I, in view of (32.16), can be written in the form

$$\operatorname{Re}\left[(a_k(s) - ib_k(s))\, \Delta^k F(t, \bar{t})\right] = c_k(s) \quad (k = 0, 1, \ldots, n - 1). \tag{32.17}$$

Making use of the representation (32.8) for polyanalytic functions we obtain by simple transformations

$$\Delta^k F(z, \bar{z}) = 4^k \sum_{p=k}^{n-1} \frac{p!}{(p-k)!}\, \bar{z}^{p-k} \left(z^p \varphi_p(z)\right)^{(k)}. \tag{32.18}$$

The main difficulty in the reduction of the boundary value problems for polyanalytic functions to boundary value problems for analytic functions, consists in the fact that the relevant representations contain a non-analytic function in $\bar{z}$. The principal task is to express the boundary value of this function by the boundary value of an analytic function; this can easily be performed for the simplest cases under consideration.

Suppose that the domain D^+ is the unit circle. Then for points on the contour $\bar{t} = 1/t$. Hence we introduce new analytic functions

$$\Phi_k(z) = \sum_{p=k}^{n-1} \frac{p!}{(p-k)!} \frac{[z^p \varphi_p(z)]^{(k)}}{z^{p-k}}. \qquad (32.19)$$

On the circumference we have

$$\Delta^k F(t, \bar{t}) = 4^k \Phi_k(t).$$

Consequently, the boundary conditions (32.17) can be written in the form

$$\mathrm{Re}\left[\frac{\Phi_k(t)}{a_k(s) + ib_k(s)}\right] = \frac{1}{4^k} c_k(s) \quad (k = 0, 1, \ldots, n-1).$$

We have arrived at n Hilbert boundary value problems for the determination of the analytic functions $\Phi_k(z)$.

The formulae of § 29.3 enable us to find the solutions of the above problems

$$\Phi_k(z) = z^{\varkappa_k} e^{i\gamma_k(z)} \left[\frac{1}{2\pi} \int_0^{2\pi} c_k(\sigma) e^{\omega_k(\sigma)} \frac{e^{i\sigma} + z}{e^{i\sigma} - z} d\sigma + Q_k(z)\right].$$

If for some k the index $\varkappa_k < 0$ we have to set $Q_k(z) \equiv 0$ and require that the following conditions of solubility be satisfied:

$$\int_0^{2\pi} c_k(\sigma) e^{\omega_k(\sigma)} e^{im_k\sigma} d\sigma = 0 \quad (m_k = 0, 1, \ldots, -\varkappa_k - 1). \quad (32.20)$$

If any one of these conditions is not satisfied the problem has no solutions.

Suppose that all conditions of solubility are satisfied; then we may regard all functions $\Phi_k(z)$ as known. To deduce the required polyanalytic function it remains to express the analytic functions $\varphi_p(z)$ entering its representation by the known functions $\Phi_k(z)$.

For $k = n - 1$ we have from (32.19)

$$[z^{n-1}\varphi_{n-1}(z)]^{(n-1)} = \frac{1}{(n-1)!} \Phi_{n-1}(z).$$

Integrating this relation $n - 1$ times we obtain the function $\varphi_{n-1}(z)$. Observe that since the function $z^{n-1}\varphi_{n-1}(z)$ and its derivatives up to the order $n - 2$ vanish for $z = 0$, all arbitrary integration constants are determined by these conditions and the function $\varphi_{n-1}(z)$ is uniquely determined by the function $\Phi_{n-1}(z)$.

When $k = n - 2$ we have

$$[z^{n-2}\varphi_{n-2}(z)]^{(n-2)} = \frac{1}{(n-2)!}\left[\Phi_{n-2}(z) - \frac{(n-1)!}{1} \cdot \frac{(z^{n-1}\varphi_{n-1}(z))^{(n-2)}}{z}\right].$$

The right-hand side is known; hence, by integration we determine $\varphi_{n-2}(z)$.

Suppose that we have determined the functions $\varphi_{k+1}(z)$, $\varphi_{k+2}(z)$, ..., $\varphi_{n-1}(z)$. Then from relation (32.19) we find that

$$[z^k\varphi_k(z)]^{(k)} = \frac{1}{k!}\left[\Phi_k(z) - \sum_{p=k+1}^{n-1} \frac{p!}{(p-k)!} \frac{(z^p\varphi_p(z))^{(k)}}{z^{p-k}}\right] = w_k(z),$$

$$(32.21)$$

where $w_k(z)$ is a known function.

Consequently, integrating k times we obtain

$$z^k\varphi_k(z) = \frac{1}{(k-1)!}\int_0^z (z-\tau)^{k-1}w_k(\tau)\,d\tau \quad (k = 1, 2, \ldots, n-1).$$

Introducing this result into formula (32.8) we arrive at the solution of the problem

$$F(z, \bar{z}) = w_0(z) + \sum_{p=1}^{n-1} \frac{\bar{z}^p}{(p-1)!}\int_0^z (z-\tau)^{p-1}w_p(\tau)\,d\tau. \quad (32.22)$$

If all indices $\varkappa_k$ are non-negative the problem is absolutely soluble and its solution depends on

$$N = \sum_{k=0}^{n-1}(2\varkappa_k + 1)$$

arbitrary real constants.

In the general case when

$$\varkappa_0 < 0, \quad \varkappa_1 < 0, \ldots, \quad \varkappa_q < 0; \quad \varkappa_{q+1} \geqq 0, \ldots, \varkappa_{n-1} \geqq 0,$$

the solution exists only if $\sum_{k=0}^{q}(-2\varkappa_k - 1)$ conditions of solubility of the type (32.20) are satisfied; if these conditions are satisfied the solution depends on $\sum_{k=q+1}^{n-1}(2\varkappa_k + 1)$ arbitrary real constants.

Let us briefly examine Problem IV. The boundary conditions of the problem (32.15) can be written in the form

$$\text{Re}\left[(a_k(s) - i b_k(s)) \frac{\partial^{n-1} F(t, \bar{t})}{\partial x^{n-k} \partial y^{k-1}}\right] = c_k(s) \quad (k = 1, 2, \ldots, n). \quad (32.23)$$

Making use of expressions (32.2) we obtain

$$\frac{\partial^{n-1}}{\partial x^{n-k} \partial y^{k-1}} = i^{k-1}\left(\frac{\partial}{\partial z} + \frac{\partial}{\partial \bar{z}}\right)^{n-k}\left(\frac{\partial}{\partial \bar{z}} - \frac{\partial}{\partial \bar{z}}\right)^{k-1}$$

$$= i^{k-1} \sum_{\alpha=0}^{n-k} \sum_{\beta=0}^{k-1} (-1)^\beta C_{n-k}^\alpha C_{k-1}^\beta \frac{\partial^{n-1}}{\partial z^{n-\alpha-\beta-1} \partial \bar{z}^{\alpha+\beta}},$$

whence, in view of the representation (32.8),

$$\frac{\partial^{n-1} F(z, \bar{z})}{\partial x^{n-k} \partial y^{k-1}} = i^{k-1} \sum_{\alpha=0}^{n-k} \sum_{\beta=0}^{k-1} (-1)^\beta C_{n-k}^\alpha \times$$

$$\times C_{k-1}^\beta \sum_{p=\alpha+\beta}^{n-1} \frac{p!}{(p-\alpha-\beta)!} \bar{z}^{p-\alpha-\beta} [z^p \varphi_p(z)]^{(n-\alpha-\beta-1)}. \quad (32.24)$$

In order to reduce the boundary conditions (32.23) to a union of Hilbert boundary value problems, let us introduce new analytic functions by replacing in the last expression $\bar{z}$ by $1/z$ and multiplying by z^{n-1} (this is done to avoid a pole at the origin).

Thus we set

$$\Phi_k(z) = \sum_{\alpha=0}^{n-k} \sum_{\beta=0}^{k-1} (-1)^\beta C_{n-k}^\alpha C_{k-1}^\beta \sum_{p=\alpha+\beta}^{n-1} \frac{p!}{(p-\alpha-\beta)!} z^{n-p+\alpha+\beta-1} \times$$

$$\times [z^p \varphi_p(z)]^{(n-\alpha-\beta-1)} \quad (k = 1, 2, \ldots, n). \quad (32.25)$$

On the contour we have

$$\frac{\partial^{n-1} F(t, \bar{t})}{\partial x^{n-k} \partial y^{k-1}} = \frac{i^{k-1}}{t^{n-1}} \Phi_k(t).$$

It follows that the boundary conditions (32.23) can be written in the form

$$\text{Re}\left[i^{k-1} \frac{a_k - i b_k}{t^{n-1}} \Phi_k(t)\right] = c_k(s).$$

Thus, we have obtained n Hilbert problems for the determination of the analytic functions $\Phi_k(z)$. Without solving these problems, suppose that they are known. We now proceed to show how to determine, in terms of the known functions $\Phi_k(z)$, the functions $\varphi_p(z)$ which enter the representation of the polyanalytic functions. We shall make use of the following three facts.

1. Derivatives of F with respect to x, y can easily be expressed by the functions $\Phi_k(z)$.

2. Derivatives of F with respect to z, $\bar{z}$ can be expressed by the derivatives with respect to x, y.

3. The functions $\varphi_p(z)$ can be expressed, on the basis of the integral representation of polyanalytic functions, by the derivatives of F with respect to z and $\bar{z}$.

We have

(i) $\left\{\dfrac{\partial^{n-1} F(z, z)}{\partial x^{n-q-\nu-1} \partial y^{q+\nu}}\right\}_{\bar{z}=\frac{1}{z}} = \dfrac{i^{q+\nu}}{z^{n-1}} \Phi_{q+\nu-1}(z);$

(ii) $\left\{\dfrac{\partial^{n-1} F(z, z)}{\partial z^{n-k} \partial \bar{z}^{k-1}}\right\}_{\bar{z}=\frac{1}{z}}$

$$= \left\{\left[\frac{1}{2}\left(\frac{\partial}{\partial x} - i\frac{\partial}{\partial y}\right)\right]^{n-k} \left[\frac{1}{2}\left(\frac{\partial}{\partial x} + i\frac{\partial}{\partial y}\right)^{k-1}\right] F(z, \bar{z})\right\}_{\bar{z}=\frac{1}{z}}$$

$$= \frac{1}{2^{n-1}} \left\{\sum_{q=0}^{n-k} \sum_{\nu=0}^{k-1} (-1)^q i^{q+\nu} C_{n-k}^q C_{k-1}^\nu \frac{\partial F(z, \bar{z})}{\partial x^{n-q-\nu-1} \partial y^{q+\nu}}\right\}_{\bar{z}=\frac{1}{z}}$$

$$= \frac{1}{2^{n-1} z^{n-1}} \sum_{q=0}^{n-k} C_{n-k}^q \sum_{\nu=0}^{k-1} (-1)^\nu C_{k-1}^\nu \Phi_{q+\nu+1}(z).$$

Introduce for brevity the notation

$$\sum_{p=k-1}^{n-1} \frac{p!}{(p-k-1)!} \cdot \frac{1}{z^{p-k-1}} [z^p \varphi_p(z)]^{(n-k)} = w_k(z).$$

We shall not dwell here on the solubility conditions due to the requirement that both sides of the last relation have the same order at the origin; we assume that these conditions are satisfied. Then setting in turn in the last relation $k = n$, $n - 1, \ldots, 1$ we determine all the functions $\varphi_p(z)$.

A complete computation of the number of linearly independent solutions is possible; it would require, however, a detailed investigation. Let us only observe that besides the arbitrary constants appearing in solving the Hilbert problems with positive indices, there also arise additional constants in view of the presence of derivatives in the boundary conditions.

Problems II and III can be solved by analogous means. Observe that, similarly to Problems I and IV, it is possible to introduce auxiliary analytic functions such that the boundary value problem for polyanalytic functions reduces to n Hilbert problems for the same functions. Then the functions $\varphi_p(z)$ entering the representation of the required polyanalytic function, are determined by the known auxiliary analytic functions.

Besides Problems I–IV formulated above, other problems may be stated. For instance, V. S. Rogozhin (1) found the solution of "the three-harmonic equation" $\Delta^3 u = 0$ with the boundary conditions

$$a_1 u + b_1 v = 0, \quad a_2 \frac{\partial u}{\partial n} + b_2 \frac{\partial v}{\partial n} = 0, \quad a_3 \Delta u + b_3 \Delta v = 0,$$

which constitute a combination of the boundary conditions of Problems I and III. Similarly, various other combinations may be investigated.

If the domain D^+ is the upper semi-plane the problem is solved in exactly the same way as for the circle. The only difference is that instead of the condition $\bar{t} = 1/t$ used in the preceding subsection we shall employ the relation $\bar{t} = t$. Consequently, in constructing the auxiliary functions, instead of replacing $\bar{z}$ by $1/z$ as was done in the case of the circle, $\bar{z}$ should be replaced simply by z.

For the solution of the Hilbert problem for the semi-plane, see the problems at the end of this chapter.

32.4. Boundary value problems for domains mapped onto the circle by means of rational functions

The closed form of the solution given in the preceding subsection for the boundary value problems of polyanalytic functions in the case of the circle, can also be deduced, by slightly altering the method, in the cases of domains which can be mapped onto the circle by means of rational functions.

This fact is well known in connection with the basic problems of the theory of elasticity, i.e. as concerns the boundary value problems for the biharmonic equation $\Delta^2 u = 0$ (see N. I. Muskhelishvili [18], p. 304); it is due to the degeneracy of the kernels of the integral equations to which the indicated problems are reduced. P. K. Segarya (1), making use of the same method (reduction to an integral equation with degenerate kernel), deduced the solution in a closed form for the boundary value problem

$$\frac{\partial^{n-1}u}{\partial x^{n-k}\,\partial y^{k-1}} = f_k(s) \quad (k = 1, 2, \ldots, n)$$

of the polyharmonic equation (Problem IV for $a_k = 1$, $b_k = 0$).

The subsequent results are taken from the unpublished paper of M. P. Ganin mentioned at the end of § 32.2.

Suppose that D is a finite or infinite domain of the plane of the complex variable z, bounded by a simple, sufficiently smooth closed contour L, and suppose that the rational function

$$z = \omega(\zeta)$$

maps conformally the domain D onto the circle $|\zeta| < 1$. The inverse function is denoted by $\zeta = \gamma(z)$. Then all boundary value problems formulated in § 32.2 can be solved for the domain D in a closed form. Consider as an example the boundary value Problem IV. In view of (32.23), (32.24) the boundary condition on the contour L can be written in the form

$$\text{Re}\left\{(a_k - ib_k)i^{k-1}\sum_{\alpha=0}^{n-k} C_{n-k}^{\alpha} \sum_{\beta=0}^{k-1}(-1)^{\beta} \times\right.$$

$$\left. \times\, C_{k-1}^{\beta} \sum_{p=\alpha+\beta}^{n-1} \frac{p!}{(p-\alpha-\beta)!}\, \bar{t}^{p-\alpha-\beta}\left(t^p \varphi_p(t)\right)^{(n-\alpha-\beta-1)}\right\} = c_k(t). \quad (32.26)$$

Setting in this relation $t = \omega(\tau)$ we transform the boundary condition on the contour L into a boundary condition on the circle $|\tau| = 1$.

Introduce n new functions

$$\Phi_k[\omega(\zeta)] = \sum_{\alpha=0}^{n-k} C_{n-k}^{\alpha} \sum_{\beta=0}^{k-1}(-1)^{\beta}\, C_{k-1}^{\beta} \sum_{p=\alpha+\beta}^{n-1} \frac{p!}{(p-\alpha-\beta)!}\, \bar{\omega}^{p-\alpha-\beta}\left(\frac{1}{\zeta}\right) \times$$

$$\times\, \left(z^p \varphi_p(z)\right)_{z=\omega(\zeta)}^{(n-\alpha-\beta-1)}. \quad (32.27)$$

These functions can have in the circle $|\zeta| < 1$ singularities only at the points at which the function $\bar{\omega}(1/\zeta)$ has singularities. Let us separate them out.

Let us set

$$\omega(\zeta) = \frac{\alpha_q \zeta^q + \alpha_{q-1}\zeta^{q-1} + \cdots + \alpha_0}{\beta_r \zeta^r + \beta_{r-1}\zeta^{r-1} + \cdots + \beta_0}.$$

Then

$$\bar{\omega}\left(\frac{1}{\zeta}\right) = \zeta^{r-q}\, \frac{\bar{\alpha}_0 \zeta^q + \bar{\alpha}_1 \zeta^{q-1} + \cdots + \bar{\alpha}_q}{\bar{\beta}_0 \zeta^r + \bar{\beta}_1 \zeta^{r-1} + \cdots + \bar{\beta}_r}.$$

For definiteness suppose that $r > q$ and denote

$$f(\zeta) = (\bar{\beta}_0 \zeta^r + \bar{\beta}_1 \zeta^{r-1} + \cdots + \bar{\beta}_r)^{n-1}. \quad (32.28)$$

The functions

$$f(\zeta)\,\Phi_k[\omega(\zeta)]$$

are analytic in the unit circle. Denote

$$\Phi_k[\omega(\zeta)] = \Phi_k^1(\zeta).$$

In terms of the new functions the boundary condition on the unit circle can be written in the form

$$\left.\begin{array}{c}\text{Re}\left\{\dfrac{a_k[\omega(\tau)] - ib_k[\omega(\tau)]}{f(\tau)}\, [i^{k-1}f(\tau)\, \Phi_k^1(\tau)]\right\} = c_k[\omega(\tau)] \\[2mm] (k = 0, 1, \ldots, n-1);\end{array}\right\} \quad (32.29)$$

hence we arrive at n Hilbert boundary value problems for the determination of the functions $f(\zeta)\Phi_k^1(\zeta)$ analytic in the unit circle.

Assume that we have solved all boundary value problems (32.29) and, consequently, we have determined all functions $\Phi_k^1(\zeta)$. Then we may also regard as known the functions $\Phi_k(z) = \Phi_k^1[\gamma(z)]$. It remains to determine the functions $\varphi_p(z)$ entering the representation of poly-analytic functions. This is performed in a complete analogy to the same problem for the circle, considered in the preceding subsection. We shall therefore omit the details quoting only the fundamental relation corresponding to relations (1), (2) of the preceding subsection:

$$\sum_{p=k-1}^{n-1} \frac{p!}{(p-k-1)!} \bar{z}^{p-k-1} \big(z^p \varphi_p(z)\big)^{(n-k)}_{\substack{z=\omega(\zeta) \\ \bar{z}=\omega\left(\frac{1}{\zeta}\right)}}$$

$$= \frac{1}{2^{n-1}} \sum_{q=0}^{n-k} C_{n-k}^q \sum_{\nu=0}^{k-1} (-1)^\nu C_{k-1}^\nu \Phi_{q+\nu+1}[\omega(\zeta)].$$

Setting in these relations in turn $k = n, n - 1, \ldots, 1$ we determine all functions $\varphi_p(z)$.

We do not consider here the problems of calculating the number of solutions and deducing the conditions of solubility. Let us only observe that they can be found exactly.

For $a_k = 1$, $b_k = 0$ we obtain the solution of the boundary value problem for polyharmonic functions, which was examined by P. K. Zeragiya (1).

32.5. Solution of the basic problem of the theory of elasticity for the domain bounded by Pascal's limaçon

Consider as an example the basic problem of the theory of elasticity for the domain D_z^+ bounded by Pascal's limaçon, which in the parametric form is given by the equations

$$x = R(\cos\theta + m\cos 2\theta), \quad y = R(\sin\theta + m\sin 2\theta)$$
$$(R > 0, \ 0 < m < \tfrac{1}{2}, \ 0 \le \theta < 2\pi).$$

It is known (see e.g. N. I. Muskhelishvili [18], p. 141) that the basic problem of the theory of elasticity is reduced to the determination of a biharmonic function $U(x, y)$ in accordance with the two conditions on the contour L:

$$\frac{\partial U}{\partial x} = g(t), \quad \frac{\partial U}{\partial y} = h(t), \tag{32.30}$$

where $g(t)$, $h(t)$ are known real functions of position on the contour, which satisfy the Hölder condition.

Because of the general representation of the polyharmonic functions (32.5) we have for the biharmonic function

$$U(x, y) = \operatorname{Re}[z\bar{z}\varphi(z) + \psi(z)], \tag{32.31}$$

where $\varphi(z)$, $\psi(z)$ are arbitrary analytic functions.

With the help of this expression, in view of formulae (32.2) we can write the boundary conditions in the form

$$\begin{rcases} \operatorname{Re}\{\bar{t}\,[t\varphi(t)]' + \psi'(t) + t\varphi(t)\} = g(t), \\ \operatorname{Re}\{i[\bar{t}\,(t\varphi(t))' + \psi'(t) - t\varphi(t)]\} = h(t). \end{rcases} \quad (32.32)$$

The function $\omega(\zeta)$ which maps conformally the domain D_z^+ onto the interior D_ζ of the unit circle Γ of the complex plane ζ and satisfies the conditions $\omega(0) = 0$, $\omega'(0) > 0$, is the following:

$$z = \omega(\zeta) = R(\zeta + m\zeta^2).$$

The inverse function is

$$\zeta = \gamma(z) = -\frac{1}{2m} + \sqrt{\left(\frac{1}{4m^2} + \frac{z}{Rm}\right)}.$$

The branch of the radical is determined by the condition $\gamma(0) = 0$. On the unit circle Γ we have

$$\bar{t} = \overline{\omega(\xi)} = R\left(\bar{\xi} + m\bar{\xi}^2\right) = \frac{R(\xi + m)}{\xi^2}.$$

Consequently, the boundary conditions assume the form

$$\begin{rcases} \operatorname{Re}\left\{\dfrac{R(\xi + m)}{\xi^2}\,\varphi_1(\xi) + \psi_1(\xi) + R(\xi + m\xi^2)\,\varphi[\omega(\xi)]\right\} = g[\omega(\xi)], \\[2mm] \operatorname{Re}\left\{i\left[\dfrac{R(\xi + m)}{\xi^2}\,\varphi_1(\xi) + \psi_1(\xi) - R(\xi + m\xi^2)\,\varphi[\omega(\xi)]\right]\right\} = h[\omega(\xi)], \end{rcases} \quad (32.33)$$

where

$$\varphi_1(\zeta) = [z\varphi(z)]'_{z=\omega(\zeta)}, \quad \psi_1(\zeta) = \psi'(z)_{z=\omega(\zeta)}.$$

Each of the boundary conditions (32.33) is the condition of Problem A for the unit circle (for $n = 2$) investigated in § 27.3. Applying formula (27.9) and replacing in it the Schwarz operator by the Schwarz integral (see § 6.1) we obtain

$$\begin{rcases} \dfrac{R(\zeta + m)}{\zeta^2}\,\varphi_1(\zeta) + \psi_1(\zeta) + R(\zeta + m\zeta^2)\,\varphi[\omega(\zeta)] \\[2mm] = \dfrac{A}{\zeta^2} - \bar{A}\zeta^2 + \dfrac{B}{\zeta} - \bar{B}\zeta + i\alpha + \dfrac{1}{2\pi i}\int\limits_\Gamma g[\omega(\xi)]\,\dfrac{\xi + \zeta}{\xi - \zeta}\,\dfrac{d\xi}{\xi}, \\[4mm] i\left\{\dfrac{R(\zeta + m)}{\zeta^2}\,\varphi_1(\zeta) + \psi_1(\zeta) - R(\zeta + m\zeta^2)\,\varphi[\omega(\zeta)]\right\} \\[2mm] = \dfrac{C}{\zeta^2} - \bar{C}\zeta^2 + \dfrac{D}{\zeta} - \bar{D}\zeta + i\beta + \dfrac{1}{2\pi i}\int\limits_\Gamma h[\omega(\xi)]\,\dfrac{\xi + \zeta}{\xi - \zeta}\,\dfrac{d\xi}{\xi}, \end{rcases} \quad (32.34)$$

where A, B, C, D are complex and α and β are real arbitrary constants.

Let us multiply the second relation by i and add it to the first. Since the function $\varphi[\omega(\zeta)]$ is analytic in the unit circle the coefficients of ζ^{-2} and ζ^{-1} vanish;

replacing ζ by $\gamma(z)$ we obtain

$$2z\varphi(z) = -(\bar{A} + i\bar{C})\gamma^2(z) - (\bar{B} + i\bar{D})\gamma(z) + i\alpha - \beta +$$
$$+ \frac{1}{2\pi i} \int_L [g(t) + ih(t)] \frac{\gamma(t) + \gamma(z)}{\gamma(t) - \gamma(z)} \frac{\gamma'(t)}{\gamma(t)} dt, \tag{32.35}$$

$$A + iC = 0, \quad B + iD = 0. \tag{32.36}$$

Setting $z = 0$ in (32.35) and taking into account that $\gamma(0) = 0$ we have

$$\alpha + i\beta = \frac{1}{2\pi} \int_L [g(t) + ih(t)] \frac{\gamma'(t)}{\gamma(t)} dt. \tag{32.37}$$

To determine $\psi(z)$ introduce into either of equations (32.34) the values of $\varphi[\omega(\zeta)]$ and $\varphi_1(\zeta)$ taken from formula (34.25) and the relation derived by differentiating it. Taking into account (32.36) we obtain

$$\psi'(z) = \frac{A}{\gamma^2(z)} + \frac{B}{\gamma(z)} + \frac{i\alpha + \beta}{2} +$$
$$+ \frac{1}{4\pi i} \int_L [g(t) - ih(t)] \frac{\gamma(t) + \gamma(z)}{\gamma(t) - \gamma(z)} \frac{\gamma'(t)}{\gamma(t)} dt +$$
$$+ \frac{R\gamma'(z)[\gamma(z) + m]}{\gamma^2(z)} \left[2\bar{A}\gamma(z) + \right.$$
$$\left. + \bar{B} - \frac{1}{2\pi i} \int_L [g(t) + ih(t)] \frac{\gamma'(t)}{[\gamma(t) - \gamma(z)]^2} dt \right]. \tag{32.38}$$

Let us equate (in view of the analyticity of the function $\psi'(z)$) the coefficients of the negative powers of $\gamma(z)$ to zero. To this end observe that

$$R\gamma'(z) = \frac{1}{1 + 2m\gamma(z)} = \sum_{n=0}^{\infty} (-1)^n [2m\gamma(z)]^n$$

and

$$\frac{1}{[\gamma(t) - \gamma(z)]^2} = \sum_{n=0}^{\infty} \frac{n[\gamma(z)]^{n-1}}{[\gamma(t)]^{n+1}}.$$

After simple transformations we arrive at the system of equations for the determination of A and B:

$$\left. \begin{array}{l} A + m\bar{B} = \dfrac{m}{2\pi i} \int_L [g(t) + ih(t)] \dfrac{\gamma'(t)}{\gamma^2(t)} dt, \\[2ex] 2m\bar{A} + B + (1 - 2m^2)\bar{B} \\[1ex] \qquad = \dfrac{1 - 2m^2}{2\pi i} \int_L [g(t) + ih(t)] \dfrac{\gamma'(t)}{\gamma^2(t)} dt + \\[2ex] \qquad + \dfrac{m}{\pi i} \int_L [g(t) + ih(t)] \dfrac{\gamma'(t)}{\gamma^3(t)} dt. \end{array} \right\} \tag{32.39}$$

The system (32.39) has solutions only if the following condition is satisfied:

$$\text{Im}\left\{\frac{1}{2\pi i}\int_L [g(t)+ih(t)]\left[\frac{1}{\gamma^2(t)}+\frac{2m}{\gamma^3(t)}\right]\gamma'(t)\,dt\right\}$$

$$= \text{Re}\left\{\frac{1}{2\pi R}\int_L [g(t)+ih(t)]\,\overline{dt}\right\} = 0. \qquad (32.40)$$

Now, this condition is the condition of vanishing of the principal moment of external forces and should necessarily be satisfied for mechanical reasons.

Setting $A = a_1 + ia_2$, $B = b_1 + ib_2$ and transforming the system (32.39) we obtain three equations

$$\left.\begin{array}{l}
a_1 + mb_1 = \text{Re}\left\{\dfrac{m}{2\pi i}\displaystyle\int_\Gamma \{g[\omega(\xi)]+ih[\omega(\xi)]\}\dfrac{d\xi}{\xi^2}\right\}, \\[3mm]
2ma_1 + b_1 = \dfrac{1}{4\pi i}\displaystyle\int_\Gamma \{g[\omega(\xi)]+ih[\omega(\xi)]\}\left[\dfrac{1}{\xi^2}+\dfrac{2m}{\xi^3}\right]d\xi, \\[3mm]
a_2 - mb_2 = \text{Im}\left\{\dfrac{m}{2\pi i}\displaystyle\int_\Gamma \{g[\omega(\xi)]+ih[\omega(\xi)]\}\dfrac{d\xi}{\xi^2}\right\}.
\end{array}\right\} \qquad (32.41)$$

in solving the system (32.41) one of the quantities a_2 or b_2 may be taken arbitrarily.

Substituting into formula (32.31) the values of $z\varphi(z)$ and $\psi(z)$ obtained by integrating relation (32.38), we find the required biharmonic function

$$U(x, y) = \text{Re}\left\{\frac{\overline{A}-m\overline{B}}{m}\,\overline{z\gamma}(z)-\frac{\overline{A}(x^2+y^2)}{Rm}+\right.$$

$$+\frac{\overline{z\gamma}(z)}{2\pi Ri}\int_L \frac{[g(t)+ih(t)]\,dt}{\gamma(t)[1+2m\gamma(t)][\gamma(t)-\gamma(z)]}+$$

$$+\frac{1}{2\pi Ri}\int_0^z d\tau \int_L \frac{[g(t)-ih(t)]\,dt}{[1+2m\gamma(t)][\gamma(t)-\gamma(\tau)]}+$$

$$+ 2R(\overline{A}+mB)\,\gamma(z)-\frac{R(A+m\overline{B})}{\gamma(z)}+$$

$$+ 2R(2ma_1+b_1)\ln\gamma(z)-$$

$$\left.-\frac{1}{2\pi Ri}\int_0^z \frac{[\gamma(\tau)+m]\,d\tau}{\gamma^2(\tau)[1+2m\gamma(\tau)]}\int_L \frac{g(t)+ih(t)}{[1+2m\gamma(t)][\gamma(t)-\gamma(\tau)]^2}\,dt\right\} + K.$$

It can easily be verified using relations (3.39) or (32.41) that the function $U(x,y)$ is analytic at the origin of the coordinate system.

The solution contains one arbitrary constant K, which corresponds to the fact that the quantities describing the state of stress depend on the derivatives of the biharmonic function $U(x, y)$ and, consequently, the presence of K has no influence on the state of stress.

The problem is soluble only under the condition that the given functions $g(t)$, $h(t)$ satisfy condition (32.40). Observe that the last condition possessing the mechanical meaning indicated above, follows also from the mathematical condition of the absence of a logarithmic singularity at the origin of the coordinate system.

§ 33. The inverse boundary value problem for analytic functions

The inverse boundary value problem is that of determining the contour of a domain from prescribed quantities on the contour. Usually the problem posed is that of finding a function belonging to a given class (an analytic function, or one which is the solution of some given equation). In this case the independent conditions defined on the contour are one more than are required for solving the usual (for the given domain) boundary value problem in the class of functions in question. The extra boundary condition is for determining the contour of the domain. Inverse boundary value problems can be posed in many ways. Here we shall consider the problem for analytic functions, which may be formulated as follows: *find the contour from boundary values of the analytic function defined thereon.*

33.1 Formulation of the problem

Two functions of the parameters s are given ($0 \leqq s \leqq l$):

$$u = u(s), \quad v = v(s), \tag{33.1}$$

which are periodic with a period l, which have first derivatives which satisfy Hölder's condition, which do not vanish simultaneously and which satisfy the condition that with two parameter values $s_1 \neq s_2$ $[u(s_1) - u(s_2)]^2 + [v(s_1) - v(s_2)]^2 \neq 0$. Equations (33.1) define in the plane of the complex variable $w = u + iv$ a simple closed Lyapounov curve which divides the plane into an interior domain D_w^+ and an exterior domain D_w^-.

"*Interior*" *problem. Find in plane z the curve L_z which bounds the finite domain D_z^+ (generally a multi-sheet domain), such that, regarding the parameter s as the length of an arc of the curve L_z, the expression $w(s) = u(s) + iv(s)$ is the boundary value of an analytic function which is a conformal mapping of the domain D_z^+ onto the domain D_w^+ or D_w^-.*

"*Exterior*" problem. *If the finite domain D_z^+ in the formulation of the interior problem is replaced by a domain D_z^- containing an infinitely remote point, the exterior inverse boundary value problem would be posed in the same way.*

If we had to deal with the ordinary boundary value problem (with the contour given), to define the analytic function within the domain to within a constant term), it would be sufficient to determine only one function ($u(s)$ or $v(s)$). Determination of the other function enables the unknown contour L_z to be defined.

In combining the domains D_z^+, D_w^+ a total of four problems could be obtained. But the only essentially different problems are those which differ in the signs of D_z, since the domain D_w^- may first be conformally mapped onto some domain of type D^+. We shall specifically assume that the domain D_w^+ has been defined. The symbol $+$ associated with D_w^+ will be omitted and this domain will be designated simply as D_w. In the z plane where the contour is required, it is impossible to reduce the exterior to the interior domain and the problems in differing signs of D_z have each to be solved independently.

We assume that the positive directions of the given contour L_w and of the required contour L_z are such that in traversing them the domains D_z^+ and D_w^+ remain on the left. Then, by virtue of the properties of a conformal mapping, the necessary solubility condition requires that with increase of s from 0 to l the contour L_w is traversed *in the positive direction for the interior problem, and in the negative direction for the exterior problem.* We shall assume throughout that these conditions are fulfilled.

32.2. Solution of the interior problem

We proceed from the well-known property of conformal mappings that the ratio of the lengths of corresponding linear elements is equal to the modulus of the derivative of the mapping function. The length of the element of the arc of the contour L_z is already given—the differential of the parameter ds. From equations (33.1) the length of the element of the curve L_w is found by the formula of differential geometry:

$$d\sigma = \sqrt{[u'^2(s) + v'^2(s)]}\, ds.$$

Hence

$$\left|\frac{dw}{dz}\right| = \frac{d\sigma}{ds} = \sqrt{[u'^2(s) + v'^2(s)]}.$$

Thus the modulus of the derivative of the function $w = w(z)$ which conformally maps the domain D_z^+ onto D_w, is known to be a function of arc s. Since L_z is unknown, the resulting formula cannot be a source for the solution of the problem. To obtain a suitable formula, we change over to the inverse function $z = z(w)$ and regard L_w as the "data-bearing" contour.

The expression $\sigma = \int_0^s \sqrt{(u'^2 + v'^2)} \, ds$ defines σ as a monotonically increasing function of s. By inversion we obtain $s = s(\sigma)$. We then have

$$\left| \frac{dz}{dw} \right| = \frac{1}{\sqrt{\{u'^2[s(\sigma)] + v'^2[s(\sigma)]\}}}. \tag{33.2}$$

On the contour L_w we therefore know the modulus of the derivative dz/dw of the function which conformally maps the domain D_w onto the domain D_z^+.

We thus arrive at the *fundamental* problem in the theory of inverse boundary value problems.

Find the function which is analytic and non-vanishing† *in the domain D_w from defined values of its modulus on the contour.*

This problem may easily be reduced to the Schwarz problem. The logarithm of the function analytic and non-vanishing in the domain is a harmonic function, being the real part of the logarithm of the analytic function itself. Consequently

$$\ln \frac{dz}{dw} = -\frac{1}{2} S \ln \{u'^2[s(\sigma)] + v'^2[s(\sigma)]\} + i\alpha, \tag{33.3}$$

where S is the Schwarz operator for the domain D_w, and α is an arbitrary real constant.

By powering and then integrating we get the function $z(w)$. One may in this way obtain a complete solution, but in practice it is more convenient to handle a domain for which an explicit expression is known for the Schwarz operator. First of all, we change from the domain D_w to some standard domain. The procedure is explained using a unit circle as the standard domain.

Suppose that $w = w(\zeta)$ is the analytic function which conformally maps the unit circle γ in plane ζ onto the domain D_w. Crossing to the contour, we have

$$u(s) = iv(s) = w(e^{i\theta}).$$

† This follows from the required conformality of the mapping.

Owing to the one-to-one correspondence between the points of the contours L_w and γ, this last relation defines s as a single-valued function of the polar angle θ

$$s = \omega(\theta). \tag{33.4}$$

We may now leave plane w and solve the inverse boundary value problem by studying the correspondence between the plane z and the auxiliary plane ζ. Considering that the polar angle is numerically equal to the arc length of the unit circle, owing to the property of conformal mappings which we have already used, we have

$$\left| \frac{dz}{d\zeta} \right| = \frac{ds}{d\theta} = \omega'(\theta).$$

Taking logarithms on both sides, we again return to the Schwarz problem for the function $\ln dz/d\zeta$. The latter is solved now by the Schwarz integral

$$\ln \frac{dz}{d\zeta} = \frac{1}{2\pi} \int_0^{2\pi} \ln \omega'(\theta) \frac{e^{i\theta} + \zeta}{e^{i\theta} - \zeta} d\theta + i\alpha = \chi(\zeta) = i\alpha.$$

By powering, we obtain

$$\frac{dz}{d\zeta} = e^{i a} e^{\chi(\zeta)},$$

whence

$$z = f(\zeta) = e^{i a} \int e^{\chi(\zeta)} d\zeta + C, \tag{33.5}$$

where C is an arbitrary complex constant.

By putting the contour values of the argument $\zeta = e^{i\theta}$ into the expression for the function $f(\zeta)$, we obtain a parametric equation for the required contour in complex form

$$z = f(e^{i\theta}) = x(\theta) + iy(\theta).$$

As shown by formula (33.5), the required domain D_z^+ is defined to within a motion (the parameter α defines the rotation and C the translatory motion). Arbitrariness due to the possibility of motion is natural and implied in the formulation of the problem. Solutions which differ only in the position of the domain are therefore regarded as the same solutions.

To establish that the motion has natural arbitrariness allowed by the solution of the problem, it is first necessary to agree as to how two different required contours carry identical data. We adopt the following definition.

Two contours L_z and L_{z_1} carry the identical data $w(s)$ $w_1(s_1)$ if, in a conformal mapping of the domains bounded by these contours, for corresponding points the values of the defined functions and their derivatives are equal, i.e. if

$$w(s) = w_1(s_1), \quad w'(s) = w_1'(s_1).$$

Suppose now that the two contours L_z and L_{z_1} are the solutions of the inverse problem and let $z_1 = z_1(z)$ be the function which effects the conformal mapping of the domains bounded by these contours.

Differentiating the first equality with respect to s, we have $w'(s)$ $= w'_1(s_1) \dfrac{ds_1}{ds}$. Bearing in mind the equality of the derivatives, we then have

$$\left| \frac{dz_1}{dz} \right| = \frac{ds_1}{ds} = 1,$$

and, furthermore,

$$\ln \left| \frac{dz_1}{dz} \right| = 0, \quad \ln \frac{dz_1}{dz} = i\alpha, \quad z_1 = e^{i\alpha} \cdot z + C.$$

The contour L_{z_1} is therefore obtainable from L_z by motion, as required.

33.3. Other methods of prescribing boundary values

In the problem under consideration the boundary values of the analytic function have been prescribed as functions of the arc length of the required contour. This defined the method of solution (finding the logarithm of the modulus of the derivative). The method of solution is changed if the parameter relating the prescribed boundary values u, v to the required contour is some other geometric quantity, e.g. an abscissa or ordinate of the required contour, its radius-vector, or the polar angle and so on. The logarithm of the derivative is no longer taken as the initial function, but the analytic function which most closely related to the selected parameter.

Suppose, for example, that the parameter is the abscissa x,

$$u = u(x), \quad v = v(x).$$

Mapping D_w onto the circle $|\zeta| < 1$ and following the same reasoning as above, the abscissa is a defined function of the polar angle θ,

$x = x(\theta)$. Considering that $x = \mathrm{Re}\, z$, we obtain

$$z = \frac{1}{2\pi} \int x(\theta) \frac{e^{i\theta} + \zeta}{e^{i\theta} - \zeta}\, d\theta + i\alpha.$$

The contour L_z is then determined to within a shift along the imaginary axis. The problem is likewise solved if the parameter is the ordinate y. If the parameter is the radius-vector ϱ, or the polar angle θ, it is necessary to take $\ln z$ as the initial function.

We shall not consider the problem of overcoming the difficulties associated with the possible non-single-valuedness of the representation of u and v in terms of these parameters [see M. T. Nuzhin (1)].

33.4. Solution of the exterior problem

In this case some point of the domain D_w must be at infinity and the mapping function $z(w)$ must therefore have a pole within D_w. Owing to the requirement of conformality of the mapping, this pole must be simple. Thus, in contrast with the foregoing, it is necessary to find a function which is analytic in D_w except at one point† where it has a simple pole. However, we may assume that we have again mapped the domain D_w onto a unit circle in the auxiliary plane ζ and that s has been evaluated as a function of the polar angle θ

$$s = \omega\,(\theta).$$

Suppose that $F(\zeta)$ is the function effecting the conformal mapping of the circle $|\zeta| < 1$ onto D_z^- and that the point at infinity is ζ_0. Then

$$\left| \frac{d\zeta}{dz} \right| = |F'(\zeta)| = \omega'(\theta).$$

The function $F'(\theta)$ has a second-order pole at ζ_0. We can reduce the problem of finding $F'(\theta)$ to the case for an analytic function by the known method of eliminating zeros and poles as in the theory of complex-variable functions.

The function $(\zeta - \zeta_0)/(1 - \bar{\zeta}_0 \zeta)$ effects a mapping of the circle onto itself for an arbitrary ζ_0. On the contour of the circle its modulus is therefore equal to unity.

---

† Here we make the additional assumption that the domain D_z^- is a single-sheet domain in the neighbourhood of the infinitely remote point.

Consequently, the function

$$\varphi(\zeta) = \left(\frac{\zeta - \zeta_0}{1 - \bar{\zeta}_0 \zeta}\right)^2 F'(\zeta)$$

is analytic inside the unit circle and has the same modulus as $F'(\zeta)$ on the contour of the circle. Hence, using the same notation, and also following the same reasoning as in the foregoing section, we have

$$\varphi(\zeta) = e^{i\alpha} e^{\chi(\zeta)}, \quad z = F(\zeta) = e^{i\alpha} \int \left(\frac{1 - \bar{\zeta}_0 \zeta}{\zeta - \zeta_0}\right)^2 e^{\chi(\zeta)} \, d\zeta + C. \quad (33.6)$$

This latter formula defines a function which, generally, also has a logarithmic singularity at ζ_0 besides the simple pole. The function is only a solution of the problem if no such singularity is present. Accordingly, it is necessary and sufficient for the residue of the function $\left(\frac{1 - \bar{\zeta}_0}{\zeta - \zeta_0}\right)^2 \varphi(\zeta)$ to be zero in relation to ζ_0. Let us evaluate this residue.

Taking, in the vicinity of ζ_0, expansions of the functions

$$(1 - \zeta_0 \zeta)^2 = (|\zeta_0|^2 - 1)^2 + 2\zeta_0(|\zeta_0|^2 - 1) + \bar{\zeta}_0^2(\zeta - \zeta_0)^2,$$

$$\varphi(\zeta) = \varphi(\zeta_0) + \varphi'(\zeta_0)(\zeta_0)(\zeta - \zeta_0) + \cdots$$

and multiplying them, we find that the required residue, equal to the coefficient of the first power of this product, is

$$c_{-1} = (|\zeta_0|^2 - 1) \left[2\bar{\zeta}_0 \varphi(\zeta_0) + (|\zeta_0|^2 - 1) \varphi'(\zeta_0) \right].$$

Equating this coefficient to zero, we find that for solubility of the exterior problem, the complex coordinate of the pole ζ_0 must satisfy the equation

$$\frac{\varphi'(\zeta_0)}{\varphi(\zeta_0)} = \frac{2\bar{\zeta}_0}{1 - |\zeta_0|^2}. \quad (33.7)$$

The solubility of the exterior problem is proved if the solubility of the last equation is established. In this respect the reasoning is somewhat artificial and the reader may wish to know that the first half of the calculations is similar to those used in the theory of single-sheet analytic functions for proving the so-called theorems of distortion.

10*

We set $|\zeta_0| = r$. Multiplying both sides of (33.7) by ζ_0 and availing ourselves of the identity† $r\dfrac{\partial}{\partial r} = \zeta_0\dfrac{\partial}{\partial\zeta_0}$, the equation becomes

$$r\frac{\partial \ln\varphi}{\partial r} = \frac{2r}{1 - r^2}$$

or

$$\frac{\partial}{\partial r}\ln|\varphi| + i\frac{\partial}{\partial r}\arg\varphi = \frac{2r}{1 - r^2}.$$

Our aim is to reduce the complex equation to a set of two real equations which are the extremum conditions of some real function.

Using one of the Cauchy–Riemann conditions in polar coordinates $\dfrac{\partial v}{\partial r} = -\dfrac{1}{r}\dfrac{\partial u}{\partial\theta}$, and bearing in mind that

$$\frac{\partial}{\partial r}\ln(1 - r^2) = \frac{-2r}{1 - r^2}, \quad \frac{\partial}{\partial\theta}\ln(1 - r^2) = 0,$$

the equation becomes

$$\frac{\partial}{\partial r}\ln[(1 - r^2)|\varphi|] - i\frac{1}{r}\frac{\partial}{\partial\theta}\ln[(1 - r^2)|\varphi|] = 0;$$

the latter is equivalent to a set of two real equations

$$\frac{\partial}{\partial r}\ln[(1 - r^2)|\varphi|] = 0, \quad \frac{\partial}{\partial\theta}\ln[(1 - r^2)|\varphi|] = 0, \qquad (33.8)$$

which is the necessary extremum condition of the surface

$$t = \Phi(r, \theta) = \ln[(1 - r^2)|\varphi(r, \theta)|].$$

Thus, to prove the solubility of equation (35.7), it is sufficient to establish that the function $\Phi(r, \theta)$ attains the extremum inside the unit circle. This can be done without difficulty. Since $u'(s)$ and $v'(s)$ belong, by definition, to class H, it may readily be inferred that $\omega'(s)$, and therefore $\ln\omega'(s)$, satisfies Hölder's condition. Hence, by virtue of the properties of the Schwarz integral, the contour values $\theta(\zeta)$ also satisfy the H condition and they are therefore bounded. Hence

$$\lim_{r\to 1}(1 - r^2)|\varphi| = 0, \quad \lim_{r\to 1}\ln[(1 - r^2)|\varphi|] = -\infty.$$

---

† We proceed from $\zeta_0 = re^{i\theta}$ and bear in mind that the argument θ needs to be regarded as constant in changing from differentiation with respect to r to differentiation with respect to ζ_0.

On a circle of radius $r = 1 - \varepsilon$, with sufficiently small ε, the function $\Phi(r, \theta)$ may assume arbitrarily large negative values. Since $\Phi(r, \theta)$ is bounded within the circle, the upper bound which it attains in a closed circle cannot be on the contour. Consequently, this upper bound, being the maximum of the function, is achieved inside the circle. Hence equations (33.8), and therefore the initial equation (33.7), are satisfied, as required.

33.5. The number of solutions of the exterior problem

If ζ_0, the root of equation (33.7), is a definite number, then from formula (33.6) the exterior problem admits the same arbitrariness over motion as the interior problem, i.e. it has a unique solution according to the particular condition. The question of the number of solutions for the exterior problem therefore merely entails finding the number of solutions for (33.7). This is generally a transcendental equation and its complete investigation is often impracticable. However, the fundamental fact that the solution need not be unique has been established. This was shown by Rogozhin (4) with the following example.

Let the domain D_w be a unit circle defined by the equations

$$u = \cos\theta, \quad v = \sin\theta, \quad \theta = \theta(s), \quad 0 \leqq s \leqq l, \quad 0 \leqq \theta(s) \leqq 2\pi,$$

the function $\theta(s)$ being defined by the differential equation

$$\frac{d\theta}{ds} = e^{\frac{\sin 2\theta}{a}}, \quad 0 < a < 1.$$

Here the domain D_w is itself standard so that the problem may be solved directly without resort to an auxiliary mapping.

Let $z = F(w)$ be the function which carries out the conformal mapping of the circle D_w onto the domain D_z^-. We then have

$$|F'_{(w)}| = \left| \frac{dz}{dw} \right| = \frac{ds}{d\theta} = e^{\frac{\sin 2\theta}{a}}.$$

Hence

$$\ln |\varphi(w)| = \ln \left| \left(\frac{w - w_0}{1 - \overline{w}_0 w} \right)^2 F'(w) \right| = -\frac{\sin 2\theta}{a}.$$

The right-hand side is the boundary value of the real part of the function iw^2/a; consequently,

$$\varphi(w) = \left(\frac{w - w_0}{1 - \bar{w}_0 w}\right)^2 F'(w) = e^{ia} e^{\frac{iw^2}{a}}.$$

Hence, according to (33.6),

$$z = F(w) = e^{ia} \int \left(\frac{1 - \bar{w}_0 w}{w - w_0}\right)^2 e^{\frac{iw^2}{a}} dw + C. \tag{33.6'}$$

The point w_0 whose image is the point at infinity is found from equation (33.7), which has the form

$$i\frac{w_0}{a} = \frac{\bar{w}_0}{1 - |w_0|^2}.$$

Putting $w_0 = r e^{i\beta}$, we arrive at the set of two equations:

$$r(1 - r^2) \cos 2\beta = 0,$$
$$r[(1 - r^2) \sin 2\beta + a] = 0.$$

The latter equation is satisfied by three values of w_0 within the circle

$$w_0 = 0, \quad w_0^* = \sqrt{(1 - a)} \, e^{\frac{\pi}{4}i},$$
$$w_0^{**} = - \sqrt{(1 - a)} \, e^{\frac{\pi}{4}i}$$

(roots $\pm \sqrt{(1 + a)} e^{\frac{\pi}{4}i}$ are discarded).

Substituting these values of w_0 into (33.6), we obtain three solutions of the problem which we denote by F, F^* and F^{**} respectively. It now reamins to establish that these solutions are essentially different. It has already been pointed out that contours which differ only in their position in the plane, correspond to one solution of the inverse boundary value problem. Moreover, it is obvious that the conformal mapping of the circle $|w| < 1$ does not itself alter the solution.

We prove that F and F^* cannot satisfy the identity

$$F^*(w) = e^{i\beta} F\left(\frac{w - b}{1 - bw}\right) + C.$$

Differentiating the latter, and having regard to (33.6') and the fact that $w_0 = 0$, we get:

$$F^*(w) = e^{i\beta} F\left(\frac{w - \bar{b}}{1 - \bar{b} w}\right) + C,$$

$$\left(\frac{1 - w_0^* w}{w - w_0}\right)^2 e^{i\frac{w^2}{a}} = e^{i\beta} \left(\frac{1 - \bar{b} w}{w - b}\right)^2 e^{\frac{i}{a}\left(\frac{w-b}{1-\bar{b} w}\right)^2} \frac{1 - |b|^2}{(1 - \bar{b} w)^2},$$

or

$$e^{\frac{i}{a}\left[w^2 - \left(\frac{w-b}{1-\bar{b} w}\right)^2\right]} = e^{i\beta} \left(\frac{w - w_0^*}{1 - \overline{w}_0^* w}\right)^2 \left(\frac{1 - \bar{b} w}{w - b}\right) \frac{1 - |b|^2}{(1 - \bar{b} w)^2}.$$

But since the right-hand side is a rational function and the left-hand side is a transcendental function, the last identity cannot be satisfied for any choice of $b\,(|b| < 1)$ for fixed $a\,(0 < a < 1)$. It may likewise be proved that the third solution is essentially different from the first two. Thus, in this example, the exterior inverse boundary value problem has three essentially different solutions. Therefore the solution cannot be unique.

33.6. The Schwarz problem with a logarithmic singularity on the contour

In the solution of the Schwarz problem we assumed a defined real part on the contour which satisfied Hölder's condition. Later we shall meet with the case when singularities of logarithmic nature can appear on the contour. We commence with a special case and consider the boundary value of a function which is analytic in the unit circle $|\delta| < 1$,

$$\varphi(\zeta) = \ln(\zeta - \zeta_0) = \ln(\zeta - e^{i\theta_0}).$$

We have

$$\varphi(e^{i\theta}) = \ln(e^{i\theta} - e^{i\theta_0}) = \ln\left[e^{i\frac{\theta+\theta_0}{2}}\left(e^{i\frac{\theta+\theta_0}{2}} - e^{-\frac{\theta-\theta_0}{2}}\right)\right]$$

$$= \ln\left[ie^{i\frac{\theta+\theta_0}{2}} \cdot 2 \sin\frac{\theta - \theta_0}{2}\right] = \ln\left|2\sin\frac{\theta - \theta_0}{2}\right| + i\gamma,$$

where

$$\gamma = \begin{cases} \dfrac{\theta + \theta_0 + \pi}{2} & \text{if } \theta > \theta_0, \\[2mm] \dfrac{\theta + \theta_0 - \pi}{2} & \text{if } \theta < \theta_0. \end{cases}$$

Hence the solution of the Schwarz problem in the class of functions which have logarithmic singularities on the contour when the boundary value of the real part is $u(\theta) = \ln\left|2\sin\dfrac{\theta - \theta_0}{2}\right|$, is the function

$$\varphi(\zeta) = \ln(\zeta - \zeta_0) + i\alpha.$$

To establish the uniqueness of the solution in the class in question, it is merely necessary to prove that there are no solutions other than an imaginary constant for the following homogeneous problem.

It is required to find the function which is analytic in a unit circle (logarithmic singularities on the contour are possible) having a zero real part on the contour.

The unknown function represents the unit circle on the imaginary axis. We therefore apply the principle of symmetry. Continuing the required function symmetrically into the exterior of the circle, we obtain a function which is analytic throughout the plane except at individual isolated points where the logarithmic singularities are possbile. Because of the requirement of single-valuedness, the function should have cuts which connect logarithmic branch points. But owing to analyticity (and therefore continuity) cuts are impossible; the singularities are therefore eliminated. The function is analytic throughout the plane: owing to Liouville's theorem, it is a constant function, which, obviously, will be purely imaginary. The uniqueness is thereby established †.

If the real part of the required function is defined as the sum

$$u(\theta) = \ln\left|2\sin\frac{\theta - \theta_0}{2}\right| + u_0(\theta),$$

where $u_0(\theta)$ is a function which satisfies Hölder's condition, by subtracting the Schwarz integral having density $u_0(\theta)$ from the required function, we revert to the case under consideration.

Considering that the function

$$\ln\left|2\sin\frac{\theta - \theta_0}{2}\right| - \ln|\theta - \theta_0| = \ln\left|\frac{2\sin\dfrac{\theta - \theta_0}{2}}{\theta - \theta_0}\right|,\cdot$$

† This reasoning also remains valid for power singularities of order less than unity instead of logarithmic singularities. But further generalization is impossible. If polar singularities are admitted on the contour, then non-zero functions exist with their real part vanishing on the contour. A good example is the function $1 - 2/z$, analytic within the circle, defined by the equation $x^2 + y^2 - 2x = 0$.

which is analytic in the neighbourhood of the point $e^{i\theta_0}$ and, therefore, satisfies Hölder's condition, we find that if the boundary value of the real part of the required analytic function is represented as the sum $\ln|\theta - \theta_0| + u_0(\theta)$, the solution is†

$$\ln(\zeta - \zeta_0) + \varphi_0(\zeta),$$

where φ_0 is bounded at ζ_0.

33.7. Singular points of the contour

The following conditions are imposed on the functions $u(s)$, $v(s)$: (1) $u'(s)$ and $v'(s)$ belong to Hölder's class, and (2) they have no common zeros; this ensures that the required contour L_z is smooth (more exactly, that it belongs to the class of Lyapounov curves). Non-observance of either of these conditions upsets the smoothness, i.e. the contour generally has singularities. Let us investigate the nature of such singularities on the assumption that the second condition is disturbed. It is also assumed that the prescribed functions in the neighbourhood of the point under consideration are expanded as by Taylor's formula.

Suppose that u and v are given by the expansions

$$u = (s - s_0)^p[u_0 + (s - s_0)\,a(s)], \quad v = (s - s_0)^p[v_0 + (s - s_0)\,b(s)],$$
$$(33.9)$$

where u_0 and v_0 are constants which do not vanish simultaneously, whilst $a(s)$ and $b(s)$ have derivatives which satisfy Hölder's condition.

From differential geometry, if p is odd, the origin ($s = s_0$) is a regular point of the contour L_w, but if p is even, we have a cusp. In the latter case one has to recognise the possiblity that the angle may be zero or 2π. We consider the cases schematically.

1. *Regular point on L_w*. Suppose that the function $w = w(\zeta)$ conformally maps the circle $|\zeta| < 1$ onto the domain D_w, the value of the angle θ_0 ($\zeta_0 = e^{i\theta_0}$) here corresponding to s_0. In the neighbourhood of ζ_0 the mapping function is

$$w(\zeta) + (\zeta - \zeta_0)\,w_1(\zeta), \quad w_1(\zeta_0) \neq 0. \qquad (33.10)$$

For the relation on the contour $|u(s) + iv(s)| = |w(e^{i\theta})|$, having regard to (33.9) and (33.10), we get:

$$|s - s_0|^p\varphi(s) = |\theta - \theta_0|\,\psi(\theta), \quad \varphi(s_0) > 0, \quad \psi(\theta_0) > 0,$$

† By the use of conformal mapping the result can be transferred to the case of any closed contour.

or, taking the root,

$$|s - s_0| \overset{p}{V}[\varphi(s)] = |\theta - \theta_0|^{\frac{1}{p}} \overset{p}{V}[\psi(\theta)].$$

Using the theory of implicit functions, we obtain the relationship:

$$\frac{ds}{d\theta} = |\theta - \theta_0|^{\frac{1}{p}-1} \psi_1(\theta), \quad \psi_1(\theta_0) > 0,$$

where $\varphi_1(\theta)$ satisfies Hölder's condition. Hence

$$\ln \left| \frac{dz}{d\zeta} \right| = \ln \frac{ds}{d\theta} = \left(\frac{1}{p} - 1 \right) \ln |\theta - \theta_0| + \psi_2(\theta).$$

According to the results in the foregoing section, we have

$$\ln \frac{dz}{d\zeta} = \left(\frac{1}{p} - 1 \right) \ln(\zeta - \zeta_0) + X_0(\zeta).$$

Hence

$$z = (\zeta - \zeta_0)^{\frac{1}{p}} F(\zeta).$$

From this latter expression and the properties of conformal mappings, the contour L_z for $\theta = \theta_0$ has a corner point with angle π/p.

2. *Cusp with angle* 2π. The function which maps the unit circle on the domain D_w, in the neighbourhood of ζ_0 is $w = \omega(\zeta) = (\zeta - \zeta_0)^2 \omega_1(\zeta) \ (\omega(\zeta_0) \neq 0)$. Hence

$$\omega(e^{i\theta}) = (\theta - \theta_0)^2 \psi(\theta), \quad (\psi(\theta_0) \neq 0).$$

By operations similar to those used in the previous case, we get

$$\frac{ds}{d\theta} = |\theta - \theta_0|^{\frac{2}{p}-1} f(\theta) \quad (f(\theta_0) > 0),$$

$$z = (\zeta - \zeta_0)^{\frac{2}{p}} F(\zeta) \quad (F(\zeta_0) \neq 0).$$

Hence the point of the contour L_z is a corner point with angle $2\pi/p$. If, in particular, $p = 2$, the corner point becomes a regular point.

3. *Cusp with zero angle.* The mapping function is

$$w \omega(\zeta) = \frac{c}{\ln(\zeta - \zeta_0) + c_0 + \varphi(\xi)}.$$

The investigations here are much more complicated than in the other two cases which we have considered, we merely point out that the required contour L_z has a cusp at the point under investigation. The details are to be found in the work of Gakhov and Mel'nik (1). Cases are studied there in which the prescribed function has singularities of an elementary nature.

EXAMPLE.
$$u = u(s) = a\cos s\,(1 + \cos s),$$
$$v = v(s) = a\sin s\,(1 + \cos s),\ \ 0 \le s \le 2\pi.$$

The contour L_w is a cardioid.

We have $u'(\pi) = v'(\pi) = 0$. *The expansion in the neighbourhood of $s = \pi$ is*

$$u(s) = (s - \pi)^2\left[-\frac{a}{2} + \alpha(s)\right],\ \ v(s) = (s - \pi^3)\beta(s).$$

Here $p = 2$. The origin $(s = \pi)$ is a cusp with† angle 2π.

The required contour L_z has a corner point with angle $2\pi/p = 2\pi/2 = \pi$, i.e. a regular point.

The required contour L_z may be found from elementary reasoning. The function

$$w = w(\zeta) = \frac{a}{2}(\zeta + 1)^2$$

maps the unit circle onto the interior of the cardioid. On the boundary we have

$$u(s) + iv(s) = w(e^{i\theta}),\ \ ae^{is}(1 + \cos s) \doteq \frac{a}{2}(e^{i\theta} + 1)^2.$$

Hence we get:

$$s = \theta,\ \ \left|\frac{dz}{d\zeta}\right| = \frac{ds}{d\theta} = 1,\ \ \ln\frac{dz}{d\zeta} = e^{i\alpha},$$
$$z = z(\zeta) = e^{i\alpha}\zeta + C.$$

The latter transformation is a motion. The contour L_z is therefore also a unit circle.

33.8. The single-sheet nature of the solution

The requirement of conformality in mapping the domain D_z onto D_w excludes zeros of the derivative of the mapping function and therefore branch points are also excluded. But this does not imply a one-sheeted representation. The contour L_w is simple and D_w is a single-sheet domain, but the function $z(w)$ resulting from the solution of the inverse boundary value problem may not be single-sheeted; in this case the contour L_z is self-intersecting and not simple.

We will explain this by an example using a transformation by the exponential function $z = e^w$.

Given in plane z the circle

$$u = a\cos\theta,\ \ v = a\sin\theta,$$

---

† This could have been established by analysis of the expansion of $u(s)$, $v(s)$ in the neighbourhood of $s = \pi$ [Gakhov and Mel'nik (1)].

10a*

where $\theta = \theta(s)$ is defined by the equation

$$\frac{d\theta}{ds} = \frac{1}{a}\, e^{-a\cos\theta} \quad (\theta(0) = 0),$$

we have

$$d\tau = a\, d\theta, \quad \left|\frac{dz}{dw}\right| = \frac{ds}{d\sigma} = \frac{1}{a}\frac{ds}{d\theta} = e^{a\cos\theta},$$

$$\ln\left|\frac{dz}{dw}\right| = a\cos\theta = \operatorname{Re} w.$$

Hence

$$\ln\frac{dz}{dw} = w + i\alpha, \quad \frac{dz}{dw} = e^{i\alpha}e^{w}, \quad z = e^{i\alpha}e^{w} + C.$$

We arrange the coordinate axes in plane z so that $\alpha = 0$, $C = 0$. The function which maps D_w onto D_z is now $z = e^{w}$. Setting $w = a\cos\theta + ia\sin\theta$, the parametric equation of the required contour L_z in Cartesian coordinates is

$$x = e^{a\cos\theta}\cos(a\sin\theta), \quad y = e^{a\cos\theta}\sin(a\sin\theta), \quad 0 \leq \theta \leq 2\pi.$$

The parametric equation of the same curve in polar coordinates is

$$\varrho = e^{a\cos\theta}, \quad \varphi = a\sin\theta.$$

The contour may easily be plotted from this last equation. The shape of the curve L_z essentially depends on the quantity a, the radius of the circle defined in plane z. If $a < \pi$, the function $z = e^{w}$ gives a single-sheet mapping of the circle and the contour L_z is simple; if, however, $a > \pi$, the mapping is multi-sheeted and the contour is self-intersecting; in the boundary case $a = \pi$ the contour has a point of

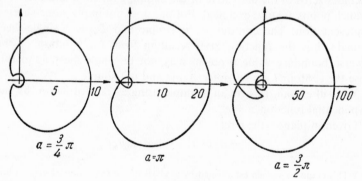

Fig. 14.

tangency†. Figure 14 shows the contour L_z for the three cases $a = 3/4\pi$, $3/2\pi$, π. The reader is advised to do this construction.

Non-intersecting contours are required in all the known applications of inverse boundary problems. It is therefore important to be able to specify the characteristics of single-sheet mappings, as far as possible simply expressed in terms of the defined functions $u(s)$, $v(s)$. The answer to this problem involves the theory of single-sheeted conformal mappings which is not of direct value in this book. Being in no position to enter into the details, we will merely cite the simple characteristic of single-sheetedness as given by Krasnovidova and Rogozhin (1).

The study is based on the following result in the theory of conformal mappings.

A function $f(z)$, analytic in the convex domain D with a continuously-continuable derivative on the boundary L, is single-sheeted in D if the variations of the argument $f'(z)$ on L are less than π, i.e. if $|\arg f'(t_1) - \arg f'(t_2)| < \pi$ for any t_1, t_2 belonging to L.

It is assumed that the defined domain D_w is a circle. The following theorem then holds:

For single-sheetedness of the required domain D_z, it is sufficient for the boundary value

$$\ln\left|\frac{dz}{dw}\right| = P(\theta) = \ln\frac{1}{u'^2(s) + v'^2(s)}$$

to satisfy Lipschitz' condition with the constant $\pi/(4\ln 2)$:

$$|P(\theta_1) - P(\theta_2)| < \frac{\pi}{4\ln 2}|\theta_1 - \theta_2|.$$

Further results are to be found in the works of Rogozhin (5) and Aksent'ev (1).

33.9. Some further topics

We will briefly consider three important questions which we cannot treat in detail.

† This result can be deduced from the well-known geometric properties of the function e^w. It single-sheetedly maps the strip $-\pi < \mathrm{Im}\, w < \pi$ in the entire z plane. If the domain D_w lies entirely inside this strip, the mapping is single-sheeted and the contour is simple; otherwise the mapping is multi-sheeted with L_z intersecting.

1. *Applications*. The theory of inverse boundary value problems is used for constructing contours which possess pre-determined properties. At the present time all the known applications are largely in hydromechanics. The main problem is that of constructing the contour from the distribution of the circumfluent flow velocity of an ideal fluid as specified on the contour. The chief technical application is in specifying an aviation profile from a given pressure distribution.

The flow potential $\theta(s)$ can be found from the velocity by integration. The profile in the stream is a stream line and therefore the stream function on the profile is constant, $\psi = \text{const}$ (it may be assumed that this constant is zero). On the required contour we therefore know the boundary value of the characteristic function of the flow $w = \theta + i\psi$. Here the contour L_w is always a portion of the axis of the abscissae. Mapping the plane cut along this portion onto the exterior of a unit circle, we have an inverse boundary value problem of type D_ζ^-, D_z^-.

Actual flows are rotational and therefore, mathematically, the inverse boundary value problem corresponding to the mechanics problem consists in finding an analytic function in D_ζ with a logarithmic singularity at infinity. This introduces considerable complications. For the problem in question all the difficulties may be overcome and a complete solution is obtainable. With a more complicated formulation of the problem (assumption of sources, or vortices, in the flow, or on the required contour) the most general formulation of the inverse boundary problem is required. In this respect the theoretical difficulties are very great and as yet they have not been overcome (see below).

We shall mention no other applications here. Readers who are interested may refer to the monograph article of Tumashev and Nuzhin [28] in which all the practical applications of inverse boundary value problems are treated, an extensive list of references also being appended there.

2. *Generalization of the functional nature*. The foregoing formulation of the inverse boundary value problem ensured the existence of continuous solutions right up to (and including) the contour (the singularities of § 33.7 were each investigated separately). It is natural to pose the question as to the widest possible assumptions about the functions $u(s)$, $v(s)$ for which the formulation of the problem and the proposed method of solution is valid. A full enquiry would entail use of the theory of real-variable functions. We only give the results.

From the theory of conformal mapping, the widest class of $u(s)$ and $v(s)$ for which the problem may be posed in the sense of § 33.1, is the class of absolutely-continuous functions. For the applicability of the mathematical apparatus for use in the solution of the problem, two additional limitations have to be imposed: (1) $u'(s)$ and $v'(s)$ should have bounded derivatives almost everywhere; (2) the set of points where $u'(s)$ and $v'(s)$ vanish simultaneously, should be of measure zero. These conditions are sufficient for the proof of solubility of the inverse boundary value problem.

In this wide formulation of the problem it is necessary to bear in mind that the solution may contain the additional multiplier $e^{\gamma(\zeta)}$, where

$$\gamma'(z) = \frac{1}{2\pi} \int_0^{2\pi} \frac{e^{i\theta} + z}{e^{i\theta} - z} \, d\psi(\theta),$$

and $\psi(\theta)$ is a singular function (a non-increasing function with a zero derivative almost everywhere). Therefore, unless further limitations are imposed on the required contour L_z, the uniqueness theorem of the interior inverse boundary value problem, proved in § 33.1, no longer holds. Uniqueness may be preserved if the class of required contours is confined to Smirnov curves†.

Readers who are interested in such generalizations may refer to the papers of F. G. Gakhov (3), (4) from which these results were taken (see also Andrianov (1)).

3. *General formulations of the inverse boundary value problem.* The simplest formulation of the problem is that in § 33.1 where it is required to find the function which gives the conformal mapping without branch points and with only one simple pole admissible (in the case of the exterior problem). In practice one meets with more complicated problems. For instance, in the simplest case, the inverse problem in hydromechanics entails finding a function with a logarithmic singularity in the domain. A generalized formulation requires the solution in a class of functions with any number of logarithmic and polar singular points. A topological theory of complex-variable functions has shown that an analytic function which has a definite number of logarithmic and polar singularities in the domain, the boundary values of which are prescribed, has an exactly determined number of branch points in the domain. The appropriate formulae are given in the paper of Gakhov and Krikunov (1). In this case it is no longer necessary to require the contour L_w to be simple. The domain D_w extends over a Riemann surface and on this surface the contour L_w is simple, though its projections in the plane are generally self-intersecting.

We formulate the inverse boundary value problem in its general form. It is given that there are 2ν functions of the parameter s

$$u = u_k(s), \quad v = v_k(s), \quad 0 < s < l_k \qquad (k = 1, 2, \ldots, \nu),$$

which‡ define ν curves L_{kw} (generally self-intersecting) in the plane of the complex variable w. In the plane of the complex variable z find ν closed curves L_{kz} which bound a finite domain D_z^+ which is generally a multi-sheet domain without branch points, such that, with the parameter s regarded as the arc length of the required contour $L_z = L_{1z} + L_{2z} + \ldots + L_{\nu z}$, the expression $w_k(s) = u_k(s) + iv_k(s)$ represents the boundary value on L_{kz} of a function $w = f(z)$ which is analytic in D_z^+ except at individual points where it may have singularities of a prescribed (logarithmic or polar) type with a given principal part.

The number of points whose images are branch points (points where $f'(z) = 0$) can be found, as shown above, by the formulae in the topological theory and this also is therefore regarded as known.

At the present time there are no solutions of this general problem. Individual special cases in practice have been solved from mechanical considerations, but without mathematical justification. Information on this subject is to be obtained from the cited article of Tumashev and Nuzhin [28].

The main theoretical problems standing in the way of the solution of the general problem are as follows.

† A Smirnov curve is a curve for which the modulus of the logarithm of the derivative of the function conformally mapping the domain bounded by it onto a circle, is represented as a Poisson integral in terms of its boundary values (see Privalov [21], p. 250).

‡ See § 36* regarding the solution of the inverse boundary value problem for a multiply-connected domain.

Principally the problem is to form that Riemann surface on which the contours given by the equations $u = \theta_z(s)$, $v \psi_k(s)$ define a domain analogous to a single-sheet domain, i.e. a domain which can be in one-to-one correspondence everywhere except at branch points mapped conformally on some standard single-sheet domain, e.g. a plane with rectilinear cuts, or cut-out circles. The contour, defined by the boundary values of the required function, the number and nature of the singularities and the number of branch points are given. The following difficulties stand in the way of a solution: (1) determination of the number and nature of the admissible singularities (and correspondingly the number of branch points) for the given boundary values; (2) the admissible arrangement of singularities and branch point for which the given contour I_w can become the contour of a domain analogous to a single-sheet domain; (3) the number of essentially different domains for the various methods of securing the sheets. The last difficulty is connected with the number of solutions of the inverse boundary value problem.

We confine ourselves to these few observations. A more detailed discussion is published in the cited paper of Gakhov and Krikunov (1).

§ 33*. Historical notes

The Hilbert boundary value problem which, similarly to the Riemann problem investigated in Chapter II, belongs to the group of fundamental boundary value problems for analytic functions, is the oldest of problems of this type. This is true for both the date of the first formulation of the problem and the first elaboration, and for the date of deducing the full solution. The same is true with respect to the first applications (theoretically, to the investigation of singular integral equations and practically, to the solution of hydrodynamical problems) and the exposition of various generalizations.

The first formulation of the Hilbert problem may be regarded as the paper of Riemann [22] "Foundations of the General Theory of Functions" (1848). In Chapter XIX speaking of the possible methods of prescribing a function analytic in a domain, he indicates that instead of prescribing its real part on the contour (which leads to the solution of the Dirichlet problem) we may prescribe a relation between the real and imaginary parts. This constitutes the most general formulation of a problem of this type.

To justify the fact that the problem (27.1) examined in the present chapter (the Hilbert problem) is less connected with the name of Riemann† than the problem (14.1) of Chapter II (the Riemann problem), we should emphasize that the problem of the actual determination of an analytic function, having a relation between its real and imaginary parts on the contour, was definitely not proposed by Riemann as a practical problem for solution. The only point he made was the manner of prescribing the function. Moreover, Riemann does not quote any formulae; the remark on the possibility of determining a function in accordance with the above indicated boundary relation was made by him together with some even more general considerations, for instance the possibility of determining the function by prescribing a relation connecting the values of its

† Some authors call it the Riemann–Hilbert problem.

real and imaginary parts at n different points of the contour. (Observe that problems of this kind have not so far been examined.) In contrast to this, the boundary value problem (14.1) (the Riemann problem) was presented by Riemann as a practical problem for solution, although this was done not to investigate it as an independent boundary value problem of analytic functions but in connection with the solution of a problem of a quite different type (see the Historical notes of Chapter II, § 19). Thus, the attribution of the boundary value problem (14.1) to Riemann is from the historical viewpoint more justified than the attribution of the boundary value problem examined in the present chapter.

A clear formulation of the Hilbert boundary value problem† (27.1) and the first attempts at its solution appeared apparently for the first time in a paper of Volterra (1) in 1883. This paper, however, for some reason, was isolated from other papers and had no important influence on subsequent investigations. Let us note a strange phenomenon, which I cannot explain, that this important paper is not quoted in any paper on the theory of boundary value problems, except for the paper [30].

A full solution of the boundary value problem considered was given by Hilbert [9] in 1905. This solution, however, was not immediately derived. The first solution given by him in 1904 in the report on the Third International Congress of Mathematicians in Heidelberg, based on a reduction to a singular integral equation with the kernel $\cot[(\sigma - s)/2]$, was erroneous (for an examination of this result see the paper by the author (1). The final solution of Hilbert constituted the foundation of all subsequent investigations of this topic, although it was not elaborated in all details. From the subsequent papers the most important is that of Noether (1); the paper of S. L. Sobolev (1) should also be noted. The latter authors investigated the Hilbert problem in all details.

Moreover, Noether's paper contains an important new development. The former authors attempted to deduce the solution of the boundary value problems (Hilbert and Riemann problems) from integral equations; here, on the contrary, the solution of the boundary value problem (Hilbert problem) was employed to investigate singular integral equations (with kernel $\cot[(\sigma - s)/2]$). The results obtained in this way turned out to be extremely fruitful; they led to fundamental properties of the singular equations (Noether theorems) and exerted a considerable influence on subsequent investigations (see the Historical notes of Chapter III, § 40).

The solution of the Hilbert problem given in the present chapter, based on the concept of regularizing factors, was first announced by the author in 1941 in his Doctor's Thesis (2) (see also [3]). The solution differs from those presented earlier only in the method, which seems to the author to be the most appropriate for the problem under consideration. In connection with the solution of the Hilbert problem the paper of I. N. Vekua should also be noted (4); in this paper an extensive exposition of the problem is given.

The Hilbert problem for the case in which the domain is the lower semi-plane is examined in the recently published book of L. A. Galin [5]. Apparently for practical reasons, he has examined only problems with zero index. A full exposition of the solution of the Hilbert problem for a unit circle is presented in the book of V. I. Smirnov [23].

† The term "Hilbert problem" was first introduced by Picard.

The general concept of a regularizing operator was introduced for the first time by the author in the papers cited above. This concept turned out to be useful in solving the Hilbert problem and also in examining inverse boundary value problems. The possibility of other applications of the regularizing factor remains still open.

Singular integral equations with the Hilbert kernel $\cot[(\sigma - s)/2]$ were first encountered in the paper of Hilbert † (1). Apparently, these were the first singular equations which became known to mathematicians. A full theory of the equation with the Hilbert kernel was presented, as already mentioned, by Noether in her frequently quoted paper. The theory of the dominant equation with Hilbert kernel in a simpler form than Noether's was presented by the author in (2). The exposition of § 31 follows essentially that of V. K. Natalevitch who considered this problem in connection with developing a theory of non-linear singular equations.

The methods of reduction of boundary value problems for polyharmonic and polyanalytic functions in the case of simplest contours (straight line and circle) to the Hilbert boundary value problem, were first announced by V. S. Rogozhin (1). A detailed elaboration of this problem was carried out by M. P. Ganin part of whose results is published in (1) and (2); the greater part, however, remained unpublished. The exposition of § 32 is all based on the papers of M. P. Ganin, including those which have not been published.

The inverse boundary value was formulated in France by Riaboushinsky (1) as a problem in harmonic functions in which it was required to find the contour from given contour values of the harmonic function and its normal derivative in the domain. Demtchenko (1) also solved the problem posed in this way. Since no "data-bearing" contour was included, he defined the data-carrier as some auxiliary circle. The formulation and solution of the problem in § 33 was given by Nuzhin (1). Nuzhin's real contribution was the introduction of a "natural" data-carrier—the contour L_w, defined by the data of the problem itself. Proof of the existence of a solution for the exterior problem (§ 33.4) is given in Gakhov's paper (4). Many technical applications have been given in mechanics papers of the Kazan scientists. An account of these papers and a list of reference up to 1957 are given in the paper of Tumashev and Nuzhin [28]. Other mathematical papers on boundary value problems are cited in the body of the text.

Problems on Chapter IV

1. Prove that the Schwarz operator for the domain exterior to the unit circle is the ordinary Schwarz integral taken with negative sign.

2. Prove that the general expression for the Schwarz operator for the upper semi-plane has the form

$$Su = \frac{1}{\pi i} \int\limits_{-\infty}^{\infty} u(\xi)\left[\frac{1}{\xi - z} + C(\xi)\right] d\xi,$$

† Simultaneously in the paper of Poincaré [20].

where $C(\xi)$ is an arbitrary real function of the real variable ξ. Hence derive two different expressions for the Schwarz operator:

(i) $\quad Su = \dfrac{1}{\pi i} \displaystyle\int\limits_{-\infty}^{\infty} \dfrac{u(\xi)}{\xi - z}\, d\xi;$ $\qquad$ (ii) $\quad Su = \dfrac{1}{\pi i} \displaystyle\int\limits_{-\infty}^{\infty} u(\xi)\, \dfrac{1 + \xi z}{\xi - z}\, \dfrac{d\xi}{1 + \xi^2}.$

3. Prove that the function $Q(z)$ which is a solution of Problem A_0 (§ 27.3) for the upper semi-plane, can be taken in the form

$$Q(z) = \Sigma \left[c_k \left(\frac{z+i}{z-i} \right)^k - \bar{c}_k \left(\frac{z-i}{z+i} \right)^k \right] + i\beta_0.$$

4. Derive an expression for the function $Q_1(z)$ which has the following properties:

(i) It is analytic in the unit circle except at the prescribed points $z_1, z_2, \ldots, z_m$ where it has simple poles.

(ii) On the unit circle the real part of $Q_1(z)$ vanishes.

Answer. $\quad Q_1(z) = i\beta_0 + \displaystyle\sum_{k=1}^{m} \left[c_k \frac{z - z_k}{1 - \bar{z}_k z} - \bar{c}_k \frac{1 - \bar{z}_k z}{z - z_k} \right].$

5. Generalize the results of the preceding problem to the case when the poles have the multiplicities $p_1, p_2, \ldots, p_m$, respectively.

Answer. $\quad Q_1(z) = i\beta_0 + \displaystyle\sum_{k=1}^{m} \sum_{j=1}^{p_k} \left[c_j \left(\frac{z - z_k}{1 - \bar{z}_k z} \right)^j - \bar{c}_j \left(\frac{z - z_k}{1 - \bar{z}_k z} \right)^{-j} \right].$

6. Solve the Hilbert problems for the unit circle

(a) $e^{-\sin 2s} \cos(\cos 2s)\, u(s) - e^{-\sin 2s} \sin(\cos 2s)\, v(s) = e^{\cos s} \cos(\sin s);$

(b) $(\cos s + \tfrac{1}{2})\, u(s) + (\sin s - \tfrac{1}{3})\, v(s) = \tfrac{1}{2}(\tfrac{49}{36} + \cos s - \tfrac{2}{3}\sin s) \ln(5 - 4\cos s).$

Answer.

(a) $F(z) = e^{-iz^2}(e^z + i\beta_0);$

(b) $F(z) = \left(z + \dfrac{1}{2} - \dfrac{1}{3}i \right) \left[\ln(z - 2) + i\beta_0 + c\, \dfrac{6z + 3 - 2i}{6 + (3 + 2i)z} - \bar{c}\, \dfrac{6 + (3 + 2i)z}{6z + 3 - 2i} \right]$

(β_0 is a real and c a complex parameter, both being arbitrary).

7. Prove that the Hilbert problem for the unit circle

$$\cos s .\, u(s) - (\sin s + \tfrac{1}{2})\, v(s) = \cos 2s + h$$

is soluble only when $h = \tfrac{1}{4}$. Derive the solution for this case.

Answer. $\quad F(z) = z - \tfrac{1}{2}i.$

8. Prove that the non-homogeneous singular integral equation with zero index

$$(\cos s + 2)\, u(s) - \frac{\sin s}{2\pi} \int\limits_{0}^{2\pi} u(\sigma) \cot \frac{\sigma - s}{2}\, d\sigma = \sin s$$

is absolutely soluble. Derive the solution.

Answer. $\quad u(s) = \tfrac{1}{2} \sin s.$

9. Prove that the non-homogeneous integral equation with zero index

$$u(s)\cos s - \frac{\sin s - 2}{2\pi} \int_0^{2\pi} u(\sigma)\cot\frac{\sigma - s}{2}\,d\sigma = (5 + 4\sin s)(\cos s + h)$$

is soluble only when $h = 0$. Derive the solution for this case.

Answer. $u(s) = \cos 2s - 2\sin s - \beta_0(\sin s + 2)$.

10. Prove that for the complex function $a(s) + ib(s)$ given on the contour L the determination of the regularizing factor $R(t) = \alpha(s) + i\beta(s)$, the image of which is the curve Γ of equation $\beta = f(\alpha)$, can be reduced to the solution of the boundary value problem

$$bu - av = -(a^2 + b^2)f\left(\frac{au + bv}{a^2 + b^2}\right),$$

where $u + iv$ is an analytic function defined by the relation

$$R(t)[a(s) + ib(s)] = u(s) + iv(s).$$

11. Suppose that $G(t) = a(s) + ib(s)$ is a function of index $\varkappa$ given on the contour L and $R(t)$ is its regularizing factor the index of which by assumption is v. Let, moreover, $\Phi^+(z)$ be a function which satisfies on L the condition $R(t)G(t) = \Phi^+(t)$ and is analytic in the domain D^+ except for, possibly, the given points $z_1, z_2, \ldots, z_m$ where it has orders $n_1, n_2, \ldots n_m$, all n_k having the same sign, and $\sum_{k=1}^m n_k = \varkappa + v$.

Prove that the function $\Phi^+(t)$ and the regularizing factor $R(t)$ are determined to within a factor which is the boundary value of a function analytic in D^+ and having there zero order.

12*. Solve the Hilbert boundary value problem for the semi-plane.
Hint. Make use of the results of Problems 2 and 3.

13. Solve the Hilbert boundary value problem for the domain exterior with respect to the unit circle.

14. Let D^- be a domain bounded by a simple contour L. Prove that the general solution of the Hilbert boundary value problem

$$\text{Re}[t'F^-(t)] = 0,$$

where

$$t' = \frac{dt}{ds},$$

is the function

$$F^-(z) = \alpha_0 \frac{\omega'(z)}{\omega(z)},$$

where $\omega(z)$ is the function which maps conformally the domain D^- onto the exterior of the unit circle (I. N. Vekua [30*], p. 272).

15. What are the conditions to be satisfied by the function $c(s)$ in order that the Hilbert problem
$$\text{Re}[t'(s) F^+(t)] = c(s)$$
be soluble?

16. Prove that the singular integral equation
$$a(s) u(s) + \frac{b(s)}{l} \int_L u(\sigma) \frac{\partial G^{\pm}(s, \sigma)}{\partial \sigma} d\sigma = c(s) \quad (a^2 + b^2 = 1),$$
where G and l are the same as in Problem 23 of Chapter I, can be reduced to the Hilbert boundary value problem
$$\text{Re}\left[\frac{F^{\pm}(t)}{a(s) \pm i b(s)}\right] = c(s)$$
with the additional condition $\int_L v\pm(\sigma) d\sigma = 0$ for the domains D^+ and D^-, respectively (Hilbert (1)).

17. Making use of the connection between the Hilbert and Cauchy kernels, reduce the solution of the dominant singular equation with Hilbert kernel to the solution of the Riemann boundary value problem. On this basis carry out a complete investigation of this equation.

18*. Derive the Noether theorems directly for the equation with Hilbert kernel (Noether (1)).

19*. Prove that for the singular integral operator
$$Ku \equiv a(s) u(s) - \frac{b(s)}{2\pi} \int_0^{2\pi} u(\sigma) \cot \frac{\sigma - s}{2} d\sigma + \int_0^{2\pi} K(s, \sigma) u(\sigma) d\sigma,$$
the adjoint operator
$$K'\xi = a(s) \xi(s) + \frac{1}{2\pi} \int_0^{2\pi} b(\sigma) \xi(\sigma) \cot \frac{\sigma - s}{2} d\sigma + \int_0^{2\pi} K(\sigma, s) \xi(\sigma) d\sigma$$
is the equivalent regularizator (V. D. Kupradze (3)).

20*. Making use of the method of § 32.2 solve the boundary value Problems II and III for polyanalytic functions (§ 32.2), for the circle.

21*. Solve the boundary value Problems I–IV for polyanalytic functions in the case of a semi-plane.

22*. Solve the basic problem of the theory of elasticity (§ 32.5) for the domain D^- bounded by the ellipse $x^2/a^2 + y^2/b^2 = 1$, making use of the fact that the domain D^- can be mapped conformally onto the interior of the unit circle by means of the function $z = \omega(\zeta) = - R(\zeta + m/\zeta)$, and onto the exterior of the unit circle by means of the function $z = R(\zeta + m/\zeta)$, where
$$R = \tfrac{1}{2}(a + b), \quad Rm = \tfrac{1}{2}(a - b).$$

23*. Prove that the boundary value Problem I for polyanalytic functions for an arbitrary contour, can be reduced to n Hilbert problems and, consequently, can be solved in a closed form (M. P. Ganin (1)).

CHAPTER V

VARIOUS GENERALIZED BOUNDARY VALUE PROBLEMS†

THE basic boundary value problems of Riemann and Hilbert, investigated in Chapters II and III, may be generalized in various ways. In this chapter the following generalizations will be considered.

1. Inclusion in the boundary conditions (of the types of Hilbert and Riemann problems) besides the values of functions and their derivatives, of terms of integral nature.

2. Examination of multiply-connected domains.

3. Formulation of boundary value problems for functions of a more general form which contain as particular cases harmonic and analytic functions, for instance for solutions of elliptic type equations and of systems of this type.

It should be observed that the generalization of the Riemann boundary value problem to the case of multiply-connected domains does not lead to significant difficulties and has already been considered. For the Hilbert problem, however, the transition from a simply-connected to a multiply-connected domain is connected with essential difficulties and from the viewpoint of the method of solution and the results it is closer to the problems of the present chapter.

Evidently, the possibilities of generalizations are not confined to those enumerated above. For instance, in the next chapter we shall investigate the problem of extension of the class of admissible coefficients and contours (discontinuous coefficients and open contours). A natural generalization is to increase the number of unknown functions determined by a system of coupled boundary conditions; these problems, however, will not be dealt with here. It is also possible to include the values of functions at various points of the contour

---

† The main contents of Chapters VI and VII are independent of the present chapter. Therefore, the reader may omit the latter, proceed to the former chapters, and return to Chapter V afterwards.

in the boundary condition, as occurs for instance in the generalized Riemann problem considered in § 18. The generalization of this kind of the Hilbert problem had not been worked out until recently†.

§ 34. Boundary value problem of Hilbert type, with the boundary condition containing derivatives

34.1. Formulation of the problem and various forms of the boundary conditions

Various authors dealing with this problem assumed the boundary condition in various forms, although essentially they are equivalent. To acquaint the reader with these forms of the boundary condition we shall present them below.

The author investigated the problem in the paper [5], published in 1938, formulating it as follows.

It is required to determine the function $\Phi(z) = u(x, y) + iv(x, y)$, analytic in the domain D^+ bounded by a simple smooth contour L, which satisfies on the contour the boundary condition

$$\sum_{k=0}^{m} \left[a_k(s) \frac{d^k u}{ds^k} + b_k(s) \frac{d^k v}{ds^k} \right] = f(s), \qquad (34.1)$$

where the coefficients $a_k(s)$, $b_k(s)$, $f(s)$ are real continuous functions, and $a_m(s)$, $b_m(s)$, $f(s)$ satisfy the Hölder condition.

I. N. Vekua (4) took the boundary condition in the form

$$\text{Re} \sum_{k=0}^{m} \left[a_k(t) \, \Phi^{(k)}(t) + \int_L h_k(t, \tau) \, \Phi^{(k)}(\tau) \, d\tau \right] = f(t), \qquad (34.2)$$

where $a_k(t)$, $h_k(t,\tau)$ are complex and $f(t)$ is a real function, of the same nature as the coefficients of the boundary condition (34.1); $h_k(t,\tau)$ are Fredholm kernels. By $\Phi^{(k)}(t)$ we understand the boundary value of the derivative of order k of the function $\Phi(z)$.

D. I. Sherman (1) formulated essentially the same problem as a problem for harmonic functions, as follows.

It is required to determine in the domain D^+ the function $u(x, y)$ which satisfies the boundary condition

$$\sum_{k=0}^{m} \sum_{j=0}^{k} a_{kj}(s) \frac{\partial^k u}{\partial x^{k-j} \partial y^j} = f(s), \qquad (34.3)$$

where $a_{kj}(s)$, $f(s)$ are coefficients of the same nature as before.

---

† Now, this problem has been investigated. See for instance the following contributions in *Doklady Akademii Nauk*: I. M. Melnik, v, 138, No. 3, 1961; L. G. Mikhailov, v, 139, No. 2, 1961; G. S. Litvinchuk and E. G. Khasabov, v, 140, No. 1, 1961 and v, 142, No. 2, 1962.

The most general of the above three boundary conditions is condition (34.2), since it contains integral terms. The non-integral terms of conditions (34.1) and (34.2) are identical; condition (34.3) contains no term with $v(s)$, i.e. if this condition be transformed to the form (34.1) it would turn out that $b_0(s) \equiv 0$.

Essentially the boundary value problems (34.1) – (34.3) are equivalent from the viewpoint of both the methods applicable to them and the results thus deduced. The investigation, as will be shown later, is confined to the non-integral terms containing the highest derivatives, which constitute the dominant part of the appropriate singular integral equation (see § 34.6). A change in the other terms leads only to a change in the regular part of the kernel which is irrelevant, since the latter contains all the same arbitrary functions— the coefficients of the boundary condition.

Since we shall solve the boundary value problem mostly in the form (34.2), we now indicate the way of reducing the other boundary conditions to it. It is very simple for (34.3); in fact, it is sufficient to use the relations

$$\frac{\partial}{\partial x} = \frac{d}{dt}, \qquad \frac{\partial}{\partial y} = i\frac{d}{dt},$$

following from the rule of differentiation of implicit functions ($t = x + iy$).

We have

$$\sum_{k=0}^{m} \sum_{j=0}^{k} a_{kj} \frac{\partial^k u}{\partial x^{k-j} \partial y^j} = \mathrm{Re}\left[\sum_{k=0}^{m} \sum_{j=0}^{k} a_{kj} \frac{\partial^k \Phi}{\partial x^{k-j} \partial y^j}\right]$$

$$= \mathrm{Re}\left[\sum_{k=0}^{m} \sum_{j=0}^{k} i^j a_{kj} \Phi^{(k)}(t)\right];$$

denoting $\sum_{j=0}^{k} i^j a_{kj}$ by $a_k(t)$ we arrive at the form (34.2).

To pass from the boundary condition (34.1) to (34.2) we express the derivatives with respect to the arc s by derivatives with respect to the complex coordinate t of the contour. Thus we obtain

$$\frac{d\Phi}{ds} = \Phi'(t)t' \qquad \frac{d^2\Phi}{ds^2} = \frac{d}{ds}[\Phi'(t)t'] = \Phi''(t)t'^2 + \Phi'(t)t'',$$

$d^m\Phi/ds^m = \Phi^{(m)}(t)t'^m +$ terms containing derivatives, of an order less than m, of the function $\Phi(t)$.

Hence we have

$$\sum_{k=0}^{m} \left[a_k(s)\frac{d^k u}{ds^k} + b_k(s)\frac{d^k v}{ds^k} \right] = \mathrm{Re}\left[\sum_{k=0}^{m} (a_k - ib_k)\frac{d^k \Phi}{ds^k} \right]$$

$$= \mathrm{Re}\left[t'^m (a_m - ib_m)\, \Phi^{(m)}(t) \right] +$$

$$+ \text{ terms containing derivatives of } \Phi \text{ of lower order.}$$

In an analogous way it is easy to pass from any of the three considered forms of the boundary condition to the other two.

Let us demonstrate with a particular example the transition from the form (34.1) to the forms (34.2) and (34.3).

Consider an equation of the form (34.1)

$$\sin s\, \frac{du}{ds} + (2 - \cos s)\frac{dv}{ds} - 2u(s) = f(s).$$

Transforming it by means of the above formulae we obtain

$$\mathrm{Re}\left\{ [\sin s - i(2 - \cos s)]\, t'(s)\, \Phi'(t) - 2\Phi(t) \right\} = f(s),$$

i.e. the boundary condition in the form of I. N. Vekua, if we assume that s is expressed by t by means of the equation of the contour. Finally, setting $t'(s) = x'(s) + iy'(s)$ we obtain the condition in the form of D. I. Sherman:

$$[x'(s)\sin s + y'(s)(2 - \cos s)]\frac{\partial u}{\partial y} +$$

$$+ [y'(s)\sin s - x'(s)(2 - \cos s)]\frac{\partial u}{\partial y} - 2u(s) = f(s).$$

Let us observe here that the particular case $m = 1$ of the problem under consideration was examined by Poincaré in connection with the theory of tides, in 1902. It was exactly this problem that prompted this outstanding mathematician to investigate the theory of singular integral equations (see the Historical notes of Chapter III, § 26).

Below (§ 34.3 – § 34.6) we shall expose a method of the solution of D. I. Sherman (4) who reduced the boundary value problem directly to a Fredholm integral equation, avoiding a singular equation†.

---

† It turns out then that the kernels of the derived equations can be expressed by elementary functions (without integrals), which is evidently convenient in numerical methods of solution of integral equations.

34.2. Representation of an analytic function by a Cauchy type integral with a real density

It is well known that representations of solutions of differential equations by means of potentials of various types play an important role in solving boundary value problems of mathematical physics. By means of such representations the boundary value problems for functions constituting solutions of differential equations of the type under consideration, are reduced to integral equations for the density of the potential. The simplest potentials are the ordinary logarithmic potentials of simple and double layers, applied to solving boundary value problems for harmonic functions †.

Bearing in mind the close connection between the integrals of Cauchy type and the potentials (§ 10) it is natural to expect that in the theory of boundary value problems for analytic functions an important role will be played by representation of analytic functions by means of Cauchy type integrals and their generalizations.

We begin by considering the simplest example, passing to more complicated ones later.

Let $\Phi^+(z)$ be an analytic function given in the domain D^+ bounded by a simple contour L. Consider the problem of representation of $\Phi^+(z)$ by the Cauchy type integral

$$\Phi(z) = \frac{1}{2\pi i} \int\limits_L \frac{\mu(\sigma)}{\tau - z} \, d\tau, \tag{34.4}$$

the density $\mu(\sigma)$ not being subject to any additional conditions. Is such a representation possible and to what extent is the function μ determined in terms of the known function $\Phi^+(z)$?

It is known that formula (34.4) defines simultaneously two analytic functions: the given function $\Phi^+(z)$ for $z \in D^+$ and, moreover, another function $\Phi^-(z)$ for $z \in D^-$. Applying the usual notation for the limiting values of the functions on the contour L we obtain in accordance with the Sokhotski formulae (4.9)

$$\mu(t) = \Phi^+(t) - \Phi^-(t). \tag{34.5}$$

---

† Thermal and wave potentials are also well investigated. This topic also includes a discussion of the operational method involving the representation of the unknown functions by Laplace–Fourier integrals and various generalizations of them.

This formula indicates that the density $\mu(t)$ of the Cauchy type integral is determined by the given function $\Phi^+(z)$ to within a term which is the boundary value of a function analytic in D^-, subject to the natural condition $\Phi^-(\infty) = 0$. Conversely, substituting into formula (34.4) the value (34.5), we obtain, in view of the Cauchy formula and the Cauchy theorem for $z \in D^+$, the given function $\Phi^+(z)$.

In order that the problem of representation of a given analytic function by a Cauchy type integral be definite, the density $\mu(\sigma)$ should be subjected to some additional conditions. These conditions can be of various types†. We first consider the simplest one, namely we demand that $\mu(\sigma)$ is a real function.

Similarly to Chapter IV we shall consider here various quantities either as functions of the complex coordinates of the points t, τ, or as functions of real variables (arc coordinates) s, σ. They are connected by the equation of the contour $t = t(s)$; thus $\mu(t) = \mu[t(s)] = \mu(s)$.

Returning to the problem under consideration let us eliminate from (34.5) the unknown function $\mu(t)$ by taking the imaginary parts. We obtain

$$\operatorname{Im} \Phi^-(t) = \operatorname{Im} \Phi^+(t).$$

The right-hand side of this relation is known. To determine $\Phi^-(z)$ we arrive at a Dirichlet problem for the domain D^- with the additional condition $\Phi^-(\infty) = 0$, the latter rendering the problem insoluble in the general case. Thus, a representation of an arbitrary function $\Phi^+(z)$ by a Cauchy type integral with a real density is in general impossible.

To make the representation possible let us introduce into the right-hand side of (34.4) a purely imaginary term which is for the time being arbitrary. In other words we seek the representation in the form

$$\Phi(z) = \frac{1}{2\pi i} \int\limits_L \frac{\mu(\sigma)}{\tau - z}\, d\tau + iC. \tag{34.6}$$

Here and in all that follows, in investigating integral representations by the function $\Phi(z)$ defined in the complementary domain we shall understand a function defined only by the Cauchy type integral without the additional terms entering the integral represen-

† There are many similarities between this problem and the problem of determination of the regularizing factor.

tation. For instance, in the present case

$$\Phi^-(z) = \frac{1}{2\pi i} \int_L \frac{\mu(\sigma)}{\tau - z} d\tau \quad \text{when} \quad z \in D^-.$$

Let us write formula (34.6) in the form

$$\Phi(z) - iC = \frac{1}{2\pi i} \int_L \frac{\mu(\sigma)}{\tau - z} d\tau$$

and let us apply to the Cauchy type integral the Sokhotski formulae. Hence we obtain

$$\mu(t) = [\Phi^+(t) - iC] - \Phi^-(t). \tag{34.7}$$

Consequently,

$$\operatorname{Im}[\Phi^-(t) + iC] = \operatorname{Im}\Phi^+(t).$$

Thus we have arrived at the Dirichlet problem for the exterior domain, without any limitations; this problem is uniquely soluble. Let

$$\operatorname{Im}[\Phi^-(z) + iC] = v(x, y).$$

Then the condition $\operatorname{Im}\Phi^-(\infty) = 0$ determines the constant C uniquely. The real part of $\Phi^-(z)$ is found from the Cauchy–Riemann equations to within a constant term. The latter is found from the condition $\operatorname{Re}\Phi^-(\infty) = 0$. Thus the real part of $\Phi^-(z)$ is also fully determined. Consequently, formula (34.7) defines the density $\mu(t)$ in an unique way.

Thus, we have the following result.

The representation of an arbitrary analytic function $\Phi^+(z)$ in the domain D^+ in the form (34.6) is possible, C and $\mu(\sigma)$ being determined by $\Phi^+(z)$ uniquely.

Suppose now that $\Phi^-(z)$ is a function analytic in D^-. Then, taking into account that the Cauchy type integral vanishes at infinity we seek the representation in the form

$$\Phi(z) = \frac{1}{2\pi i} \int_L \frac{\mu(\sigma)}{\tau - z} d\tau + \Phi(\infty). \tag{34.8}$$

Reasoning analogous to that carried out above, leads, for the determination of $\mu(t)$, to the relation

$$\mu(t) = \Phi^+(t) - [\Phi^-(t) - \Phi(\infty)]. \tag{34.9}$$

Now we obtain for the determination of the imaginary part of $\Phi^+(z)$ the uniquely soluble Dirichlet problem

$$\text{Im}\,\Phi^+(t) = \text{Im}\,[\Phi^-(t) - \Phi(\infty)].$$

The function $\Phi^+(z)$ itself is determined to within a real term, and then *the density $\mu(t)$ is calculated by means of formula* (34.9) *also to within a real term.*

34.3. The Cauchy type integral the density of which is the product of a given complex function and a real function

We now proceed to investigate a more general representation of analytic functions, by integrals of Cauchy type the densities of which are products of a prescribed function, which will be written in the form $1/[a(s) + ib(s)]$, and an unknown real function $\mu(s)$.

Thus, we seek the representation of the analytic function $\Phi(z)$ in the form

$$\Phi(z) = \frac{1}{2\pi i} \int_L \frac{\mu(\sigma)}{a(\sigma) + ib(\sigma)} \frac{d\tau}{\tau - z}. \qquad (34.10)$$

We assume that $a + ib \neq 0$. Without any loss of generality we may also assume that $a^2 + b^2 = 1$.

Consider the case $z \in D^+$, i.e. we regard the functions $\Phi^+(z)$ as known and $\Phi^-(z)$ as arbitrary. It was shown in the preceding subsection that the problem of representing an analytic function in the form of a Cauchy type integral with a real density, is reduced to the solution of the Dirichlet problem for the complementary domain. Similarly, the problem of representing it in the form (34.10) is reduced to the solution of the Hilbert problem. Since the latter essentially depends on the index let us consider the following various cases.

1. $\varkappa = \text{Ind}\,(a + ib) > 0$. As before we have

$$\frac{\mu(\sigma)}{a(s) + ib(s)} = \Phi^+(t) - \Phi^-(t). \qquad (34.11)$$

Multiplying by the denominator and taking the imaginary parts we obtain

$$\text{Im}\,\{[a(s) + ib(s)]\,\Phi^-(t)\} = \text{Im}\,\{[a(s) + ib(s)]\,\Phi^+(t)\}.$$

This relation is the boundary condition of the Hilbert problem for the exterior domain D^- (see § 29.4), the only difference being that the factor $1/(a + ib)$ is replaced by $a + ib$† (see the end of § 27.3).

Multiplying the last relation throughout by the regularizing factor $p(s)$ (for the domain D^-) of the function $a + ib$ and denoting

$$p(s) \operatorname{Im} [(a + ib) \, \Phi^+(t)] = f(s),$$

we arrive at the relation

$$\operatorname{Im} [t^\varkappa e^{i\gamma(t)} \, \Phi^-(t)] = f(s).$$

Making use of the results of § 29.4 and formula (27.10) we obtain

$$\Phi^-(z) = z^{-\varkappa} e^{-i\gamma(z)} [iSf + iQ(z)]. \tag{34.11'}$$

Next we determine the function $\mu(s)$ in accordance with formula (34.11). Since $\Phi^-(\infty) = 0$ the function $Q(z)$ defined by formula (29.27) contains only $2\varkappa - 1$ arbitrary real constants. Consequently, for $\varkappa > 0$ the representation (34.10) is possible and the function $\mu(s)$ depends on $2\varkappa - 1$ arbitrary real constants.

2. $\varkappa \leqq 0$. For a non-positive index, taking the representation in the form (34.10) we would arrive at the Hilbert problem which is, in this case, in general insoluble. To ensure the solubility let us introduce into the representation an additional polynomial and seek the function $\Phi(z)$ in the form

$$\Phi(z) = \frac{1}{2\pi i} \int_L \frac{\mu(\sigma)}{a(\sigma) + ib(\sigma)} \frac{d\tau}{\tau - z} + \sum_{k=0}^{-\varkappa} c_k z^k, \tag{34.12}$$

where c_k are for the time being arbitrary constants.

Proceeding as before we first obtain

$$\mu(s) = [a(s) + ib(s)] \left[\Phi^+(t) - \Phi^-(t) - \sum_{k=0}^{-\varkappa} c_k t^k \right], \tag{34.13}$$

and then we are led to the problem

$$\operatorname{Im} \left\{ t^\varkappa e^{i\gamma(t)} \left[\Phi^-(t) + \sum_{k=0}^{-\varkappa} c_k t^k \right] \right\} = f(s),$$

the solution of which has the form

$$\Phi^-(z) = z^{-\varkappa} e^{i\gamma(z)} [iSf + \alpha_0] - \sum_{k=0}^{-\varkappa} c_k z^k.$$

---

† The indicated difference implies that the results of the present subsection differ from those of § 29.4 in the sign of the index.

Expanding the right-hand side into series in the vicinity of $z = \infty$ denoting

$$z^{-\varkappa} e^{-i\gamma(z)} [iSf + \alpha_0] = \sum_{k=-\infty}^{-\varkappa} A_k z^k.$$

By assumption we have

$$\Phi^-(\infty) = 0.$$

This condition is satisfied if we set

$$c_k = A_k \quad (k = 0, 1, \ldots, -\varkappa).$$

The coefficients $c_0, c_1, \ldots, c_{-\varkappa-1}$ are determined by the last relations in a unique way. Since the coefficient $A_{-\varkappa}$ contains the real arbitrary quantity α_0, the relation $c_\varkappa = A_{-\varkappa}$ does not completely determine the complex coefficient $c_{-\varkappa}$ but yields two real equations with three real unknowns. Let us consider these equations in more detail.

Suppose that

$$iSf + \alpha_0 = \alpha_0 + i\beta_0 + \frac{B}{z} + \cdots,$$

$$e^{-i\gamma(z)} = e^{\omega_1} e^{-i\omega} = \alpha_1 - i\beta_1 + \frac{D}{z} + \cdots$$

are the expansions of the corresponding functions in the vicinity of $z = \infty$, and

$$\alpha_1 = e^{\omega_1(\infty)} \cos \omega(\infty), \quad \beta_1 = e^{\omega_1(\infty)} \sin \omega(\infty).$$

Then from the relation

$$c_{-\varkappa} = c_{-\varkappa}^{(1)} + i c_{-\varkappa}^{(2)} = (\alpha_1 \alpha_0 + \beta_1 \beta_0) + i(\alpha_1 \beta_0 - \beta_1 \alpha_0)$$

for the three arbitrary quantities $c_{-\varkappa}^{(1)}, c_{-\varkappa}^{(2)}, \alpha_0$ we obtain two equations

$$\alpha_1 \alpha_0 + \beta_1 \beta_0 = c_{-\varkappa}^{(1)}, \quad \alpha_1 \beta_0 - \beta_1 \alpha_0 = c_{+\varkappa}^{(2)}. \tag{34.14}$$

The known coefficients α_1, β_1 cannot simultaneously vanish. If both of them are different from zero the system (34.14) can be satisfied by setting one of the constants $c_{-\varkappa}^{(1)}, c_{-\varkappa}^{(2)}$ equal to zero, i.e. by taking $c_{-\varkappa}$ real or purely imaginary (no matter which). If one of the numbers α_1, β_1 is zero (i.e. $\cos\omega(\infty)$ or $\sin\omega(\infty)$ vanishes) $c_{-\varkappa}$ can be taken real as $\beta_1 \neq 0$ and purely imaginary as $\alpha_1 \neq 0$. Thus, in all cases $c_{-\varkappa}$ can be regarded as real or purely imaginary.

Consequently, for $\varkappa \leq 0$ representation (34.12) is possible, and the coefficients $c_0, c_1, \ldots, c_{-\varkappa-1}$ are some complex constants and $c_\varkappa$ is a real or purely imaginary number. Consequently, representation

(34.12) *contains* $2\varkappa + 1$ *real parameters which must necessarily be introduced in order that the representation be possible.*

In an analogous way we solve the problem for the case when the function $\Phi(z)$ is prescribed in the domain D^-. In this case the problem of representing $\Phi^-(z)$ in the indicated form is reduced to the solution of the Hilbert problem for the interior domain D^+. Without considering the details let us present the results.

1. $\varkappa = \mathrm{Ind}\,(a + ib) \leqq 0$. The function $\Phi(z)$ analytic in the domain D^- can now be represented in the form

$$\Phi(z) = \frac{1}{2\pi i}\int\limits_{L} \frac{\mu(\sigma)}{a(\sigma) + ib(\sigma)}\,\frac{d\tau}{\tau - z} + \Phi(\infty), \qquad (34.15)$$

where $\mu(s)$ contains $-2\varkappa + 1$ real arbitrary constants.

2. $\varkappa > 0$. In this case the representation can have the form

$$\Phi(z) = \frac{1}{2\pi i}\int\limits_{L} \frac{\mu(\sigma)}{a(\sigma) + ib(\sigma)}\,\frac{d\tau}{\tau - z} + \Phi(\infty) + \sum_{k=1}^{\varkappa} \frac{c_k}{(z - z_0)^k}, \qquad (34.16)$$

where z_0 is a fixed point inside L and c_k are some constants determined uniquely by $\Phi(z)$; $c_\varkappa$ can be taken either real or purely imaginary.

34.4. Integral representation of an analytic function the mth derivative of which is representable by the Cauchy type integral

In solving boundary value problems of the type (34.1)–(34.3) it is useful to have an integral representation of analytic functions in which the highest derivative entering the boundary condition is representable by the Cauchy type integral of the form considered in the preceding subsection.

The kernel of such an integral representation can obviously be obtained by integrating the Cauchy kernel $1/(\tau - z)$ m times. However, we shall for simplicity of the exposition confine ourselves to the verification of the result by differentiating m times†. We first prove the validity of the identity

$$\frac{(-1)^m}{(m - 1)!}\,\frac{d^m}{dz^m}\left[(\tau - z)^{m-1}\ln\left(1 - \frac{z}{\tau}\right)\right] = \frac{1}{\tau - z}. \qquad (34.17)$$

† We are not interested here in additional terms arising in integration, since it is sufficient for us to have just one expression for the kernel.

The proof can be carried out by a direct application of the Leibniz formula for differentiating a product. However, bearing in mind a later more complicated case in which an application of the Leibniz formula is difficult, we shall prove the identity (34.17) by the method of mathematical induction, taking into account that it is obviously valid for $m = 1$. For $m = n + 1$ we have

$$\frac{(-1)^{n+1}}{n!} \frac{d^{n+1}}{dz^{n+1}} \left[(\tau - z)^n \ln \left(1 - \frac{z}{\tau} \right) \right]$$

$$= \frac{(-1)^{n+1}}{n!} \frac{d^n}{dz^n} \left[-n(\tau - z)^{n-1} \ln \left(1 - \frac{z}{\tau} \right) - (\tau - z)^{n-1} \right].$$

Since, in view of the assumption of induction,

$$\frac{d^n}{dz^n} \left[(\tau - z)^{n-1} \ln \left(1 - \frac{z}{\tau} \right) \right] = \frac{(n-1)!}{(-1)^n} \frac{1}{\tau - z},$$

and $[d^n/dz^n](\tau - z)^{n-1} = 0$, identity (34.17) is proved.

We now proceed to derive the integral representation. Its form will depend on $\text{Ind}(a + ib) = \varkappa$, as was also the case in the simpler problem examined in the preceding subsection. Moreover, it will depend on the order of the derivative m. *The form of the representation will change, depending on whether $m > \varkappa$ or $m \leqq \varkappa$.*

1. $m > \varkappa$. We prove that in this case the function $\Phi^+(z)$ is representable in the form

$$\Phi(z) = \frac{(-1)^m}{(m-1)!} \frac{1}{2\pi i} \int_L \frac{\mu(\sigma)}{a(\sigma) + ib(\sigma)} (\tau - z)^{m-1} \ln \left(1 - \frac{z}{\tau} \right) d\tau +$$

$$+ \sum_{k=0}^{m-\varkappa} c_k z^k, \quad (34.18)$$

where we take the branch of the logarithm which vanishes for $z = 0$.

In proving the assertion we shall have to differentiate between two cases, depending on the sign of $\varkappa$.

(a) $\varkappa \leqq 0$. Differentiating the relation m times (34.18) we obtain

$$\Phi^{(m)}(z) = \frac{1}{2\pi i} \int_L \frac{\mu(\sigma)}{a(\sigma) + ib(\sigma)} \cdot \frac{d\tau}{\tau - z} +$$

$$+ \sum_{k=m}^{m-\varkappa} k(k-1) \ldots (k - m + 1) c_k z^{k-m}. \quad (34.19)$$

We have arrived at representation (34.12) for the function $\Phi^{(m)}(z)$ if $k - m$ is replaced by k and $(k + m)(k + m - 1) \ldots (k + 1) c_{k+m}$

by c_k). In view of the investigation of the preceding subsection representation (34.19) is possible and the coefficients $c_m, c_{m+1}, \ldots, c_{m-\varkappa-}$ are definite complex constants while $c_{m-\varkappa}$ can be taken as either a real or a purely imaginary constant. The function $\mu(s)$ is determined in a unique way.

To determine the first m coefficients of the sum let us compare the first m terms of the expansion of the both sides of formula (34.18) into the Taylor series. (The remaining terms are identical, on account of relation (34.19).) We have

$$\left. \frac{1}{k!} \, \Phi^{(k)}(0) = - \frac{(-1)^m}{(m-1)!} \, \frac{\alpha_k}{2\pi i} \int_L \frac{\mu(\sigma) \, \tau^{m-k-1}}{a(\sigma) + ib(\sigma)} \, d\tau + c_k \atop (k = 0, 1, \ldots, m-1), \right\} \quad (34.20$$

where the coefficients a_k obtained by the multiplication of the series for $(\tau - z)^{m-1}$ and $\ln(1 - z/\tau)$ are the following:

$$\alpha_0 = 0; \quad \alpha_k = \frac{1}{k} - \frac{C_{m-1}^1}{k-1} + \cdots + \frac{(-1)^{k-1} C_{m-1}^{k-1}}{1}$$
$$(k = 1, 2, \ldots, m-1).$$

We take into account here that the function $\mu(s)$ is already known.

Relations (34.20) determine the coefficients $c_0, c_1, \ldots, c_{m-1}$ uniquely. Thus, representation (34.18) is proved. It has incidentally been established that *the sum entering this representation contains* $2(m - \varkappa) + 1$ *real parameters*.

(b) $\varkappa > 0$. Differentiating (34.18) we arrive at the relation

$$\Phi^{(m)}(z) = \frac{1}{2\pi i} \int_L \frac{\mu(\sigma)}{a(\sigma) + ib(\sigma)} \, \frac{d\tau}{\tau - z}. \quad (34.21$$

For $\Phi^{(m)}(z)$ we have a representation of the type (34.10). The investigation of the preceding subsection implies that representation (34.21) is possible in the case under consideration, and in accordance with formulae (34.11) and (34.11') $\mu(s)$ depends on $2\varkappa - 1$ arbitrary real constants.

Similarly to the case $\varkappa \leqq 0$ we equate the first m terms of the expansion of both sides of relation (34.18) in the vicinity of the origin which leads again to relation (34.20) where we have to set

$$c_{m-\varkappa+1} = c_{m-\varkappa+2} = \cdots = c_{m-1} = 0;$$

the m complex equations (34.20) (i.e. $2m$ real ones) contain altogether

$$2(m - \varkappa + 1) + 2\varkappa - 1 = 2m + 1 \quad (34.22$$

real parameters. It can be proved (we shall not do so here) that all these equations are independent. They completely determine the coefficients $c_0, c_1, \ldots, c_{m-\varkappa-1}$ and the arbitrary constants entering the expression for $\mu(s)$; $c_{m-\varkappa}$ can be taken as real or purely imaginary.

Thus, in this case also we have proved representation (34.18) and it has also been established that *the representation contains* $2(m - \varkappa) + 1$ *real parameters and the density* $\mu(s)$ *is determined uniquely by the given function* $\Phi^+(z)$†.

Finally let us consider the last case.

2. $m \leqq \varkappa$. Now we could expect that the non-integral sum in representation (34.18) to be absent; however, in this case, taking into account the choice of the branch of $\ln(1 - z/\tau)$, we would obtain that the function represented by the integral vanishes at $z = 0$. Consequently, to represent an arbitrary function $\Phi(z)$, it is necessary to retain in the sum the constant term c_0, i.e. to seek the representation in the form

$$\Phi(z)$$
$$= \frac{(-1)^m}{(m-1)!} \frac{1}{2\pi i} \int_L \frac{\mu(\sigma)}{a(\sigma) + ib(\sigma)} (\tau - z)^{m-1} \ln\left(1 - \frac{z}{\tau}\right) d\tau + c_0, \quad (34.23)$$

where

$$c_0 = \Phi(0). \qquad (34.24)$$

The proof is carried out in exactly the same way as before. First we establish that $\mu(s)$ is determined by $\Phi^{(m)}(z)$ and contains linearly $2\varkappa - 1$ real constants. (It is to be borne in mind that here it is necessary that $\varkappa > 0$.) Further, the expansion in the vicinity of the origin yields relation (34.24) and also $m - 1$ complex equations which are satisfied on account of the presence of the constants in $\mu(s)$. These equations make it possible to determine only $2(m - 1)$ constants, the remaining ones being arbitrary; consequently, the function $\mu(s)$ is *determined by the given function* $\Phi^+(z)$ *not uniquely, but to within an expression which depends linearly on*

$$2\varkappa - 1 - 2(m - 1) = 3(\varkappa - m) + 1 \qquad (34.25)$$

arbitrary real constants.

† Let us note that although in (a) and (b) the number of the real parameters is given by the same formula, their role is somewhat different. In the case (a) all of them enter only the additional sum, whereas in the case (b) some of them also enter the expression for $\mu(s)$.

11*

34.5. Reduction of the boundary value problem to a Fredholm integral equation

We now take the boundary condition in the form (34.2)

$$\operatorname{Re} \sum_{k=0}^{m} \left[a_k(t)\, \Phi^{(k)}(t) + \int_L h_k(t,\tau)\, \Phi^{(k)}(\tau)\, d\tau \right] = f(s) \quad (34.26)$$

and assume that $a_m(t) \neq 0$. Moreover, to make it possible to apply the integral representation of the preceding subsection, let us subject the contour L to the additional restriction that it constitutes a Lyapounov curve (see § 18.1).

To solve the problem we apply the integral representation (34.18) taking for the known function $a + ib$ in this representation, the coefficient of the highest derivative of the boundary condition, $a_m(t)$:

$$\Phi(z) = \frac{(-1)^m}{(m-1)!} \frac{1}{2\pi i} \int_L \frac{\mu(\sigma)}{a_m(\tau)} (\tau - z)^{m-1} \ln\left(1 - \frac{z}{\tau}\right) d\tau + \sum_{k=0}^{m-x} c_k z^k. \quad (34.27)$$

Differentiating this formula in turn m times we obtain the integral representations for all derivatives of $\Phi(z)$ up to mth order. Then the kernels of the representations of the function and its derivatives up to $(m-2)$th order are continuous up to the contour and their limiting values may be derived by a direct substitution for the variable z of its contour value t. The kernel of the $(m-1)$th derivative has on the contour a singularity of logarithmic type and it is a Fredholm kernel; its limiting value is also obtained by replacing z by t. Only the mth derivative represented by the Cauchy type integral (34.19), in passing to the limiting values yields in accordance with the Sokhotski formula the singular Cauchy kernel.

Let us examine in more detail the term containing the highest derivative; we shall prove that our choice of the integral representation $(a + ib = a_m(t))$ leads to the reduction of the singular integral to an ordinary integral

$$\operatorname{Re}\left[a_m(t)\, \Phi^{(m)}(t) \right] = \operatorname{Re}\left[\frac{1}{2}\mu(s) + \frac{1}{2\pi i} \int_L \frac{a_m(t)}{a_m(\tau)} \frac{\mu(\sigma)}{\tau - t}\, d\tau \right].$$

Transform the last integral as follows:

$$\int_L \frac{a_m(t)}{a_m(\tau)} \frac{\mu(\sigma)}{\tau - t}\, d\tau = \int_L \frac{a_m(t) - a_m(\tau)}{a_m(\tau)(\tau - t)}\, \mu(\sigma)\, d\tau + \int_L \frac{\mu(\sigma)}{\tau - t}\, d\tau. \quad (34.28)$$

Since $a_m(t)$ satisfies the Hölder condition, the first integral is ordinary. Consequently, the problem is reduced to investigating the integral

$$\text{Re}\left[\frac{1}{2\pi i}\int_L \frac{\mu(\sigma)}{\tau - t}\,d\tau\right].$$

Let us remember the resolution of the Cauchy type integral into the sum of two potentials, given in § 10; by formulae (10.3), (10.4) we have

$$\text{Re}\left[\frac{1}{2\pi i}\int_L \frac{\mu(\sigma)}{\tau - t}\,d\tau\right] = \frac{1}{2\pi}\int_L \mu\,d\theta = \frac{1}{2\pi}\int_L \mu(\sigma)\frac{\cos(r, n)}{r}\,d\sigma, \quad (34.29)$$

the notation being the same as in § 10.

Under the conditions imposed upon the contour L (it is a Lyapounov curve) the potential of double layer in the right-hand side of the last relation is an ordinary integral, i.e. the kernel $[\cos(r, n)]/r$ is a Fredholm one.

Let us now pass to real variables, using the equation of the curve to replace t, τ by the arc coordinates s, σ. As a consequence we obtain for the determination of $\mu(s)$ a Fredholm integral equation

$$\mu(s) + \int_L K(s, \sigma)\,\mu(\sigma)\,d\sigma = 2f(s) - \text{Re}\sum_{k=0}^{m-\varkappa} c_k\omega_k(s), \quad (34.30)$$

where $K(s, \sigma)$ is a Fredholm kernel which can be written explicitly. Since we shall not need this explicit expression anywhere, we do not quote it. The $\omega_k(s)$ are some definite complex functions which are obtained by substituting into the boundary condition the sum $\sum_{k=0}^{m-\varkappa} c_k t^k$ and its derivatives, and collecting the terms containing the same constant c_k. For $m \leq \varkappa$ only the term $c_0\omega_0(s)$ should be preserved in the sum.

Solving the integral equation and applying formula (34.27) we arrive at the solution of the original boundary value problem (34.26). Consequently, *the investigation of problems of solubility of the boundary value problem reduces to an investigation of the solubility of the integral equation* (34.30).

Prior to proceeding to the investigation of the integral equation let us note the dual role played by the constants c_k in deriving the integral representation on one hand and in investigating the integral equation on the other hand.

In deriving the integral representation the function $\Phi(z)$ was regarded as known, and, consequently, the constants c_k entering the representation and determined by this function were also regarded as known. In contrast to this, in making use of the integral representation to solve boundary value problems the function $\Phi(z)$ is the unknown to be found. Hence, the constants entering the integral representation are also to be regarded as unknown and, consequently, in considering the questions of solubility they should be regarded as arbitrary parameters.

Thus, the problem is reduced to an investigation of the solubility of the Fredholm integral equation the right-hand side of which contains a number of arbitrary parameters.

34.6. Problems of solubility of the boundary value problem

Introducing into consideration the real and imaginary parts of the parameters c_k and the functions $\omega_k(s)$ and taking into account that $c_{m-\varkappa}$ is real or purely imaginary, we can write the integral equation (34.30) in the form

$$\mu(s) + \int_L K(s, \sigma)\mu(\sigma)\,d\sigma = 2f(s) + \sum_{k=1}^{2m-2\varkappa+1} \alpha_k g_k(s), \quad (34.31)$$

where α_k are arbitrary real parameters and $g_k(s)$ are some definite real functions.

Before the investigation, let us make some remarks of a general nature concerning the solution of boundary value problems by the method of integral equations.

An investigation of boundary value problems by the method of integral equations can be regarded as complete only when without solving the integral equation it is possible to count the number of its linearly independent solutions and to express this result by a quantity characterizing the boundary value problems. An investigation of this kind can be performed in two cases.

1. When the kernel of the equation is a completely defined function, independent of any arbitrary functions appearing in the boundary condition. This case is, for instance, encountered in solving the Dirichlet and Neumann problems where the kernels of the equations are the potential of the double layer and the normal derivative of the potential of simple layer, respectively.

2. When the kernel contains arbitrary functions, but, in view of certain facts of a physical or analytic nature inherent in the boundary value problem under investigation, it is possible to prove the uniqueness of the solution (the insolubility of the homogeneous problem). This case occurs for instance in the boundary value problems of hydrodynamics and the theory of elasticity where the investigation is carried out on the basis of an earlier proof of uniqueness of solutions of the corresponding mechanical problems. We have met a similar situation in § 17 where, although the kernel contained an arbitrary function $\alpha(t)$, analytic facts enabled us to prove beforehand the insolubility of the problem with zero jump (§ 17.2).

In the present case the kernel of the equation $K(s, \sigma)$ contains the coefficients of the boundary condition which are arbitrary functions. On the other hand, on account of a great generality of the boundary value problem under consideration it is impossible to draw any conclusions concerning the number of solutions, directly from the boundary condition. Consequently, in this case we do not carry out a complete investigation of the boundary value problem by calculating the number of its solutions, but derive some statements concerning its solubility only, without solving the integral equation.

(i) Consider in detail the case $m > \varkappa$, which in accordance with the results of § 34.3 is characterized by the fact that the density $\mu(s)$ and the coefficients c_k (consequently also α_k) are uniquely expressible in terms of the function $\Phi(z)$, the relation being homogeneous, i.e. when $\Phi(z) \equiv 0$ we also have $\mu \equiv 0$ and $\alpha_k \equiv 0$. It is evident that the converse statement follows from the integral representation (34.27).

It is easy to deduce from this property that if the solutions of the boundary value problem, $\Phi_1(z)$ and $\Phi_2(z)$ are linearly independent, then it is impossible to find constants p_1, p_2 different from zero, such that the following relations are satisfied:

$$p_1\mu_1 + p_2\mu_2 \equiv 0; \quad p_1\alpha_k^{(1)} + p_2\alpha_k^{(2)} = 0 \qquad (34.32)$$
$$(k = 1, 2, \ldots, 2(m - \varkappa) + 1).$$

To investigate the integral equation (34.31) we shall make use of the third Fredholm theorem (§ 20.2). Assume that the homogeneous equation corresponding to the given equation (34.31) has q solutions. Then the equation
$$v(s) + \int_L K(\sigma, s) v(s) \, ds = 0,$$

adjoint to the one investigated has also q solutions. Suppose that

$$v_1(s), \ v_2(s), \ \ldots, \ v_q(s)$$

is a complete system of these solutions.

Then for the equation (34.31) to be soluble, it is necessary and sufficient that the conditions

$$\int_0^l \left[2f(s) + \sum_{k=1}^{2(m-\varkappa)+1} \alpha_k g_k(s) \right] v_j(s) \, ds = 0 \quad (j = 1, 2, \ldots, q). \quad (34.33)$$

are satisfied. Denoting

$$2 \int_0^l f(s) \, v_j(s) \, ds = -f_j, \quad \int_0^l g_k(s) \, v_j(s) \, ds = g_{jk}, \quad (34.34)$$

we find that the solubility of the equation considered is equivalent to the solubility of the following system of linear algebraic equations for the arbitrary constants α_k:

$$\sum_{k=1}^{2(m-\varkappa)+1} g_{jk} \alpha_k = f_j \quad (j = 1, 2, \ldots, q). \quad (34.35)$$

Consider first the homogeneous boundary value problem ($f(s) \equiv 0$). Now all $f_j = 0$ and system (34.35) is also homogeneous. Denote the rank of the matrix $\| g_{jk} \|$ of the system, i.e. the number of independent equations it contains, by r. Obviously, $r \leqq \min \{q, 2(m - \varkappa) + 1\}$. The homogeneous system of algebraic equations is always soluble; therefore, our homogeneous system is absolutely soluble and has $2(m - \varkappa) + 1 - r$ linearly independent solutions.

Examining the general solution of the integral equation (34.31) (for $f(s) \equiv 0$) expressed in terms of the resolvent,

$$\mu(s) = \sum_{k=1}^{2(m-\varkappa)+1} \alpha_k \left[g_k(s) + \int_L R(s, \sigma) g_k(\sigma) \, d\sigma \right] + \sum_{k=1}^{q} \beta_k \mu_k(s), \quad (34.36)$$

where the last sum represents the general solution of the homogeneous equation (34.31), we observe that this solution depends on $2(m - \varkappa) + + 1 - r + q$ linearly independent parameters. Introducing this value of μ into the integral representation (34.27) we find that the solution of the boundary value problem contains the same number of parameters. Taking into account that $r \leqq q$ we obtain the following result.

For $m > \varkappa$ the homogeneous problem is absolutely soluble and has not less than $2(m - \varkappa) + 1$ [exactly $2(m - \varkappa) + 1 + q - r$] solutions.

We now proceed to the non-homogeneous boundary value problem $(f(s) \not\equiv 0)$. It is known from linear algebra that if the rank of the extended matrix derived by adding to it the column of free terms is also r, then the non-homogeneous system (34.35) is soluble. This implies the following conclusion.

THEOREM. *In order that the non-homogeneous boundary value problem* (34.26) *be soluble, it is necessary and sufficient that the rank r of the matrix $\| g_{jk} \|$ of the system* (34.35) *does not increase on adding to the matrix the column of the free terms.* If this condition is satisfied, i.e. the non-homogeneous problem is soluble, then the solution depends on $2(m - \varkappa) + 1 + q - r$ real constants.

(ii) Assume now that $m \leqq \varkappa$. The difference between the present case and the preceding one consists in the fact that in the integral representation the expression for $\mu(s)$ in terms of $\Phi(z)$ contains arbitrary constants, the number of which is, in accordance with formula (34.25), equal to $2(\varkappa - m) + 1$. Hence, there also exist non-zero solutions of the integral equation to which there correspond zero solutions of the boundary value problem. This implies at once that the number of solutions of the boundary value problem is $2(\varkappa - m) + 1$ less than the number of solutions of the integral equation. This fact being taken into account, the investigation is entirely analogous to that of the preceding case. We shall not dwell on the details here.

The above reasoning is typical for many problems. It is encountered in all boundary value problems of the general type considered in the present chapter, and, moreover, in numerous other problems (for instance in the theory of Fredholm integro-differential equations on which an extensive literature exists). An outline of this reasoning follows.

To solve the problem, an integral equation is constructed, the right-hand side of which depends linearly on a number of parameters. An assumption is made concerning the number of solutions of the adjoint integral equation, and on the basis of the third Fredholm theorem the conditions of solubility are deduced. Thus, we arrive at a system of linear algebraic equations for the parameters. Making assumptions concerning its rank a conclusion is obtained about the solubility and the number of solutions of the original problem.

Thus, the investigation is carried out according to the scheme "if . . ., then . . .", there being no way of indicating exactly in the general case when this "if" occurs. More precise results can be obtained only in some particular cases.

34.7. Other methods of investigation

It was already mentioned at the beginning of the section that there are other methods of solving the problem under consideration, worked out by F. D. Gakhov, and I. N. Vekua. The method of I. N. Vekua consists in making use of an integral representation deduced by him in connection with this problem; using our notation we may write this representation in the form

$$\Phi(z) = \frac{(-1)^m}{(m-1)!} \frac{1}{2\pi i} \int_L \mu(\sigma) \left(1 - \frac{z}{\tau}\right)^{m-1} \ln\left(1 - \frac{z}{\tau}\right) d\sigma + c_0. \quad (34.37)$$

It is readily observed that this representation can be derived from the integral representation of D. I. Sherman (34.23), if we set there

$$a + ib = \tau^{m-1}\tau'.$$

The index of the last expression is m which corresponds exactly to formula (34.23). Making use of this representation[†] I. N. Vekua reduces the boundary value problem to a singular integral equation with Cauchy kernel (it would be even simpler to obtain the Hilbert kernel), the index of which is $2(m - \varkappa)$ where $\varkappa = \text{Ind } a_m(t)$. Applying to the latter the theory of singular integral equations I. N. Vekua established the results deduced in the preceding subsection[‡].

The method of F. D. Gakhov consists in employing the formulae of Hilbert which connect the boundary values of the real and imaginary parts of an analytic function. For the circle these formulae, which are in this case very simple, were derived in § 6.2. In the case of an arbitrary contour they are more complicated (see Problem 23 of Chapter I) and contain the Green function.

This method leads to integral equations with more complicated kernels than the two indicated above, which causes difficulties in

---

[†] The representation of I. N. Vekua was announced before that of D. I. Sherman.

[‡] I. N. Vekua states the conditions of solubility also in another form, by means of the concept of completeness of a kernel. An actual determination of the completeness of a kernel can be carried out by solving the appropriate integral equation. Therefore, the exact results given by I. N. Vekua concern only the cases when the kernel does not contain arbitrary functions (the Dirichlet and Hilbert problems).

practical solutions of problems. However from the viewpoint of foundations of the theory—investigation of solubility problems—this method has advantages over others since here difficulties connected with the elaboration of integral representations do not arise and, moreover, an interpretation of the deduced results is simpler because the unknown in the integral equation is the required function of the boundary value itself, not an auxiliary function (the density of the integral representation) as is the case of other methods.

Taking into account these circumstances let us also describe this method of solution. We shall confine ourselves to the case when the domain is the unit circle. Incidentally, the general case can be reduced to this particular case by a conformal mapping.

Consider the boundary condition in the form which is most suitable for the application of our method

$$\sum_{k=0}^{m} \left[a_k(s) \frac{d^k u}{ds^k} + b_k(s) \frac{d^k v}{ds^k} + \int_0^{2\pi} M_k(s, \sigma) \frac{d^k u}{d\sigma^k} d\sigma + \right.$$
$$\left. + \int_0^{2\pi} N_k(s, \sigma) \frac{d^k v}{d\sigma^k} d\sigma \right] = f(s). \quad (34.38)$$

For the basic unknown we take the highest derivative $d^m u/ds^m$ and we express it in terms of all remaining unknown quantities entering the boundary condition.

The quantities $d^{m-1}u/ds^{m-1}, \ldots, du/ds, u(s)$ can be expressed in terms of $d^m u/ds^m$ by successive integration. Applying the known form of transformation of a multiple integral into an ordinary one†, we have

$$\frac{d^{m-1}u}{ds^{m-1}} = \int_0^s \frac{d^m u}{d\sigma^m} d\sigma + \alpha_0, \quad \frac{d^{m-2}u}{ds^{m-2}} = \int_0^s (s - \sigma) \frac{d^m u}{d\sigma^m} d\sigma + \frac{\alpha_0}{1} s + \alpha_1, \ldots,$$

$$u(s) = \frac{1}{(m-1)!} \int_0^s (s - \sigma)^{m-1} \frac{d^m u}{d\sigma^m} d\sigma +$$

$$+ \frac{\alpha_0}{(m-1)!} s^{m-1} + \cdots + \alpha_{m-2} s + \alpha_{m-1}. \quad (34.39)$$

† See for instance V. I. Smirnov, *Course of higher mathematics* (Kurs vysshei matematiki), vol. II, ed. 10, p. 58.

The constants $\alpha_0, \alpha_1, \ldots, \alpha_{m-2}$ are not arbitrary; they should so be chosen that the functions $d^{m-1}u/ds^{m-1}, \ldots, u(s)$ are periodic. It is known that an integral of a periodic function is a periodic function if and only if the expansion of the integrand into the Fourier series (we have in mind functions which admit such an expansion) does not contain the free term, i.e. if

$$F(s) = \int_0^s f(\sigma)\, ds$$

then the condition of periodicity is the following:

$$\int_0^{2\pi} f(\sigma)\, d\sigma = 0. \tag{34.40}$$

On this basis we shall assume that the function $d^m u/ds^m$ satisfies the condition

$$\int_0^{2\pi} \frac{d^m u}{d\sigma^m}\, d\sigma = 0, \tag{34.41}$$

and the constants $\alpha_0, \alpha_1, \ldots, \alpha_{m-2}$ will be chosen in such a way that condition (34.40) is satisfied. For instance, in view of (34.40) α_0 is given by the formula

$$\alpha_0 = \frac{1}{2\pi} \int_0^{2\pi} \sigma \frac{d^m u}{d\sigma^m}\, d\sigma.$$

(In deriving it we employed the formula for the change of the order of integration in a double integral with a variable upper bound; see the preceding footnote.) Since the function $u(s)$ is no longer subject to the condition that its integral is a periodic function, the free term of its expansion into the Fourier series may be arbitrary. Therefore the parameter α_{m-1} in formulae (34.39) remains arbitrary.

Thus, all lower derivatives are expressible by $d^m u/ds^m$, the expression for $u(s)$ containing one arbitrary parameter and the function $d^m u/ds^m$ itself being subject to condition (34.41).

To derive expressions for the derivatives of v we start from the Hilbert formula (6.6)

$$v(s) = -\frac{1}{2\pi} \int_0^{2\pi} u(\sigma) \cot \frac{\sigma - s}{2}\, d\sigma + v_0. \tag{34.42}$$

Integrating by parts we obtain

$$v(s) = \frac{1}{2\pi} \int_0^{2\pi} \frac{du}{d\sigma} \ln\left(\sin^2 \frac{\sigma - s}{2}\right) d\sigma + v_0. \qquad (34.43)$$

Differentiating with respect to s we have

$$\frac{dv}{ds} = -\frac{1}{2\pi} \int_0^{2\pi} \frac{du}{d\sigma} \cot \frac{\sigma - s}{2} d\sigma. \qquad (34.44)$$

Repeating this operation $m - 1$ times more we arrive at the relation

$$\frac{d^m v}{ds^m} = -\frac{1}{2\pi} \int_0^{2\pi} \frac{d^m u}{d\sigma^m} \cot \frac{\sigma - s}{2} d\sigma. \qquad (34.45)$$

All lower derivatives can be deduced from this formula by integrating with respect to s:

$$\frac{d^{m-k} v}{ds^{m-k}} = -\frac{1}{2\pi} \int_0^{2\pi} \frac{d^m u}{d\sigma^m} A_k(s, \sigma) d\sigma \quad (k = 1, 2, \ldots, m - 1), \qquad (34.46)$$

where $A_k(s, \sigma)$ is the integral of multiplicity k of the function $\cot[(\sigma - s)/2]$, with respect to s. The arbitrary constants appearing in the integration are chosen in such a way that the following conditions are satisfied:

$$\int_0^{2\pi} A_k(s, \sigma) ds = 0 \quad (k = 1, 2, \ldots, m - 1).$$

The function $v(s)$ possesses the same properties as have been stated for the function $u(s)$. Its expression, therefore, in terms of $d^m u/ds^m$ also contains one arbitrary constant

$$v(s) = -\frac{1}{2\pi} \int_0^{2\pi} \frac{d^m u}{d\sigma^m} A_m(s, \sigma) d\sigma + v_0. \qquad (34.47)$$

Denote for simplicity

$$\frac{d^m u}{ds^m} = g(s), \quad \alpha_{m-1} = \alpha.$$

Introducing into the boundary condition (34.38) the appropriate quantities in accordance with formulae (34.39)†, (34.45)–(34.47) we arrive at the singular integral equation

$$a_m(s)\,g(s) - \frac{b_m(s)}{2\pi} \int_0^{2\pi} g(\sigma) \cot \frac{\sigma - s}{2}\, d\sigma + \int_0^{2\pi} K(s,\sigma)\,g(\sigma)\,d\sigma$$

$$= f(s) - \alpha\gamma(s) - v_0\delta(s), \qquad (34.48)$$

where

$$\gamma(s) = a_0(s) + \int_0^{2\pi} M_0(s,\sigma)\,d\sigma, \quad \delta(s) = b_0(s) + \int_0^{2\pi} N_0(s,\sigma)\,d\sigma,$$

and $K(s,\sigma)$ is a completely determined Fredholm kernel.

It may turn out that in equation (34.48) the functions $\gamma(s)$, $\delta(s)$ are linearly dependent, in particular one of the functions, $\gamma(s)$ or $\delta(s)$ (or both) vanish. Set for definiteness $\gamma(s) = 0$ and $\delta(s) = 0$. Since $\gamma(s)$ and $\delta(s)$ arise as the result of substitution in the left-hand side of relation (34.48) the values $u = 1$, $v = 0$ and $v = 1$, $u = 0$, it follows that the latter functions are solutions of the homogeneous problem (34.48). Evidently, there correspond to these solutions only trivial solutions of the homogeneous integral equation (34.48). Consequently, problem (34.48) has two solutions more than the integral equation (34.38) with the additional condition (34.41).

It is known from the general theory (§ 32) that the number of its solutions (without the additional condition) is not less than the number of solutions of the corresponding dominant equation ($K(s,\sigma) \equiv 0$).

Denote the index of the function $a_m(s) + ib_m(s)$ by $\varkappa_1$. Then, taking into account the additional condition (34.41) we find that *the number of solutions of equation* (34.48) *is not less than* $2\varkappa_1 - 1$. Therefore, *the number of solutions of the boundary value problem is not less than* $2\varkappa_1 + 1$. In the remaining cases analogous reasoning leads to the same result. This is the only result which can be deduced without solving the integral equation. Moreover, it is evident that here we could also carry out an analysis analogous to that of the preceding subsection,

---

† All integrals in formulae (34.39) are written in the form $\int_0^{2\pi}$, assuming that the factors of $g(\sigma)$ in the interval $(s, 2\pi)$ vanish.

the starting point of which was the assumption concerning the number of solutions of the adjoint equation.

Analysing the above obtained results we discover that essentially they reduce to the following two statements.

(1) *The problem can be reduced to a linear Fredholm integral equation.* (Singular solutions derived by the methods of F. D. Gakhov and I. I. N. Vekua constitute only an intermediate stage.)

(2) If $\varkappa \geqq 0$ *the number of solutions of the boundary value problem is not less than the number of solutions of the simplest problem obtained by omitting all terms except the terms containing the highest derivatives* (*non-integral*†), *i.e.*

$$a_m(s)\frac{d^m u}{ds^m} + b_m(s)\frac{d^m v}{ds^m} = 0 \quad \text{(in the form of F. D. Gakhov),} \quad (34.49)$$

$$\text{Re}[c_m(t)\,\Phi^{(m)}(t)] = 0 \qquad \text{(in the form of I. N. Vekua).} \quad (34.40)$$

The second result is especially apparent in solving the problem by the method exposed here as the last. In solving the problems by using an integral representation the deduced results are somewhat disguised by the intermediate integral representation which stands between the solutions of the integral equation and the solutions of the boundary value problem.

To avoid misunderstanding observe that our remarks concern only the general formulation of the problem. In considering particular problems it may be possible to obtain more precise results without solving the integral equation, and even to establish the exact number of solutions (see Problem 7 at the end of this chapter).

Finally let us establish the identity of the results obtained in this and in the preceding subsections (see § 34.6). On the basis of equation (34.2′) we see that to establish the correspondence we have to set

$$t'^m(a_m - ib_m) = c_m(t)$$

or, passing to the complex quantities

$$a_m + ib_m = t'^m \bar{c}_m(t).$$

Since

$$\text{Ind}\,(a_m + ib_m) = \varkappa_1; \quad \text{Ind}\,t'^m = m; \quad \text{Ind}\,\bar{c}_m(t) = -\varkappa,$$

---

† If $\varkappa < 0$ this result may be incorrect since the simplest problem has m solutions $\Phi_k(z) = z^k$ ($k = 0, 1, \ldots, m-1$), whilst the general problem may not have them at all.

we have

$$\varkappa_1 = m - \varkappa.$$

It follows that the number of solutions $2\varkappa_1 + 1$ derived here, corresponds to the number $2(m - \varkappa) + 1$ of solutions obtained in the preceding subsection.

§ 35. Boundary value problem of Riemann type with the boundary condition containing derivatives

The Riemann boundary value problem can be generalized in the same sense as was done in the preceding section for the Hilbert problem. The generalized problem will be formulated as follows.

A simple smooth closed contour L is given, which divides the plane into two domains D^+, D^-. It is required to determine two functions $\Phi^+(z)$, $\Phi^-(z)$ analytic in the domains D^+, D^-, respectively (or one sectionally analytic function $\Phi(z)$), which satisfy, on the contour, the boundary condition

$$\sum_{k=0}^{n} \left[a_k(t) \frac{d^k \Phi^+(t)}{dt^k} + \int_L A_k(t, \tau) \frac{d^k \Phi^+(\tau)}{d\tau^k} d\tau \right] -$$

$$- \sum_{k=0}^{p} \left[b_k(t) \frac{d^k \Phi^-(t)}{dt^k} + \int_L B_k(t, \tau) \frac{d^k \Phi^-(\tau)}{d\tau^k} d\tau \right] = f(t), \quad (35.1)$$

where $a_k(t)$, $b_k(t)$ are known continuous functions and $a_n(t)$, $b_p(t)$ satisfy the Hölder condition and do not vanish. $A_k(t, \tau)$, $B_k(t, \tau)$ are Fredholm kernels.

The method of solution of the above formulated boundary value problem, similarly to that solved in the preceding section, consists in reducing it to an integral equation; this was first performed by L. G. Magnaradze (2). We shall describe a simpler method of solution, based on a special integral representation derived by Y. M. Krikunov (1).

35.1. An integral representation for sectionally analytic functions

The purpose of the integral representation described below is to express the unknown functions and all their derivatives entering the boundary condition (35.1), in terms of one unknown function. Moreover, the kernel of the integral representation is chosen in such a way that the highest derivatives are expressed by a Cauchy type integral.

We assume that the limiting values of the derivative of order n of the function $\Phi^+(z)$, and of the derivative of order p of the function $\Phi^-(z)$, satisfy the Hölder condition on L. For definiteness we suppose that the origin of the coordinate system belongs to D^+.

We shall prove *that the functions $\Phi^+(z)$, $\Phi^-(z)$ can be represented by the formulae*

$$\Phi^+(z) = \frac{(-1)^n}{(n-1)!}\frac{1}{2\pi i}\int_L \mu(\tau)(\tau-z)^{n-1}\ln\left(1-\frac{z}{\tau}\right)d\tau + \sum_{k=0}^{n-1}\frac{c_k}{k!}z^k,$$

$$\tag{35.2}$$

$$\Phi^-(z) = \frac{(-1)^p}{(p-1)!}\frac{1}{2\pi i}\int_L \frac{\mu(\tau)}{\tau^p}\left[(\tau-z)^{p-1}\ln\left(1-\frac{\tau}{z}\right)+\right.$$

$$\left. + \sum_{k=0}^{p-2}\beta_k\tau^{p-k-1}z^k\right]d\tau + \Phi^-(\infty), \quad (35.3)$$

where $\mu(\tau)$ is a complex function which satisfies the Hölder condition and c_k are complex constants; $\mu(\tau)$ and c_k are uniquely determined by $\Phi^+(z)$, $\Phi^-(z)$ and

$$\beta_k = \frac{(-1)^{k+1}C_{p-1}^{k+1}}{1} + \frac{(-1)^{k+2}C_{p-1}^{k+2}}{2} + \cdots + \frac{(-1)^{p-1}C_{p-1}^{p-1}}{p-k-1}$$

$(C_{p-1}^{k+1}, C_{p-1}^{k+2}, \ldots, C_{p-1}^{p-1}$ *are the binomial coefficients). By $\ln(1-z/\tau)$ we understand the branch which vanishes as $z=0$, and by $\ln(1-\tau/z)$ the branch which vanishes as $z=\infty$.*

Proof. In deriving a representation which satisfies the conditions formulated above, it is necessary to remember that the derivative of order p of an analytic function has at infinity a zero of order $p+1$. Therefore, in deriving formulae (35.2), (35.3) we shall use the fact that the function $d^n\Phi^+(z)/dz^n$ is analytic in D^+ and the function $z^p\,d^p\Phi^-(z)/dz^p$ is analytic in D^- and vanishes at infinity; therefore they can be represented by Cauchy type integrals with the same density (§ 14.2)

$$\frac{d^n\Phi^+(z)}{dz^n} = \frac{1}{2\pi i}\int_L \frac{\mu(\tau)}{\tau-z}d\tau, \tag{35.4}$$

$$\frac{d^p\Phi^-(z)}{dz^p} = \frac{z^{-p}}{2\pi i}\int_L \frac{\mu(\tau)}{\tau-z}d\tau. \tag{35.5}$$

The density $\mu(t)$ is determined by the functions $\Phi^+(z)$, $\Phi^-(z)$ in accordance with the Sokhotski formula

$$\mu(t) = \frac{d^n \Phi^+(t)}{dt^n} - t^p \frac{d^p \Phi^-(t)}{dt^p}. \tag{35.6}$$

The representations for the functions $\Phi^+(z)$, $\Phi^-(z)$ are deduced from relations (35.4), (35.5) by integrating, with respect to z, n and p times respectively.

It follows from identity (34.17)

$$\frac{d^n}{dz^n} \left[\frac{(-1)^n}{(n-1)!} (\tau - z)^{n-1} \ln \left(1 - \frac{z}{\tau} \right) \right] = \frac{1}{\tau - z},$$

established in § 34.4, that the function obtained by integrating relation (35.4) n times with respect to z can be represented in the form (35.2).

The arbitrary constants of integration c_k can be determined by expanding both sides of relation (35.2) into series in the vicinity of the origin of the coordinate system

$$\frac{1}{k!} [\Phi^+(0)]^{(k)} = - \frac{(-1)^n}{(n-1)!} \frac{\alpha_k}{2\pi i} \int_L \mu(\tau) \tau^{n-k-1} d\tau + \frac{c_k}{k!}, \tag{35.7}$$

where

$$\alpha_0 = 0, \quad \alpha_k = \frac{1}{k} - \frac{C_{n-1}^1}{k-1} + \frac{C_{n-1}^2}{k-2} - \cdots + \frac{(-1)^{k-1} C_{n-1}^{k-1}}{1}$$

$$(k = 1, 2, \ldots, n-1)$$

(compare with relations (34.20)).

Relations (35.7) determine all the constants. Consequently, relation (35.2) has been proved. Introducing into relation (35.2) the values of the coefficients determined from equation (35.7) we can write the representation of $\Phi^+(z)$ in a different form, namely

$$\Phi^+(z) = \frac{(-1)^n}{(n-1)!} \frac{1}{2\pi i} \int_L \mu(\tau) \left[(\tau - z)^{n-1} \ln \left(1 - \frac{z}{\tau} \right) + \right.$$

$$\left. + \sum_{k=1}^{n-1} \alpha_k \tau^{n-k-1} z^k \right] d\tau + \sum_{k=0}^{n-1} \frac{1}{k!} [\Phi^+(0)]^{(k)} z^k. \tag{35.8}$$

This formula is convenient when the values of $\Phi^+(0)$, $[\Phi^+(0)]'$, $\ldots$, $[\Phi^+(0)]^{(n-1)}$ are given beforehand. Then representation (35.8) does not contain any arbitrary constants. If the initial values of the function and its derivatives are unknown it is more convenient to use the representation in the form (35.2).

Applying the method of complete induction, as was done in deriving identity (34.17), we obtain the identity

$$\frac{(-1)^p}{(p-1)!} \tau^{-p} \frac{d^p}{dz^p} \left[(\tau - z)^{p-1} \ln \left(1 - \frac{\tau}{z} \right) \right] = \frac{1}{z^p (\tau - z)}. \quad (35.9)$$

Hence we find that the function obtained by integrating both sides of relation (35.5) p times has the form

$$\Phi^-(z) = \frac{(-1)^p}{(p-1)!} \frac{1}{2\pi i} \int_L \frac{\mu(\tau)}{\tau^p} (\tau - z)^{p-1} \ln \left(1 - \frac{\tau}{z} \right) d\tau + \sum_{k=0}^{p-1} d_k z^k. \quad (35.10)$$

To determine the integration constants d_k let us make use of the expansions of both sides of the last relation in the vicinity of $z = \infty$. Taking into account that the expansion of the function $\Phi^-(z)$, since it is analytic in D^-, does not contain positive powers, we obtain

$$d_0 = \frac{(-1)^p \beta_0}{(p-1)! \, 2\pi i} \int_L \mu(\tau) \tau^{-1} d\tau + \Phi^-(\infty),$$

$$d_k = \frac{(-1)^p}{(p-1)!} \frac{\beta_k}{2\pi i} \int_L \mu(\tau) \tau^{-k-1} d\tau \quad (k = 1, 2, \ldots, p - 2),$$

$$d_{p-1} = 0,$$

where

$$\beta_k = \sum_{j=k+1}^{p-1} \frac{(-1)^j C_{p-1}^j}{j - k}.$$

Substituting the known values of d_k in formula (35.10) we arrive at formula (35.3).

Thus, the integral representations (35.2), (35.3) have been completely proved.

35.2. Solution of the boundary value problem

To solve the boundary value problem (35.1) let us make use of the integral representations (35.2) and (35.3) just established.

Differentiating successively relation (35.2) n times and relation (35.3) p times we obtain integral representations for all derivatives of the functions $\Phi^+(z)$, $\Phi^-(z)$ up to the required orders. As z tends to a point t of the contour, the kernels of these derivatives behave as the kernels of the derivatives of the integral representation (34.27), i.e. the kernels of the derivatives up to $(n-1)$th and $(p-1)$th order, respectively, are Fredholm kernels, whereas the derivatives of the highest orders (n and p, respectively) are given by the Cauchy type integrals (35.4), (35.5) and in accordance with the Sokhotski formulae their limiting values have the form

$$[\Phi^+(t)]^{(n)} = \frac{1}{2}\mu(t) + \frac{1}{2\pi i}\int_L \frac{\mu(\tau)}{\tau - t}\, d\tau,$$

$$[\Phi^-(t)]^{(p)} = -\frac{1}{2}\frac{\mu(t)}{t^p} + \frac{t^{-p}}{2\pi i}\int_L \frac{\mu(\tau)}{\tau - t}\, d\tau.$$

Substituting the boundary values of the functions $\Phi^+(t)$, $\Phi^-(t)$ and their derivatives into boundary condition (35.1) we obtain for the determination of the function $\mu(t)$ the singular integral equation

$$a(t)\,\mu(t) + \frac{b(t)}{\pi i}\int_L \frac{\mu(t)}{\tau - t}\, d\tau + \int_L K(t, \tau)\mu(\tau)\, d\tau = f(t) - \sum_{k=0}^{n-1} c_k \omega_k(t),$$

$$(35.11)$$

where

$$a(t) = \tfrac{1}{2}[a_n(t) + t^{-p} b_p(t)], \quad b(t) = \tfrac{1}{2}[a_n(t) - t^{-p} b_p(t)]. \quad (35.12)$$

$K(t,\tau)$ is a completely determined Fredholm kernel and $\omega_k(t)$ are functions which are obtained by collecting the terms containing the constant c_k.

Since

$$a(t) + b(t) = a_n(t) \neq 0 \quad \text{and} \quad a(t) - b(t) = t^{-p} b_p(t) \neq 0,$$

equation (35.11) is a singular integral equation with Cauchy kernel of the normal type and the general theory can be applied to it (Chapter III).

An investigation of the solubility problems can be carried out according to the same scheme as in the preceding section. The investigation can be carried out directly for the singular equation by using the Noether theorems, as was done by I. N. Vekua for the case of the problem considered in the preceding section; we may also first perform the regularization and then make use of the Fredholm theorems, as was done in § 34.6. Without going into details let us indicate the main results of the investigation.

The coefficient of the Riemann problem corresponding to equation (35.11) has according to formula (21.5) the form

$$G(t) = \frac{a(t) - b(t)}{a(t) + b(t)} = t^{-p} \frac{b_p(t)}{a_n(t)}. \tag{35.13}$$

Denoting

$$\mathrm{Ind}\, \frac{b_p(t)}{a_n(t)} = \mathrm{Ind}\, b_p(t) - \mathrm{Ind}\, a_n(t) = \varkappa_1, \tag{35.14}$$

we shall call this expression *the index of the boundary value problem* (35.1). Then the index of the singular integral equation has the form

$$\varkappa = \varkappa_1 - p. \tag{35.15}$$

Taking into account the presence in the right-hand side of the equation of n arbitrary parameters and the property of a complete singular integral equation (§ 23.3) that the number of its solutions is not less than the number of solutions of the corresponding dominant equation, we arrive at the following conclusion: *the number of linearly independent solutions of the boundary value problem* (35.1) *is not less than* $\varkappa_1 - p + n$ *where* $\varkappa_1$ *is the index of the problem and* n *is the highest order of the derivative of the function* $\Phi^+(z)$ *present in the boundary condition, p being the corresponding order of the derivative of the function* $\Phi^-(z)$.

This is the result which can be deduced without solving the equation (or, which is the same, its adjoint).

Similarly to the preceding section the results of the investigation deduced without solving the equation can be summarized as follows.

1. *The boundary value problem* (35.1) *can be reduced to a linear integral equation* (*first singular and then Fredholm's*).

2. *If* $\varkappa \geqq 0$ *the number of solutions of the problem is not less than the number of solutions of the simplest problem obtained by omitting in the boundary condition all terms except those* (*non-integral*) *which contain the highest derivatives*†.

† See previous footnote on p. 315.

The first statement is obvious; let us explain the second one. Consider the simplest problem

$$a_n(t) [\Phi^+(t)]^{(n)} - b_p(t) [\Phi^-(t)]^{(p)} = 0 \qquad (35.16)$$

or

$$[\Phi^+(t)]^{(n)} = \frac{b_p(t)}{a_n(t)} [\Phi^-(t)]^{(p)}. \qquad (35.16')$$

Regarding $[\Phi^+(z)]^{(n)}$, $[\Phi^-(z)]^p$ as unknown functions we may consider the last boundary value problem as an ordinary Riemann problem. The fact that the "interior" unknown function is a derivative of a function analytic in D^+ does not impose any additional conditions upon this function. The fact that the "exterior" function is the pth derivative of a function analytic in D^- is expressed by the condition that it must have at infinity a zero of $(p + 1)$th order.

Consequently, the Riemann boundary value problem (35.16') has to be solved in the class of functions which have at infinity a zero of order $(p + 1)$.

It is readily observed that the number of such solutions is $\varkappa_1 - p$ where, as before, $\varkappa_1 = \text{Ind} [b_p(t)/a_n(t)]$ is the index of the problem. In integrating the "interior" function there arise n arbitrary constants; the integration of the "exterior" functions yields no new arbitrary constants, since all functions obtained by integration should vanish at infinity. Consequently, problem (35.16) has $\varkappa_1 - p + n$ solutions, which is in agreement with the result deduced before. The reader is advised to perform the investigation independently according to § 34.6, using in this case not the Fredholm theorems but the Noether theorems.

35.3. New method of solving the problem

Rogozhin (3) has recently constructed a new integral representation by which the boundary value problem (35.1) can be reduced directly, without the singular integral equation, to the Fredholm equation in the same way as in § 34.3–§ 34.6 for the generalized Hilbert problem. It is done by introducing into Krikunov's integral representation (35.2) an additional multiplier which is correspondingly chosen.

It has been established as follows:

Suppose that $c(t)$ is a prescribed function on L which satisfies Hölder's condition and does not vanish. If $n \geqq x = \text{Ind } c(t)$, the following integral representation holds good for the sectionally-analytic function

$\Phi^{\pm}(z)$ with limit values of the derivatives of $\Phi^{+(n)}(t)$, $\Phi^{-(p)}(t)$ on the contour satisfying Hölder's condition

$$\Phi^{+}(z) = \frac{(-1)^n}{(n-1)!} \frac{1}{2\pi i} \int_L \frac{\mu(\tau)}{c(\tau)} (\tau - z)^{p-1} \ln\left(1 - \frac{z}{\tau}\right) d\tau +$$

$$+ \sum_{k=0}^{n-\varkappa-1} g_k z^k, \qquad (35.17)$$

$$\Phi^{-}(z) = \frac{(-1)^p}{(p-1)!} \frac{1}{2\pi i} \int \frac{\mu(\tau)}{\tau^p} (\tau - z)^{p-1} \ln\left(1 - \frac{\tau}{z}\right) +$$

$$+ \sum_{k=0}^{p-2} \beta_k \tau^{p-k-1} z^k \Bigg] d\tau + \Phi^{-}(\infty), \qquad (35.18)$$

where $\mu(\tau)$ is a complex function which satisfies Hölder's condition, g_k are constants found from $\Phi^{\pm}(t)$, and β_k are given by the expressions

$$\beta_k = \sum_{j=k+1}^{p-1} \frac{(-1)^j C_{p-1}^j}{j-k}.$$

If $n \geq \varkappa$, it is necessary to replace $\Sigma g_k z^k$ in formula (35.17) by $g_0 = \Phi^{+}(0)$.

For the proof, as in § 34.4, we must take three cases into account. We give the reasoning for the cases when $n > \varkappa > 0$. Differentiating formulae (35.17) and (35.18) n and p times respectively and using the notation

$$\Phi^{+(n)}(z) = F^{+}(z), \quad z^p \Phi^{-(p)}(z) = F^{-}(z),$$

we get

$$F^{+}(z) = \frac{1}{2\pi i} \int_L \frac{\mu(\tau)}{c(\tau)} \frac{d\tau}{\tau - z}, \qquad (35.19)$$

$$F^{-}(z) = \frac{1}{2\pi i} \int_L \frac{\mu(\tau)}{\tau - z} d\tau. \qquad (35.20)$$

Putting $\Psi^{-}(z)$ and $\Psi^{+}(z)$ respectively for the functions defined by the integrals in the complementary domains, from the Sokhotski formulae we have

$$F^{+}(t) + \Psi^{-}(t) = \frac{\mu(t)}{c(t)}, \quad \Psi^{+}(t) - F^{-}(t) = \mu(t). \qquad (35.21)$$

Eliminating $\mu(t)$, for the determination of $\Psi^{\pm}(z)$ and g_k we arrive at the Riemann problem:

$$\Psi^{+}(t) = c(t)\Psi^{-}(t) + c(t) F^{+}(t) + F^{-}(t), \qquad (35.22)$$

which is solved on the condition that $\Psi^-(\infty) = 0$. Its free term only vanishes if $F(z) = 0$. In fact, by equating it to zero we would obtain the homogeneous boundary value problem having the coefficient $-1/c(t)$ with negative index. It would then be found that $F(z) = 0$. According to § 14.5 problem (35.22) is solved unconditionally and its solution (the formula (14.18′)) contains linearly $\varkappa$ arbitrary constants.

The constants g_k and β_k in the integral representation, and also the $\varkappa$ arbitrary constants in $\mu(t)$, are found by equating the initial n terms of the expansion of both sides of (35.17) in the neighbourhood of the coordinate origin and the first p terms of (35.18) in the neighbourhood of the point at infinity. We have already performed this operation in the foregoing section and § 34.4, so there is no need to repeat it here.

The case $\varkappa < 0$ is different in that the Riemann problem with negative index is solved to prove the representation. The $|\varkappa|$ arbitrary constants which are left after differentiating (35.17) are chosen for this problem to be soluble. Otherwise the procedure is similar.

By use of the integral representation the boundary value problem (35.1) can be reduced to an integral equation. If we take $c(t) = a_n(t)/t^p b_p(t)$, it is easy to see that the terms containing higher derivatives give an integral with the kernel

$$\left[\frac{b_p(\tau)}{\tau^p a_n(\tau)} - \frac{b_p(t)}{t^p a_n(t)}\right]\frac{1}{\tau - t},$$

and the integral equation is therefore a Fredholm equation.

The integral equation can be investigated in the same way as in § 34.6.

35.4. Singular integro-differential equation

The boundary value problem (35.1) is closely connected with an integro-differential equation with Cauchy kernel of the form

$$\sum_{k=0}^{p} \alpha_k(t)\, v^{(k)}(t) + \sum_{k=0}^{q} \frac{1}{\pi i}\int_L \frac{P_k(t, \tau)}{\tau - t}\, v^{(k)}(\tau)\, d\tau = f(t), \quad (35.23)$$

where $\alpha_k(t)$, $f(t)$, $P_k(t,\tau)$ are known functions satisfying the Hölder condition and $v(t)$ is the unknown function belonging to the class of functions the rth derivative ($r = \max(p, q)$) of which satisfies the Hölder condition.

Let us write equation (35.23) in a form with the dominant part separated out

$$\sum_{k=0}^{p} \alpha_k(t)\, v^{(k)}(t) + \sum_{k=0}^{q} \frac{\beta_k(t)}{\pi i} \int_L \frac{v^{(k)}(\tau)}{\tau - t}\, d\tau +$$

$$+ \sum_{k=0}^{q} \int_L Q_k(t, \tau)\, v^{(k)}(\tau)\, d\tau = f(t), \quad (35.24)$$

where

$$\beta_k(t) = P_k(t, t), \quad Q_k(t, \tau) = \frac{1}{\pi i} \frac{P_k(t, \tau) - P_k(t, t)}{\tau - t}.$$

Obviously $\beta_k(t)$ satisfies the Hölder condition and $Q_k(t,\tau)$ are Fredholm kernels.

We also assume that the quantities $\alpha_p(t)$ for $p > q$, $\beta_p(t)$ for $q > p$ and $\alpha_p(t) \pm \beta_q(t)$ for $q = p$ do not vanish anywhere.

Introduce an analytic function by means of the Cauchy type integral the density of which is the required solution of the integro-differential equation

$$\Phi(z) = \frac{1}{2\pi i} \int_L \frac{v(\tau)}{\tau - z}\, d\tau. \quad (35.25)$$

According to the Sokhotski formulae (4.14), (4.15) we have

$$\left. \begin{array}{l} v^{(k)}(t) = \dfrac{d^k \Phi^+(t)}{dt^k} - \dfrac{d^k \Phi^-(t)}{dt^k}, \\[2ex] \dfrac{1}{\pi i} \displaystyle\int_L \dfrac{v^{(k)}(\tau)}{\tau - t}\, d\tau = \dfrac{d^k \Phi^+(t)}{dt^k} + \dfrac{d^k \Phi^-(t)}{dt^k}. \end{array} \right\} \quad (35.26)$$

Substituting the last expressions into equation (35.24) we obtain

$$\sum_{k=0}^{r} \left[a_k(t) \frac{d^k \Phi^+(t)}{dt^k} + \int_L Q_k(t, \tau) \frac{d^k \Phi^+(\tau)}{d\tau^k}\, d\tau \right] -$$

$$- \sum_{k=0}^{r} \left[b_k(t) \frac{d^k \Phi^-(t)}{dt^k} + \int_L Q_k(t, \tau) \frac{d^k \Phi^-(\tau)}{d\tau^k}\, d\tau \right] = f(t) \quad (35.27)$$

where

$$a_k(t) = \alpha_k(t) + \beta_k(t), \quad b_k(t) = \alpha_k(t) - \beta_k(t) \quad (k = 0, 1, \ldots, r),$$

and we suppose that $\alpha_k \equiv 0$ for $k > p$ and $p < q$, $\beta_k(t) \equiv 0$ for $k > q$ and $p > q$.

It is evident that the boundary value problem (35.27) constitutes a particular case of the problem examined above (35.1); we may therefore omit the reasoning connected with its solution.

Computing the solution of the problem (35.27) in accordance with the Sokhotski formula

$$v(t) = \Phi^+(t) - \Phi^-(t)$$

we obtain the solution of the original integro-differential equation (35.23).

§ 36. The Hilbert boundary value problem for multiply-connected domains

An analytic function defined in a multiply-connected domain has the following peculiarity: although it is analytic in the domain, i.e. it may be expanded in the vicinity of every point into the Taylor series, it may not be single-valued. For the Riemann problem, when the boundary condition is given in the complex form, this fact has no significant influence on the solution. A single-valued prescription of the boundary condition leads necessarily in this case to a single-valued solution. The case is different when the analytic function is prescribed by a boundary condition in the real form as it occurs in the Hilbert problem. Here, naturally, there occur solutions which are analytic but not single-valued. If according to the conditions of the problem it is required to deduce single-valued solutions, there arise serious difficulties in separating out such solutions. These difficulties in the general case have not as yet been overcome.

Our exposition will consist of two independent parts. In the first part, following the method of solution of the Hilbert problem in a simply-connected domain (§ 29), we shall present a complete solution of the homogeneous problem in the class of analytic many-valued functions. In the second part, using methods similar to those applied in the preceding sections of this chapter for solving generalized problems, we shall give results obtained so far concerning the solution of the problem in the class of analytic single-valued functions.

As in Chapter IV we begin by presenting certain auxiliary results.

36.1. The Dirichlet problem for multiply-connected domains

Suppose that D^+ is a $(m + 1)$-connected domain bounded by simple non-intersecting smooth curves $L_0, L_1, \ldots, L_m$, where L_0 encloses all the others (Fig. 13, § 16.1). The contours will be subjected to the additional condition that they are Lyapounov curves (see § 18.1).

The classical Dirichlet problem can be stated as follows.

THE DIRICHLET PROBLEM. *It is required to find the function $u(x, y)$ which is harmonic in D^+ and continuous in $D^+ + L$, in accordance with the boundary condition*

$$u = f(s), \tag{36.1}$$

where $f(s)$ is a continuous function prescribed on the contour.

If the contour L_0 is absent, i.e. D^+ is infinite, we impose also the condition that u tends to a definite limit at infinity.

In the case of a simply-connected domain we have frequently used the fact that the Dirichlet problem is uniquely soluble. The same is true in the case of the above problem for a multiply-connected domain, if only the unknown function be subject to no additional conditions. In some applications (including also our further investigations) it is required that the unknown function is not simply harmonic but, moreover, it constitutes the real part of a single-valued analytic function. (Generally speaking it is the real part of a many-valued analytic function.) Under this additional assumption the Dirichlet problem is in the general case insoluble. To make it soluble it is necessary to introduce certain arbitrary parameters into the prescribed value on the contour; this leads to the so-called modified Dirichlet problem. We now present the precise formulation.

THE MODIFIED DIRICHLET PROBLEM†. *It is required to find the function $u(x, y)$ constituting the real part of a function $\Phi(z)$ analytic and single-valued in the domain D^+, which is continuous on the contour and satisfies the boundary condition*

$$u = f(s) + h(s), \tag{36.2}$$

where $h(s)$ assumes on the contours L_j arbitrary constant values h_j which are not prescribed beforehand.

An investigation proves that the constants h_j are determined by the condition of the problem (the requirement of single-valuedness of

† This term was introduced by N. I. Muskhelishvili.

$\Phi(z)$), if only one of them be prescribed. We shall hereafter assume that $h_0 = 0$.

The uniqueness of the solution of the classical Dirichlet problem follows from the well-known assertion that a non-constant harmonic function attains the minimum and maximum on the contour.

To prove the uniqueness of the solution of the modified problem it is sufficient to establish that the homogeneous problem ($f(s) \equiv 0$) has only zero solution.

We proceed from the well-known formula of the theory of harmonic function

$$\iint_D \left[\left(\frac{\partial u}{\partial x}\right)^2 + \left(\frac{\partial u}{\partial y}\right)^2\right] dx\,dy = \int_L u \frac{\partial u}{\partial n}\,ds.$$

Applying the Cauchy–Riemann equation[†] and the boundary condition we have

$$\int_L u \frac{\partial u}{\partial n}\,ds = -\int_L u \frac{\partial v}{\partial s}\,ds = -\sum_{j=1}^m h_j \int_{L_i} dv = 0,$$

since by assumption the function v is single-valued. Hence $\partial u/\partial x = \partial u/\partial y = 0$ and consequently $u = \text{const}$. But since on L_0 we have $u = 0$, it follows also that $u(x, y) = 0$ everywhere.

If the condition $h_0 = 0$ be discarded the constant h_0 is the solution of the homogeneous problem and two solutions of the modified problem can differ by a constant.

We now present a brief account of the solution of the Dirichlet problem. The reader may find a detailed exposition in the book of N. I. Muskhelishvili [17*], §§ 60 – 65.

We begin by solving the modified Dirichlet problem with the help of the potential of double layer. The given function $f(s)$ is assumed to satisfy the Hölder condition[‡].

The analytic function $\Phi(z)$ is represented by the Cauchy type integral

$$\Phi(z) = u(x, y) + iv(x, y) = \frac{1}{2\pi i} \int_L \frac{\mu(\sigma)}{\tau - z}\,d\tau + iC \qquad (36.3)$$

[†] We do not attempt here to justify the validity of these relations on the contour.

[‡] We could subsequently dispose of the requirement that the function $f(s)$ satisfies the Hölder condition and prove that the deduced solution is also valid when $f(s)$ is simply continuous. We shall not carry out the relevant reasoning here (see N. I. Muskhelishvili [17*], § 61).

ssuming that the unknown real density $\mu(s)$ satisfies the Hölder condition.

Taking, in accordance with the Sokhotski formulae, the limiting values on the contour, separating the real part (see § 10) and inserting t in the boundary condition we arrive at the Fredholm integral equation

$$\mu(s) + \frac{1}{\pi} \int_L \mu(\sigma) \frac{\cos(r, n)}{r} \, d\sigma = 2 [f(s) + h(s)]. \tag{36.4}$$

The homogeneous integral equation

$$\mu(s) + \frac{1}{\pi} \int_L \mu(\sigma) \frac{\cos(r, n)}{r} \, d\sigma = 0 \tag{36.4'}$$

has the solutions

$$\mu_j(s) = \begin{cases} c_j & \text{for } t \in L_j \ (j = 1, 2, \ldots, m), \\ 0 & \text{on the remaining contours.} \end{cases} \tag{36.5}$$

Let us prove this assertion.

Equation (36.4') can be written in the form

$$\text{Re} \lim_{z \to t(s)} \left[\frac{1}{2\pi i} \int_L \frac{\mu(\sigma)}{\tau - z} \, d\tau \right] = 0.$$

If $\mu = \mu_j$ the integral over L is reduced to the integral over L_j. Taking into account that

$$\int_{L_j} \frac{\mu(\sigma)}{\tau - z} \, d\tau = 0$$

for z located outside L_j, we find that the functions (36.5) do in fact satisfy equation (36.4').

The equation can have no other solutions, in view of the following reasoning. According to a familiar property of the Cauchy type integral, if the function $\mu(s)$ constitutes a solution of equation (36.4'), then it is the boundary value of a function analytic in D^-. But $\mu(s)$ is real. Consequently, the boundary value of the imaginary part of the function analytic in D^- is zero. Now, the analytic function $\mu(z)$ is constant in D^-, and in every connected part of D^- (the domains $D_0^-, D_1^-, \ldots, D_m^-$) different values of this constant are possible. Thus, only such functions $\mu(s)$ can satisfy equation (36.4') which take on every contour L_j certain constant values. To satisfy relations (36.4') it is necessary to equate to zero the constant on L_0. Taking all possible combinations of the constants we arrive at solutions (36.5).

The general solution of equation (36.4') is a linear combination of the following m linearly independent solutions:

$$\mu_j(s) = \begin{cases} 1 & \text{for} \quad t \in L_j \quad (j = 1, 2, \ldots, m), \\ 0 & \text{on the remaining contours.} \end{cases} \tag{36.5'}$$

According to the Fredholm theorems the homogeneous equation adjoint to (36.4') also has m solutions and the non-homogeneous equation (36.4) is soluble only when the m conditions of orthogonality are satisfied

$$\int_L [f(s) + h(s)] \, v_j(s) \, ds = 0 \quad (j = 1, 2, \ldots, m). \tag{36.6}$$

Here $v_j(s)$ are linearly independent solutions of the adjoint homogeneous equation.

Relations (36.6) yield a system of m equations for the arbitrary constants h_j. This system is uniquely soluble and thus all constants are determined. Thus, the modified Dirichlet problem is soluble and all constants h_j acquire completely definite values.

We may infer from the above established solubility of the modified Dirichlet problem that a harmonic function constituting the real part of an analytic and single-valued function $\Phi(z)$ can be represented uniquely by the potential of a double layer

$$u(x, y) = \mathrm{Re} \left[\frac{1}{2\pi i} \int_L \frac{\mu(\sigma)}{\tau - z} \, d\tau \right].$$

Hence, the following important corollary follows. *A function $\Phi(z)$ analytic and single-valued in a finite multiply-connected domain D^+ can be represented by the Cauchy type integral with a real density*

$$\Phi(z) = \frac{1}{2\pi i} \int_L \frac{\mu(\sigma)}{\tau - z} \, d\tau + iC, \tag{36.7}$$

where C is an arbitrary real constant.

If the contour L_0 is absent and, consequently, the domain becomes infinite, we could prove the validity of the representation

$$\Phi(z) = \frac{1}{2\pi i} \int_L \frac{\mu(\sigma)}{\tau - z} \, d\tau + \Phi(\infty). \tag{36.8}$$

The density $\mu(s)$ is determined in terms of the given function $\Phi(z)$ to within an arbitrary constant on every interior contour L_j (formula (36.5)); the latter are the values of μ for which the Cauchy type integral for $z \in D^+$ vanishes.

Thus, *in the case of a multiply-connected domain, all representations of analytic functions by Cauchy type integrals, which were derived in § 34.2 for the case of a simply-connected domain (formulae (34.6), (34.8)), still hold; however, there is a difference, namely the density μ is not determined uniquely by the function $\Phi(z)$.*

On the basis of the investigation of the modified Dirichlet problem it is easy to deduce a method of solving the classical Dirichlet problem.

We seek the solution of the latter problem in the form

$$u(x, y) = U(x, y) + \sum_{k=1}^{m} A_k \ln |z - z_k|, \tag{36.9}$$

where $U(x, y)$ is the real part of an analytic single-valued function, z_k are some arbitrary fixed points in the domains $D_1^-, D_2^-, \ldots, D_m^-$ and A_k are for the time being arbitrary constants.

The function $U(x, y)$ should satisfy the boundary condition

$$U = f(s) - \sum_{k=1}^{m} A_k \ln |t - z_k|.$$

Let us add to the right-hand side of these relations on the contours L_j some arbitrary constants h_j, i.e. we consider the modified Dirichlet problem

$$U = f(s) - \sum_{k=1}^{m} A_k \ln |t - z_k| + h_j \quad \text{on} \quad L_j,$$

which, as we know, is soluble. Multiplying the right-hand side by an arbitrary solution of the adjoint homogeneous equation $v_j(s)$, integrating over the contour and taking into account (36.6) we obtain the relation

$$h_j = -f_j + \sum_{k=1}^{m} \gamma_{jk} A_k \quad (j = 1, 2, \ldots, m).$$

We deduce the solution of the original problem if we set $h_j = 0$ in the last relations; this leads to a system of linear equations for A_k:

$$\sum_{k=1}^{m} \gamma_{jk} A_k = f_j \quad (j = 1, 2, \ldots, m). \tag{36.10}$$

The homogeneous system obtained from (36.10) by setting $f_j = 0$ has only the trivial solution, since if we assume that it has solutions different from zero this means that the original problem has solutions when $f(s) \equiv 0$, which is impossible on account of the uniqueness theorem. Consequently, the determinant of the system (36.10) does not vanish and the system has a unique solution.

Introducing the obtained values of $U(x, y)$ and A_k into expression (36.9) we obtain the solution of the Dirichlet problem.

If the contour L_0 is absent, the above proved theorem remains valid, the only difference being that now the constants A_k must be subjected to the condition

$$\sum_{k=1}^{m} A_k = 0,$$

following from the requirement of boundedness of the solution at infinity.

D. I. Sherman (5) proved that the Dirichlet problem for a multiply-connected domain may also be solved by another method not connected with the solution of the modified problem. He sought the function $u(x, y)$ in the form

$$u(x, y) = \frac{1}{\pi} \int_L v(s) \frac{\partial \ln r}{\partial n} \, ds + \sum_{j=1}^{m} B_j \ln r_j, \quad B_j = \int_{L_j} v(s) \, ds,$$

where the density $v(s)$ is real and $r_j = |z - z_j|$. The integral equation for $v(s)$ (it is constructed in the same way as equation (36.4)), turns out in this case to be always uniquely soluble. In applying this method, there is no necessity to construct and to solve system (36.10).

36.2. The Schwarz operator for a multiply-connected domain

The Schwarz operator for a multiply-connected domain is defined in exactly the same way as was done in § 27.2 for a simply-connected domain. We shall here employ the notations and the concepts of that section.

Suppose that

$$G(z, \zeta) = \ln \frac{1}{r} + g(x, y; \xi, \eta)$$

is the Green function. It is known that its determination consists in solving a particular case of the Dirichlet problem, which establishes its existence.

Let $H(z, \zeta)$ be the function adjoint to G with respect to the variable z and let

$$M(z, \zeta) = G + iH$$

be the complex Green function.

In all that follows the positive direction on the contour will be taken with respect to the domain D^+. This means that the contour L_0 is traversed anticlockwise and the contours L_j $(j = 1, 2, \ldots, m)$ clockwise. The positive direction of the normal is that inside the domain D^+. Thus, the positive directions of the tangent and the normal on every curve of the contour are situated relatively as the x and y axes.

If $u + iv$ is analytic and by $\partial/\partial s$ and $\partial/\partial n$ on the contour we understand the limiting values of the derivatives with respect to the corresponding directions in the domain, then the Cauchy–Riemann equations

$$\frac{\partial u}{\partial s} = \frac{\partial v}{\partial n}, \quad \frac{\partial u}{\partial n} = -\frac{\partial v}{\partial s}, \tag{36.11}$$

are satisfied on the contour. Let us also note that ds in the integrand is always a positive quantity.

Returning to the complex Green function let us examine its analytic nature.

The function $M(z, \zeta)$ is many-valued for two reasons: first, it contains the term $-\ln(\zeta - z)$ and, secondly, since the domain is multiply-connected, in general, on describing in the domain D^+ any interior contour L_j, the function acquires an increment. By definition the Green function $G(z, \zeta)$ is single-valued. Denote the increment of $H(z, \zeta)$ on describing an interior contour L_j anticlockwise, by $2\pi\beta_j(\zeta)$. According to relation (36.11) these quantities are given by the formulae

$$\beta_j(\zeta) = -\frac{1}{2\pi} \int\limits_{L_j} \frac{\partial H}{\partial s} \, ds = \frac{1}{2\pi} \int\limits_{L_j} \frac{\partial G}{\partial n} \, ds, \tag{36.12}$$

ζ being an interior point of the domain D^+ and z a point of the contour. On describing a contour L_j anticlockwise the function $M(z, \zeta)$ acquires the increment $i2\pi\beta_j(\zeta)$. The same increment has the function $\beta_j(\zeta) \ln(z - z_j)$, where z_j is a point inside L_j. Consequently, the complex Green function has the representation

$$M(z, \zeta) = M_0(z, \zeta) - \ln(\zeta - z) + \sum_{j=1}^{m} \beta_j(\zeta) \ln(z - z_j),$$

where $M_0(z, \zeta)$ is a function analytic and single-valued in z.

Consider the nature of the function β_j. Rewriting formula (36.12) with a change of notation for the variables, taking into account the symmetry of the Green function with respect to the variables z, ζ we have

$$\beta_j(z) = \frac{1}{2\pi} \int\limits_{L_j} \frac{\partial G(z, \tau)}{\partial v} \, d\sigma. \tag{36.13}$$

Comparing the last formula with (27.2) yielding the solution of the Dirichlet problem in a closed form by means of the Green function we observe that *formula (36.13) determines $\beta_j(z)$ as functions harmonic in D^+*. On the contour they assume the following values:

$$\beta_j(\tau) = \begin{cases} 1 & \text{for } \tau \in L_j \quad (j = 1, 2, \ldots, m), \\ 0 & \text{on the remaining contours.} \end{cases} \tag{36.14}$$

Let us now construct the Schwarz kernel taking the derivative of M with respect to the positive direction of the normal

$$T(z, \zeta) = \frac{\partial M(z, \zeta)}{\partial \nu} = \frac{\partial M_0(z, \zeta)}{\partial \nu} - \frac{1}{\zeta - z} \frac{\partial \zeta}{\partial \nu} + \sum_{j=1}^{m} \alpha_j(\zeta) \ln(z - z_j), \tag{36.15}$$

where

$$\alpha_j(\zeta) = \frac{\partial \beta_j(\zeta)}{\partial \nu}. \tag{36.14'}$$

On the basis of the same reasoning as in § 27.2 we find that the formula

$$F(z) = \frac{1}{2\pi} \int_L T(z, \tau) u(\sigma) \, d\sigma + iv(z_0) \tag{36.16}$$

yields an analytic (but in general not single-valued) function, the real part of which takes on the contour the prescribed value $u(s)$.

Consequently, similarly to § 27.2, the Schwarz operator is given by the formula

$$Su \equiv \frac{1}{2\pi} \int_L T(z, \tau) u(\sigma) \, d\sigma. \tag{36.17}$$

The analytic function $F(z)$ defined by the Schwarz operator

$$F(z) = Su + iv_0,$$

has the form

$$F(z) = F_0(z) + \sum_{j=1}^{m} \frac{v_j}{2\pi} \ln(z - z_j), \tag{36.18}$$

where

$$F_0(z) = \frac{1}{2\pi} \int_L \left[\frac{\partial M_0(z, \tau)}{\partial \nu} + \frac{1}{\tau - z} \frac{\partial \tau}{\partial \nu} \right] u(\sigma) \, d\sigma$$

is an analytic single-valued function and

$$v_j = \int_L u(\sigma) \alpha_j(\tau) \, d\sigma \tag{36.19}$$

are some definite constants.

It is readily observed in view of formula (36.18), that the constants v_j are equal to the increments of the imaginary part of the function $F(z)$ on describing the contours L_j.

In order that the Schwarz operator should yield a single-valued function it is evidently necessary to require that all v_j vanish. Hence we have the following important result.

In order that the arbitrary function $u(s)$ prescribed on the contour be the real part of an analytic and single-valued function, it is necessary and sufficient that it satisfies the m relations

$$\int_L u(s)\,\alpha_j(t)\,ds = 0 \quad (j = 1, 2, \ldots, m). \tag{36.20}$$

In the case of an infinite domain D^+ where the contour L_0 is absent, all above formulae hold, provided the following conditions for β_j and α_j following from the boundedness at infinity, are satisfied:

$$\sum_{j=1}^{m} \beta_j(\zeta) = 1, \quad \sum_{j=1}^{m} \alpha_j(\zeta) = 0. \tag{36.21}$$

For a circular ring the complex Green function can be expressed by the function ϑ_1 known in the theory of elliptic functions, and the Schwarz kernel by the Weierstrass function ζ. The appropriate formulae the reader may find for instance in the book of N. I. Akhiyezer *Elements of the theory of elliptic functions*, pp. 221–9.

36.3. Problem A for a multiply-connected domain

The non-homogeneous Problem A and the homogeneous Problem A_0 are formulated for a multiply-connected domain in the same manner as for a simply-connected domain (see § 27.3). The unknown analytic function which can have a pole at one point z_0 is subjected to no additional restrictions; consequently, we admit as solutions many-valued functions.

To solve the problem we make use of the conformal mapping of the multiply-connected domain D^+ onto the unit circle. It is known that such a mapping cannot be carried out by means of a single-valued function. The mapping, however, becomes possible if we also admit many-valued functions. By such a many-valued function we understand a certain initial element obtained by expanding the function into power series in the vicinity of a point belonging to D^+, and its analytic continuations along all possible paths located wholly in D^+.

The following theorem holds. *Any domain D^+ of the plane z having more than two boundary points, can be conformally mapped onto the circle $|\zeta| < 1$ in such a way that to a prescribed point $z_0 \in D^+$ and a prescribed direction at this point, there correspond (for the fundamental element of the mapping function) the point $\zeta = 0$ and the positive direction of the real axis. This mapping is unique.*

It can be proved that all other branches of this function can be obtained from one of them (for instance from the fundamental element) by means of a linear fractional mapping which maps a unit circle onto itself. Details of such a mapping the reader may find for example in the book of G. M. Goluzin [6], Chapter VI.

Suppose that $\zeta = \omega(z)$ is the above mentioned function. Then the reasoning analogous to that in § 27.3 proves that the following function is a solution of Problem A_0:

$$Q(z) = i\beta_0 + \sum_{k=1}^{n} \left\{ C_k [\omega(z)]^k - \overline{C}_k [\omega(z)]^{-k} \right\}, \tag{36.22}$$

where n is the order of the pole.

Obviously, the solution of Problem A is given by the formula

$$F(z) = Su + Q(z). \tag{36.23}$$

Similarly to the case of a simply-connected domain the solution contains $2n + 1$ arbitrary real constants.

36.4. The regularizing factor

As in § 28.2 we introduce the concept of regularizing factor. Consider the complex function $a(s) + ib(s)$ given on the contour L and non-zero on it. Let us examine the particular case when the increment of the argument of $a(s) + ib(s)$ in describing any interior contour, vanishes; the problems which we shall be interested in reduce to this case.

Set

$$\text{Ind}\,[a(s) + ib(s)] = \frac{1}{2\pi} \left\{ \arg(a + ib) \right\}_L = \frac{1}{2\pi} \left\{ \arg(a + ib) \right\}_{L_0} = \varkappa,$$

where $\varkappa$ is an integer. As before we suppose that the origin belongs to the domain D^+.

DEFINITION. *We shall call the regularizing factor of the complex function $a(s) + ib(s)$ a real positive function $p(s)$ (in general many-*

*valued), such that the product $p(s)[a(s) + ib(s)]$ is the boundary value
of an analytic (in general many-valued) function having zero order
everywhere in the domain D^+, except at the origin of the coordinate
system where its order is equal to the index $\varkappa$.*

By definition we have

$$p(s)[a(s) + ib(s)] = t^{\varkappa} e^{i\gamma(t)}, \tag{36.24}$$

where $\gamma(z) = \omega(x, y) + i\omega_1(x, y)$ is a function analytic in D^+.

Equating the moduli and the arguments we obtain

$$p(s)\sqrt{[a^2(s) + b^2(s)]} = |t|^{\varkappa} e^{-\omega_1(s)}, \quad \arg\left[\frac{a(s) + ib(s)}{t^{\varkappa}}\right] = \omega(s). \tag{36.25}$$

It follows that

$$\gamma(z) = S \arg\left[\frac{a + ib}{t^{\varkappa}}\right], \tag{36.26}$$

where, for definiteness, the function $\gamma(z)$ has been subjected to the
additional condition

$$\operatorname{Im}\gamma(z_0) = 0, \tag{36.27}$$

where z_0 is a fixed point belonging to the domain D^+.

In view of formulae (36.18) the function $\gamma(z)$ can be represented
in the form

$$\gamma(z) = \gamma_0(z) + \sum_{j=1}^{m} \frac{v_j}{2\pi} \ln(z - z_j), \tag{36.28}$$

where $\gamma_0(z)$ is a single-valued analytic function and v_j are constants
defined by the relations

$$v_j = \int_L \alpha_j(s)\,\omega(s)\,ds. \tag{36.29}$$

The regularizing factor $p(s)$ is given by the formula

$$p(s) = \frac{|t|^{\varkappa}}{\sqrt{[a^2(s) + b^2(s)]}} e^{-\operatorname{Im}\gamma_0(t)} e^{-\sum_{j=1}^{m} \frac{v_j}{2\pi} \arg(t - z_j)}. \tag{36.30}$$

In view of presence of the factors $\exp[-(v_j/2\pi)\arg(t - z_j)]$ the
regularizing factor $p(s)$ is a many-valued function. In describing
interior contours L_j in the positive direction (clockwise) it acquires
the factors e^{v_j}; in describing the exterior contour L_0 it acquires the
factor $\exp(-\sum_{j=1}^{m} v_j)$. Evidently, the regularizing factor is a single-

valued function if and only if

$$v_j = \int_L \omega(s)\,\alpha_j(t)\,ds = 0 \quad (j = 1, 2, \ldots, m), \tag{36.31}$$

where $\alpha_j(t)$ are given by formula (36.14′).

By virtue of (36.28) relation (36.24) can be written in the form

$$p(s)\,[a(s) + ib(s)] = t^\varkappa e^{i\gamma(z)} \prod_{j=1}^{m} (t - z_j)^{\frac{iv_j}{2\pi}}. \tag{36.32}$$

Conditions (36.31) are also conditions of single-valuedness of the last function.

The above reasoning has established the existence of the regularizing factor. By considerations similar to those of §.28.2 we can easily prove its uniqueness (z_j being prescribed).

36.5. The solution of the homogeneous Hilbert problem in the class of many-valued functions

Let us write the boundary condition of the Hilbert problem in the form

$$\mathrm{Re}\left[\frac{F(t)}{a(s) + ib(s)}\right] = 0. \tag{36.33}$$

Denote

$$\frac{1}{2\pi}\left\{\arg(a + ib)\right\}_{L_j} = \varkappa_j,$$

all contours being described in the positive direction with respect to the domain D^+ (the interior contours clockwise and the exterior in the opposite direction). The functions $(t - z_j)^{-\varkappa_j}$ $(j = 1, 2, \ldots, m)$ when z_j is located inside L_j acquire on the contour L_j the same increment in the argument as $a + ib$. The function

$$[a(s) + ib(s)]\prod_{j=1}^{m}(t - z_j)^{\varkappa_j}$$

acquires on all interior contours a zero increment.

Let us multiply the numerator and the denominator of the left-hand side of relation (36.33) by $\prod_{j=1}^{m}(t - z_j)^{\varkappa_j}$. Then in the denominator we obtain a function of the same class as that considered in the preceding subsection, since its indices on all interior contours are zero. The function $F(z)\prod_{j=1}^{m}(z - z_j)^{\varkappa_j}$, however, is of the same nature as the original function $F(z)$. We assume that we have already carried out the appropriate transformation, so that in the boundary condition (36.33) the function $a + ib$ has on all interior contours an increment equal to zero.

Multplying the denominator of relation (36.33) by its regularizing factor we reduce the boundary condition to the form

$$\text{Re}\left[\frac{F(t)}{t^{\varkappa} e^{i\gamma(t)}}\right] = 0. \tag{36.34}$$

Let us consider the following cases.

1. $\varkappa \geqq 0$. Relation (36.34) constitutes Problem A_0. Hence

$$F(z) = z^{\varkappa} e^{i\gamma(z)} Q(z). \tag{36.35}$$

2. $\varkappa < 0$. The solution of the problem in the class of analytic functions is impossible, since the function satisfying the boundary condition has at the origin of the coordinate system a pole of order not lower than $-\varkappa$.

The above obtained results can be formulated as follows.

THEOREM. *If $\varkappa = \text{Ind}(a + ib) \geqq 0$ the Hilbert problem for a multiply-connected domain, in the class of analytic (not single-valued) functions has $2\varkappa + 1$ linearly independent solutions given by formula (36.35), $\gamma(z)$ being defined by formula (36.28) and $Q(z)$ by (36.22).*

If $\varkappa < 0$ a solution of the problem in the class of analytic functions is impossible.

Since the class of single-valued analytic functions constitutes only a part of the whole class of functions analytic in the domain D^+ we may state the following important corollary. *The Hilbert problem with a negative index is insoluble in the class of single-valued analytic functions.*

This corollary has important applications in investigating problems of solubility of the Hilbert problem in the class of single-valued functions.

Thus, we observe that the homogeneous Hilbert problem for a multiply-connected domain in the class of many-valued functions, can be investigated as for a simply-connected domain where only single-valued functions are considered.

We do not consider here the non-homogeneous problem, since in view of many-valuedness of the regularizing factor we would encounter difficulties connected with the computation of the Schwarz operator for many-valued functions. By its nature this problem is closely related to the contents of the next chapter.

The present formulation of the Hilbert problem characterized by the fact that many-valued functions are admitted as solutions, is not generally accepted. Other authors (D. A. Kveselava, I. N. Vekua) admit only single-valued functions. To obtain such solutions it would be necessary to solve Problem A_0 in the class of many-valued functions and subsequently to take only these solutions of Problem A_0 for which conditions (36.31) are satisfied.

To date no results have been obtained on the latter topic; therefore, to derive single-valued solutions of the Hilbert problem we abandon our method of investigation and proceed to another, based on representing the required functions by the Cauchy type integrals and on an examination of the resulting integral equations.

36.6. Integral equation of the Hilbert problem

We seek the solution of the Hilbert problem

$$\mathrm{Re}\left\{[a(s) - ib(s)]\, F(t)\right\} = c(s) \qquad (36.36)$$

in the class of single-valued functions. The required function is represented in the form of a Cauchy type integral with a real density

$$F(z) = \frac{1}{2\pi i} \int_L \frac{\mu(\sigma)\, d\tau}{\tau - z} + iC. \qquad (36.37)$$

In accordance with the investigations of § 36.1 (formula (36.7)) this representation is possible and C is given uniquely by the function $F(z)$, while the density $\mu(s)$ is given to within a constant term on every interior contour.

To simplify the further investigation let us represent the term C in the form of an integral containing the density μ. Let us select any of the interior contours, for instance L_m, and select the constant α so that the relation

$$C = \int_{L_m} [\mu(\sigma) + \alpha]\, d\sigma$$

is satisfied. Then, replacing on the contour L_m the density $\mu(\sigma) + \alpha$ by $\mu(\sigma)$ without altering $\mu(\sigma)$ on the other contours, we arrive at a representation of the form

$$F(z) = \frac{1}{2\pi i} \int_L \frac{\mu(\sigma)}{\tau - z}\, d\tau + i \int_{L_m} \mu(\sigma)\, d\sigma, \qquad (36.37')$$

the density μ entering the latter representation being completely defined on L_m and defined on the other contours to within a constant term of the form

$$c_1\mu_1 + c_2\mu_2 + \cdots + c_{m-1}\mu_{m-1},$$

where c_j are real arbitrary constants and $\mu_1, \ldots, \mu_{m-1}$ are given by the formulae

$$\mu_j(s) = \begin{cases} 1 & \text{for} \quad t \in L_j \quad (j = 1, 2, \ldots, m-1), \\ 0 & \text{on the remaining contours.} \end{cases} \tag{36.38}$$

Taking in accordance with the Sokhotski formulae the limiting values of the function (36.37′) and substituting into the boundary condition (36.36) we obtain for the determination of $\mu(s)$ the integral equation

$$a(s)\mu(s) + \mathrm{Re}\left\{\frac{a(s) - ib(s)}{\pi i}\int_L \frac{\mu(\sigma)}{\tau - t}d\tau\right\} + 2b(s)\int_{L_m}\mu(\sigma)\,d\sigma = 2c(s), \tag{36.39}$$

which will be examined later (§ 36.8).

36.7. The adjoint integral equation and the adjoint Hilbert problem

To investigate the solubility of the above integral equation let us construct the adjoint equation. To this end we pass to the real coordinates, setting

$$t = t(s), \quad \tau = \tau(\sigma), \quad d\sigma = \tau'(\sigma)\,d\sigma,$$

and we represent the last integral of the left-hand side of equation (36.39) in the form

$$\int_L h_m(s, \sigma)\,\mu(\sigma)\,d\sigma,$$

where

$$h_m(s, \sigma) = \begin{cases} 2b(s) & (s \in L, \ \sigma \in L_m), \\ 0 & (s \in L, \ \sigma \in L_k, \ k \neq m). \end{cases}$$

The kernel transposed with respect to $h_m(s, \sigma)$ has the form

$$h_m(s, \sigma) = \begin{cases} 2b(\sigma) & (s \in L_m, \ \sigma \in L), \\ 0 & (s \in L_k, \ k \neq m, \ \sigma \in L). \end{cases}$$

The homogeneous equation adjoint to (36.39) can now be written as follows:

$$a(s)\nu(s) - \mathrm{Re}\left\{\frac{t'(s)}{i\pi}\int_L \frac{a(\sigma) - ib(\sigma)}{\tau - t}\nu(\sigma)\,d\sigma\right\} = h_m, \tag{36.39′}$$

where h_m denotes the constant equal to $-2\int_L b(\sigma)\nu(\sigma)\,d\sigma$ on L_m and to zero on all remaining curves L_k.

The last integral equation can be regarded as constructed for the solution of a Hilbert problem, in the same way as equation (36.39) was constructed for the solution of the problem (36.36).

Consider the Cauchy type integral defined in $\sum\limits_{k=0}^{m} D_j^-$

$$\Phi(z) = -\frac{1}{\pi i}\int_L \frac{a(\sigma)-ib(\sigma)}{\tau - z}\,\nu(\sigma)\,d\sigma = -\frac{1}{\pi i}\int_L \frac{a(\sigma)-ib(\sigma)}{\tau'(\sigma)}\cdot\frac{\nu(\sigma)}{\tau - z}\,d\tau.$$

(36.40)

On the basis of the Sokhotski formula applied to all domains D_j^- equation (36.39) can be written in the form

$$\operatorname{Re}\left[t'(s)\,\Phi(t)\right] = h_m.$$

(36.41)

By assumption the function $\Phi(z)$ given by formula (36.40) is defined in the domain supplementing D^+. Consequently, relations (36.41) represent the boundary conditions of the Hilbert problem for $m + 1$ simply-connected domains $D_1^-, D_2^-, \ldots, D_m^-$ lying inside the contours $L_1, \ldots, L_m$, and D_0^- lying outside the curve L_0. The problems for the domains $D_1^-, \ldots, D_{m-1}^-$ are interior homogeneous while for the domain D_m^- the problem is interior non-homogeneous and for D_0 exterior homogeneous. Since the vector dt is tangential to the contour and $|dt| = ds$, therefore $t' = e^{i\alpha}$, where α is the angle between the tangent to the contour and the axis of the abscissae. In going round each contour L_k in an anticlockwise direction $\alpha(s)$ acquires the increment 2π; therefore $\operatorname{Ind} t' = 1$ on all the contours L_k. By definition (see §§ 29.1, 29.4) the indices of the problems (36.41) are $\operatorname{Ind} \bar{t}' = -\operatorname{Ind} t' = -1$. Consequently, all boundary value problems have the negative index $\varkappa = -1$.

It follows in accordance with the properties of the Hilbert problem that for the domains $D_1^-, \ldots, D_{m-1}^-$ we have $\Phi(z) = 0$.

To solve the problem for the domain D_m^- let us perform the conformal mapping onto the unit circle of the plane ζ. Denote the mapping function by $z = \omega(\zeta)$ and the complex coordinate of the circle by $\xi = e^{i\theta}$. Then we have

$$t'(s) = \frac{dt}{d\xi}\frac{d\xi}{d\theta}\frac{d\theta}{ds} = \omega'(\xi)\,i\xi\,\frac{d\theta}{ds},$$

and the boundary condition takes the form

$$\operatorname{Re}\left[\xi \Phi_1(\xi)\right] = h_m \frac{ds}{d\theta},$$

where

$$\Phi_1(\xi) = i\omega'(\xi)\,\Phi[\omega(\xi)].$$

The solution is given by the expression

$$\Phi_1(\zeta) = \zeta^{-1}\left[h_m \frac{1}{2\pi}\int\limits_0^{2\pi} \frac{ds}{d\theta}\,\frac{e^{i\theta}+\zeta}{e^{i\theta}-\zeta}\,d\theta + iC\right],$$

where for solubility we have to set $C = 0$ and, moreover, to require that the condition

$$h_m \frac{1}{2\pi i}\int\limits_0^{2\pi} \frac{ds}{d\theta}\,d\theta = 0$$

is satisfied. The last relation implies that $h_m = 0$.

Consequently, we have found that in the domain D_m^-, $\Phi(z) = 0$ and, moreover,

$$h_m = -2\int_L b(\sigma)\,v(\sigma)\,d\sigma = 0. \tag{36.42}$$

Let us finally proceed to the domain D_0^-. According to representation (36.40) $\Phi(z)$ should satisfy the condition $\Phi(\infty) = 0$. A reasoning similar to that used before leads to the solution of the last problem in the form

$$\Phi(z) = C\,\frac{\omega'(z)}{\omega(z)}, \tag{36.43}$$

where C is a real constant and $\zeta = \omega(z)$ is the function mapping the domain D_0^- conformally onto the exterior of the unit circle; it can be taken in the form

$$\omega(z) = z + \sum_{k=0}^{\infty} c_k z^{-k}.$$

To determine C we make use of formula (36.40) and set $z = \infty$ in relation (36.43). Taking into account that

$$\lim_{z\to\infty}\left[\frac{\omega(z)}{\omega'(z)}\,\frac{1}{t-z}\right] = -1,$$

12 a*

we find in view of (36.42)

$$C = \text{Re}\left\{ \frac{1}{2\pi i} \int_L [a(\sigma) - ib(\sigma)]\, v(\sigma)\, d\sigma \right\}$$

$$= -\frac{1}{2\pi} \int_L b(\sigma)\, v(\sigma)\, d\sigma = \frac{1}{4\pi} h_m = 0.$$

Consequently, for the domain D_0^-, we also have $\Phi(z) \equiv 0$.

Thus, we have proved that $\Phi(z) \equiv 0$ at an arbitrary point of the domain supplementing D^+ to the entire plane.

This result implies that the density of the Cauchy type integral (36.40) is the boundary value of a function $\varphi^+(z)$ analytic in D^+, i.e.

$$\frac{a(s) - ib(s)}{t'(s)}\, v(s) = \varphi^+(t). \tag{*}$$

Eliminating $v(s)$ we obtain

$$\text{Re}\left\{ it'(s)\, [a(s) + ib(s)]\, \varphi^+(t) \right\} = 0. \tag{36.44}$$

This relation is the boundary condition of the Hilbert problem for the function $\varphi^+(z)$. This problem will be called the adjoint to the Hilbert problem (36.36).

36.8. Investigation of the solubility problems

Solving the integral equation (36.39) we find with the help of formula (36.37) all solutions of the Hilbert boundary value problem (36.36). Consequently, the investigation of the solubility problems of the boundary value problem is essentially reduced to the investigation of the solubility of the integral equation and then to the question whether to every solution of the integral equation there corresponds a (non-zero) solution of the boundary value problem.

Consider the homogeneous integral equation

$$a(s)\,\mu(s) + \text{Re}\left\{ [a(s) - ib(s)] \frac{1}{\pi i} \int_L \frac{\mu(\sigma)}{\tau - t}\, d\tau \right\} +$$

$$+ 2b(s) \int_{L_m} \mu(\sigma)\, d\sigma = 0, \tag{36.45}$$

corresponding to the homogeneous Hilbert boundary value problem $(c(s) \equiv 0)$. First we prove that equation (36.45) has $m - 1$ non-zero solutions to which there correspond the zero $(F(z) \equiv 0)$ solutions of the boundary value problem.

In fact, equation (36.45) can be written in the form

$$\lim_{z \to t} \operatorname{Re} \left\{ [a(s) - ib(s)] \left[\frac{1}{2\pi i} \int_L \frac{\mu(\sigma)}{\tau - z} \, d\tau + i \int_{L_m} \mu(\sigma) \, d\sigma \right] \right\} = 0.$$

A consideration similar to that in § 36.1 in solving equation (36.4), shows that the functions

$$\mu_j(s) = \begin{cases} 1 & \text{for} \quad t \in L_j \quad (j = 1, 2, \ldots, m - 1) \\ 0 & \text{on the remaining contours} \end{cases} \tag{36.38}$$

are solutions of equation (36.45).

For these values of $\mu(s)$ formula (36.37) yields the function $F(z)$ identically equal to zero. It follows that *the number of solutions of the Hilbert boundary value problem is $m - 1$ less than the number of solutions of the corresponding integral equation.*

Denote by k the number of solutions of the integral equation (36.45) and by l the number of solutions of the corresponding (homogeneous) Hilbert problem. In view of the above

$$k - l = m - 1. \tag{36.46}$$

Denote by k', l' the corresponding numbers for the adjoint integral equation and Hilbert problem. According to relation (*) to every solution of the adjoint integral equation (36.39') there corresponds a definite (non-zero) solution of the adjoint Hilbert boundary value problem. Consequently, we have

$$k' = l'. \tag{36.47}$$

Let us calculate the index of the integral equation (36.45). To this end let us separate out its dominant part. Since (see § 10)

$$\frac{d\tau}{\tau - t} = \frac{dr}{r} + i \, d\theta,$$

the integral equation (36.45) can be written in the form

$$a(s) \mu(s) - \frac{b(s)}{\pi} \int_L \mu(\sigma) \frac{dr}{r} + \frac{a(s)}{\pi} \int_L \mu(\sigma) \, d\theta + 2b(s) \int_{L_m} \mu(\sigma) \, d\sigma = 0.$$

The integral $\int_L \mu(\sigma) \, d\theta$ constitutes a potential of a double layer and under the assumption concerning the contour (Lyapounov curves) made above, it is not singular. The singular integral is the first one, i.e. $\int_L \mu(\sigma)\, dr/r$. We could prove on a geometric basis that $dr/r = \frac{1}{2}\cot\left[(\sigma - s)/2\right]$ + a Fredholm kernel, and thus reduce equation (36.45) to an equation with Hilbert kernel (real). It is simpler, however, to separate out the dominant part with the Cauchy kernel. To this end add and subtract the term

$$\frac{ib(s)}{\pi}\int_L \mu(\sigma)\, d\theta\,;$$

whence

$$a(s)\,\mu(s) - \frac{b(s)}{\pi}\int_L \frac{\mu(\sigma)}{\tau - t}\, d\tau + \frac{a(s) + ib(s)}{\pi}\int_L \mu(\sigma)\, d\theta +$$

$$+ 2b(s)\int_{L_m} \mu(\sigma)\, d\sigma = 0.$$

The coefficient of the corresponding Riemann problem has the form (§ 21.1)

$$G(t) = \frac{a(s) + ib(s)}{a(s) - ib(s)}.$$

Consequently, the index of the equation has the value

$$\text{Ind}\, \frac{a(s) + ib(s)}{a(s) - ib(s)} = 2\varkappa. \tag{36.48}$$

Let us also calculate the index $\varkappa'$ of the adjoint Hilbert problem (36.44)

$$\varkappa' = \text{Ind}\left\{\bar{t}'\left[a(s) - ib(s)\right]\right\} = -\text{Ind}\, t' - \varkappa.$$

As shown in § 36.7, $t' = dt/ds = e^{i\alpha}$ where α is the angle between the tangent to the contour and the x-axis. When the point t describes the exterior contour in the positive direction the argument of t' acquires the increment 2π and when it describes the interior contour the increment is -2π; hence $\text{Ind}\, t' = 1 - m$ and, consequently,

$$\varkappa' = -\varkappa + m - 1. \tag{36.49}$$

DEFINITION. *We shall call* $\text{Ind}\, t'$ (*i.e. the number of full revolutions of the tangent to the contour in describing the contour in the positive direction*), *the corner order*† *of the domain.*

† See M. Morse, *Topological methods in the theory of functions of a complex variable*, Princeton, 1947, § 18.

The corner order of the domain under consideration is equal to $1 - m$.

Relation (36.49) can also be stated as follows: *the index of the adjoint Hilbert problem is equal to the sum of the index of the Hilbert problem considered and the corner order of the domain, both taken with negative signs.*

On the basis of the third Noether theorem (§ 23.2) we have

$$k - k' = 2\varkappa.$$

Taking into account relations (36.46), (36.47), (36.49) we obtain

$$l - l' = k - k' - (m - 1) = 2\varkappa - (m - 1) = \varkappa - \varkappa', \quad (36.50)$$

i.e. *the difference between the numbers of solutions of the adjoint Hilbert problems is equal to the difference between their indices.*

This is the fundamental result which can be deduced from the theory of integral equations.

We now make use of a result deduced in a direct examination of the boundary value problem (§ 36.5, the corollary).

The boundary value problem with a negative index is insoluble, i.e. for $\varkappa < 0$, $l = 0$ and for $\varkappa' < 0$, $l' = 0$.

It follows that for $\varkappa < 0$ the homogeneous Hilbert problem considered is insoluble, and for $\varkappa > m - 1$ the adjoint problem is insoluble (and consequently, also the adjoint singular integral equation).

Making use of the second Noether theorem on the solubility of the non-homogeneous singular integral equation we arrive at the following result.

THEOREM. *If the index of the Hilbert problem is negative, then the homogeneous problem is insoluble and the non-homogeneous is soluble if and only if the $-2\varkappa + m - 1$ conditions of solubility are satisfied (the orthogonality of the free term to all solutions of the adjoint integral equation).*

If the index of the Hilbert problem $\varkappa > m - 1$, then the non-homogeneous problem is absolutely soluble and the homogeneous problem has $2\varkappa - (m - 1)$ solutions.

Consequently, the number of solutions $l_\varkappa$ in the cases $\varkappa < 0$ and $\varkappa > m - 1$ may be expressed by the following formula:

$$l_\varkappa = \min(0, 2\varkappa - m + 1). \quad (36.51)$$

The last result can also be stated as follows:

If the index of the problem is greater than the corner order of the domain taken with negative sign, then the number of solutions of the problem is equal to twice the index of the problem plus the corner order of the domain.

The result deduced in Chapter IV concerning the number of solutions of the Hilbert problem for a simply-connected domain ($m = 0$) constitutes a particular case of the above result.

Thus, we have fully investigated the cases $\varkappa < 0$ and $\varkappa > m - 1$. It still remains to examine the problem† for $0 \leqq \varkappa \leqq m - 1$. The intermediate cases $\varkappa = 0$ and $\varkappa = m - 1$ can also be investigated and we now proceed to do so.

36.9. Investigation of the cases $\varkappa = 0$ and $\varkappa = m - 1$

We now have to modify the methods applied so far. First we make a remark. In § 36.5 (formula (36.35)) it was proved that in the case $\operatorname{Ind}(a + ib) \geqq 0$ the existence of a unique regularizing factor is a sufficient condition of solubility of the homogeneous Hilbert problem $\operatorname{Re}[(a - ib) F(t)] = 0$. It is readily observed that since $\operatorname{Ind}(a + ib) = 0$ the existence of this factor is also a necessary condition of solubility of the problem considered. In fact, if the factor does not exist, the left-hand side of the boundary condition cannot be represented as the real part of a single-valued function analytic in the domain and, consequently, the solution of the Hilbert problem in the class of single-valued analytic functions is impossible.

LEMMA. *The Hilbert boundary value problem*

$$\operatorname{Re}\left[e^{ih(s)} F(t)\right] = 0, \tag{36.52}$$

where

$$h(s) = 0 \ \text{on} \ L_0, \quad h(s) = h_j = \text{const on} \ L_j \quad (j = 1, 2, \ldots, m), \tag{36.53}$$

and

$$-\frac{\pi}{2} \leqq h_j \leqq \frac{\pi}{2}, \tag{36.54}$$

is soluble if and only if $h(s) \equiv 0$.

---

† Vekua called such cases singular (see [30*]).

Proof. Suppose that $p(s)$ is the regularizing factor of the problem so that
$$p(s)\,e^{ih(s)} = e^{i\gamma(t)}.$$

According to the considerations of § 36.5 the determination of the regularizing factor is equivalent to the determination of a single-valued analytic function $\gamma(z)$ according to the boundary condition $\mathrm{Re}\,\gamma(t) = h(s)$. In view of the uniqueness theorem of the modified Dirichlet problem (§ 36.1) $\mathrm{Re}\,\gamma(z) \equiv 0$. Hence also $h(s) \equiv 0$; this completes the proof.

Observe that by changing, if necessary, the sign in the boundary condition (36.52), which is equivalent to multiplying the relation by $e^{\pm\pi i}$, we can always make h_j satisfy condition (36.54).

We now proceed to investigate the main problem.

A. Let $\varkappa = \mathrm{Ind}\,[a(s) + ib(s)] = 0$. We seek the regularizing factor in the class of single-valued functions. We know that the determination of the factor is equivalent to the determination of the analytic function $\gamma(z)$ according to the boundary value of its real part $\omega(s)$. Since the solution of this problem in the class of single-valued functions is in general impossible, we shall reduce the problem to the modified Dirichlet problem.

Set
$$a(s) + ib(s) = e^{i\omega(s)} = e^{-ih(s)}\,e^{i[\omega(s)+h(s)]},$$

$h(s)$ being given by expressions (36.53). According to the considerations of § 36.1 it is possible to find a single-valued analytic function $\gamma(z)$ the real part of which assumes on the contour the value $\omega(s) + h(s)$ (the h_j are not prescribed beforehand). Consequently, the boundary condition of the homogeneous Hilbert problem can be reduced to the form
$$\mathrm{Re}\,[e^{ih(s)}\,e^{-i\gamma(t)}\,F(t)] = 0, \qquad (36.55)$$

all the h_j being uniquely determined and satisfying condition (36.54).

If in the last expression $h(s) \equiv 0$, we say that *the conditions of uniqueness are satisfied*. If any one of the h_j does not vanish we say that the conditions of uniqueness are not satisfied. Let us consider both cases.

1. *The conditions of uniqueness are not satisfied*, $h(t) \not\equiv 0$. In view of the lemma the homogeneous Hilbert problem has no solutions, i.e. $l = 0$. Hence, according to formula (36.51) $l' = k' = m - 1$. Consequently, in this case *the non-homogeneous problem is soluble if and only if $m - 1$ conditions of orthogonality are satisfied*.

2. *The conditions of uniqueness are satisfied,* $h(t) \equiv 0$. Now the homogeneous boundary value problem has one solution, $l = 1$. Hence $l' = k' = m$ and it follows that the non-homogeneous problem is soluble if m conditions are satisfied.

B. Suppose now that $\varkappa = m - 1$. Then, in accordance with formula (36.49), $\varkappa' = 0$.

Let us follow analogous reasoning for the adjoint Hilbert problem. We again have two cases.

1. The conditions of uniqueness for the adjoint problem are not satisfied. Then $l' = k' = 0$, $l = m - 1$. In this case the homogeneous and non-homogeneous Hilbert problems are absolutely soluble and the first has $m - 1$ linearly independent solutions.

2. The conditions of uniqueness for the adjoint problem are satisfied. Now $l' = k' = 1$, $l = m$. The homogeneous Hilbert problem has m solutions. For the solubility of the non-homogeneous problem it is necessary and sufficient that one condition of orthogonality is satisfied.

To appreciate fully the results obtained in the present subsection it should be taken into account that the problem, whether the conditions of uniqueness are satisfied or not, can be solved independently of the integral equations. Let us elucidate this question in detail.

The vanishing of all the h_j means that the known function $\omega(s) = \arctan [b(s)/a(s)]$ is the boundary value of the real part of a function $\gamma(z)$ analytic and single-valued in D^+. If the function $\gamma(z)$ be sought by means of the Schwarz operator, in view of the uniqueness we obtain the same function as from the solution of the modified Dirichlet problem, provided the conditions of uniqueness (36.31) are satisfied. It is evident that the fulfilment of the latter conditions is equivalent to the vanishing of all h_j. The functions $\alpha_j(s)$ entering conditions (36.31) are independent of the conditions of the boundary value problem and depend only on the form of the contour L (see formula (36.13) and (36.14$'$)), and therefore can be determined without solving the integral equations of the problem.

Let us illustrate what has been said above by an example.

Suppose that the given domain D^+ is a ring bounded by two concentric circles $q < |z| < 1$. Observe that any doubly-connected domain can be reduced to this particular case by means of a conformal mapping (with the appropriate choice of q)†.

† See for instance M. V. Keldysh (1).

Consider the Hilbert boundary value problem,

$$\text{Re}\,[(a - ib)\,F(t)] = c$$

with zero index.

First let us determine the functions $\beta_j(z)$ according to conditions (36.14). Here $m = 1$. A direct verification shows that the function β_1, harmonic in the ring and taking the value unity on the inner circle and zero on the outer one, can be given in polar coordinates by the relation

$$\beta_1(r, \theta) = \frac{1}{\ln q}\,\ln r.$$

Further, we have

$$\alpha_1(1, \theta) = \frac{\partial \beta_1}{\partial v}\bigg|_{r=1} = -\frac{\partial \beta_1}{\partial r}\bigg|_{r=1} = -\frac{1}{\ln q} \quad \text{on } L,$$

$$\text{and} \quad \alpha_1(q, \theta) = \frac{\partial \beta_1}{\partial v}\bigg|_{r=q} = \frac{\partial \beta_1}{\partial r}\bigg|_{r=1} = \frac{1}{q\ln q} \quad \text{on } L_1.$$

Let us now write down the conditions of uniqueness (36.31). We have on L_0, $d\sigma = d\theta$ and integration is carried out from 0 to 2π, while on $L_1, \sigma = (2\pi - \theta)q$, $d\sigma = -q\,d\theta$ with the integration bounds $(2\pi, 0)$. Changing the bounds of integration in the last integral and denoting the values of $\arctan b(\sigma)/a(\sigma)$ on L_0 and L_1 by $\omega_0(\theta)$, $\omega_1(\theta)$, respectively, we obtain

$$-\int_0^{2\pi} \frac{1}{\ln q}\,\omega_0(\theta)\,d\theta + \int_0^{2\pi} \frac{1}{\ln q}\,\frac{1}{q}\,\omega_0(\theta)\,q\,d\theta = 0,$$

or

$$\frac{1}{2\pi}\int_0^{2\pi} \omega_0(\theta)\,d\theta = \frac{1}{2\pi}\int_0^{2\pi} \omega_1(\theta)\,d\theta.$$

Taking into account that the last quantities constitute the first coefficients $a_0^{(0)}$, $a_0^{(1)}$ of the expansions of the corresponding functions into the Fourier series we arrive at the following result:

If in the expansion of the function $b(\theta)/a(\theta)$ on the circles $|z| = 1$ and $|z| = q$ into the Fourier series the first coefficients $a_0^{(0)}$, $a_0^{(1)}$ of the expansions are identical, then both homogeneous adjoint Hilbert problems have one solution each; the non-homogeneous problems are soluble if one condition is satisfied.

If these conditions are different the homogeneous adjoint problems are insoluble and the non-homogeneous problems are soluble absolutely and uniquely.

Let us also state the results for a triply-connected domain

$$m = 2, \quad l - l' = 2\varkappa - 1.$$

I. $\varkappa < 0$. The homogeneous problem is insoluble. The non-homogeneous problem is soluble if $1 - 2\varkappa$ conditions are satisfied.

II. $\varkappa > 1$. The homogeneous problem has $2\varkappa - 1$ solutions. The non-homogeneous problem is absolutely soluble.

III. $\varkappa = 0, l - l' = -1$.

(1) The conditions for single-valuedness of the function $(a + ib)$ are satisfied. The homogeneous problem has one solution, while the non-homogeneous problem is soluble if two conditions are satisfied.

(2) The conditions of uniqueness for the function $(a + ib)$ are not satisfied.

The homogeneous problem is insoluble, while the non-homogeneous problem is soluble if one condition is satisfied.

IV. $\varkappa = 1, l - l' = 1$.

(1) The conditions for single-valuedness of the function $t'(a - ib)$ are satisfied.

The homogeneous problem has two solutions, while the non-homogeneous problem is soluble if one condition is satisfied.

(2) The conditions for single-valuedness of the function $t'(a - ib)$ are not satisfied.

The homogeneous problem has one solution, while the non-homogeneous problem is absolutely soluble.

Thus, for doubly-connected and triply-connected domains the investigation of the Hilbert problem may be regarded as completed. For $m > 2$, $m - 2$ cases are still not examined, where $1 \leqq \varkappa \leqq m - 2$.

36.10. Connection with the mapping onto a plane with slits

We now proceed to set out some considerations with regard to the solubility of the Hilbert problem for the remaining cases not examined above: $0 < \varkappa < m - 1$.

Suppose that the index of the problem is unity and the conditions for single-valuedness of the function $a + ib$ are satisfied. Then the boundary condition of the homogeneous problem can be reduced to the form

$$\operatorname{Re}\left[\frac{F(t)}{t}\right] = 0. \tag{36.56}$$

This problem has the obvious solution $F(z) = iCz$. Does there exist a solution of the problem, such that $F(z)/z \neq \text{const}$? If there does, in accordance with condition (36.56), the function $\omega(z) = F(z)/z$ takes purely imaginary values on the contour L. The geometrical meaning of this fact is that in this case there exists a single-valued analytic function $\omega(z)$ which maps the domain D bounded by the contour L onto a plane with straight slits located along the imaginary axis, and has a simple pole at the origin.

Conversely, if a function $\omega(z)$ establishing this mapping exists, then the Hilbert problem (36.56) has more than one solution.

Thus, the problem of the number of solutions of the boundary value problem considered turns out to be closely related to the geometric properties of the domain.

It is known† from the theory of conformal mappings that there exists a function which maps conformally any multiply-connected

† See for instance M. V. Keldysh (1).

domain onto the plane with straight slits parallel to a straight line; in general, however, these slits do not lie on one straight line.

DEFINITION. *If the multiply-connected domain D is mapped by a single-valued analytic function possessing a simple pole at the origin, onto the plane with slits located on one straight line, then we call the domain D a domain of type D^0; otherwise the domain is not of type D^0.*

It can easily be proved that there exist domains of type D^0 of any finite connectivity. For instance if the domain has an axis of symmetry dividing it into two simply-connected domains, then it is of type D^0. This can easily be verified by mapping the part of the domain lying on one side of the axis of symmetry, onto the semi-plane and by making use of the principle of symmetry. It is known† that doubly-connected and triply-connected domains belong to type D^0. This follows at once from the above mentioned fact, since any doubly-connected or triply-connected domain can be mapped onto a circular domain having an axis of symmetry.

For domains of type D^0 the following exact result concerning the number of solutions of the Hilbert problem, can be deduced.

THEOREM. *If the domain is of type D^0 and the conditions for single-valuedness (36.31) are satisfied, then the homogeneous Hilbert boundary value problem with the index $0 \leqq \varkappa \leqq m$ has $l = \varkappa + 1$ linearly independent solutions.*

Proof. The foregoing theory implies that in the case under consideration the boundary condition can be reduced to the form

$$\operatorname{Re}[t^{-\varkappa}F(t)] = 0. \tag{36.57}$$

The problem expressed by this condition can be formulated as follows: to determine the function $Q(z)$ which is single-valued and analytic in D, except at the origin of the coordinate system where a pole of order not exceeding $\varkappa$ is admitted, and the real part of which function vanishes on the contour. This is the problem which in §36.3 was called Problem A_0, but with the additional restriction of single-valuedness of the solution. It turns out that for domains of type D^0 the function $Q(z)$ can easily be constructed.

Let $\omega(z) = u + iv$ map the domain D onto the plane with straight slits along the real axis, the zero being mapped onto the point at

---

† S. A. Chaplygin, On the theory of triplane [in Russian], *Collected Works*, vol II, 1948.

infinity. For definiteness we require also that the contour L_0 is mapped onto the segment of the straight line $(0, 1)$.

Then the functions

$$W_k(z) = i[\omega(z)]^k \qquad (k = 0, 1, \ldots, \varkappa) \qquad (36.58)$$

are solutions of the problem. In fact, these functions have at the origin of the coordinate system a pole of order not exceeding $\varkappa$ and, moreover, in view of the property of the function $\omega(z)$, we have on the contour $\mathrm{Im}\,\omega(t) = v(t) = 0$. Hence

$$\mathrm{Re}[W_k(t)] = \mathrm{Re}\{i[u(t) + iv(t)]^k\} = \mathrm{Re}\{i[u(t)]^k\} = 0.$$

We readily observe that, in view of the formulation of the problem (36.57), if a function $\Phi(z)$ is a solution for a certain $\varkappa = \varkappa_0$, then it is a solution also for any $\varkappa > \varkappa_0$.

We now prove that besides the $\varkappa + 1$ solutions which, obviously, are linearly independent, there are no other solutions. Let us assume the converse, i.e. that for some $\varkappa_0 (0 \leqq \varkappa_0 \leqq m)$ there exists another solution $\Phi(z)$. Then in view of the above remark it is also a solution for $\varkappa = m$. It follows that the problem for $\varkappa = m$ has $m + 2$ linearly independent solutions

$$W_0(z), W_1(z), \ldots, W_m(z), \quad \Phi(z).$$

But this statement contradicts the fact established above (§ 36.8) that for $\varkappa = m$ the problem has $2m - (m - 1) = m + 1$ solutions. This completes the proof of the theorem.

The general solution $Q(z)$ has the form

$$Q(z) = i \sum_{k=0}^{\varkappa} \beta_k \omega^k(z), \qquad (36.59)$$

where β_k are arbitrary real constants.

According to formula (36.51) we have

$$l' = l - 2\varkappa + (m - 1) = m - \varkappa.$$

Consequently, the non-homogeneous Hilbert problem is in the case under consideration soluble if $m - \varkappa$ conditions are satisfied.

Let us now consider some general features of the above deduced results.

It has been established that for the values $\varkappa < 0$ and $\varkappa > m - 1$ the problems of solubility of the Hilbert problem depend only on the value of this index. In the cases $\varkappa = 0$ and $\varkappa = m - 1$ the problem is more complicated, and to investigate the solubility we have to introduce another characteristic property—the condition of single-valuedness. For both cases under consideration two sub-cases are possible, depending on whether the conditions of single-valuedness are satisfied or not.

Both quantities characterizing the solubility (the index and the condition of single-valuedness) depend themselves on the form of the contour and therefore the problems of solubility of the Hilbert problem are in some way connected with the geometric properties of the domain. But for $\varkappa \leqq 0$ and $\varkappa \geqq m - 1$ only those properties are relevant to the domain which determine the index of $a + ib$ and the values of v_j (formulae (36.19)).

The case is different when $0 < \varkappa < m - 1$. It has been proved in the present subsection that when investigating the problems of solubility of the Hilbert problem, a third characteristic property occurs, namely whether or not the domain is of the type D^0. For domains of type D^0, when the conditions of single-valuedness are satisfied, the problems of solubility have completely been solved.

For $0 < \varkappa < m - 1$ the following three cases are still not examined.

I. *The domain does not belong to type D^0.*

(a) The conditions of single-valuedness are satisfied.

(b) The conditions of single-valuedness are not satisfied.

II. *The domain belongs to type D^0.* The conditions of single-valuedness are not satisfied.

It is not clear at this time whether or not the domain is of the type D^0 is the only additional geometric characteristic influencing the· solubility of the Hilbert problem, or there exist also other characteristics. The complete results deduced for doubly-connected and triply-connected domains indicate that for these domains the property of being of the type D^0 is of primary importance.

36.11. Concluding remarks

The reader who knows, by the results of Chapter IV, the complete form of the solution of the Hilbert problem in the case of a simply-connected domain, after reading this section experiences a disappointment. The variety of the methods of investigation and the cases $0 < \varkappa < m - 1$ left unexplored indicate the incompleteness of the solution.

The question may arise, whether we should speak of an incompleteness of the solution or of an impossibility to deduce more definite results. The results concerning the generalized problems derived in the first sections of the present chapter have been incomparably less definite, but then we regarded the situation as suiting the nature of the problem. Is not the case now the same? I think it is not. There is an essential difference between the problems considered in the first sections of this chapter and the contents of the present section. It consists in the fact that in the case of boundary value problems of the general type, exactly in view of their generality there are no other methods of investigation (in the general case) except a reduction to integral equations. The results obtained there contain all that can be derived by this method; therefore, these results, notwithstanding their serious indefiniteness, can be regarded as complete. A further improvement of these results can only be expected when considering particular cases. For applications it can turn out to be very important but it does not change the nature of our general statement.

The case is different for the Hilbert problem. It belongs to the simplest problems and in the case of a simply-connected domain a complete solution is deduced by a method which is not connected with integral equations. In the case of a multiply-connected domain this method is still applicable, but in view of the possibility of an appearance of many-valued solutions it does not lead to complete results. Nevertheless, also in this case the first complete result (the insolubility of the problem with a negative index) has been derived by this method. Further complete results ($\varkappa = 0$, $\varkappa \geqq m - 1$) have been deduced by combining this method with the method of integral equations.

The existence of this second method which is independent of integral equations, entitles us to consider the indefiniteness of the results for $0 < \varkappa < m - 1$ as following not from the nature of the problem but from the imperfectness of the investigation, and enables us to expect that further research will lead to a complete solution of the problem.

36.12. Some new results

The investigation of cases with the index $0 < \varkappa < m - 1$ has been developed in the works of Vekua and Boyarski. Being unable to include the details, we give the main findings from Vekua's book [30*]. (This book contains a supplement by Boyarski.)

1. In § 36.10 it was shown that if the conditions of single-valuedness are fulfilled, the number of linearly independent solutions of the homogeneous problem cannot decrease with increasing index, i.e. $l_{\varkappa+1} \geqq l_\varkappa$. It has now been established that this also holds in the general case. The proof is straightforward. It is deduced from the fact that the number of solutions of the boundary problem Re $[e^{ih(s)} F(t)] = 0$ in the class of functions with a permissible pole of order $\varkappa + 1$ is in every case not less than the number for a permissible pole of order $\varkappa$. But, on the other hand, the number of solutions should not increase too quickly. It can easily be shown that for a unit increase in the index, no more than two new independent solutions can be added, namely those with the principal parts beginning with the terms $1/z^{\varkappa+1}$ and $i/z^{\varkappa+1}$. Consequently, an estimate of the number of solutions is

$$l_\varkappa \leqq l_{\varkappa+1} \leqq l_\varkappa + 2. \tag{36.60}$$

2. The exact upper bound of the number of solutions is

$$l_\varkappa \leqq \varkappa + 1. \tag{36.61}$$

This was obtained by Boyarski by a simple, but ingenious, method of calculating the number of solutions of the boundary value problem by multiplying the boundary conditions of two mutually conjugate boundary value problems. Firstly the inequality $l_\varkappa + l'_\varkappa \leqq m + 1$ was derived, and (36.61) followed from this by the use of (36.50).

The accuracy of the estimate follows the problems in which it is made. One example is the theorem in § 36.10.

3. The investigation of the solubility of a boundary value problem can be reduced to an investigation of the solubility of a set of p linear algebraic equations with $2\varkappa$ unknowns. This has also been shown by Kveselav (4). Calculating the number of solutions of the boundary value problem is equivalent to establishing the rank of the corresponding matrix. Whenever the rank is 2π, the number of solutions is calculated from the formula (36.51) $l_\varkappa = \min(0, 2\varkappa - m + 1)$, which holds for non-exceptional cases of the index (§ 36.8). Boyarski established that the sum of the squares of moduli of minors of order $2\varkappa$ of the matrix is an analytic function of the constants $h_1, \ldots, h_m$ and of the point z_0 at which the pole of the solution is assumed to occur (in our case the pole was at the origin). Thus the rank of the matrix is not equal to $2\varkappa$ only at the zeros of some analytic function. Hence the normal form of (36.51) holds in all but exceptional cases.

4. As in § 36.10, if the conditions of single-valuedness are fulfilled a relationship is established with the mapping onto a plane with cuts along the real axis; in the general case, a relationship is established with mappings onto a plane with radial cuts.

§ 36*. Inverse boundary value problem for a multiply-connected domain

Consider the problem as in § 33 except that the prescribed domain in plane w is multiply-connected. The required domain is accordingly multiply-connected also.

The method of solution is the same as in § 33.2 except that here in the general case there is no canonical domain for which an explicit expression of the Schwarz operator could be known; the Schwarz problem can therefore only be solved directly in the prescribed domain D_w after overcoming the difficulties associated with the possible multi-valued nature of the solution.

36*.1. Formulation of the problem

We are given $(bn + 1)$ pairs of functions of the parameters s

$$u = u_k(s), \quad v = v_k(s), \quad 0 \leqq s \leqq l_k, \quad k = 0, 1, \ldots, n, \quad (36^*.1)$$

which are periodic with the period l_k, and which have derivatives that satisfy Hölder's condition and do not vanish. Equations (36*.1) define $(n + 1)$ closed curves $L_{0w}, L_{1w}, \ldots, L_{nw}$ in the plane $w = u + iv$.

It is assumed that the functions $u_k(s)$ and $v_k(s)$ are such that the curves L_{kw} do not intersect themselves or each other. They determine an $(n + 1)$-connected domain D_w in plane w, being the contour of the domain $(L_w = L_{0w} + L_{1w} + \ldots + L_{nw})$.

Interior problem. Find in the z-plane $(n + 1)$ curves $L_{0z}, L_{1z}, \ldots, L_{nz}$ $(L_z = L_{0z} + L_{1z} + \ldots + L_{nz})$ such that, in regarding s as the length of the arcs of the curves L_{kn}, the complex functions $w_k(s) = u_k(s) + iv_k(s)$ are the boundary value of the analytic function $w = w(z)$ which conformally maps the domain D_z^+ onto the domain D_w.

Exterior problem. The interior inverse problem is formulated in the same way as above except that the finite domain D_z^+ is replaced by D_z^- containing the point at infinity.

To be specific we assume that D_w is a finite domain and that the contour L_{0w} embraces all the rest. If the prescribed domain had been an infinite domain outside the contours L_k, by taking the point w_0 inside the contour L_{0w} and carrying out the transformation $w = 1/(w - w_0)$, we would revert to the formulated case.

36*.2. Solution of the interior problem

For the length of the arc σ of the contour L_w, we have

$$\sigma = \sigma(s) = \int_0^s \sqrt{[u'^2(s) + v'^2(s)]}\,ds. \qquad (36^*.2)$$

The function $\sigma(s)$ and its inverse $s(\sigma)$ are monotonic and positive, having a derivative which satisfies Hölder's condition.

For the function $z = z(w)$ which carries out the conformal mapping of the domain D_w onto D_z^+, by virtue of the geometric properties of the mapping, the following relationship holds good in the contour

$$\left|\frac{dz}{dw}\right| = \frac{ds}{d\sigma} = \frac{1}{\sqrt{\{u'^2[s(\sigma)] + v'^2[s(\sigma)]\}}} = \omega(\sigma). \qquad (36^*.3)$$

The solution of the inverse problem thus reduces to the solution of the "direct" boundary value problem.

Find in the multiply-connected domain D_w the analytic and single-valued function dz/dw which has no zeros in this domain if its modulus on the contour is equal to a given positive function $w(\sigma)$ which satisfies Hölder's condition. The function $\ln|dz/dw|$ is harmonic and constitutes the real part of the analytic function $\ln dz/dw$. Let us elucidate the nature of the latter in the domain D_w.

Since dz/dw has no zeros, the function $\ln dz/dw$ is analytic in the domain D_w; but it need not necessarily be single valued. In traversing the closed contour in D_w around some interior contour L_{kw}, it can acquire an increment which is a multiple of $2\pi i$.

The function $\ln(dz/dw)$ may therefore be represented as

$$\ln\frac{dz}{dw} = \Omega(w) + \sum_{j=1}^n m_j \ln(w - w_j), \qquad (36^*.4)$$

where $\Omega(w)$ is an analytic and single-valued function in D_w, the m_j are integers and w_j are fixed points inside the contours L_{jw}.

The problem reduces to finding the function $\ln dz/dw$, as defined by (36*.4), which is analytic in D_w with its real part on the contour equal to the prescribed function $\ln \omega(s)$. This problem is solved by the Schwarz operator S (see § 36.2)

$$\ln\frac{dz}{dw} = S \ln \omega + iv_0 = \gamma(w) + \sum_{j=1}^{n} \frac{p_j}{2\pi i} \ln(w - w_j) + iv_0. \qquad (36*.5)$$

The constants p_j are determined by formulae (36.19):

$$p_j = \int_L \alpha_j(\sigma) \ln \omega(\sigma) \, d\sigma; \qquad (36*.6)$$

$\gamma(w)$ is an analytic single-valued function.

36*.3. Solubility conditions

By comparing (36*.4) and (36*.5) it is seen that (36*.5) only determines the solution of the problem when the constants p_j are multiples of 2π. Hence

THEOREM 1. *For the inverse boundary value problem to be soluble in the case of a multiply-connected domain, the following conditions must necessarily be fulfilled:*

$$\int_L \alpha_j(\sigma) \ln \omega(\sigma) \, d\sigma = 2\pi m_j \qquad (j = 1, 2, \ldots n), \qquad (36*.7)$$

where m_j are integers (positive or negative). Assuming that these conditions are in fact fulfilled, let us consider the sufficient conditions for solubility.

Powering and then integrating in (36*.5), we obtain for the required function $z(w)$:

$$z = e^{iv_0} \int e^{\gamma(w)} \prod_{j=1}^{n} (w - w_j)^{m_j} \, dw + C. \qquad (36*.8)$$

If all the integers m_j are positive, this latter formula defines $z(w)$ as an analytic and single-valued function. But if some of the m_j are negative, the expression for z can contain logarithmic terms. For none to be present, it is necessary and sufficient for the residues of the integrands to vanish at those points w_j which correspond to negative m_j.

Let us now calculate these residues.

Let $m_k = -q_k < 0$ for some $j = k$. The corresponding residue, equal to the coefficient of $(w - w_k)^{q_k-1}$ of the expansion of the function $e^{\gamma(\omega)} \prod\limits_{j \neq k} (w - w_j)^{m_j}$ in the neighbourhood of w_k, is then found by the formula

$$\frac{1}{(q_k - 1)!} \frac{d^{q_k-1}}{dw^{q_k-1}} \left[e^{\gamma(w)} \prod_{j \neq k} (w - w_j)^{m_j} \right]_{w-w_k}.$$

Hence:

THEOREM 2. *If all the integers m_j in the expresison (36*.7) are positive, the inverse boundary value problem is soluble. If negative integers $m_k = -q_k$ are found amongst them, then for the solubility of the problem the additional condition*

$$\frac{d^{q_k-1}}{dw^{q_k-1}} \left[e^{\gamma(z)} \prod_{j \neq k} (w - w_j)^{m_j} \right]_{w=w_k} = 0 \tag{36*.9}$$

must be fulfilled for each integer.

With these two theorems the problems of the solubility conditions for the interior inverse boundary value problem are solved.

The required contour can be obtained by changing over in (36*.8) to the contour values.

EXAMPLE. Given that the domain D_w^+ is a circular ring $q < |w| < 1$, by the same reasoning as in the example of § 36.9 we arrive at the condition:

$$\frac{1}{2\pi} \int_0^{2\pi} \ln \omega_0(\theta)\, d\theta - \frac{1}{2\pi} \int_0^{2\pi} \ln \omega_1(\theta)\, d\theta = m \ln \frac{1}{q}$$

or

$$a_0^0 - a_0^1 = m \ln \frac{1}{q},$$

where a^0, a_0^1 are the initial terms of the Fourier expansion of the function $\ln \omega(\theta)$ on the circles $|w| = 1$ and $|w| = q$ respectively.

If $m > 0$, the latter condition is also sufficient, but if $m < 0$, then according to (36*. 9) the following condition must also be satisfied

$$\frac{d^{-m-1}}{d\omega^{-m-1}} \left[e^{\gamma(w)} \right]_{w=0} = 0,$$

where $\gamma(w)$ is a function definable in terms of $\omega(\theta)$ by the Schwarz operator which, in the case in question, can be expressed explicitly in terms of elliptic functions according to Vill's formula.

The material of this section is taken from the author's article (5).

§ 37. General boundary value problem of Riemann type for multiply-connected domains

In this subsection we shall extend the solution of the general boundary value problem (35.1) to the case of a multiply-connected domain. As in § 35 where the solution was given for a simply-connected domain, the device employed is an integral representation of the unknown sectionally analytic function. We begin the exposition by deriving this representation.

37.1. The integral representation

The integral representation is based on the same principles as in § 35.1. However, now, as always in problems concerning multiply-connected domains, complications arise due to the possibility of many-valuedness in the analytic functions defined by integral representations. To explain and justify the complexity of the integral representations applied below, let us first consider a simple example.

Suppose that in a doubly-connected domain bounded by the exterior contour L_0 and the interior L_1 the derivative of an analytic function is given by the Cauchy type integral

$$[\Phi^+(z)]' = \frac{1}{2\pi i} \int_L \frac{\mu(\tau)}{\tau - z} \, d\tau.$$

Integrating from 0 to z we obtain for the function itself the expression

$$\Phi^+(z) = -\frac{1}{2\pi i} \int_L \mu(\tau) \ln\left(1 - \frac{z}{\tau}\right) d\tau + \Phi^+(0). \tag{37.1}$$

This formula represents in general a many-valued function. The many-valuedness is due to the integral over the interior contour, because for the function defined by this integral any closed contour in the domain D^+, which contains the contour L_1 separates the two branch points of the kernel $z = \tau$ and $z = \infty$.

The many-valuedness can be established by a direct computation. In fact, expression (37.1) can be written in the form

$$\Phi^+(z) = -\frac{1}{2\pi i} \int_{L_0} \mu(\tau) \ln\left(1 - \frac{z}{\tau}\right) d\tau - \frac{1}{2\pi i} \int_{L_1} \mu(\tau) \ln\left(1 - \frac{\tau - z_1}{z - z_1}\right) d\tau -$$

$$- \ln(z - z_1) \frac{1}{2\pi i} \int_{L_1} \mu(\tau) \, d\tau + \frac{1}{2\pi i} \int_{L_1} \mu(\tau) \ln(-\tau) \, d\tau + \Phi^+(0),$$

where z_1 is a point belonging to D^-.

All terms of the right-hand side, except the third are single-valued. It follows that for the single-valuedness of the integral representation (37.1) the density $\mu(\tau)$ should satisfy the condition

$$\int_{L_1} \mu(\tau) \, d\tau = 0.$$

This simple example indicates that it is impossible to achieve the single-valuedness of a representation of an analytic function by strictly following the way which led to representations (35.2), (35.3).

We now proceed to the general case of an $(m + 1)$-connected domain (Fig. 13, p. 114). D^+ is a connected domain and $D^- = D_0^- + D_1^- + \ldots + D_m^-$ is the sum of separate domains, which for brevity will also be called a domain.

Let us find the integral representation for a function $\Phi(z) = \{\Phi^+(z), \Phi^-(z)\}$, sectionally analytic and single-valued in the domains D^+, D^-, assuming that $[\Phi^+(z)]^{(n)}$ and $[\Phi^-(z)]^{(p)}$ are known.

Some considerations which are outside the scope of our exposition prove that the required representation for $\Phi(z)$ can be derived by using the following representation for the derivatives:

$$[\Phi^+(z)]^{(n)} = \sum_{j=1}^{m} \frac{1}{(z - z_j)^n} \frac{1}{2\pi i} \int_{L_j} \frac{\mu(\tau)}{\tau - z} d\tau + \frac{1}{2\pi i} \int_{L_0} \frac{\mu(\tau)}{\tau - z} d\tau, \qquad (37.2)$$

$$\frac{z^{p+mn}}{\prod\limits_{j=1}^{m} (z - z_j)^n} [\Phi^-(z)]^{(p)}$$

$$= \sum_{j=1}^{m} \frac{1}{(z - z_j)^n} \frac{1}{2\pi i} \int_{L_j} \frac{\mu(\tau)}{\tau - z} d\tau + \frac{1}{2\pi i} \int_{L_0} \frac{\mu(\tau)}{\tau - z} d\tau, \qquad (37.3)$$

where $\mu(t)$ is the unknown complex function and z_j are some fixed points of the domains D_j^-.

In accordance with the Sokhotski formulae the density $\mu(t)$ is expressed in terms of the known $[\Phi^+(z)]^{(n)}$, $[\Phi^-(z)]^{(p)}$ by the relations

$$\mu(t) = [\Phi^+(t)]^{(n)} - \frac{t^{p+mn}}{\prod\limits_{j=1}^{m} (t - z_j)^n} [\Phi^-(t)]^{(p)} \qquad (t \in L_0), \qquad (37.4)$$

$$\mu(t) = (t - z_q)^n \left\{ [\Phi^+(t)]^{(n)} - \frac{t^{p+mn}}{\prod\limits_{j=1}^{m} (t - z_j)^n} [\Phi^-(t)]^{(p)} \right\} \qquad (37.5)$$

$$(t \in L_q; \quad q = 1, 2, \ldots, m).$$

The derivation of the representations for the functions $\Phi^+(z)$, $\Phi^-(z)$ consists of integrating formulae (37.2), (37.3). To this end let us employ the expansion of the kernels $1/[(z - z_j)^n (\tau - z)]$ into simple fractions

$$\frac{1}{(z - z_j)^n (\tau - z)} = \frac{-1}{(\tau - z_j)^n (z - \tau)} + \sum_{k=1}^{n} \frac{1}{(\tau - z_j)^{n-k+1} (z - z_j)^k}.$$

Integrating n times both sides of relation (37.2) from 0 to z, after rather cumbersome transformations we obtain

$$\Phi^+(z) = \frac{(-1)^n}{(n - 1)!} \sum_{j=1}^{m} \frac{1}{2\pi i} \int_{L_j} \frac{\mu(\tau)}{(\tau - z_j)^n} (\tau - z)^{n-1} \ln\left(1 - \frac{\tau - z_j}{z - z_j}\right) d\tau +$$

$$+ \frac{(-1)^n}{(n - 1)!} \frac{1}{2\pi i} \int_{L_0} \mu(\tau) (\tau - z)^{n-1} \ln\left(1 - \frac{z}{\tau}\right) d\tau + P_{n-1}(z), \qquad (37.6)$$

where $P_{n-1}(z)$ is a polynomial of degree $n-1$ the coefficients of which can be computed by comparing the first n terms of the expansions of both sides of relation (37.6) into power series.

By $\ln(1 - z/\tau)$, $\ln[1 - (\tau - z_j)/(z - z_j)]$ we understand the branches which vanish for $z = 0$, $z = \infty$, respectively. Since all terms of the right-hand side of formula (37.6) are single-valued when z lies in D^+ it follows that the function $\Phi^+(z)$ is analytic and single-valued in the domain D^+.

For the function $\Phi^-(z)$ we obtain two distinct representations, depending on whether z belongs to one of the finite domains D_q^- or to the infinite domain D_0^-. Omitting the transformations we quote the final results of integrating relation (37.3) p times with respect to z from z_j to z; for $z \in D_q^-$ $(q = 1, 2, \ldots, m)$ we have

$$\Phi^-(z) = \frac{(-1)^p}{(p-1)!} \sum_{j=1}^{m} \frac{1}{2\pi i} \int_{L_j} \mu(\tau) \frac{\prod\limits_{k=1}^{m}{}'(\tau - z_k)^n}{\tau^{p+mn}} (\tau - z)^{p-1} \ln\left[1 - \frac{z - z_j}{\tau - z_j}\right] d\tau +$$

$$+ \frac{(-1)^p}{(p-1)!} \frac{1}{2\pi i} \int_{L_0} \mu(\tau) \frac{\prod\limits_{k=1}^{m}(\tau - z^k)^n}{\tau^{p+mn}} (\tau - z)^{p-1} \ln\left(1 - \frac{z}{\tau}\right) d\tau +$$

$$+ \sum_{k=0}^{p-1} A_k z^k \ln z + \sum_{k=1}^{mn} \frac{B_k}{z^k} + \sum_{k=0}^{p-1} d_{qk} z^k, \tag{37.7}$$

the prime denoting that in the product the factor corresponding to $k = j$ is to be omitted. The constants A_k, B_k are expressed in a definite manner by the function $\mu(\tau)$, and d_{qk} are constants to be computed by expanding both sides of relation (37.7) into power series in the vicinity of the point $z = z_q$.

For $z \in D_0^-$ the representation of $\Phi^-(z)$ has the form

$$\Phi^-(z) = \frac{(-1)^p}{(p-1)!} \frac{1}{2\pi i} \int_{L} \mu(\tau) \frac{\prod\limits_{k=1}^{m}{}'(\tau - z_k)^n}{\tau^{p+mn}} (\tau - z)^{p-1} \ln\left(1 - \frac{\tau}{z}\right) d\tau +$$

$$+ \sum_{k=1}^{mn} \frac{B_k}{z^k} + \sum_{k=0}^{m-1} d_{0k} z^k. \tag{37.8}$$

the prime denoting that in the product the term for $k = j$ is to be omitted when $\tau \in L_j (j = 1, 2, \ldots, m)$. The constants B_k are the same as in formula (37.7) and the constants d_{0k} are determined by expanding both sides of relation (37.8) into series in the vicinity of the point $z = \infty$.

37.2. Boundary value problem and integro-differential equation

Let us formulate the following boundary value problem: to determine the function $\Phi^+(z)$ which is analytic and single-valued in the $(m + 1)$-connected domain D^+ and the function $\Phi^-(z)$ which is analytic† in "the domain" D^-, in accordance with

---

† Here analyticity is necessarily single-valuedness as well.

the boundary conditions on the contour

$$\sum_{k=1}^{n}\left[a_k(t)\frac{d^k\Phi^+(t)}{dt^k}+\int_L M_k(t,\tau)\frac{d^k\Phi^+(\tau)}{d\tau^k}\,d\tau\right]-$$

$$-\sum_{k=1}^{p}\left[b_k(t)\frac{d^k\Phi^-(t)}{dt^k}+\int_L N_k(t,\tau)\frac{d^k\Phi^-(\tau)}{d\tau^k}\,d\tau\right]=f(t),\qquad(37.9)$$

where $a_k(t)$, $b_k(t)$, $M_k(t,\tau)$, $N_k(t,\tau)$ are functions of the same kind as in the boundary value problem (35.1).

With the help of the integral representations (37.6), (37.7), (37.8) we can express all unknown functions entering the boundary condition by the density of the representation $\mu(t)$. Then all derivatives up to $(n-1)$th and $(p-1)$th orders, respectively, are expressed by integrals in which the boundary value is obtained by simply replacing z by t. The highest derivatives are given by the Cauchy type integrals according to formulae (37.2), (37.3). Passing to the limit we have

$$\left.\begin{array}{l}[\Phi^+(t)]^n=\tfrac{1}{2}(t-z_j)^{-n}\mu(t)+\psi(t)\quad(t\in L_j),\\[2mm]\dfrac{t^{p+mn}}{\displaystyle\prod_{j=1}^{m}(t-z_j)^n}[\Phi^-(t)]^{(p)}=-\dfrac{1}{2}(t-z_j)^{-n}\mu(t)+\psi(t)\\[6mm]\hspace{3cm}(j=1,2,\ldots,m),\end{array}\right\}\qquad(37.10)$$

$$\left.\begin{array}{l}[\Phi^+(t)]^n=\tfrac{1}{2}\mu(t)+\psi(t),\\[2mm]\dfrac{t^{p+mn}}{\displaystyle\prod_{j=1}^{m}(t-z_j)^n}[\Phi^-(t)]^{(p)}=-\dfrac{1}{2}\mu(t)+\psi(t),\quad t\in L_0\end{array}\right\}\qquad(37.11)$$

where $\psi(t)$ denotes the right-hand side of formulae (37.2), (37.3), z being replaced by t.

After replacing all unknown functions by their expressions in terms of $\mu(t)$, the boundary condition (37.9) is reduced to a singular integral equation with Cauchy kernel of the form

$$a(t)\,\mu(t)+\frac{b(t)}{\pi i}\int_L\frac{\mu(\tau)}{\tau-t}\,d\tau+\int_L K(t,\tau)\,\mu(\tau)\,d\tau=g(t),\qquad(37.12)$$

where

$$\left.\begin{array}{l}a(t)=a_n(t)+t^{-p-mn}\displaystyle\prod_{j=1}^{m}(t-z_j)^n b_p(t),\\[4mm]b(t)=a_n(t)-t^{-p-mn}\displaystyle\prod_{j=1}^{m}(t-z_j)^n b_p(t),\end{array}\right\}\qquad(37.13)$$

$K(t,\tau)$ is a completely definite Fredholm kernel, $g(t)$ is a function which can be expressed by the known functions and the constants of the integral representation.

The further investigation is carried out as before; it has been considered in detail in §34 and §35.

The singular integro-differential equation of type (35.17) deduced on the contour $L=L_0+L_1+\ldots+L_m$ in the same way as in §35.3 can be reduced to the boundary value problem (37.9).

We shall not consider here the problem of Hilbert type of general form similar to problem (34.2), referring the reader to the paper of D. I. Sherman (1). A full and simplified exposition is given in the paper by Khasabov (1).

§ 38. Boundary value problems for equations of elliptic type

38.1. General information

For the reader's convenience we shall give here some well-known properties of linear differential equations of elliptic type. The differential equations of second order with two independent variables

$$A\frac{\partial^2 u}{\partial x^2} + 2B\frac{\partial^2 u}{\partial x\,\partial y} + C\frac{\partial^2 u}{\partial y^2} + D\frac{\partial u}{\partial x} + E\frac{\partial u}{\partial y} + Fu = \varphi \quad (38.1)$$

can be classified in accordance with the so-called characteristics, i.e. the solutions of two ordinary differential equations of first order

$$A\,dy - [B \pm \sqrt{(B^2 - AC)}]\,dx = 0. \quad (38.2)$$

If the expression $B^2 - AC$ is negative (the characteristics are complex) the equation is of elliptic type; when $B^2 - AC > 0$ the equation is of hyperbolic type and when $B^2 - AC = 0$ of parabolic type.

Making use of the solutions of the equation of characteristics, the general equation (38.1) can be reduced to the simplest (canonical) form by a change of the independent variables. The canonical form of an equation of elliptic type is the following:

$$\Delta u + a\frac{\partial u}{\partial x} + b\frac{\partial u}{\partial y} + cu = \varphi, \quad (38.3)$$

where Δ is the Laplace operator, a, b, c, φ are known functions of x, y, which we assume to be continuously differentiable in the domain under consideration, including the boundary.

The simplest equation of the elliptic type is the Laplace equation ($a = b = c = \varphi = 0$) which we have frequently encountered previously.

A typical property of equations of the elliptic type which results in their essential difference from equations of other types, is the fact that their solutions are completely determined by prescribing one boundary condition. The simplest is that arising from the prescription on the contour of the values of the function itself, or of its normal derivative; this constitutes the classical boundary value problems which are usually called, as for the Laplace equation, the first and the second boundary value problems, or else the Dirichlet and the Neumann problems. Another classical problem is that of determining the solution of an equation in terms of a linear combination of the function and its normal derivative on the contour; this problem is called the mixed problem. The modern investigations are characterized by formulating boundary value problems of general type, such as the problem considered in § 34.

In the next subsection we shall present an outline of the classical methods of solution of the first boundary value problem.

38.2. Classical methods of solution

THE DIRICHLET PROBLEM. *To determine a function $u(x, y)$ possessing in the domain D bounded by a simple smooth closed contour, continuous derivatives up to the second order, continuous on the contour and satisfying the differential equation of elliptic type* (38.3) *and the boundary condition*

$$u = f(s). \tag{38.4}$$

The device used for the solution of the problem is the Green function for the Laplace equation, which has already been made use of in § 27.2. The basis is the following fact (Hilbert's theorem). The solution of the Poisson equation

$$\Delta u = \varphi(x, y), \tag{38.5}$$

which vanishes on the contour, is the function

$$u(x, y) = -\frac{1}{2\pi} \int\int_D G(x, y; \xi, \eta) \, \varphi(\xi, \eta) \, d\xi \, d\eta. \tag{38.6}$$

To derive the solution of equation (38.5) which takes on the contour the given value $f(s)$ it is sufficient to add to the solution (38.6) a harmonic function $v(x, y)$ which satisfies the boundary condition (38.4). Since the solution of this problem is well known we may assume that this function has already been found.

Formula (38.6) enables us to reduce the boundary value problem (38.4) to an integro-differential equation. Writing equation (38.3) in the form

$$\Delta u = -a\frac{\partial u}{\partial x} - b\frac{\partial u}{\partial y} - cu + \varphi,$$

13*

we have in view of formula (38.6)

$$u(x, y) = \frac{1}{2\pi} \int\int_D G(x, y; \xi, \eta) \left[a(\xi, \eta) \frac{\partial u}{\partial \xi} + \right.$$

$$\left. + b(\xi, \eta) \frac{\partial u}{\partial \eta} + c(\xi, \eta) u - \varphi(\xi, \eta) \right] d\xi\, d\eta + v(x, y). \tag{38.7}$$

Integrating by parts and taking into account that in view of the properties of the Green function (on the contour $G \equiv 0$) the integrated terms vanish, we arrive at the integro-differential equation

$$u = \frac{1}{2\pi} \int\int_D \left[Gc - \frac{\partial(aG)}{\partial \xi} - \frac{\partial(bG)}{\partial \eta} \right] u\, d\xi\, d\eta + f_1, \tag{38.8}$$

where

$$f_1 = \frac{1}{2\pi} \int\int_D G(x, y; \xi, \eta)\, \varphi(\xi, \eta)\, d\xi\, d\eta + v(x, y)$$

is a known function.

We have here only formally indicated a method of reduction of the Dirichlet boundary value problem to an integral equation, omitting the derivation itself. The difficulty of the latter consists in the justification of the legitimacy of integration by parts and then in proving that the function determined by equation (38.8) is twice differentiable. All proofs are significantly simplified if the coefficients of the equation and the contour be subjected to sufficient limitations. Without going into details let us observe that in deriving equation (38.8) in the most general class of prescribed functions and contours, we make an essential use of the approximation of the known functions by uniformly convergent sequences of analytic functions and the approximation of the contour by sequences of analytic contours. A detailed exposition of the problem is presented in two papers of Lichtenstein published in v. 142 and 143 (1913) of *Journ. für reine u. angew. Mathem.*

On the basis of the Fredholm theorems, equation (38.8) implies the following statement.

If the homogeneous problem ($\varphi \equiv 0$, $f \equiv 0$) is insoluble, the non-homogeneous problem is absolutely soluble and has a unique solution. If the homogeneous problem is soluble, the non-homogeneous problem is soluble only if certain conditions are satisfied. If these conditions are satisfied the solution of the non-homogeneous problem is determined to within the general solution of the homogeneous problem.

In the general case of arbitrary coefficients of the equation, more precise results cannot be obtained. There is one important particular case when independently of the integral equation a rigorous result on the solubility of the Dirichlet problem can be deduced.

THEOREM. *If, in equation* (38.3),

$$c(x, y) \leqq 0 \tag{38.9}$$

everywhere in the domain, then the homogeneous Dirichlet problem is insoluble and the nonhomogeneous problem is uniquely soluble.

The proof may be found for instance in the course of analysis of V. I. Smirnov, vol. IV, § 234.

Let us mention here a remarkable property of equations of the elliptic type, which was first established in the case considered above by Picard, and in the general case by S. N. Bernstein.

A differential equation of the elliptic type with analytic coefficients is satisfied by an analytic function.

38.3. Integral representation of solutions

Consider the differential equation (38.3) the coefficients of which are analytic functions. It was indicated above that all solutions of such an equation are also analytic functions. Consequently, the equation can be investigated also in the case of complex values of the argument. This constitutes the basis of the following considerations.

It was mentioned at the beginning of the section that the division of equations into hyperbolic and elliptic types is carried out in accordance with the characteristics which can be either real or imaginary. This division in the case of real equations is essential and of primary importance. However, in the case of equations defined also for complex values of the arguments (analytic) the difference between the elliptic and hyperbolic types vanishes, and equations of one type can be reduced to equations of the other type by formal transformations. This constitutes the basis of the derivation presented below of an integral representation of solutions of equations of elliptic type, by analytic functions.

First, using the method of § 32.1 we shall formally reduce the considered equation of elliptic type to an equation of a hyperbolic type. Next, making use of the property of the latter equations that their solutions can be deduced by successive approximations, we construct the required integral representation.

We proceed to the derivation of the representation.

We suppose for the time being that in equation (38.3) the independent variables x, y and the function u are complex, and we transform the equation to the new variables

$$z = x + iy, \quad \bar{z} = x - iy.$$

Observe that if, as indicated above, the variables are regarded as complex, the variables z, $\bar{z}$ are to be regarded as two independent complex variables. If in particular x, y are real, z, $\bar{z}$ are complex conjugate values of one complex variable.

We have (see § 32.1)

$$\frac{\partial}{\partial x} = \frac{\partial}{\partial z} + \frac{\partial}{\partial \bar{z}}, \quad \frac{\partial}{\partial y} = i\left(\frac{\partial}{\partial z} - \frac{\partial}{\partial \bar{z}}\right), \quad \Delta = 4\frac{\partial^2}{\partial z\,\partial \bar{z}}.$$

In the new variables equation (38.3) takes the form

$$4\frac{\partial^2 u}{\partial z\,\partial \bar{z}} + (a + ib)\frac{\partial u}{\partial z} + (a - ib)\frac{\partial u}{\partial \bar{z}} + cu = \varphi.$$

Introducing the notations

$$A(z, \bar{z}) = \frac{a + ib}{4}, \quad \overline{A(z, \bar{z})} = \frac{a - ib}{4}, \quad \frac{C}{4} = c(z, \bar{z}), \quad F = \frac{\varphi}{4},$$

equation (38.3) can be written in the form

(38.10)

$$\frac{\partial^2 u}{\partial z\,\partial \bar{z}} + A\frac{\partial u}{\partial z} + \bar{A}\frac{\partial u}{\partial \bar{z}} + Cu = F. \tag{38.11}$$

Here, if x, y are real, A and $\bar{A}$ are complex conjugate functions and C and F are real.

Equation (38.11) is of hyperbolic type in the canonical form, and the whole theory of these equations can be applied here. It is known that the solutions of such equations, similarly to the case of ordinary differential equations, can be derived by the method of successive approximations. To apply this method conveniently we first reduce the differential equation (38.11) to an integral equation of Volterra type. To this end let us integrate equation (38.11) with respect to the variables z and $\bar{z}$, if necessary eliminating by integration by parts the derivatives $\partial u/\partial z$ and $\partial u/\partial \bar{z}$.

For simplification of the transformations we replace the variables z, $\bar{z}$ in equation (38.11) by ζ, $\bar{\zeta}$, respectively, and we regard the variables z, $\bar{z}$ as the upper bounds of integration; for the lower bounds of integration we take some constants z_0 and $\bar{z}_0$.

Integrating with respect to ζ we obtain

$$\frac{\partial u}{\partial \bar{\zeta}} + \int\limits_{z_0}^{z} A(\zeta, \bar{\zeta}) \frac{\partial u}{\partial \zeta} d\zeta + \int\limits_{z_0}^{z} \overline{A(\zeta, \bar{\zeta})} \frac{\partial u}{\partial \bar{\zeta}} d\zeta + \int\limits_{z_0}^{z} C(\zeta, \bar{\zeta}) u \, d\zeta$$

$$= \int\limits_{z_0}^{z} F(\zeta, \bar{\zeta}) d\zeta + \omega(\bar{\zeta}),$$

where $\omega(\bar{\zeta})$ is an arbitrary analytic function in the variable $\bar{\zeta}$. Integrating the first integral by parts

$$\int\limits_{z_0}^{z} A(\zeta, \bar{\zeta}) \frac{\partial u}{\partial \zeta} d\zeta = A(\zeta, \bar{\zeta}) u(\zeta, \bar{\zeta}) \Bigg|_{z_0}^{z} - \int\limits_{z_0}^{z} \frac{\partial A}{\partial \zeta} u(\zeta, \bar{\zeta}) d\zeta.$$

and including the term $A(z_0, \bar{\zeta}) u(z_0, \bar{\zeta})$ in the arbitrary function ω_1 we can write the equation in the form

$$\frac{\partial u}{\partial \bar{\zeta}} + \int\limits_{z_0}^{z} \overline{A(\zeta, \bar{\zeta})} \frac{\partial u}{\partial \bar{\zeta}} d\zeta + A(z, \bar{\zeta}) u(z, \bar{\zeta}) + \int\limits_{z_0}^{z} \left[C(\zeta, \bar{\zeta}) - \frac{\partial A}{\partial \zeta} \right] u \, d\zeta$$

$$= \int\limits_{z_0}^{z} F(\zeta, \bar{\zeta}) d\zeta + \omega_1(\bar{\zeta}).$$

Integrating this expression with respect to $\bar{\zeta}$ from $\bar{z}_0$ to $\bar{z}$ and performing for the integral

$$\int\limits_{\bar{z}_0}^{\bar{z}} \overline{A(\zeta, \bar{\zeta})} \frac{\partial u}{\partial \zeta} d\zeta$$

the same transformation as above, we arrive at the following integral equation of the Volterra type:

$$u(z, \bar{z}) + \int\limits_{z_0}^{z} \overline{A(\zeta, \bar{z})} u(\zeta, \bar{z}) d\zeta + \int\limits_{\bar{z}_0}^{\bar{z}} A(z, \bar{\zeta}) u(z, \bar{\zeta}) d\bar{\zeta} +$$

$$+ \int\limits_{\bar{z}_0}^{\bar{z}} \int\limits_{z_0}^{z} H(\zeta, \bar{\zeta}) u(\zeta, \bar{\zeta}) d\zeta \, d\bar{\zeta} = F_1(z, \bar{z}) + \varphi(z) + \psi(\bar{z}), \quad (38.12)$$

where

$$H = C - \frac{\partial A}{\partial \zeta} - \frac{\partial \bar{A}}{\partial \bar{\zeta}}, \quad F_1 = \int_{\bar{z}_0}^{\bar{z}} \int_{z_0}^{z} F(\zeta, \bar{\zeta}) \, d\zeta \, d\bar{\zeta};$$

φ and ψ are arbitrary analytic functions of their arguments.

Observe that if x, y be regarded as real and, consequently, z, $\bar{z}$ as complex conjugate, in view of the fact that the whole equation is real, the arbitrary functions φ and ψ are also complex conjugate.

Equation (38.12) belongs to a somewhat unusual type of the Volterra equations, since it contains both ordinary and double integrals. We are dealing with functions of two variables, so the double integral therefore is normal, whereas the ordinary integrals should be eliminated. The solution of equations of the above obtained type is known. The detailed exposition, in connection with the problem under consideration, is for instance given in the book of G. M. Müntz, Integral Equations (in Russian), p. 158.

To reduce equation (38.12) to the ordinary form in which the ordinary integrals are absent, we apply the following method. Write the equation in the form

$$u(z, \bar{z}) + \int_{z_0}^{z} \overline{A(\zeta, \bar{z})} \, u(\zeta, \bar{z}) \, d\zeta = v(z, \bar{z}),$$

where v is the sum of the terms of the equation (except the first two) taken to the right-hand side.

The latter equation we consider as an equation with one variable z (the variable $\bar{z}$ is considered as a parameter). Writing the solution of this equation in the usual form, making use of the resolvent, we have

$$u(z, \bar{z}) = v(z, \bar{z}) + \int_{z_0}^{z} R(z, \bar{z}, \zeta) \, v(\zeta, \bar{z}) \, d\zeta,$$

and we obtain an equation of the same type as (38.12) without the ordinary integral with respect to ζ. By means of the same transformation in the derived equation we eliminate the ordinary integral with respect to $\bar{\zeta}$ and the equation takes the usual form of the Volterra type of equation

$$u(z, \bar{z}) + \int_{\bar{z}_0}^{\bar{z}} \int_{z_0}^{z} K(z, \bar{z}; \zeta, \bar{\zeta}) \, u(\zeta, \bar{\zeta}) \, d\zeta \, d\bar{\zeta}$$
$$= \alpha(z, \bar{z}) \varphi(z) + \alpha_1(z, \bar{z}) \psi(\bar{z}) + F_2(z, \bar{z}), \quad (38.13)$$

where K, α, α_1, F_2 are completely definite functions for which an explicit expression in terms of A, $\bar{A}$, H can be written. If x, y are real, then α and α_1 are complex conjugate functions.

Observe that the resolvents used for the elimination of the ordinary integrals can, in view of an exceptional simplicity of the kernels (A is independent of $\bar{z}$ and $\bar{A}$ of z), be computed not by means of the general formulae as infinite series of iterative kernels, but as solutions of simple linear differential equations of the first order (see Problem 19 at the end of this chapter).

Equation (38.13) can be solved in the usual way, by the method of successive approximations. Writing the solution in terms of the resolvent we obtain the expression for u in the form

$$u(z, \bar{z}) = \alpha(z, \bar{z})\,\varphi(z) + \alpha_1(z, \bar{z})\,\psi(\bar{z}) +$$

$$+ \int_{z_0}^{z} \beta(z, \bar{z}, \zeta)\,\varphi(\zeta)\,d\zeta + \int_{\bar{z}_0}^{\bar{z}} \beta_1(z, \bar{z}, \bar{\zeta})\,\psi(\bar{\zeta})\,d\bar{\zeta} +$$

$$+ \int_{\bar{z}_0}^{\bar{z}} \int_{z_0}^{z} \gamma(z, \bar{z}; \zeta, \bar{\zeta})\,F(\zeta, \bar{\zeta})\,d\zeta\,d\bar{\zeta}, \quad (38.14)$$

where β, β_1, γ are completely definite functions obtained from the kernel K by the method of successive approximations. Ultimately, all these functions are expressed in a completely definite manner by the coefficients of the equation. If the original equation is homogeneous the last term drops out; in the following, only this case will be considered.

When the original equation (38.3) is real the functions φ, ψ; α, α_1; β, β_1 represent pairs of complex conjugate functions and the expression for u can be written in the form†

$$u(x, y) = \text{Re}\left[\alpha(z, \bar{z})\,\varphi(z) + \int_{z_0}^{z} \beta(z, \bar{z}, \zeta)\,\varphi(\zeta)\,d\zeta\right]. \quad (38.15)$$

This is the formula for the required integral representation of the solutions of an equation of elliptic type with analytic coefficients

$$\Delta u + a\frac{\partial u}{\partial x} + b\frac{\partial u}{\partial y} + cu = 0 \quad (38.3')$$

† We replace, for simplicity, 2α and 2β again by α and β.

by the arbitrary analytic function $\varphi(z)$. If the latter function is known the function u is uniquely determined in terms of it. It can be proved, which we shall not do here, that if the solution u of the elliptic equation is given, then it determines the analytic function $\varphi(z)$ in a unique way, in accordance with the formula

$$\varphi(z) = 2u\left(\frac{z}{2}, \frac{z}{2i}\right) - u(0, 0)\,\alpha(z, 0) \qquad (38.16)$$

(we have set here $z_0 = \bar{z}_0 = 0$).

The derivation of this formula is given in the book of I. N. Vekua [30], § 12.

In representation (38.15), since it follows from the derivation, the function α is expressed by integrals of the coefficients of the equation, while the function β is found with the help of the procedure of successive approximations. However, important particular cases can be indicated, when the infinite series of successive approximations can be summed; then the function β is also expressed by some known functions (see Problems 16, 17, 18 at the end of this chapter).

The above derivation of the integral representation, based directly on the method of successive approximations, was presented by I. N. Vekua in his first papers on this topic. In his later papers, the summary of which is given in the book [30], to derive the integral representation the theory of hyperbolic equations is essentially employed, in particular Riemann's method of integration of such equations. Observe that, except for some particular cases when the application of this theory makes it possible to simplify the determination of the representation, in general the basis of all other methods is the method of successive approximations and the other methods are only a form of the latter procedure.

38.4. The general boundary value problem

For a general equation of elliptic type the same boundary value problems can be formulated as for the Laplace equation. Let us state the general boundary value problem first announced and solved by I. N. Vekua [30] (§ 21).

In the domain D bounded by a simple smooth closed contour L it is required to determine a function satisfying in D the differential equation of elliptic type

$$\Delta u + a\frac{\partial u}{\partial x} + b\frac{\partial u}{\partial y} + cu = 0,\qquad(38.17)$$

where a, b, c are functions analytic in the domain D, continuous in D + L, and satisfying the boundary condition

$$\sum_{\substack{j,k=0,\ldots,n}}^{j+k\le n}\left[a_{jk}(s)\,u_{jk}(s) + \int b_{jk}(s,\sigma)\,u_{jk}(\sigma)\,d\sigma\right] = f(s).\quad(38.18)$$

The functions a_{jk}, f satisfy the Hölder condition, b_{jk} are Fredholm kernels and the notation

$$u_{jk} = \frac{\partial^{j+k}u}{\partial x^j\,\partial y^k}\quad(u_{00} = u)$$

is employed.

Inserting into the boundary condition the value of u_{jk} from the integral representation (38.15) we arrive at the boundary value problem (34.2) for the analytic function $\varphi(z)$. We have already described the methods of solution of this problem in § 34.

In the book of I. N. Vekua [30] which has frequently been quoted, the above problem is formulated for the case of a multiply-connected domain. The solution is based a given integral representation for solutions of equation of an elliptic type, in the case of a multiply-connected domain.

§ 39. Boundary value problems for systems of elliptic equations

39.1. Various forms of the system and general remarks

Consider the elliptic system† of differential equations of first order

$$\left.\begin{array}{l}\dfrac{\partial u}{\partial x} - \dfrac{\partial v}{\partial y} = a(x,y)\,u + b(x,y)\,v + f(x,y),\\[2mm]\dfrac{\partial u}{\partial y} + \dfrac{\partial v}{\partial x} = c(x,y)\,u + d(x,y)\,v + g(x,y).\end{array}\right\}\qquad(39.1)$$

† The system of equations $a_{11}u_x + a_{12}v_y = F_1(x,y;u,v)$, $a_{21}u_y + a_{22}v_x = F_2(x,y;u,v)$ is said to be elliptic if the determinant $\begin{vmatrix} a_{11} & a_{12} \\ a_{21} & a_{22} \end{vmatrix}$ is a positive-definite or negative-definite function.

13 a*

It constitutes a natural generalization of the Cauchy–Riemann system which is obtained when $a = b = c = d = f = g = 0$.

It is convenient to investigate system (39.1) in the complex form. Multiply, therefore, the second equation by i and add it to the first. Introducing the notation

$$\frac{\partial}{\partial \bar{z}} = \frac{1}{2} \left(\frac{\partial}{\partial x} + i \frac{\partial}{\partial y} \right), \tag{39.2}$$

we write system (39.1) in the form of one complex equation

$$\frac{\partial U}{\partial \bar{z}} = AU + B\bar{U} + F, \tag{39.3}$$

where

$$\left. \begin{array}{l} U = u + iv, \quad A = \tfrac{1}{4}(a + d + ic - ib), \\ B = \tfrac{1}{4}(a - d + ic + ib), \quad F = \tfrac{1}{2}(f + ig). \end{array} \right\} \tag{39.4}$$

Conversely, separating in (39.3) the real and imaginary parts we return to system (39.1). The functions a, b, c, d, f, g and consequently also A, B, F are hereafter regarded as continuous in the considered domain.

If $A = B = F = 0$ we obtain the equation

$$\frac{\partial U}{\partial \bar{z}} = 0, \tag{39.5}$$

which constitutes a system of two Cauchy–Riemann equations in complex form. Thus, the function

$$U = u + iv,$$

which satisfies equation (39.5) is an ordinary analytic function of the complex variable $z = x + iy$. If A, B, F do not simultaneously vanish, equation (39.3) defines complex functions of a much more general class. It will be indicated later that these functions possess many essential properties of analytic functions.

Let us make some general remarks concerning the contents of the present section. Problems constituting parts of the main topic of this book—the solution of boundary value problems—will be described in detail, using the above theory of boundary value problems for analytic functions. For the reader's convenience we shall also present the required auxiliary material (integral representations). Sometimes we shall indicate only the logical connection, without presenting a detailed proof. Always in these cases references will be indicated, in which the relevant proofs may be found; mainly it will be a paper of I. N. Vekua [30*].

39.2. Functions of class $C_{\bar{z}}$

According to formula (39.2) the operation $\partial/\partial z$ is determined in terms of the derivatives with respect to x and y and hence it is meaningful only when the latter derivatives exist. We now define $\partial/\partial z$ as an independent operation with no reference to formula (39.2); it may turn out that it is meaningful also when the derivatives $\partial/\partial x$, $\partial/\partial y$ do not exist.

To generalize the notion of the operation $\partial/\partial \bar{z}$ we shall apply the device familiar in mathematics, which consists in making use of integral identities. Those identities are closely connected with physical ideas.

Let us agree on the notation. Suppose that D is the domain in which the equation (39.3) is investigated, and L is its contour. For the domain D we shall take a finite domain bounded by $m + 1$ simple non-intersecting contours (in § 16 it was called D^+).

As before, the complex variables on the contour will be denoted by the letters t, τ and the same variables in the domain D by $z = x + iy, \zeta = \xi + i\eta$.

If we first define the operation $\partial/\partial \bar{z}$ in accordance with formula (39.2) it is easy to derive the celebrated integral identity (Ostrogradski's formula) relating the integral over the domain to the integral over the contour

$$\int\int_{D_1} \frac{\partial U}{\partial \bar{\zeta}} \, d\xi \, d\eta = \frac{1}{2i} \int_{L_1} U(\tau) \, d\tau, \qquad (39.6)$$

where D_1 is the domain located wholly in the domain D and L_1 is its boundary. Suppose now that the curve L_1 contracts to the point $z = x + iy$. Applying to the left-hand side of relation (39.6) the theorem of the mean value and passing to the limit we obtain

$$\frac{\partial U}{\partial \bar{z}} = \lim_{L_1 \to z} \frac{1}{2i\,|D_1|} \int_{L_1} U(\tau) \, d\tau, \qquad (39.7)$$

where $|D_1|$ is the area inside the contour L_1.

Let us now abandon the original assumption of the existence of the derivatives $\partial/\partial x$, $\partial/\partial y$, on the basis of which formulae (39.6), (39.7) have been derived. We regard the above reasoning only as heuristic and define the operation $\partial/\partial \bar{z}$ by relation (39.7).

Functions for which the limit in the right-hand side of relation (39.7) *exists and is continuous will be called the functions of class $C_{\bar{z}}$.*

By means of examples it can be proved that the class $C_{\bar{z}}$ is wider than the class of functions possessing continuous partial derivatives (see Problem 23 at the end of this chapter).

The above defined class of functions $C_{\bar{z}}$ is important, since it contains solutions of equation (39.3) the coefficients of which are only continuous; the latter assumption would be insufficient if we required the continuity of $\partial u/\partial x$, $\partial u/\partial y$.

I. N. Vekua called the solutions of equation (39.3) for which the operation $\partial/\partial\bar{z}$ exists only in the sense of definition (39.7) *regular solutions*, and these for which partial derivatives $\partial u/\partial x$, $\partial u/\partial y$ exist *completely regular solutions*.

39.3. The fundamental solution

Consider a double integral over the domain D, with Cauchy kernel

$$U(z) = -\frac{1}{\pi}\iint\limits_{D}\frac{f(\xi,\eta)}{\zeta-z}\,d\xi\,d\eta, \tag{39.8}$$

where $f(x, y)$ is continuous in D.

Since the kernel of the integral is a function analytic everywhere if $\zeta - z$ does not vanish, the integral represents a function analytic outside D and vanishing at infinity. In the domain D it is evident that the function $U(z)$ is continuous. It can be proved that it obeys the condition

$$|U(z_1) - U(z_2)| \leqq M|z_1 - z_2|\,|\ln\alpha|z_1 - z_2||, \tag{39.9}$$

where M, α are positive constants.

It follows from inequality (39.9) that $U(z)$ satisfies the Hölder condition with exponent arbitrarily close to unity.

In the theory of the equations considered, function (39.8) plays the same role as the plane logarithmic potential for the Laplace equation.

THEOREM. *Function* (39.8) *belongs to the class $C_{\bar{z}}$ and satisfies the equation*

$$\frac{\partial U}{\partial \bar{z}} = f(x, y). \tag{39.10}$$

To prove the theorem let us divide integral (39.8) into two: $U_1(z)$ taken over a subdomain containing the point z and bounded by the contour L_1, and U_2 over the remaining part of the domain. We have

$$\frac{1}{2i|D_1|} \int_{L_1} U(z) \, dz = \frac{1}{2i|D_1|} \int_{L_1} U_1(z) \, dz + \frac{1}{2i|D_1|} \int_{L_1} U_2(z) \, dz.$$

By virtue of the Cauchy theorem the second integral vanishes. The first integral does not alter when the contour outside the domain D_1 is deformed. Hence we obtain

$$\frac{1}{2i|D_1|} \int_{L_1} U(z) \, dz = \frac{1}{2i|D_1|} \int_{\Gamma} U_1(z) \, dz$$

$$= \frac{1}{2\pi i |D_1|} \iint_{D_1} f(\xi, \eta) \, d\xi \, d\eta \int_{\Gamma} \frac{dz}{z - \zeta} = \frac{1}{|D_1|} \iint_{D_1} f(\xi, \eta) \, d\xi \, d\eta,$$

where Γ is a circle of a sufficiently large radius.

Passing to the limit, when $D_1 \to z$, we find in view of the mean value theorem, that the limit of the last integral is $f(x, y)$. Consequently, formula (39.7) implies that relation (39.10) holds. The solution given by formula (39.8) will be called *the fundamental solution* of equation (39.10).

Formula (39.10) is very important on account of its further applications. The above proved theorem is completely analogous to the theorem of the potential theory asserting that the plane logarithmic potential is a particular solution of the Poisson equation $\Delta u = f$.

It is known that the general solution of a non-homogeneous equation is composed of the general solution of the homogeneous equation and a particular solution of the non-homogeneous one. Since the solutions of the homogeneous equation $\partial U/\partial \bar{z} = 0$ corresponding to the non-homogeneous equation (39.10) is given by an arbitrary analytic function $\Phi(z)$, it follows that the general solution of equation (39.10) is given by the formula

$$U(z) = \Phi(z) - \frac{1}{\pi} \iint_D \frac{f(\xi, \eta)}{\zeta - z} \, d\xi \, d\eta, \tag{39.11}$$

which is analogous to the representation of a solution of the Poisson equation in the form of a sum of a harmonic function and the plane logarithmic potential.

It can be proved that under the condition of continuity of U in $D + L$ the analytic function $\Phi(z)$ can be represented in the form of a Cauchy type integral the density of which is the function U itself,

$$\Phi(z) = \frac{1}{2\pi i} \int_L \frac{U(\tau)}{\tau - z}\, d\tau. \qquad (39.12)$$

Replacing f by its value from equation (39.10) we obtain the following representation of functions of class $C_{\bar{z}}$:

$$U(z) = \frac{1}{2\pi i} \int_L \frac{U(\tau)}{\tau - z}\, d\tau - \frac{1}{\pi} \int\!\!\int_D \frac{\partial U}{\partial \bar{\zeta}}\, \frac{d\xi\, d\eta}{\zeta - z}. \qquad (39.13)$$

A rigorous derivation of the latter formula under the general definition of the class $C_{\bar{z}}$ is of considerable difficulty and to save space we do not quote it (see I. N. Vekua [29] p. 225–6).

39.4. The normal form of the system

Let us consider a particular case of system (39.1)

$$\frac{\partial u}{\partial x} - \frac{\partial v}{\partial y} = au + bv, \quad \frac{\partial u}{\partial y} + \frac{\partial v}{\partial x} = -bu + av, \qquad (39.14)$$

or, in the complex form

$$\frac{\partial U}{\partial \bar{z}} = AU, \quad A = \frac{1}{2}(a - ib), \quad U = u + iv. \qquad (39.15)$$

We shall now prove that the function

$$U_0 = e^{\omega_0(z)}, \quad \omega_0(z) = -\frac{1}{\pi} \int\!\!\int_D \frac{A(\xi, \eta)}{\zeta - z}\, d\xi\, d\eta \qquad (39.16)$$

is a solution of equation (39.15).

The proof consists in direct differentiation†. According to the rule of differentiation of an implicit function we have

$$\frac{\partial U_0}{\partial \bar{z}} = e^{\omega_0} \frac{\partial \omega_0}{\partial \bar{z}} = U_0 A,$$

which completes the proof.

---

† It is first proved that for the operation $\partial/\partial \bar{z}$ in the class $C_{\bar{z}}$ the ordinary formulae of differentiation of a product and quotient hold, and moreover the theorem stating that if $\Phi(z)$ is a known analytic function and U is a function of class $C_{\bar{z}}$, then the following formula for differentiation of an implicit function holds:

$$\frac{\partial \Phi(U)}{\partial \bar{z}} = \Phi'(U) \frac{\partial U}{\partial \bar{z}}.$$

Take the solution of equation (39.15) in the form

$$U(z) = \varphi(z)\, U_0(z),$$

where U_0 is the solution just found and $\varphi(z)$ is a function at present unknown. Substituting into equation (39.15) and applying the rule of differentiation of a product and of an implicit function we find that $\varphi(z)$ is a solution of the equation

$$\frac{\partial \varphi}{\partial \bar{z}} = 0,$$

i.e. it is an ordinary analytic function of the complex variable z. It follows that any solution of equation (39.15) can be represented in the form

$$U(z) = \varphi(z)\, e^{\omega_0(z)}, \tag{39.17}$$

where $\varphi(z)$ is an arbitrary analytic function, and $\omega_0(z)$ is given by formula (39.16).

Making use of the above solution of the particular equation, the general equation (39.3) can be reduced to a simpler form without the term AU.

We seek the solution of the equation

$$\frac{\partial U}{\partial \bar{z}} = AU + B\bar{U} + F \tag{39.3}$$

in the form

$$U = V e^{\omega(z)}, \quad \omega(z) = -\frac{1}{\pi} \iint\limits_{D} \frac{A(\xi, \eta)}{\zeta - z}\, d\xi\, d\eta. \tag{39.18}$$

Applying the rule of differentiation of a product and an implicit function we find that V satisfies the equation

$$\frac{\partial V}{\partial \bar{z}} = C(z)\bar{V} + H(z), \tag{39.19}$$

where

$$C = B e^{\bar{\omega} - \omega}, \quad H = F e^{-\omega}. \tag{39.20}$$

Equation (39.19) is equivalent to the system of the form

$$\frac{\partial u}{\partial x} - \frac{\partial v}{\partial y} = au + bv + f, \quad \frac{\partial u}{\partial y} + \frac{\partial v}{\partial x} = bu - av + g \tag{39.21}$$

which is said to be *normal*. We shall hereafter consider the system in its *normal* form assuming that the term AU has been eliminated by means of the method given above.

39.5. An auxiliary representation of the solutions and its implications

Consider the homogeneous equation in its normal form

$$\frac{\partial U}{\partial \bar{z}} = A\bar{U}. \tag{39.22}$$

Denoting

$$A\frac{\bar{U}}{U} = A_0, \tag{39.23}$$

we can write it in the form

$$\frac{\partial U}{\partial \bar{z}} = A_0 U. \tag{39.22'}$$

This equation is of the form (39.15) of the particular case, the only difference being the fact that the coefficient A_0 depends on U. Consider the function

$$\omega(z) = -\frac{1}{\pi} \iint\limits_{D} \frac{A_0(\xi, \eta)}{\zeta - z} \, d\xi \, d\eta. \tag{39.24}$$

It belongs to the class $C_{\bar{z}}$ everywhere in D except for the zeros of U. Let us enclose the zeros in small circles of radii ε and denote the remaining domain by D_ε. On account of formula (39.17) we have in the domain D_ε

$$U(z) = \varphi_\varepsilon(z) \, e^{\omega_\varepsilon(z)}, \tag{39.25}$$

where $\varphi_\varepsilon(z)$ is an arbitrary analytic function and $\omega_\varepsilon(z)$ is a function of class $C_{\bar{z}}$ defined in the domain D_ε by the integral (39.24).

We now pass to the limit, the radii ε of the circles tending to zero. Denoting $U = \varrho e^{i\theta}$ we have $\bar{U}/U = e^{-2i\theta}$. Hence, at the points where $U = 0$ the function A_0 is bounded although it does not tend to any definite limit.

Integral (39.24) is a function continuous in the entire domain D†. Consequently, in the limit, (39.25) is reduced to the following:

$$U(z) = \varphi(z) \, e^{\omega(z)}. \tag{39.26}$$

Here $\varphi(z)$ is a function analytic in the domain D and $\omega(z)$ is a function of class $C_{\bar{z}}$ defined by formula (39.24)‡.

Formula (39.26) yields a representation of the solution of equation (39.22). Although the right-hand side of it contains the unknown

---

† The reasoning is also valid for points at which the function U has poles.

‡ I. N. Vekua (5) extended formula (39.26) to the case of an infinite domain, assuming that $A(z)$ is a continuous function.

function U, it is of a great importance in a qualitative investigation of boundary value problems. We shall now present an application of this formula to an examination of the properties of the function U.

PRINCIPLE OF THE ARGUMENT. *Suppose that U is a solution of equation (39.22) regular everywhere in D, except at a finite number of points where it has poles. If U is continuous and does not vanish on the contour L the relation*

$$N - P = \frac{1}{2\pi} \{\arg U(t)\}_L, \qquad (39.27)$$

is valid. Here N and P are the numbers of the zeros and poles of the function U in D, their multiplicities being taken into account.

To prove this assertion we employ the representation (39.26). Since the function $e^{\omega(z)}$ is regular and has no zeros, the zeros and poles of U coincide with the zeros and poles of $\varphi(z)$. In view of the principle of argument for an analytic function, we have

$$N - P = \frac{1}{2\pi} \{\arg \varphi(t)\}_L.$$

But, in accordance with formula (39.26), the following relation holds:

$$\{\arg U(t)\}_L = \{\arg \varphi(t)\}_L + \{\operatorname{Im} \omega(t)\}_L.$$

In view of the single-valuedness of $\operatorname{Im} \omega(t)$ the last term of the right-hand side vanishes. This completes the proof of the principle of the argument (39.27).

Finally, we shall prove an auxiliary statement playing the same role in the theory of functions of class $C_{\bar{z}}$ as the Liouville theorem for analytic functions.

BASIC LEMMA. *If U is a solution of equation (39.22) regular in D and continuous on the contour, and it takes on the contour the values of a function analytic in D + L and vanishing at infinity, then U = 0 everywhere in D.*

We again employ representation (39.26). Since, according to formula (39.24) the function $\omega(z)$ is analytic outside $D + L$, then the function $\varphi(z) = Ue^{-\omega(z)}$ takes on L the values of a function analytic outside $D + L$. On the other hand, according to the definition, it constitutes a function analytic in D. Hence, in view of the theorem of analytic continuation and the Liouville theorem for the analytic functions we find that $\varphi(z) \equiv 0$. Consequently, also $U = 0$ in the domain of validity of representation (39.26), i.e. in D.

Let us now state one more assertion of the same kind, established by I. N. Vekua (5), which to an even greater degree resembles the Liouville theorem.

COUNTERPART OF THE LIOUVILLE THEOREM. *Suppose that $U(z)$ is a solution of equation* (39.22), *regular in the entire domain, except at a finite number of rectifiable Jordan curves and a finite number of points. If $U(z)$ is continuous and bounded in the entire domain and vanishes at one point, then it vanishes identically.*

As before, the proof is based on formula (39.26), the only difference being that this formula is now applied to an infinite domain.

We note without proof one more important property.

A solution of equation (39.22) *which is regular in D, cannot have a limit point of zeros inside the domain.*

Similarly for the case of analytic functions, this implies the principle of uniqueness.

If two solutions of equation (39.22) *regular in the domain D coincide on an infinite set of points, having in D a limit point, then these solutions are identical.*

39.6. Integral representation of solutions

For simplicity we take the homogeneous equation and assume that it has already been reduced to the normal form

$$\frac{\partial U}{\partial \bar{z}} = A\bar{U}. \tag{39.22}$$

Let us construct an integral equation for the function $U(z)$. For this purpose we make use of formula (39.13), $\partial U/\partial \bar{z}$ being given by relation (39.22). Thus, we obtain the integral equation

$$U(z) + \frac{1}{\pi} \int\int_D \frac{A(\xi, \eta)\,\bar{U}(\xi, \eta)}{\zeta - z}\, d\xi\, d\eta = \Phi(z), \tag{39.28}$$

where $\Phi(z)$ is an arbitrary function analytic in D.

We now prove that the non-homogeneous integral equation (39.28) is absolutely soluble. To this end consider the corresponding homogeneous equation

$$U(z) + \frac{1}{\pi} \int\int_D \frac{A(\zeta)\,\overline{U(\zeta)}}{\zeta - z}\, d\xi\, d\eta = 0, \tag{39.28'}$$

which we can write in the form

$$U(z) = -\frac{1}{\pi} \int \int_D \frac{A(\zeta)\,\overline{U(\zeta)}}{\zeta - z}\, d\xi\, d\eta,$$

and suppose that U is one of its solutions. We know that the integral

$$-\frac{1}{\pi} \int\int \frac{A(\zeta)\,\overline{U(\zeta)}}{\zeta - z}\, d\zeta\, d\eta$$

defines a function which is continuous in the entire domain, analytic in D^- and vanishing at infinity. Consequently, in view of equation (39.28′) itself, $U(z)$ takes on L the values of a function analytic outside $D + L$ and vanishing at infinity. Moreover, it is clear from the derivation of equation (39.28′) that the function $U(z)$ which satisfies integral equation (39.28′) is also a solution of equation (39.22).

According to the basic lemma $U(z) \equiv 0$.

Thus, the homogeneous equation (39.28′) is insoluble and this implies the absolute solubility of the non-homogeneous integral equation (39.28), for an arbitrary analytic function $\varPhi(z)$.

Replacing A by λA we obtain the same result for the solubility of this equation. Consequently, the equation has no eigenfunctions which implies that it can be solved by the method of successive approximations. Applying this method we arrive at the formula

$$U(z) = \varPhi(z) + \int\int_D \varGamma_1(z,\zeta)\,\varPhi(\zeta)\, d\xi\, d\eta + \int\int_D \varGamma_2(z,\zeta)\,\overline{\varPhi(\zeta)}\, d\xi\, d\eta,$$

$$\tag{39.29}$$

where

$$\varGamma_1(z,\zeta) = \sum_{j=1}^{\infty} K_{2j}(z,\zeta), \quad \varGamma_2(z,\zeta) = \sum_{j=1}^{\infty} K_{2j+1}(z,\zeta), \quad (39.30)$$

$$K_1(z,\zeta) = -\frac{A(\xi,\eta)}{\pi(\zeta - z)}; \quad K_n(z,\zeta) = \int\int_D K(z,\zeta_1)\,K_{n-1}(\zeta_1,\zeta)\, d\xi_1\, d\eta_1.$$

$$\tag{39.31}$$

The functions $\varGamma_1$, $\varGamma_2$ are called the *resolvents* of the integral equation (39.28).

Formula (39.39) yields an integral representation of all solutions of equation (39.22) by an arbitrary analytic function $\varPhi(z)$. It constitutes the basic device in solving boundary value problems for the considered type of equations.

The above derived integral representation leads directly to an integral representation of the solutions of the non-homogeneous equation

$$\frac{\partial U}{\partial \bar{z}} = A\bar{U} + F.$$

For this purpose it is sufficient to replace in formula (39.29) $\Phi(z)$ by

$$\Phi(z) - \frac{1}{\pi} \iint\limits_{D} \frac{F(\xi, \eta)}{\zeta - z} \, d\xi \, d\eta.$$

Let us present one more formula, constituting the counterpart of the Cauchy formula for analytic functions. Introducing into formula (39.29) the Cauchy type integral (39.12) instead of the function $\Phi(z)$ we obtain

$$U(z) = \frac{1}{2\pi i} \int\limits_{L} \Omega_1(z, \tau) \, U(\tau) \, d\tau - \frac{1}{2\pi i} \int\limits_{L} \Omega_2(z, \tau) \, \overline{U(\tau)} \, d\tau, \quad (39.32)$$

where $\Omega_1(z,\tau)$, $\Omega_2(z,\tau)$ are completely definite functions which we do not quote here. The function $U(z)$ unknown in the domain D is expressed in terms of its values on the contour of the domain; hence formula (39.32) is the counterpart of the Cauchy formula for analytic functions.

39.7. Boundary value problems

For functions of class $C_{\bar{z}}$ which satisfy system (39.1) or, which is the same, equation (39.3), we can formulate the same boundary value problems as for analytic functions of the complex variable z. We assume that by means of substitution (39.18) we have already reduced the system to the normal form, so that in what follows we shall deal only with the latter system. For simplicity of formulation we confine ourselves to the homogeneous equation.

Consider the boundary value problem of Hilbert type.

It is required to determine in the domain D a solution regular in this domain and continuous in $D + L (L = L_0 + L_1 + \cdots + L_m)$, $U = u + iv$, of the equation

$$\frac{\partial U}{\partial \bar{z}} = A\bar{U}, \quad (39.22)$$

which satisfies the boundary condition

$$\operatorname{Re}\left[(\alpha - i\beta)\, U\right] \equiv au + \beta v = \gamma, \quad (39.33)$$

where α, β, γ are functions given on L and satisfying the Hölder condition. The curves L_j are assumed to be the Lyapounov curves.

The solution of the problem is carried out by reducing it to boundary value problems for analytic functions. For this purpose we make use of the representations of solutions (39.26) and (39.29) deduced in this section.

Making use of the integral representation (39.29) we reduce the above formulated boundary value problem to the boundary value problem for the analytic function $\Phi(z)$:

$$\text{Re}\left\{(\alpha - i\beta)\left[\Phi(t) + \int\int_D \Gamma_1(t, \zeta)\,\Phi(\zeta)\,d\xi\,d\eta + \right.\right.$$
$$\left.\left. + \int\int_D \Gamma_2(t, \zeta)\,\overline{\Phi(\zeta)}\,d\xi\,d\eta\right]\right\} = \gamma. \quad (39.34)$$

This boundary value problem of analytic functions is then reduced to an integral equation. To this end it is necessary to employ one of the integral representations for analytic functions. We may then, as does I. N. Vekua, making use of the integral representation (36.7)

$$\Phi(z) = \frac{1}{2\pi i}\int_L \frac{\mu(\sigma)}{\tau - z}\,d\tau + iC,$$

reduce the problem to the singular integral equation

$$\text{Re}\left\{(\alpha - i\beta)\left[\mu(s) + \frac{1}{\pi i}\int \frac{\mu(\sigma)}{\tau - t}\,d\tau\right]\right\} + \int_L K(s, \sigma)\,\mu(\sigma)\,d\sigma = \gamma_1,$$
$$(39.35)$$

where $K(s, \sigma)$ is a completely defined Fredholm kernel.

Observe that the dominant part of this equation is identical with that in equation (39.39) obtained in § 36.6 for the solution of the ordinary Hilbert problem for a multiply-connected domain.

Making use of the integral representation of D. I. Sherman

$$\Phi(z) = \frac{1}{2\pi i}\int_L \frac{\mu(\sigma)}{a(\tau) - ib(\sigma)}\frac{d\tau}{\tau - z} + P(z),$$

we can also reduce the problem to a Fredholm integral equation. The considerations required for the justification of this fact are the same as in § 34.5, in deriving equation (34.20).

We now proceed to investigate the problems of solubility of the boundary value problem (39.33). The simplest way is to employ representation (39.26) and to reduce the problem to one already solved—to an investigation of the solubility of the ordinary Hilbert problem.

Substituting into the boundary condition (39.33) the expression for U from formula (39.26) we obtain

$$\text{Re}\,[e^{\omega}(\alpha - i\beta)\,\varphi(t)] = \gamma.$$

Writing $e^{\omega}(\alpha - i\beta) = h(t)$ we arrive at the Hilbert boundary value problem

$$\text{Re}\,[h(t)\,\sigma(t)] = \gamma \tag{39.36}$$

for the analytic function $\varphi(z)$.

Since

$$\omega(z) = -\frac{1}{\pi}\iint\limits_{D} A\,\frac{\overline{U}}{U}\,\frac{d\xi\,d\eta}{\zeta - \zeta}$$

depends itself on U, the boundary value problem (39.36) can only serve as a device for a qualitative examination of the original problem. Although the function $h(t)$ itself is unknown, its index can be calculated exactly. In fact,

$$\text{Ind}\,h(t) = \text{Ind}\,(\alpha - i\beta) + \text{Ind}\,e^{\omega} = \text{Ind}\,(\alpha - i\beta)$$

(see the derivation of the principle of the argument in § 39.5).

Thus, the index of the Hilbert problem (39.36) is equal to the index of the original problem (39.33). Let us establish a connection between the solutions of these problems.

To every solution U of the given problem (39.33) there corresponds in accordance with formula (39.26) a definite solution of the Hilbert boundary value problem (39.36). We now prove the converse, i.e. that to every solution $\varphi(z)$ of the Hilbert problem (39.36) there corresponds one solution of the original problem (39.33). The question is to prove that the formula $U = \varphi e^{\omega}$ is uniquely invertible with respect to U.

We now proceed to prove this assertion.†

Denote the supplement of $D + L$ to the entire plane by D^{-}. First note that the formula

$$\omega(z) = -\frac{1}{\pi}\iint\limits_{D} A(\zeta)\,\frac{\overline{U(\zeta)}}{U(\zeta)}\,\frac{d\xi\,d\eta}{\zeta - z} \tag{39.37}$$

† The proof was announced by L. G. Mikhailov (2); its concept was given by I. N. Vekua (5).

determines the function $\omega(z)$ not only in the domain D where the function $U(z)$ is defined, but in the entire plane. It was already mentioned in § 39.3 that $\omega(z)$ is analytic in D^- and vanishes at infinity. On passing through the contour L it is continuous and in the domain D it satisfies the condition

$$\frac{\partial_\omega}{\partial \bar{z}} = A \frac{\overline{U}}{U}. \tag{39.38}$$

After this observation let us invert formula (39.26). It is sufficient to find $e^{\omega(z)}$. On the basis of the above properties of the function $\omega(z)$ we should require that the unknown function $e^{\omega(z)} = V(z)$ satisfies the following conditions:
(1) it is continuous in the entire plane,
(2) it is analytic in D^- and satisfies the condition $V(\infty) = 1$,
(3) it satisfies in the domain D the equation

$$\frac{\partial V}{\partial \bar{z}} = A(z) \frac{\overline{\varphi(z)}}{\varphi(z)} \overline{V} \tag{39.39}$$

(this latter equation is derived by differentiating the relation $U = \varphi V$ with respect to $\bar{z}$ and taking into account equation (39.22)).

Equation (39.39) is of the same kind as the original equation (39.22), the coefficient however being different.

Thus, we have arrived at the following problem: to find a solution of equation (39.39) in the domain D^+, which is continuously continuable into the domain D^- as an analytic function and has the value unity at infinity.

It follows from the considerations of § 39.6 that the determination of $V(z)$ is equivalent to the solution of the integral equation

$$V(z) + \frac{1}{\pi} \iint\limits_D A(\zeta) \frac{\overline{\varphi(\zeta)}}{\varphi(\zeta)} \overline{V(\zeta)} \frac{d\xi\, d\eta}{\zeta - z} = \Phi(z),$$

where $\Phi(z)$ is a function analytic in D^+.

Since the left-hand side of the equation is analytic in D^-, $\Phi(z)$ is also analytic in this domain. Further, substituting $z = \infty$ in both sides of the equation we obtain $\Phi(\infty) = 1$. Hence, according to the Liouville theorem $\Phi(z) \equiv 1$.

Consequently, $V(z)$ is a solution of the following integral equation:

$$V(z) + \frac{1}{\pi} \iint\limits_D A(\zeta) \frac{\overline{\varphi(\zeta)}}{\varphi(\zeta)} \overline{V(\zeta)} \frac{d\xi\, d\eta}{\zeta - z} = 1.$$

Solving the latter equation by the method of successive approximations we obtain

$$V(z) = e^{\omega(z)} = 1 + \iint_D \Gamma_1^{\varphi}(z, \zeta)\, d\xi\, d\eta + \iint_D \Gamma_2^{\varphi}(z, \zeta)\, d\xi\, d\eta, \quad (39.40)$$

where $\Gamma_1^{\varphi}, \Gamma_2^{\varphi}$ are resolvents deduced from formulae (39.30) by replacing A by $A\bar{\varphi}/\varphi$. Substituting the solution into formula (39.26) we have

$$U(z) = \varphi(z)\left[1 + \iint_D \Gamma_1^{\varphi}(z, \zeta)\, d\xi\, d\eta + \iint_D \Gamma_2^{\varphi}(z, \zeta)\, d\xi\, d\eta\right]. \quad (39.41)$$

The last formula associates with every analytic function $\varphi(z)$ a definite function $U(z)$. Thus, there exists a one to one correspondence between the solutions of the original boundary value problem (39.33) and the Hilbert problem (39.36).

On this basis, we can apply to the original boundary value problem (39.33) all results concerning the Hilbert problem (39.36) which follow from the knowledge of its index, i.e. the results of § 36.8 for $\varkappa = \mathrm{Ind}(\alpha + i\beta) < 0$ and $\varkappa > m - 1$. This is exactly the result derived by I. N. Vekua [29] on the basis of a direct investigation of the integral equation (39.35). There is no purpose in considering the application to the original problem of the properties of the Hilbert problem concerning the cases $\varkappa = 0$ and $\varkappa = m - 1$, because the question of whether or not the conditions of single-valuedness are satisfied for the problem (39.36) cannot be answered; in fact we do not know the function $h(t)$. To investigate the given boundary value problem (39.33) in these cases we have to consider the integral equation (39.35).

39.8. Riemann boundary value problem. Formulation and auxiliary relationships

As before we shall here consider the homogeneous equation (39.22) in which we assume $A(z)$ to be continuous in the entire plane except on passing through the contour L, the discontinuities of the first kind then being possible. We assume also that it satisfies the condition $|A(z)| < M/|z|^{\alpha}$, $\alpha > 1$ in the vicinity of infinity. The contour L and the domains D^+, D^- are the same as in § 16 (Fig. 12).

In accordance with the formulation of the Riemann problem, it is required *to find a pair of solutions of equation* (39.22) $U^+(z)$, $U^-(z)$ *which are regular in the domains* D^+, D^-, *respectively, continuous up*

to the contour L, and the boundary values of which, on L, satisfy the linear relation
$$U^+(t) = G(t) U^-(t) + g(t), \qquad (39.42)$$

where $G(t)$, $g(t)$ are known functions, of the position on the contour, which satisfy the Hölder condition and, moreover, $G(t) \neq 0$.

The above pair of functions will for brevity be called the *sectionally regular solution.*

We are faced with a homogeneous or a non-homogeneous problem, depending on whether $g(t)$ vanishes or not.

The method of solution of this problem is in many respects analogous to that of the Riemann problem for analytic functions (see § 14). As in that case the basis is the solution of the two simplest problems, namely: $U^+ = U^-$, the problem of a continuous continuation, and $U^+ - U^- = g$, the problem of the jump. To solve these problems we shall need some auxiliary relations, to the derivation of which we now proceed. In contrast to the theory described above we shall have to consider the solutions of equation (39.22) defined in the infinite domain.

1. *The counterpart of the Cauchy type integral.* If in formula (39.29) the analytic function $\Phi(z)$ be replaced by the Cauchy type integral
$$\Phi(z) = \frac{1}{2\pi i} \int_L \frac{\varphi(\tau)}{\tau - z} d\tau,$$

where $\varphi(\tau)$ is a function which satisfies the Hölder condition, we arrive at the formula
$$U(z) = \frac{1}{2\pi i} \int_L \Omega_1(z, \tau) \varphi(\tau) d\tau - \frac{1}{2\pi i} \int_L \Omega_2(z, \tau) \overline{\varphi(\tau)} \, \overline{d\tau} \quad (39.43)$$

(see formula (39.32)). The kernels Ω_1, Ω_2 are completely defined functions.

Taking into account the presence in formulae (39.32) of the term $\Phi(z)$, the last relation can be written in the form
$$U(z) = \frac{1}{2\pi i} \int_L \frac{\varphi(\tau)}{\tau - z} d\tau + \Omega(z), \qquad (39.44)$$

$\Omega(z)$ being continuous in the entire plane, and $\Omega(\infty) = 0$. The last form justifies the name the counterpart of the Cauchy type integral, given to formula (39.43).

It can easily be deduced from equation (39.44) that for a function $U(z)$ represented·by the counterpart of the Cauchy type integral, the following formulae analogous to the Sokhotski formulae for the Cauchy type integral are valid:

$$\left.\begin{aligned}
& U^+(t) - U^-(t) = \varphi(t), \\
& U^+(t) + U^-(t) = \\
& \qquad \frac{1}{2\pi i} \int_L \Omega_1(t,\tau)\,\varphi(\tau) - \frac{1}{2\pi i} \int_L \Omega_2(t,\tau)\,\overline{\varphi(\tau)}\,d\bar{\tau}.
\end{aligned}\right\} \quad (39.45)$$

2. *The counterpart of powers.* Let us take for $\varphi(z)$ in formula (39.41) the powers z^k and iz^k. Then we obtain

$$\left.\begin{aligned}
U_{2k}(z) &= z^k \left[1 + \iint_E \Gamma_{1,k}(z,\zeta)\,d\xi\,d\eta + \right. \\
& \qquad\qquad\qquad \left. + \iint_E \Gamma_{2,k}(z,\zeta)\,d\xi\,d\eta \right], \\
U_{2k+1}(z) &= iz^k \left[1 + \iint_E \Gamma^i_{1,k}(z,\zeta)\,d\xi\,d\eta + \right. \\
& \qquad\qquad\qquad \left. + \iint_E \Gamma^i_{2,k}(z,\zeta)\,d\xi\,d\eta \right],
\end{aligned}\right\} \quad (39.46)$$

where $\Gamma_{1,k}$, $\Gamma_{2,k}$ denote the resolvents of the equation $\partial U/\partial \bar{z} = A(\bar{z}^k/z^k)\,U$, $\overline{\Gamma}^i_{1,k}$, $\overline{\Gamma}^i_{2,k}$ the resolvents of the equation $\partial U/\partial \bar{z} = -A(\bar{z}^k/z^k)\,U$, the integrals being taken over the entire plane E.

It can easily be proved that the functions U_k given by formulae (39.46) are linearly independent in the sense of combinations with real coefficients. To prove this assertion, assuming the converse it is sufficient to consider the appropriate linear relation for $z = \infty$. There exists a one to one correspondence between the powers z^k, iz^k, and the functions U_{2k}, U_{2k+1}. Formulae (39.46) determine linearly independent solutions of equation (39.22) which are continuous in the entire domain, except at infinity where they are of order $-k$. It is natural to call these solutions the counterparts of the powers.

3. *The counterpart of the polynomial.* Construct the linear combination of the powers with real coefficients

$$U_P(z) = \sum_{k=0}^{2n+1} A_k U_k(z); \quad (39.47)$$

the function $U_P(z)$ will be called the counterpart of the polynomial. There exists a one to one correspondence between the functions $U_P(z)$ and the polynomials $P(z) = \sum_{k=0}^{n} c_k z^k$ with the coefficients $c_k = A_{2k} + i A_{2k+1}$.

The counterpart of the polynomial is completely described by two properties: the continuity in the entire plane and the prescription of its order at infinity. The validity of this statement can easily be established making use of formula (39.26). On this basis the following theorem can be proved.

The solution of equation (39.22) *continuous in the entire plane and having a finite order at infinity, is identically equal to* $U_P(z)$.

This theorem is the counterpart of the generalized Liouville theorem (see § 13) and constitutes the principal device in deducing the general solution of the Riemann boundary value problem. Particular cases of the theorem are the lemma and the counterpart of the Liouville theorem, which were established in § 39.5.

We now proceed to solve the Riemann problem.

39.9. The solution of the Riemann boundary value problem

We begin by solving the simplest problem of auxiliary nature.

It is required to determine a sectionally regular solution of equation (39.22) *which has at infinity a prescribed order and which satisfies on the contour L the boundary condition*

$$U^+(t) - U^-(t) = g(t), \qquad (39.48)$$

where $g(t)$ is a given function satisfying the Hölder condition.

The general solution of the above formulated problem can be represented as the sum of a particular solution vanishing at infinity and the general solution of the corresponding homogeneous problem. On the basis of the first formula (39.45) and the counterpart of the generalized Liouville theorem the solution is represented in the form of the sum of the counterpart of the Cauchy type integral (39.43) with density $g(t)$ and the counterpart of polynomial (39.47)

$$U(z) = \frac{1}{2\pi i} \int_L \Omega_1(z, \tau) g(\tau) \, d\tau - \frac{1}{2\pi i} \int_L \Omega_2(z, \tau) \overline{g(\tau)} \, d\overline{\tau} + U_P(z).$$
$$(39.49)$$

The Homogeneous Problem.

$$U^+(t) = G(t) U^-(t). \qquad (39.50)$$

According to formula (39.26), taking into account that $\omega(z)$ is continuous in the entire plane we have

$$U^+(t) = \varphi^+(t)\, e^{\omega(t)}, \quad U^-(t) = \varphi^-(t)\, e^{\omega(t)}.$$

Inserting these expressions into the boundary condition we arrive at the Riemann boundary value problem for the sectionally analytic function $\varphi(z)$:

$$\varphi^+(t) = G(t)\, \varphi^+(t). \tag{39.51}$$

In view of the one to one correspondence between the sectionally analytic function $\varphi(z)$ and the sectionally regular function, the considered boundary value problem (39.50) has the same number of solutions as the ordinary Riemann boundary value problem (39.51). Consequently, all results of § 14.3 concerning the solubility of problem (39.51) can also be applied to the boundary value problem under consideration.

Let $X(z)$ be the canonical function of the boundary value problem (39.51). *The index $\varkappa$ of the coefficient $G(t)$ will be called the index of the considered problem (39.50), and $X(z)$ its canonical function.*

To derive an explicit solution of the problem under consideration we employ the inversion formula (39.41). Replacing there the analytic function $\varphi(z)$ by the linearly independent solutions $z^k X(z)$, $iz^k X(z)$ $(k = 0, 1, \ldots, \varkappa)$ (in the sense of combinations with real coefficients) of problem (39.51), we obtain $2\varkappa + 2$ linearly independent solutions of problem (39.50)

$$\left.\begin{aligned}
U_{2k}(z) &= z^k X(z)\left[1 + \iint_E \Gamma_{1,k,\mathrm{x}}(z, \zeta)\, d\xi\, d\eta +\right.\\
&\qquad\qquad \left.+ \iint_E \Gamma_{2,k,\mathrm{x}}(z, \zeta)\, d\xi\, d\eta\right],\\
U_{2k+1}(z) &= iz^k X(z)\left[1 + \iint_E \Gamma^i_{1,k,\mathrm{x}}(z, \zeta)\, d\xi\, d\eta +\right.\\
&\qquad\qquad \left.+ \iint_E \Gamma^i_{2,k,\mathrm{x}}(z, \zeta)\, d\xi\, d\eta\right],
\end{aligned}\right\} \tag{39.52}$$

where $\Gamma_{1,k,\mathrm{x}}$, $\Gamma_{2,k,\mathrm{x}}$ are the resolvents of the equation

$$\frac{\partial U}{\partial \bar{z}} = A(z)\, \frac{\overline{z^k X(z)}}{z^k X(z)}\, \overline{U};$$

$\Gamma^i_{2,k,\mathrm{x}}$, $\Gamma^i_{2,k,\mathrm{x}}$ are the resolvents of the equation

$$\frac{\partial U}{\partial \bar{z}} = -A(z)\, \frac{\overline{z^k X(z)}}{z^k X(z)}\, \overline{U}.$$

THE NON-HOMOGENEOUS PROBLEM.

$$U^+(t) = G(t)\,U^-(t) + g(t). \tag{39.53}$$

To solve this problem we shall make use of the method employed in § 14.5 for the solution of the ordinary Riemann problem. Replacing in the boundary condition the coefficient $G(t)$ by the ratio of the canonical functions we reduce the boundary condition to the form

$$\frac{U^+(t)}{X^+(t)} - \frac{U^-(t)}{X^-(t)} = \frac{g(t)}{X^+(t)}. \tag{39.54}$$

We have arrived at a boundary value problem of the form (39.48). To solve the boundary value problem (39.54) we can use formula (39.54), replacing g by g/X^+ and U by U/X. Taking into account these transformations we have to bear in mind that $V(z) = U(z)/X(z)$ is a solution of the new differential equation

$$\frac{\partial V}{\partial \bar{z}} = A(z)\,\frac{\overline{X(z)}}{X(z)}\,\overline{V}, \tag{39.55}$$

derived by differentiating $V = U/X$ with respect to $\bar{z}$.

Thus, the solution of the boundary value problem (39.53) can be written in the form

$$U(z) = X(z)\,[W(z) + V_P(z)], \tag{39.56}$$

where

$$W(z) = \frac{1}{2\pi i}\int_L \omega_1(z,\tau)\,\frac{g(\tau)}{X^+(\tau)}\,d\tau - \frac{1}{2\pi i}\int_L \omega_2(z,\tau)\,\frac{\overline{g(\tau)}}{X^+(\tau)}\,d\bar{\tau}, \tag{39.57}$$

$$V_P(z) = \sum_{k=0}^{2\varkappa+1} A_k V_k(z), \tag{39.58}$$

$$V_{2k}(z) = z^k\Big[1 + \int\!\!\int_E \Gamma_{1,k,X}(z,\zeta)\,d\xi\,d\eta + \int\!\!\int_E \Gamma_{2,k,X},(z,\zeta)\,d\xi\,d\eta\Big], \tag{39.59}$$

$$V_{2k+1}(z) = iz^k\Big[1 + \int\!\!\int_E \Gamma^i_{1,k,X},(z,\zeta)\,d\xi\,d\eta + \int\!\!\int_E \Gamma^i_{2,k,X}(z,\zeta)\,d\xi\,d\eta\Big],$$

ω_1, ω_2 are definite functions and Γ_1, Γ_2 are the resolvents of the equations

$$\frac{\partial V}{\partial z} = \pm\,A(z)\,\frac{\overline{z^k X(z)}}{z^k X(z)}\,\overline{V}.$$

If the required solution be subject to the condition $U(\infty) = 0$ we have to replace $\varkappa$ by $\varkappa - 1$ in formula (39.58). We shall carry out subsequent considerations under this condition.

If $\varkappa \leqq 0$ we have to set in formula (39.56) $V_P(z) = 0$, since the solution in this case is given by the formula

$$U(z) = X(z) W(z). \tag{39.60}$$

If $\varkappa < 0$ the last formula defines a function which does not satisfy the condition $U(\infty) = 0$. In order to satisfy this condition it is necessary to demand that $W(z)$ has at infinity a zero of order $-\varkappa + 1$. This makes it impossible to derive the condition of solubility of the problem by the same method as in § 14.5 for the ordinary Riemann problem, since $W(z)$ is a non-analytic function and cannot be expanded into power series. We reduce the problem to an investigation of an analytic function. In view of formula (39.26)

$$W(z) = \varphi(z) e^{\omega(z)},$$

where

$$\omega(z) = -\frac{1}{\pi} \iint_E A(\zeta) \frac{\overline{X(\zeta)}}{X(\zeta)} \frac{\overline{W(\zeta)}}{W(\zeta)} \frac{d\xi \, d\eta}{\zeta - z}.$$

Since $e^{\omega(\infty)} = 1$ the zero of $W(z)$ at infinity has the same multiplicity as the zero of the analytic function $\varphi(z)$. It follows from the boundary condition

$$W^+(t) - W^-(t) = \frac{g(t)}{X^+(t)}$$

and from the formula

$$\varphi^+(t) - \varphi^-(t) = \frac{g(t)}{X^+(t)} e^{-\omega(t)}$$

implied by it, that

$$\varphi(z) = \frac{1}{2\pi i} \int \frac{g(\tau)}{X^+(\tau)} e^{-\omega(\tau)} \frac{d\tau}{\tau - z}.$$

Hence we obtain the condition of solubility of the considered non-homogeneous problem, in the form

$$\int \frac{g(\tau)}{X^+(\tau)} e^{-\omega(\tau)} \tau^k \, d\tau = 0 \quad (k = 0, 1, \ldots, -\varkappa - 1).$$

This result can be stated in the form of a theorem analogous to that given at the end of § 14.5.

Thus, from the quantitative standpoint, for the Riemann boundary problem in the considered class of functions we have derived the same results as for the same problem in the class of analytic functions. The case is different as concerns an effective solution. The solution of the ordinary Riemann problem is given in a closed form in terms of Cauchy type integrals. Now, however (although formally the solution also has a closed form), the kernels and resolvents in the integrand have to be calculated from Fredholm integral equations. Thus, finally the boundary value problem is reduced to a number of integral equations which can be solved by the method of successive approximations.

39.10. Additional remarks

$1°$. The problem for the non-homogeneous equation

$$\frac{\partial U}{\partial \bar{z}} = A \bar{U} + F$$

can easily be reduced to the problem considered before for the homogeneous equation, by the substitution $U = U_0 + V$ where

$$U_0 = -\frac{1}{\pi} \int\int_E \Omega_1(z, \zeta) F(\zeta) \, d\xi \, d\eta - \frac{1}{\pi} \int\int_E \Omega_2(z, \zeta) \overline{F(\zeta)} \, d\xi \, d\eta.$$

The boundary condition for the function U is replaced by a boundary condition for the function V which satisfies a homogeneous differential equation. Then, while the initial boundary condition for U was homogeneous, the condition for V is non-homogeneous; thus we can regard the boundary value problem as homogeneous only when both the boundary condition and the differential equation are homogeneous†. In the remaining cases the boundary value problem is non-homogeneous.

$2°$. For the class of functions under consideration we can formulate the generalized boundary value problem

$$U^+ [\alpha(t)] = G(t) U^- (t) + g(t), \tag{39.61}$$

analogous to problem (17.2) for analytic functions, which we considered in § 17. The methods of solution are also similar. The qualitative examination is somewhat more complicated than in § 17, but the results are very similar (L. G. Mikhailov(2)).

$3°$. We can also formulate boundary value problems containing derivatives but their investigation presents serious difficulties. They consist firstly in deducing conditions ensuring the existence of the derivative

$$\frac{\partial U}{\partial \bar{z}} = \frac{1}{2} \left(\frac{\partial U}{\partial x} + i \frac{\partial U}{\partial y} \right)$$

† Except for one special case, when a non-homogeneous equation and a non-homogeneous boundary condition can lead to a homogeneous problem.

in the domain under consideration, where U is the regular solution, secondly in finding the class to which the above function belongs†, and thirdly in the investigation of the problem of existence of the limiting values of this function. So far this topic has been investigated only in the thesis of B. V. Bojarski *Some boundary value problems for an elliptic system with two independent variables* (M.G.U., 1954) who considered a boundary condition containing the first derivative.

§ 40. Historical notes

Historical remarks concerning the general boundary value problem examined in § 34 were made in considering the problem itself and we shall not dwell on this topic here.

The generalized boundary value problem of Riemann type (§ 35) was first investigated by L. G. Magnaradze (2). His solution (it was only published in form of a summary of formulae, with no indication of the method of their derivation) is not complete and very complicated. The solution given in § 35 is due to Yu. M. Krikunov (1), (2). A more detailed elaboration of the problem was given by this author in his thesis (3). M. P. Ganin (4), (5) generalized this problem to the case of a multiply-connected domain and formulated it directly in a very general form for many unknown functions, assuming that the limiting values from inside and from outside the contour are taken at different points of the contour. The device for solving this general problem was the same integral representation (37.6)–(37.8) which we employed in solving the case considered in § 37. A detailed exposition of the problem the reader may find in the thesis of M. P. Ganin (5).

The Hilbert problem for multiply-connected domains was first considered by D. A. Kveselava (4). Making use of the integral representation (36.37) he reduced the problem to a singular integral equation and investigated the problem along approximately the same lines as we did in § 34.6 for the generalized problem. The most important result deduced by D. A. Kveselava consisted in establishing the insolubility of the problem for the case of a negative index. This result was the basis of further investigations. The adjoint problem was not examined by D. A. Kveselava.

The investigation given in § 36.7 follows the paper of I. N. Vekua [29], in which it was applied to a more general case (boundary value problem (39.33)). The investigation of the limiting cases $\varkappa = 0$ and $\varkappa = m - 1$, and the establishing of the connection with conformal mapping was presented in a joint paper of the author and E. G. Khasabov (1).

The examination of boundary value problems for linear elliptic equations in the classical formulation has a long history and we cannot discuss it here. The modern methods of solution of boundary value problems for equations with analytic coefficients based on a reduction of the elliptic equations to hyperbolic ones are presented in the papers of N. I. Vekua. The very concept of reduction of an elliptic equation to a hyperbolic one has been known for a long time and has been carried out in various ways. It was first used in practice by Hans Levy in 1927, for proving the analiticity of the solution of an elliptic equation with analytic

† In general, it is no longer a solution of an equation of the considered type.

coefficients (see Progress in Mathematical Sciences [Uspekhi matematitcheskhikh nauk], vol. 8, 1941, pp. 100–106). I. N. Vekua systematically developed this method and applied it to solving boundary value problems for equations of not only second order but also of higher orders. A survey exposition is given in his book [30], which we have frequently quoted.

The system of equations of elliptic type (§ 39) were investigated by a great number of mathematicians, both Soviet and foreign. Among the Soviet investigators who considered the general properties, we first of all note M. A. Lavrentyev; he examined the problem from a geometric viewpoint in connection with the theory of quasi-conformal mappings. The same ideas are found in the papers of B. V. Shabat. We should also mention the papers of G. N. Polozhy who deduced many important properties of solutions of these equations. He used the term "p-analytic" functions.

The basic boundary value problem (the Dirichlet problem) for this type of equations was first investigated by Hilbert, [9]. By means of the Green's function he reduced the problem to a Fredholm integral equation but he did not examine its solubility. The boundary value problem (39.33) was first tackled by N. K. Usmanov (1), (2). To derive the integral equations he employed partly the methods of Hilbert and partly the methods of I. N. Vekua treated in § 38. He has not carried out a rigorous investigation of the solubility problems.

A survey of the problem is given in the extensive paper of I. N. Vekua, quoted frequently in § 39, [30*]. It contains the most complete consideration of the solubility problems; that presented in § 39.7 which completely reduces the problem to the Hilbert boundary value problem, belongs to the present author. The solution of the Riemann boundary value problem (§ 39.8 and 39.9) follows the paper of L. G. Mikhailov (2).

Problems on Chapter V

1. Solve the generalized Hilbert boundary value problem

$$\sin s \frac{du}{ds} - \cos s \frac{dv}{ds} + \cos s\, u(s) + \sin s\, v(s) = \cos 3s\, (\cos s - 1)$$

for the unit circle.

Answer. $F(z) = u + iv = -\frac{1}{8}z^5 + \frac{1}{3}z^4 - \frac{1}{4}z^3 + Az^2 + Bz + \overline{A},$

where A, B are arbitrary constants.

2. Prove that the generalized Hilbert boundary value problem

$$\sin s \frac{du}{ds} + (2 - \cos s) \frac{dv}{ds} - 2u(s) = 2(\cos 2s + h \cos s)$$

for the unit circle is soluble only when $h = -1$. Derive the solution for this case.

Answer. $F(z) = u + iv = z^2 + i\alpha + A(z - \frac{1}{2}),$

where A is a complex and α a real parameter.

14*

3. Solve the generalized Hilbert boundary value problem

$$\sin s\, \frac{du}{ds} + (2 - \cos s)\, \frac{dv}{ds} - 2\cos s u(s) - 2\sin s v(s)$$

$$= 2\cos 3s - 3\cos 2s + \tfrac{3}{2}\cos s - \tfrac{1}{4}$$

for the unit circle.

Answer. $F(z) = u + iv = \alpha\, i(z - 1) + \tfrac{1}{3} z(z - \tfrac{1}{2})^2$.

4. Prove that if the coefficients of the boundary value problem

$$a_1\, \frac{du}{ds} + b_1\, \frac{dv}{ds} + a_0 u + b_0 v = c$$

satisfy the condition

$$\left(a_0 - \frac{da_1}{ds} \right) b_1 - \left(b_0 - \frac{db_1}{ds} \right) a_1 = 0,$$

then the solution of the problem can be reduced to the solution of the Hilbert problem and, consequently, it can be given in a closed form (F. D. Gakhov [2]).

5. Solve the generalized Hilbert boundary value problem

$$\cos 2s \cos s\, \frac{du}{ds} + \sin 2s \cos s\, \frac{dv}{ds} + (-2\sin 2s \cos s + \cos 2s \sin s)\, u +$$

$$+ (2\cos 2s \cos s + \sin 2s \sin s)\, v = \cos^2 s (\sin 2s - 3\sin s)$$

for the unit circle.

Hint. Make use of the results of the preceding problem.

Answer. $F(z) = \tfrac{1}{2} z^6 - \tfrac{1}{4} z^5 + (A_2 + \tfrac{1}{2})\, z^4 + (A_1 + \alpha - \tfrac{1}{4})\, z^3 + i\beta z^2 - A_1 z - A_2$.

6. Solve the generalized Hilbert boundary value problem

$$\cos \varkappa s\, \frac{du}{ds} - \sin \varkappa s\, \frac{dv}{ds} - \varkappa \sin \varkappa s u - \varkappa \cos \varkappa s v = -2\sin 2s,$$

for the unit circle ($\varkappa$ is an integer).

Answer. For $\varkappa \leq 0$, $F(z) = u + iv = z^{-\varkappa}[z^2 + c + Q(z)]$ where $Q(z)$ is given by formula (27.6), for $0 < \varkappa \leq 2$, $F(z) = z^{2-\varkappa}$, and for $\varkappa \leq 2$ the problem is insoluble.

7. Prove that the homogeneous Poincaré boundary value problem

$$\frac{du}{dn} + B(s)\, \frac{du}{ds} + c(s)\, u = f(s),$$

the coefficients of which satisfy the condition

$$\tfrac{1}{2} B'(s) - c(s) \geq 0,$$

is uniquely soluble (B. V. Khvedelidze (1)).

8. Prove that the Hilbert boundary value problem

$$\mathrm{Re}\,[(a - ib)\, F(t)] = au + bv = c$$

can be reduced to a Fredholm integral equation by means of the integral representation

$$F(z) = \frac{1}{2\pi i} \int \frac{\mu(\sigma)}{a(\sigma) - ib(\sigma)} \frac{d\tau}{\tau - z} + P(z),$$

where $\mu(\sigma)$ is a real function and $P(z)$ is a suitable polynomial. Hence, carry out an investigation of the solubility problems of the Hilbert problem (D.I. Sherman (4)).

9. Solve the generalized Riemann boundary value problem

$$t^2 \frac{d^2 \Phi^+}{dt^2} + 4t \frac{d\Phi^+}{dt} + 2\Phi^+(t) = \frac{d\Phi^-}{dt} + \frac{2t}{1+t^2} \Phi^-(t) + \frac{6t^3 + 6t - 1}{t^2 + 1},$$

$$\Phi^-(\infty) = 0,$$

under the following assumptions:
   (a) the points $i, -i \in D^+$, the point $0 \in D^-$;
   (b) the points $i, -i, 0 \in D^+$.

Answer.   (a) $\Phi^+(z) = z + \dfrac{c_1}{z} + \dfrac{c_2}{z^2}, \quad \Phi^-(z) = \dfrac{z}{1+z^2} + \dfrac{c_3}{1+z^2}$;

   (b) $\Phi^+(z) = z, \quad \Phi^-(z) = \dfrac{z}{1+z^2} + \dfrac{c_3}{1+z^2}.$

10. Solve the integro-differential equation

$$t^2 \varphi''(t) + (1 + 4t) \varphi'(t) + \frac{2(t^2 + t + 1)}{t^2 + 1} \varphi(t) + \frac{t^2}{\pi t} \int_L \frac{\varphi''(\tau)}{\tau - t} d\tau +$$

$$+ \frac{4t - 1}{\pi i} \int_L \frac{\varphi'(\tau)}{\tau - t} d\tau + 2 \frac{t^2 - t + 1}{t^2 + 1} \frac{1}{\pi i} \int_L \frac{\varphi(\tau)}{\tau - t} d\tau = 2 \frac{6t^3 + 6t - 1}{t^2 + 1}$$

under the same assumptions with respect to the contour as in the preceding problem.

Answer.   (a) $\varphi(t) = \dfrac{t^3}{1+t^2} + \dfrac{c_1}{t} + \dfrac{c_2}{t^2} + \dfrac{c_3}{1+t^2}$;

   (b) $\varphi(t) = \dfrac{t^3}{1+t^2} + \dfrac{c}{1+t^2}.$

11. Prove that the boundary value problem

$$t(t-1)[\Phi^+(t)]' + (2t - 1)\Phi^+(t)$$

$$= (t+1)[\Phi^-(t)]' + 2\Phi^-(t) + 2\frac{at^4 - t^3 - 1}{t^3}, \quad \Phi^-(\infty) = 0,$$

where $0, 1 \in D^+$, is soluble only when $a = 2$. Solve the problem in this case.

Answer. $\Phi^+(z) = 2, \quad \Phi^-(z) = -\dfrac{1}{z^2}.$

12. Prove that the integro-differential equation

$$(t^2 + 1)\, \Phi'(t) + (2t + 1)\, \Phi(t) + \frac{t^2 - 2t - 1}{\pi i} \int_L \frac{\Phi'(\tau)}{\tau - t}\, d\tau +$$

$$+ \frac{2t - 3}{\pi i} \int_L \frac{\Phi(\tau)}{\tau - t}\, d\tau = 4\, \frac{a t^4 - t^3 - 1}{t^3},$$

where L is the contour of the preceding problem, is soluble only when $a = 2$. Derive the solution for this case.

Answer.　$\Phi(t) = \dfrac{2 t^2 + 1}{t^2}$.

13. Solve the boundary value problem

$$\frac{1}{t}\, [\Phi^+(t)]'' - \frac{1}{t^2}\, [\Phi^+(t)]'$$

$$= t^3 [\Phi^-(t)]'' + 6 t^2 [\Phi^-(t)]' + 6 t \Phi^-(t) + 2\, \frac{4 t^3 + 8 t^2 + 4 t + 1}{(t + 1)^2}, \quad \Phi^-(\infty) = 0,$$

in the following cases:
(a) the point $-1 \in D^+$, the points $0, 1 \in D^-$;
(b) the points $-1, 0 \in D^+$, the point $1 \in D^-$.

Answer.　(a) The problem is insoluble.

(b)　$\Phi^+(z) = z^4 + 4 c_1 z^3 + c_2 z^2 - 2 z + c_3$,

$$\Phi^-(z) = 2\, \frac{\ln(z + 1)}{z^3} + 6 c_1 z^{-1} - 2\, \frac{\ln z}{z^3} + c_4 z^{-2} + c_5 z^{-3}.$$

14. Solve the integro-differential equation

$$\frac{t^4 + 1}{t}\, \Phi''(t) + \frac{6 t^4 - 1}{t^2}\, \Phi'(t) + 6 t \Phi(t) + \frac{1 - t^4}{t}\, \frac{1}{\pi i} \int_L \frac{\Phi''(\tau)}{\tau - t}\, d\tau -$$

$$- \frac{6 t^4 + 1}{t^2}\, \frac{1}{\pi i} \int_L \frac{\Phi'(\tau)}{\tau - t}\, d\tau - \frac{6 t}{\pi i} \int_L \frac{\Phi(\tau)}{\tau - t}\, d\tau = 4\, \frac{4 t^3 + 8 t^2 + 4 t + 1}{(t + 1)^2}$$

under the same assumption with respect to the contour as in the preceding problem.

Answer. (a) There is no solution.

(b)　$\Phi(t) = \dfrac{t^7 - 2 \ln(t + 1)}{t^3} + 2 c_1\, \dfrac{2 t^4 - 3}{t} + c_2 t^2 +$

$$+ 2\, \frac{\ln t - t^4}{t^3} + c_3 - c_4 t^{-2} - c_5 t^{-3}.$$

15. Solve the following Hilbert boundary value problems for the circular ring $\frac{1}{2} < |z| < 1$:

(a)　$\cos\theta\, u(\theta) + \sin\theta\, v(\theta) = \sin\theta$　　　　on　$|z| = 1$,

　　　$2 \cos\theta\, u(\theta) + 2 \sin\theta\, v(\theta) = \cos 2\theta$　　on　$|z| = \frac{1}{2}$;

(b) $\cos\theta\, u(\theta) + \sin\theta\, v(\theta) = 1$ ⠀⠀⠀⠀⠀⠀on $|z| = 1$,

$$\cos\left(\theta + \frac{\pi}{6}\right) u(\theta) + \sin\left(\theta + \frac{\pi}{6}\right) v(\theta) = \sin\theta \quad \text{on} \quad |z| = \frac{1}{2}.$$

Answer. (a) $F(z) = u + iv = -\dfrac{4}{15} z^3 - \dfrac{4}{3} iz^2 + i\beta z - \dfrac{1}{3} i + \dfrac{4}{15} z^{-1}$;

⠀⠀⠀⠀(b) $F(z) = \left\{1 - i\sqrt{3} + \dfrac{1}{13}\left[(10 + i6\sqrt{3})\, z - \dfrac{10 - i6\sqrt{3}}{z}\right]\right\} z$.

Remark. In the case (a) the conditions of single-valuedness are satisfied. Therefore the homogeneous problem has the solution $\varphi(z) = iCz$ and the non-homogeneous problem is soluble if one condition is satisfied, which is indeed the case.

In the case (b) the homogeneous problem is insoluble and the non-homogeneous problem is uniquely and absolutely soluble.

16*. Prove that every solution of the equation

$$\Delta u + \lambda^2 u = 0$$

can be represented in the form

$$u = \operatorname{Re}\left[\Phi(z) - \int_{z_0}^{z} \Phi(t)\, \frac{\partial}{\partial t}\, I_0\left(\lambda\sqrt{\{z(z-t)\}}\right) dt \right],$$

where $\Phi(z)$ is an arbitrary analytic function and I_0 is the Bessel function of zero order (I. N. Vekua [30]).

17*. Prove that the kernel of the integral representation of the elliptic equation with constant coefficients

$$\Delta u + a\, \frac{\partial u}{\partial x} + b\, \frac{\partial u}{\partial y} + cu = 0$$

can be expressed by elementary functions and a Bessel function of zero order.

Hint. After transformation to the complex variables

$$\frac{\partial^2 u}{\partial z\, \partial \bar{z}} + A\, \frac{\partial u}{\partial z} + \bar{A}\, \frac{\partial u}{\partial \bar{z}} + Cu = 0$$

by means of the substitution $u = e^{-\bar{A}z - A\bar{z}}$ reduce the equation to that considered in the preceding problem.

18*. Prove that every solution of the equation

$$\Delta u + \frac{4n(n+1)}{(1 + x^2 + y^2)^2}\, u = 0$$

can be represented in the form

$$u = \operatorname{Re}\left[\Phi(z) - \int_{0}^{z} \Phi(t)\, \frac{\partial}{\partial t}\, P_n\left(\frac{1 - z\bar{z} + 2\bar{z}t}{1 + z\bar{z}}\right) dt \right],$$

where P_n is the Legendre polynominal (I. N. Vekua [30]).

19*. Prove that the resolvent R of the Volterra integral equation

$$u(z, \bar{z}) + \int_{z_0}^{z} \bar{A}(\zeta, \bar{z})\, u(\zeta, \bar{z})\, d\zeta = v(z, \bar{z})$$

can be derived as the solution of the ordinary linear differential equation

$$\frac{dR}{dz} = -\bar{A}(z, \bar{z})\, R(z, \bar{z}, \zeta)$$

(I. N. Vekua [30]).

20. Prove that the system of equations

$$\frac{\partial u}{\partial x} = p(x, y)\frac{\partial w}{\partial y}, \quad \frac{\partial u}{\partial y} = -p(x, y)\frac{\partial w}{\partial x}$$

(the solutions $u + i\,w$ which G. N. Polozhy called p-analytic functions) is reduced by means of the substitution $v = pw$ to the system

$$\frac{\partial u}{\partial x} - \frac{\partial v}{\partial y} = -\frac{1}{p}\frac{\partial p}{\partial y}\, v, \quad \frac{\partial u}{\partial y} + \frac{\partial v}{\partial x} = \frac{1}{p}\frac{\partial p}{\partial x}\, v.$$

21*. Prove that the equation

$$\frac{\partial U}{\partial \bar{z}} = AU + B\bar{U} + F$$

in a conformal mapping of the domain is transformed into an equation of the same type (I. N. Vekua [29]).

22*. Prove that the equation of second order of elliptic type

$$\Delta w + p\frac{\partial w}{\partial x} + q\frac{\partial w}{\partial y} = 0$$

is reduced to the equation

$$\frac{\partial U}{\partial \bar{z}} = AU + \bar{A}\bar{U}$$

by means of the substitution

$$U = \frac{\partial w}{\partial x} - i\frac{\partial w}{\partial y},$$

where $A = -\frac{1}{4}(p + iq)$ (I. N. Vekua [29]).

23. Prove that the function

$$U(z) = -\frac{1}{\pi}\int\int \frac{A(\xi, \eta)}{\zeta - z}\, d\xi\, d\eta,$$

where

$$A(x, y) = \frac{e^{i\theta}}{\ln\dfrac{1}{r}} \qquad (re^{i\theta} = z = x + iy),$$

is a function of class $C_{\bar{z}}$ and has no continuous derivatives with respect to x, y at the origin of the coordinate system.

24*. Consider the equation (1)

$$\frac{\partial U}{\partial \bar{z}} = A\bar{U} \quad (U = u + iv)$$

and suppose that $U_* = u_* + iv_*$ is the solution of the equation

$$\frac{\partial U_*}{\partial z} = -\bar{A}\bar{U}_*,$$

which does not vanish in the domain under consideration (it can be proved that such a solution always exists). Introduce the function

$$w(z) = \frac{1}{2i} \int\limits_{z_0}^{z} (U_* U \, d\zeta - \bar{U}\bar{U}_* \, d\bar{\zeta}) = \int\limits_{z_0}^{z} [(uv_* + vu_*) \, d\xi + (uu_* - vv_*) \, d\eta].$$

Prove that by means of the substitution

$$u = \frac{v_*}{u_*^2 + v_*^2} \frac{\partial w}{\partial x} + \frac{u_*}{u_*^2 + v_*^2} \frac{\partial w}{\partial y}; \quad v = \frac{u_*}{u_*^2 + v_*^2} \frac{\partial w}{\partial x} - \frac{v_*}{u_*^2 + v_*^2} \frac{\partial w}{\partial y}$$

equation (1) is reduced to the elliptic equation of second order

$$\Delta u + p \frac{\partial u}{\partial x} + q \frac{\partial u}{\partial y} = 0,$$

where p, q are known functions determined by the relation

$$p + iq = -\frac{4}{U_*} \frac{\partial U_*}{\partial \bar{z}}$$

(I. N. Vekua [29]).

25*. Prove that if the elliptic equation of second order

$$\Delta w + p \frac{\partial w}{\partial x} + q \frac{\partial w}{\partial y} + rw = 0$$

has a solution w_1 which does not vanish in the domain under consideration, then by means of the substitution $w = w_1 W$ it can be reduced to the form

$$\Delta W + p_1 \frac{\partial W}{\partial x} + q_1 \frac{\partial W}{\partial y} = 0,$$

and subsequently reduced to the equation

$$\frac{\partial U}{\partial \bar{z}} = AU + \bar{A}\bar{U}$$

(see Problem 22).

CHAPTER VI

BOUNDARY VALUE PROBLEMS AND SINGULAR INTEGRAL EQUATIONS WITH DISCONTINUOUS COEFFICIENTS AND OPEN CONTOURS

IN ALL the problems considered so far the coefficients of the boundary conditions were regarded as continuous functions † and it was assumed that the contour of the domain consists of closed curves. Consequently, solutions were made up only of functions continuous up to the contour. The continuity of the solution on the contour did not follow from the particular restrictions imposed on the solution but from the very formulation of the problem. If we analyse the investigation of the fundamental boundary value problems (Chapters II and IV) we find that an essential restriction imposed upon the solution was the requirement of integrability of its boundary values. Then the continuity of the solution follows automatically from the conditions of continuity imposed upon the coefficients (see for instance Problem 12, Chapter II).

In this chapter we shall extend the formulation of the problems, admitting coefficients possessing discontinuities of the first kind at isolated points, and contours which are composed of both closed and open curves. Consequently, we shall have to extend the class of permissible functions constituting solutions of the problem.

We preserve the requirement of integrability (an exception will be considered only in § 45); the necessity of preserving this condition follows from various considerations. First of all in practical applications only this class of functions turns out to be useful. From the theoretical viewpoint this condition makes the formulation of the problem more definite. If we admit solutions which have infinities of

† The exceptional cases examined in § 15, where the coefficient could have an infinity, are not typical, since the investigation was close to that for the case of continuous coefficients.

406

an arbitrarily high order on the contour then a simple argument proves that for any of the investigated boundary value problems there exist an arbitrary number of linearly independent solutions.

It should be observed that the extension of the problem carried out here is of great practical value, since many important problems with the so-called "mixed" boundary conditions are reduced to these problems with discontinuous coefficients.

Prior to proceeding to further considerations the reader should recall the contents of §§ 8.1–8.4 and § 14.

§ 41. Solution of the Riemann problem with discontinuous coefficients by reduction to a problem with continuous coefficients

41.1. Formulation of the problem and determination of the function in terms of a given jump

We first consider the simplest case when the contour consists of one simple closed curve. Suppose that the functions $G(t)$, $g(t)$ in the boundary condition of the Riemann problem

$$\Phi^+(t) = G(t)\,\Phi^-(t) + g(t) \tag{41.1}$$

satisfy the Hölder condition everywhere on L, except at the points t_1, $t_2, \ldots, t_m$ where they have discontinuities of the first kind. The poinst preceding t_k in describing the contour in the positive direction will be regarded as the left vicinity of this point, while the points following it will be regarded as the right vicinity. The left and the right limits at the point t_k will, as usual, be denoted by $G(t_k - 0)$ and $G(t_k + 0)$ $(G(t_k - 0) \neq G(t_k + 0))$. We shall assume that neither of these limiting values vanishes. In addition, the assumption is made that the boundary condition is satisfied everywhere except at the points of discontinuity where it is meaningless.

The solution of the problem is sought in the class of functions which are integrable on the contour. It follows that the solution is continuous, in the Hölder sense, everywhere except possibly at the points t_k. At the latter points various possibilities may occur.

1. It may be demanded that the solution is bounded at all points of discontinuity; thus a solution is sought which is bounded everywhere.

2. It may be demanded that the solution is bounded at some points of discontinuity, allowing integrable discontinuities at the remaining points of discontinuity.

3. Solutions may be admitted which have integrable infinities at all points, if this is allowed by the conditions of the problem†.

The first class of solutions is the narrowest. The second class is obtained from the first by admitting infinities at some points, and it is wider than the first. The third class is the widest. It will be proved later that the number of solutions depends on the class in which the solution is sought, and it may turn out that a problem soluble in a wider class is insoluble in a narrower class. Similarly as was done in § 14 we begin our investigation from the simplest case by the determination of a sectionally analytic function in terms of its jump. Set in (41.1) $G(t) = 1$ seeking the analytic functions $\Phi^+(z)$, $\Phi^-(z)$ in accordance with the boundary condition

$$\Phi^+(t) - \Phi^-(t) = g(t); \qquad (41.2)$$

$g(t)$ can have at some separate points discontinuities of the first kind or singularities of the form

$$g(t) = \frac{g^*(t)}{(t - t_k)^\gamma} \qquad (\operatorname{Re}\gamma < 1), \qquad (41.3)$$

where $g^*(t)$ satisfies the Hölder condition.

The solution is sought in the class of functions having on the contour the estimate

$$|\Phi^\pm(t)| < \frac{C}{|t - t_k|^\alpha}, \qquad (\alpha < 1) \qquad (41.4)$$

and vanishing at infinity.

In view of the Sokhotski formulae the problem is solved by the function

$$\Phi(z) = \frac{1}{2\pi i} \int\limits_L \frac{g(\tau)}{\tau - z} d\tau, \qquad (41.5)$$

and according to the investigation of §§ 8.1–8.4 this function has at the points of discontinuity of the first kind of the function $g(t)$, a logarithmic singularity, and, at the points where $g(t)$ is of the form (41.3) a singularity of the same nature.

---

† We shall see later that there exist points of discontinuity where the integrability automatically implies boundedness.

Let us prove that problem (41.2) cannot have other solutions in the class of functions considered. As it always occurs in linear problems the investigation of the uniqueness is reduced to the problem of the solubility of the homogeneous problem. In the case under consideration it has the form

$$\Phi^+(t) - \Phi^-(t) = 0. \qquad (41.2')$$

This relation indicates that the functions $\Phi^+(z)$, $\Phi^-(z)$ are analytically continuable through the contour L and, consequently, constitute a unified analytic function in the whole plane. This function may only have isolated singularities. In view of estimate (41.4) these singularities cannot be poles or essential singularities, and hence they can only be branch points. But a single-valued function possessing branch points should have lines of discontinuities which cannot take place in view of the continuity of the functions in the domains D^+, D^- and condition (41.2') on the contour. Therefore, problem (41.2') has only the zero solution and, consequently, solution (41.5) of problem (41.2) is unique.

Observe that reasoning of this kind constitutes the basis of the assertion that in the case of continuity of the coefficient $G(t)$, the admittance of an integrable singularity of the solution on the contour, does not extend the class of solutions.

If the solution of problem (41.2) be sought in the class of functions having at infinity a pole of an order $\varkappa$, then it can easily be proved, relying as in § 14 on the generalized Liouville theorem, that the solution is provided by the function

$$\Phi(z) = \frac{1}{2\pi i} \int\limits_L \frac{g(\tau)}{\tau - z} \, \varphi \tau + P_\varkappa(z), \qquad (41.6)$$

where $P_\varkappa(z)$ is a polynomial of degree $\varkappa$ with arbitrary coefficients.

The method of solution of the general problem (41.1) described in the present section consists in reducing it to a problem considered before in § 14. For this purpose special functions are constructed which have at the points t_k the same discontinuities as $G(t)$ and possess the property that if they are the coefficients of the Riemann problem this problem can be solved. Introducing with the help of these functions new unknown functions (see (41.12)) we arrive at the Riemann problem which already has continuous coefficients.

We now proceed to construct the above mentioned auxiliary functions.

41.2. Fundamental auxiliary functions

The concept of construction of such functions consists in the possibility of regarding a many-valued analytic function in a suitably cut plane, as a discontinuous single-valued function.

Introduce two analytic functions

$$(z - z_0)^\gamma, \quad (z - t_1)^\gamma,$$

where z_0 is a point of the domain D^+, t_1 is a point on the contour and $\gamma = \alpha + i\beta$ is a complex number. The branch points of these functions are z_0, ∞ and t_1, ∞, respectively. Let us make a cut in the plane z from the point z_0 to the point ∞ through t_1. In the plane cut in this way both functions are single-valued, the cut being for them the line of discontinuity. We agree to locate the segment of the cut $z_0 t_1$ wholly in D^+ and consider the functions

$$\omega^+(z) = (z - t_1)^\gamma, \quad \omega^-(z) = \left(\frac{z - t_1}{z - z_0}\right)^\gamma. \tag{41.7}$$

Owing to the manner of performing the cut the function $\omega^+(z)$ is analytic in D^+ and $\omega^-(z)$ is analytic in D^-. On the contour both functions are continuous (and analytic) everywhere except possibly at the point t_1. The behaviour of these functions at the point t_1 depends on the value of γ. We set aside this problem for the time being. Consider now the ratio of the boundary values. Obviously, for all points of the contour, except t_1 we have

$$\Omega(t) = \frac{\omega^-(t)}{\omega^+(t)} = (t - z_0)^{-\gamma}. \tag{41.8}$$

At the point t_1, $\Omega(t)$ is undefined; we now find the ratio of the left and right limits of $\Omega(t)$ which coincide with the corresponding limits of the function $(t - z_0)^{-\gamma}$.

Since the left and right vicinities of the point t_1 are located on different sides of the cut for the function $(z - z_0)^\gamma$, in view of a property of the power function,

$$\frac{\Omega(t_1 - 0)}{\Omega(t_1 + 0)} = e^{-2\pi i \gamma}. \tag{41.9}$$

We now investigate the behaviour of the functions $\omega^+(z)$, $\omega^-(z)$ in the vicinity of the point t_1. To this end it is sufficient to investigate the function $\omega^+(z) = (z - t_1)^\gamma$.

Denoting $z - t_1 = re^{i\theta}$ we have

$$(z - t_1)^\gamma = e^{\gamma \ln(z - t_1)} = r^\alpha e^{-\beta \theta} e^{i(\beta \ln r + \alpha \theta)}. \tag{41.10}$$

It follows from the last expression that if $\alpha \neq 0$ the order of the function $(z - t_1)^\gamma$ depends only on α. If $\alpha > 0$ it has a zero of order $+\alpha$ and if $\alpha < 0$ an infinity of order $-\alpha$. Finally when $\alpha = 0$ this quantity is bounded but does not tend, as z approaches t_1, to any limit. For $-1 < \alpha < 0$ the functions $\omega^+(z)$, $\omega^-(z)$ have at the point t_1 infinities of an integrable order.

41.3. Reduction to a problem with continuous coefficients. The simplest case

Consider first the homogeneous problem

$$\Phi^+(t) = G(t)\,\Phi^-(t), \tag{41.1'}$$

assuming that the coefficient $G(t)$ has only one point of discontinuity t_1.
Setting

$$\gamma = \frac{1}{2\pi i} \ln \frac{G(t_1 - 0)}{G(t_1 + 0)}, \tag{41.11}$$

we construct in accordance with formulae (41.7) the functions $\omega^+(z)$ and $\omega^-(z)$.
We introduce new unknown functions $\Phi_1^\pm(z)$ setting

$$\Phi^\pm(z) = \omega^\pm(z)\,\Phi_1^\pm(z). \tag{41.12}$$

Boundary condition (41.1) now takes the form

$$\Phi_1^+(t) = G_1(t)\,\Phi_1^-(t), \tag{41.13}$$

where

$$G_1(t) = \frac{\omega^-(t)}{\omega^+(t)}\,G(t).$$

Introducing the notation of the preceding subsection we obtain

$$G_1(t) = \Omega(t)\,G(t) = (t - z_0)^{-\gamma}\,G(t). \tag{41.14}$$

The new coefficient $G_1(t)$ is already everywhere continuous on L, including the point t_1. In fact, applying formulae (41.9) and (41.11) we have

$$\frac{G_1(t_1 - 0)}{G_1(t_1 + 0)} = \frac{\Omega(t_1 - 0)}{\Omega(t_1 + 0)}\,\frac{G(t_1 - 0)}{G(t_1 + 0)} = e^{-2\pi i \gamma}\,\frac{G(t_1 - 0)}{G(t_1 + 0)} = 1,$$

i.e. $G_1(t_1 - 0) = G_1(t_1 + 0)$ which means the continuity of $G_1(t)$ at the point t_1. Thus, problem (41.1') with the discontinuous coefficient

$G(t)$ has been reduced to problem (41.13) with the continuous coefficient $G_1(t)$, which was investigated in § 14.

Consider the index of the problem. The behaviour of the solution of problem (41.1′) near the point of discontinuity t_1 depends on the behaviour of the functions $\omega^+(z)$, $\omega^-(z)$—this is seen from formulae (41.12). It was shown at the end of the preceding subsection that the behaviour of the latter functions is determined by the quantity $\alpha = \mathrm{Re}\,\gamma$, i.e. by the choice of the branch of the logarithm in formula (41.11). This choice should be made depending on the admissible class of solutions, namely it is necessary to require that the condition

$$0 \leqq \mathrm{Re}\,\gamma < 1 \tag{41.15}$$

is satisfied, if solutions bounded near the point t_1 are admitted; if we seek solutions having at the point t_1 an infinity of an integrable order† we have the condition

$$-1 < \mathrm{Re}\,\gamma < 0. \tag{41.16}$$

Let us find what is the meaning of these conditions for the coefficient $G(t)$ of the original problem (41.1′).

Denote by θ the increment of a branch of the argument of $G(t)$, in describing the contour L. Let us observe at once that θ can be regarded as the jump of the argument of $G(t)$ at the point of discontinuity, taken with opposite sign. Hence we may set

$$\frac{G(t_1 - 0)}{G(t_1 + 0)} = \varrho\, e^{i\theta}. \tag{41.17}$$

Then in accordance with formula (41.11)

$$\gamma = \frac{1}{2\pi i} \ln \frac{G(t_1 - 0)}{G(t_1 + 0)} = \frac{\theta}{2\pi} - \varkappa - i\frac{\ln \varrho}{2\pi}, \tag{41.18}$$

the integer $\varkappa$ being chosen such that conditions (41.15) and (41.16) are satisfied; the latter take the form

$$0 \leqq \frac{\theta}{2\pi} - \varkappa < 1, \tag{41.15′}$$

$$-1 < \frac{\theta}{2\pi} - \varkappa < 0. \tag{41.16′}$$

† The condition $\mathrm{Re}\,\gamma > -1$ follows immediately from the preceding one. The reason for the condition $\mathrm{Re}\,\gamma < 1$ is that superfluous conditions are not imposed on the solution. The requirement $\gamma > 1$ would result in a loss of solutions.

This implies the following formulae:

for the class of bounded solutions

$$\varkappa = \left[\frac{\theta}{2\pi}\right]; \qquad (41.19)$$

for the class of solutions unbounded near the point t_1

$$\varkappa = \left[\frac{\theta}{2\pi}\right] + 1. \qquad (41.20)$$

The symbol $[x]$ denotes, as in number theory, the greatest integer not exceeding x.

Let us now compute the index of problem (41.13). Taking into account formulae (41.14), (41.17) and (41.18) (the last defines the choice of the branch of the logarithm) we obtain

$$\text{Ind } G_1(t) = \frac{1}{2\pi i}\left\{\ln G_1(t)\right\}_L$$

$$= \frac{1}{2\pi i}\ln\left(\frac{G(t_1 - 0)}{G(t_1 + 0)}e^{-2\pi i\gamma}\right) = \frac{1}{2\pi i}\ln e^{2\pi i\varkappa} = \varkappa.$$

This quantity will be called the index of the considered problem (41.1'). Thus the index is determined by formulae (41.19) or (41.20) and depends on the selected class of solutions, the index of the problem in the class of unbounded solutions being greater by unity than that in the class of bounded solutions.

If $\theta/2\pi$ is an integer, of the two conditions (41.15') and (41.16') only the first can be satisfied for $\varkappa = \theta/2\pi$.

In this case the point of discontinuity t_1 will be called *the point of automatic boundedness*. The solution of problem (41.1') is bounded near this point but does not tend, on approaching t_1, to any limit (see the end of § 41.2).

Having elucidated the principles by way of the simplest case, we proceed to solving the problem in the general case.

41.4. Solution of the homogeneous problem

Suppose that the coefficient $G(t)$ has discontinuities of the first kind at the points $t_1, t_2, \ldots, t_l$. Let us split the contour L into sections between t_k and t_{k+1} by the points of discontinuity and let us define arbitrarily at every initial point of the section the branch of $\arg G(t_k + 0)$; at the last point of the section $\arg G(t_{k+1} - 0)$ will be obtained from the selected branch of $\arg G(t_k + 0)$ by a continuous change.

As before denote

$$\frac{G(t_k - 0)}{G(t_k + 0)} = A_k = \varrho_k\, e^{i\theta_k} \quad (k = 1, 2, \ldots, l),$$

where θ_k is the jump of the argument at the point of discontinuity t_k, taken with opposite sign.

Set

$$\gamma_k = \frac{1}{2\pi i}\ln A_k = \frac{\theta_k}{2\pi} - \varkappa_k - i\,\frac{\ln \varrho_k}{2\pi} \qquad (41.21)$$

(the integers $\varkappa_k$ are given by formulae (41.19) or (41.20)),

$$\Phi^+(z) = \prod_{k=1}^{l}(z - t_k)^{\gamma_k}\, \Phi_1^+(z), \quad \Phi^-(z) = \prod_{k=1}^{l}\left(\frac{z - t_k}{z - z_0}\right)^{\gamma_k} \Phi_1^-(z), \quad (41.22)$$

the cuts for the functions $(z - z_0)^{\gamma_k}$ passing through the point t_k of the contour. Substituting into equation (41.1') we arrive at the boundary condition

$$\Phi_1^+(t) = \prod_{k=1}^{l}(t - z_0)^{-\gamma_k}\, G(t)\, \Phi_1^-(t), \qquad (41.23)$$

in which, by definition, the function $(t - z_0)^{-\gamma_k}$ has a jump in passing through the point t_k.

A consideration similar to that in the case of one discontinuity proves that the coefficient $G_1(t) = \prod_{k=1}^{l}(t - z_0)^{-\gamma_k} G(t)$ is a continuous function. Consequently, we have again arrived at the solution of the problem with a continuous coefficient. It is easy to compute the index of the coefficient $G_1(t)$. Taking the increment of the argument of $G_1(t)$ over the whole contour L as the sum of increments over the sections between the points of discontinuity and taking into account that every binomial $(t - z_0)^{\gamma_k}$ satisfies at the point t_k the condition

$$\frac{(t_k - 0 - z_0)^{\gamma_k}}{(t_k + 0 - z_0)^{\gamma_k}} = e^{2\pi i\gamma_k},$$

we obtain

$$\text{Ind } G_1(t) = \frac{1}{2\pi i}\{\ln G_1(t)\}_L = \frac{1}{2\pi i}\sum_{k=1}^{l}\{\ln G_1(t)\}_{t_k}^{t_{k-1}}$$

$$= \frac{1}{2\pi i}\sum_{k=1}^{l}\ln\left[\frac{G(t_k - 0)}{G(t_k + 0)}\frac{(t_k - 0 - z_0)^{-\gamma_k}}{(t_k + 0 - z_0)^{-\gamma_k}}\right] = \varkappa_1 + \varkappa_2 + \cdots + \varkappa_l.$$

$$(41.24)$$

The last quantity will be called the *index* of the original problem (41.1').

It has already been noted that the numbers $\varkappa_k$ have the value

$$\varkappa_k = \left[\frac{\theta_k}{2\pi}\right],$$

if the solution is sought in the class of bounded functions, and

$$\varkappa_k = \left[\frac{\theta_k}{2\pi}\right] + 1$$

in the class of unbounded integrable functions.

It follows that the index of the problem in the class of functions having at a prescribed point of discontinuity an infinity of an integrable order, is greater by unity than the index in the class of functions bounded at the considered point.

Suppose that we want to find the solution having an integrable infinity at some p points of discontinuity and bounded at the remaining $l - p$ points of this kind (all points of automatic discontinuity enter, obviously, the second group). According to the above reasoning the index of the problem in this class of functions is given by the formula

$$\varkappa = \sum_{k=1}^{l} \left[\frac{\theta_k}{2\pi}\right] + p. \tag{41.25}$$

We could express the index by the increment of the argument of $G(t)$. However, we cannot obtain here such a simple formula as for the case when there was one point of discontinuity. We shall therefore omit this expression. In solving actual problems the index is completely determined by the last formula.

To derive the general solution of the problem considered we have to find in accordance with the rules of § 14 the general solution $\Phi_1(z)$ of problem (41.23) with the continuous coefficient; then formulae (41.22) yield the general solution of the original problem. We shall not quote here the relevant formulae and we proceed to solve the non-homogeneous problem. The general solution of the latter also contains, as is well known, the general solution of the homogeneous problem; the latter will contain all required formulae.

41.5. Solution of the non-homogeneous problem

Consider the non-homogeneous problem

$$\Phi^+(t) = G(t)\,\Phi^-(t) + g(t).$$

We make the same assumptions with respect to $G(t)$ as in the preceding subsection, and we assume that $g(t)$ satisfies the Hölder condition.

Introducing new functions $\Phi_1^+(z)$, $\Phi_1^-(z)$, by means of the substitutions (41.22) we reduce the boundary condition to the form

$$\Phi_1^+(t) = \prod_{k=1}^{l}(t - z_0)^{-\gamma_k}\,G(t)\,\Phi_1^-(t) + \prod_{k=1}^{l}(t - t_k)^{-\gamma_k}\,g(t).$$

Replacing the continuous coefficient $\prod\limits_{k=1}^{k}(t - z_0)^{-\gamma_k}G(t)$ by the ratio of the canonical functions (see § 14.4) we reduce the boundary condition to the form

$$\frac{\Phi_1^+(t)}{X_1^+(t)} - \frac{\Phi_1^-(t)}{X_1^-(t)} = \frac{\prod\limits_{k=1}^{l}(t - t_k)^{-\gamma_k}\,g(t)}{X_1^+(t)}.$$

The latter problem is problem (41.2) of the determination of an analytic function in terms of prescribed jumps, the function $\Phi_1^-(z)/X_1^-(z)$ having at infinity order $-\varkappa$. According to relation (41.6) the solution is given by the formula †

$$\frac{\Phi_1(z)}{X_1(z)} = \frac{1}{2\pi i}\int_L \frac{\prod\limits_{k=1}^{l}(\tau - t_k)^{-\gamma_k}\,g(\tau)}{X_1^+(\tau)}\,\frac{d\tau}{\tau - z} + P_\varkappa(z). \quad (41.26)$$

When $\varkappa < 0$ we have to set $P_\varkappa(z) \equiv 0$.

In view of formulae (41.22) the solution of the problem under consideration has the form

$$\left.\begin{aligned}
\Phi^+(z) &= \prod_{k=1}^{l}(z - t_k)^{\gamma_k}\,X_1^+(z)\,[\Psi^+(z) + P_\varkappa(z)],\\[2mm]
\Phi^-(z) &= \prod_{k=1}^{l}\left(\frac{z - t_k}{z - z_0}\right)^{\gamma_k}X_1^-(z)\,[\Psi^-(z) + P_\varkappa(z)],
\end{aligned}\right\} \quad (41.27)$$

† It is now clear that Re γ_k should be less than unity (see the remark on p. 412 and also § 8.4).

where

$$X_1^+(z) = e^{\Gamma_+(z)}, \quad X_1^-(z) = (z - z_0)^{-\varkappa} e^{\Gamma_-(z)},$$

$$\left.\begin{aligned}
\Gamma(z) &= \frac{1}{2\pi i} \int_L \frac{\ln\left[(\tau - z_0)^{-\varkappa} \prod_{k=1}^{l} (\tau - z_0)^{-\gamma_k} G(\tau)\right]}{\tau - z} d\tau, \\[2em]
\Psi(z) &= \frac{1}{2\pi i} \int_L \frac{\prod_{k=1}^{l} (\tau - t_k)^{-\gamma_k} g(\tau)}{X_1^+(\tau)(\tau - z)} d\tau.
\end{aligned}\right\} \quad (41.28)$$

It can readily be established that the solution given by the last relations belongs to the prescribed class.

For instance suppose that $\operatorname{Re}\gamma_j > 0$. Then by virtue of formulae (8.20) the function $\Psi(z)$ has near the point t_j the order of increase of $(z - t_j)^{-\gamma_j}$ and, consequently, in view of the presence of the factor $(z - t_j)^{\gamma_j}$, $\Phi(z)$ is bounded.

If $\operatorname{Re}\gamma_j < 0$ then $\Psi(z)$ is bounded at the point t_j and $\Phi(z)$ has an infinity of order $\operatorname{Re}\gamma_j$.

If the solution be subjected to the additional condition

$$\Phi^-(\infty) = 0, \tag{41.29}$$

$P_\varkappa(z)$ in formula (41.27) should be replaced by $P_{\varkappa-1}(z)$.

The result of our investigation may be stated in the form of a theorem which is a repetition of the theorem of § 14.5.

THEOREM. *If the index $\varkappa$ of the Riemann problem in the considered class is positive, then the general solution of the non-homogeneous problem depends linearly on $\varkappa + 1$ arbitrary constants and it is given by formula (41.27); if the additional condition $\Phi^-(\infty) = 0$ is imposed, $P_\varkappa(z)$ should be replaced by $P_{\varkappa-1}(z)$.*

When $\varkappa \leqq 0$ the homogeneous problem in the considered class is insoluble and the non-homogeneous problem is soluble if and only if the $-\varkappa$ conditions

$$\int_L \frac{\prod_{k=1}^{l} (\tau - t_k)^{-\gamma_k} g(\tau)}{X_1^+(\tau)} \tau^{j-1} d\tau = 0 \quad (j = 1, 2, \ldots, -\varkappa) \tag{41.30}$$

are satisfied.

If these conditions are satisfied the non-homogeneous problem has a unique solution which is derived from formula (41.27) by setting $P_\varkappa(z) \equiv 0$.

We recall that the index in the considered class coincides with the index of problem (41.23) with the continuous coefficient, and it is given by formula (41.25).

Remark. Although formulae (41.27) contain explicitly the arbitrary point z_0 it can easily be proved (we leave this to the reader) that the general solution is independent of the choice of this point.

Let us now consider what alterations have to be made in the deduced solution, if we admit discontinuities at a finite number of points for $g(t)$.

Suppose that in the vicinities of the points $t'_1, t'_2, \ldots, t'_r$, $g(t)$ has the representation

$$g(t) = \frac{g_*(t)}{(t - t'_k)^{\gamma'_k}} \quad (\gamma'_k = \alpha'_k + i\beta'_k, \quad 0 \leqq \alpha'_k < 1), \quad (41.31)$$

where $g_*(t)$ satisfies the Hölder condition and at the points $t''_1, t''_2, \ldots, t''_s$ it has discontinuities of the first kind, $g(t''_k + 0) \neq g(t''_k - 0)$.

We shall examine separately two cases.

1. *All points of discontinuity of the free term are distinct from the ends†.* All considerations leading to the solution under the assumption that the free term satisfies the Hölder condition remain valid also in the considered case, and we again arrive at formulae (41.27), (41.28). The only difference lies in the fact that now the solutions given by these formulae, besides the singularities at the ends determined by the selected class of solutions, have singularities at the points of discontinuity of the function $g(t)$. In view of the properties of the Cauchy type integral (§ 8.2 and § 8.4) the solution has singularities of the form

$$\Phi(t) = (t - t'_k)^{-\gamma'_k} \Phi^*(t)$$

at the points $t'_1, t'_2, \ldots, t'_r$ and singularities of logarithmic type at the points

$$t''_1, t''_2, \ldots, t''_s.$$

In contrast to the ends where the behaviour of the solution depends to a certain degree on our arbitrary choice of the class, at the points

† By the ends we mean the points of discontinuity of the coefficient $G(t)$.

of discontinuity of the free term the behaviour of the solution depends wholly on the nature of the discontinuity.

2. *Some points of discontinuity of $g(t)$ coincide with the ends.* For the points of discontinuity of $g(t)$ which do not coincide with the ends, it is evident that the above reasoning remains valid. Further, it is easy to see that the coincidence of the points of discontinuity of $g(t)$ of the first kind with the ends does not introduce any new facts, as compared with the above the case of the free term satisfying the Hölder condition; this is due to the fact that the behaviour of the function

$$\frac{\Pi(\tau - t_k)^{-\gamma_k} g(\tau)}{X_1^+(\tau)}$$

which is the density of the Cauchy type integral, is the same for both $g(t)$ satisfying the Hölder condition and having discontinuities of the first kind.

For the points t_k' coinciding with the ends more complicated singularities may appear: this case therefore has to be considered separately.

Suppose that $t_k' = t_k$. Introducing into formulae (41.28) expression (41.31) for $g(t)$ we find that the density of the appropriate Cauchy type integral has at the point t_k the form

$$(t - t_k)^{-(\gamma_k + \gamma_k')} \varphi_1(t)$$

($\varphi_1(t)$ satisfies the Hölder condition), i.e. it has the order $-(\alpha_k + \alpha_k')$.

In accordance with the definition, $\alpha_k' \geqq 0$ while the number α_k vanishes for the singular end, and it is positive in the case of a non-singular end if the required solution is bounded at the end t_k, and otherwise negative. In Chapter I an investigation of the Cauchy type integral was carried out, the density of which had infinities of order smaller than unity. If we seek solutions unbounded at the point t_k, then $\alpha_k + \alpha_k' < 1$ and the applicability of the theory is ensured. A simple calculation proves that in this case the solution has at the point t_k an order not lower than $\min(\alpha_k, -\alpha_k')$.

For solutions in the class of bounded functions, when $\alpha_k + \alpha_k' < 1$ the theory is also applicable, but the solution is actually unbounded at the point under consideration (except when $\alpha_k' = 0$); its order is $-\alpha_k'$.

If $\alpha_k + \alpha_k' > 1$ an additional investigation is necessary.

§ 42. Riemann boundary value problem for open contours

42.1. Formulation and solution of the problem

Let the contour L consist of a union of m simple, smooth, non-intersecting curves $L_1, L_2, \ldots, L_m$, the ends of which are a_k and b_k (the positive direction is from a_k to b_k) and suppose that $G(t)$, $g(t)$ are prescribed on L and satisfy the Hölder condition; moreover, we have everywhere $G(t) \neq 0$.

It is required to determine the function $\Phi(z)$ analytic in the entire plane, except at the points of the contour L; its boundary values $\Phi^+(t)$, $\Phi^-(t)$ on approaching L from the left and from the right are integrable functions satisfying the boundary condition

$$\Phi^+(t) = G(t)\,\Phi^-(t) + g(t). \tag{42.1}$$

The Riemann problem for an open contour is, by its formulation, essentially different from the problem for a closed contour; in

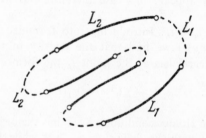

Fig. 15

fact, the entire plane with a cut along the curve L constitutes one domain and it is necessary to seek, instead of two independent analytic functions $\Phi^+(z)$, $\Phi^-(z)$, one analytic function $\Phi(z)$ for which the contour L is the line of discontinuity.

The above formulated problem can be reduced to that considered in the preceding section for a closed contour, with discontinuous coefficients, by means of the following formal method.† We complete the contour L by m arbitrary lines $L_1', L_2', \ldots, L_m'$ so that one closed curve C is constructed $(L_1 + L_1' + \cdots + L_m + L_m' = C)$ (Fig. 15).

---

† The concept of this method was first announced verbally by N. N. Meyman.

It is evident that on the additional curves, $\Phi^+(t) = \Phi^-(t)$. Consequently, problem (42.1) can be regarded as the problem for the closed contour
$$\Phi^+(t) = G_1(t)\,\Phi^-(t) + g_1(t), \qquad (42.2)$$
where

$$\left.\begin{array}{ll} G_1(t) = G(t), & g_1(t) = g(t) \quad \text{on} \;\; L_k, \\ G_1(t) = 1, & g_1(t) = 0 \quad \text{on} \;\; L_k'(k = 1, 2, \ldots, m). \end{array}\right\} \quad (42.3)$$

Now the solution constitutes an example of an application of the theory of the preceding section. Set $G(a_k) = \varrho_k e^{i\theta_k}$ and $G(b)_k = \varrho_k' e^{i(\theta_k + \varDelta_k)}$ where $\varDelta_k$ denotes the change of $\arg G(t)$ on L_k. Then

$$\left.\begin{array}{ll} G_1(a_k - 0) = 1, & G_1(a_k + 0) = G(a_k) = \varrho_k\,e^{i\theta_k}; \\ G_1(b_k - 0) = G(b_k) = \varrho_k'\,e^{i(\theta_k + \varDelta_k)}, & G_1(b_k + 0) = 1; \\ \dfrac{G_1(a_k - 0)}{G_1(a_k + 0)} = \dfrac{1}{\varrho_k}\,e^{-i\theta_k}, & \dfrac{G_1(b_k - 0)}{G_1(b_k + 0)} = \varrho_k'\,e^{i(\theta_k + \varDelta_k)}. \end{array}\right\} \quad (42.4)$$

The value of the argument θ_k† at the initial point a_k of the contour will be taken in the range
$$-2\pi < \theta_k \leqq 0 \qquad (42.5)$$
if a solution bounded at the point a_k is required, and in the range
$$0 < \theta_k < 2\pi \qquad (42.6)$$
if the solution is to be unbounded at the point a_k.

These conditions are imposed only to simplify the future formulae and have no influence on the final results.

Set (see formula (4.21))

$$\left.\begin{array}{l} \gamma_k = \dfrac{1}{2\pi i}\ln\left(\dfrac{1}{\varrho_k}\,e^{-i\theta_k}\right) = -\dfrac{\theta_k}{2\pi} + i\,\dfrac{\ln \varrho_k}{2\pi}, \\[2mm] \gamma_k' = \dfrac{1}{2\pi i}\ln\left(\varrho_k'\,e^{i(\theta_k + \varDelta_k)}\right) = \dfrac{\theta_k + \varDelta_k}{2\pi} - \varkappa_k = i\,\dfrac{\ln \varrho_k'}{2\pi}, \end{array}\right\} \quad (42.7)$$

where $\varkappa_k$ are integers which by formulae (41.16), (41.18) are

$$\varkappa_k = \left[\frac{\theta_k + \varDelta_k}{2\pi}\right] \qquad (42.8)$$

---

† It should be borne in mind that now θ_k denotes simply the jump in the argument of the function $G_1(t)$ at the point a_k and $\theta_k + \varDelta_k$ the jump at the point b_k, taken as before with negative sign.

if the solutions are bounded at the point b_k, and

$$\varkappa_k = \left[\frac{\theta_k + \varDelta_k}{2\pi}\right] + 1 \qquad (42.9)$$

for a solution unbounded at this end.

The index is given by the formula

$$\varkappa = \sum_{k=1}^{m} \varkappa_k,$$

$\varkappa_k$ being defined by formula (42.8) or (42.9), depending on whether the solution is to be bounded or unbounded at the end b_k. The initial value of the argument at the point a_k should satisfy condition (42.5) for a bounded and (42.6) for an unbounded solution at the point.

It is readily observed that the admittance of an unbounded solution at one of the ends increases the index by unity. This is a simple result of the corresponding property for the points of discontinuity.

The ends on which $G(t)$ is a real positive number (the argument is a multiple of 2π) will be called *the ends of automatic boundedness* (singular ends). If the initial point a_k is such an end, we have to set $\theta_k = 0$ while when it is the final point b_k, $\varkappa_k = (\theta_k + \varDelta_k)/2\pi$.

Introducing new functions

$$\left.\begin{array}{l}
\Phi^+(z) = \prod_{k=1}^{m} (z - a_k)^{\gamma_k}(z - b_k)^{\gamma_k'} \Phi_1^+(z), \\[2mm]
\Phi^-(z) = \prod_{k=1}^{m} \left(\dfrac{z - a_k}{z - z_0}\right) \gamma_k \left(\dfrac{z - b_k}{z - z_0}\right)^{\gamma_k'} \Phi_1^-(z),
\end{array}\right\} \qquad (42.10)$$

we arrive at the problem with the continuous coefficient $G(t)$ on the closed contour C

$$\Phi_1^+(t) = \prod_{k=1}^{m} (t - z_0)^{-\gamma_k}(t - z_0)^{-\gamma_k'} G_1(t) \Phi_1^-(t) +$$

$$+ \prod_{k=1}^{m} (t - a_k)^{-\gamma_k}(t - b_k)^{-\gamma_k'} g_1(t). \quad (42.11)$$

Now the solution is carried out in the usual way and there is no need to describe it here. It can easily be proved (see Problem 1 at the end of this chapter) that the solution is independent of the nature of the additional curve L'.

42.2. Example

As an example of the above theory let us consider a problem (encountered in numerous applications) which will frequently be quoted later.

Suppose that on the contour L composed of m open curves $G(t) = -1$, i.e. the boundary condition has the form

$$\Phi^+(t) = -\Phi^-(t) + g(t). \qquad (42.12)$$

Write down the boundary condition for the closed contour

$$\Phi^+(t) = G_1(t)\,\Phi^-(t) + g_1(t); \qquad (42.12')$$

$$G_1(t) = -1, \quad g_1(t) = g(t) \quad \text{on} \quad L,$$

$$G_1(t) = 1, \quad g_1(t) = 0 \quad \text{on} \quad L'.$$

We first seek the solution in the most general class of functions unbounded at all ends. In this case we obtain

$$G_1(a_k - 0) = 1; \qquad G_1(a_k + 0) = -1 = e^{i\pi};$$

$$\frac{G_1(a_k - 0)}{G_1(a_k + 0)} = e^{-i\pi}; \qquad G_1(b_k - 0) = -1 = e^{i\pi};$$

$$G_1(b_k + 0) = 1; \qquad \frac{G_1(b_k - 0)}{G_1(b_k + 0)} = e^{i\pi};$$

$$\gamma_k = \frac{1}{2\pi i} \ln e^{-i\pi} = -\frac{1}{2}.$$

Since $\theta_k = \pi$ and $\Delta_k = 0$ we have $\varkappa_k = [\tfrac{1}{2}] + 1 = 1$ and $\gamma_k' = -\tfrac{1}{2}$. Consequently

$$\varkappa = \sum_{k=1}^{m} \varkappa_k = m. \qquad (42.13)$$

According to substitution (42.10) the problem on the closed contour C with the continuous coefficient has the form

$$\Phi_1^+(t) = \prod_{k=1}^{m}(t - z_0)^{\frac{1}{2}}(t - z_0)^{\frac{1}{2}}\,G_1(t)\,\Phi_1^-(t) +$$

$$+ \prod_{k=1}^{m}(t - a_k)^{\frac{1}{2}}(t - b_k)^{\frac{1}{2}}\,g_1(t). \qquad (42.14)$$

The function $G_1(t)$ changes its sign at the points a_k, b_k. By definition, one of the binomials $(t - z_0)^{\frac{1}{2}}$ changes its sign at each of the above points. Hence we may set †

$$\prod_{k=1}^{m} (t - z_0)^{\frac{1}{2}} (t - z_0)^{\frac{1}{2}} G_1(t) = (t - z_0)^m.$$

For brevity we introduce the notation

$$\left. \begin{array}{l} \prod_{k=1}^{m} (z - a_k)^{\frac{1}{2}} (z - b_k)^{\frac{1}{2}} = R(z); \\[2mm] \prod_{k=1}^{m} (z - a_k)^{\frac{1}{2}} = R_a(z); \quad \prod_{k=1}^{m} (z - b_k)^{\frac{1}{2}} = R_b(z). \end{array} \right\} \tag{42.15}$$

Then condition (42.14) takes the form

$$\Phi_1^+(t) = (t - z_0)^m \Phi_1^-(t) + R(t) g_1(t). \tag{42.16}$$

Since the index of the problem $\varkappa = m$ we have in accordance with formulae (41.28)‡

$$X_1^+(z) = 1, \quad X_1^-(z) = (z - z_0)^{-m}.$$

According to (41.27) the solution of the initial problem unbounded at all ends is given by the formula

$$\Phi^+(z) = \Phi^-(z) = \Phi(z) = \frac{1}{R(z)} [\Psi(z) + P_m(z)], \tag{42.17}$$

where

$$\Psi(z) = \frac{1}{2\pi i} \int_L \frac{R(\tau) g(\tau)}{\tau - z} d\tau. \tag{42.18}$$

In applications we are frequently confronted with a solution which satisfies the additional condition

$$\Phi(\infty) = 0. \tag{42.19}$$

In this case we have to replace in formula (42.17) $P_m(z)$ by $P_{m-1}(z)$ (all following formulae concern the solution satisfying condition (42.19)).

If at any ends a_k, b_j we require the boundedness of the solution, the corresponding exponents are to be taken $\gamma_k = \gamma_j' = \frac{1}{2}$. The requirement of boundedness at any end decreases the index by unity; this fact has already been mentioned.

---

† This example enables the reader to observe clearly the process of elimination of the discontinuity.

‡ Obviously, in this case and in all following cases the functions $X_1^{\pm}(z)$ can be determined directly.

The solution bounded at the initial points a_k and unbounded at the final points b_k has the form

$$\Phi(z) = \frac{R_a(z)}{R_b(z)} \frac{1}{2\pi i} \int_L \frac{R_b(\tau)}{R_a(\tau)} \frac{g(\tau)}{\tau - z} d\tau \qquad (42.20)$$

(index $\varkappa$ is here equal to zero).

The solution bounded at all the ends has the form

$$\Phi(z) = \frac{R(z)}{2\pi i} \int_L \frac{g(\tau)}{R(\tau)} \frac{d\tau}{\tau - z}, \qquad (42.21)$$

and, since here $\varkappa = -m$ the solution exists only when the m conditions of solubility

$$\int_L \frac{g(\tau)}{R(\tau)} \tau^{j-1} d\tau = 0 \quad (j = 1, 2, \ldots, m). \qquad (42.22)$$

are satisfied.

Finally let us write the solution in the most general case. Denote uniformly all ends by the letters $c_1, c_2, \ldots, c_{2m}$. Suppose that on some of the ends $c_1, c_2, \ldots, c_p$ the solution is bounded, whereas on the remaining ends $c_{p+1}, \ldots, c_{2m}$ an integrable infinity is admitted. The ends of automatic boundedness, evidently, are here absent. Denote

$$R_1(z) = \sqrt{\left(\prod_{k=1}^{p} (z - c_k) \right)},$$
$$R_2(z) = \sqrt{\left(\prod_{k=p+1}^{2m} (z - c_k) \right)}. \qquad (42.23)$$

The index is $m - p$ and the solution has the form

$$\Phi(z) = \frac{R_1(z)}{R_2(z)} \left[\frac{1}{2\pi i} \int_L \frac{R_2(\tau)}{R_1(\tau)} \frac{g(\tau)}{\tau - z} d\tau + P_{m-p-1}(z) \right]. \qquad (42.24)$$

If $p \geqq m$, then $P_{m-p-1} \equiv 0$ and in the case $p > m$ the solution exists only when the conditions of solubility

$$\int_L \frac{R_2(\tau)}{R_1(\tau)} g(\tau) \tau^{j-1} d\tau = 0 \quad (j = 1, 2, \ldots, p - m) \qquad (42.25)$$

are satisfied.

42.3. Inversion of the Cauchy type integral

Consider the practically important problem of inversion of the Cauchy type integral

$$\frac{1}{\pi} \int_L \frac{\varphi(\tau)}{\tau - t} \, d\tau = f(t). \tag{42.26}$$

The contour L is considered to be the same as in the preceding subsection.

Equation (42.26) is a particular case of the singular integral equation with Cauchy kernel, the theory of which in the case of closed contours was described in Chapter III, and in the case of open contours will be described in the next chapter. Here we solve this equation without employing the general theory but making use of a method on which this theory is based.

Introduce the analytic function (see § 21.1)

$$\Phi(z) = \frac{1}{2\pi i} \int_L \frac{\varphi(\tau)}{\tau - z} \, d\tau. \tag{42.27}$$

In view of the Sokhotski formulae (4.10)

$$\frac{1}{\pi i} \int_L \frac{\varphi(\tau)}{\tau - t} \, d\tau = \Phi^+(t) + \Phi^-(t).$$

Introducing this relation into equation (42.26) we arrive at the boundary value problem

$$\Phi^+(t) = -\Phi^-(t) - if(t), \tag{42.28}$$

examined in the preceding subsection $(g(t) = -if(t))$ (under the condition (42.19)).

The solution of equation (42.26) can be deduced from that for the boundary value problem by means of the Sokhotski formulae (4.9)

$$\varphi(t) = \Phi^+(t) - \Phi^-(t).$$

1. *Solution unbounded at all the ends.* In order to pass in formula (42.17) to boundary values we define more exactly the branches of the many-valued functions entering this formula. For $R(t)$ we take any definite branch of this function. The branch of $R(z)$ is for instance chosen in such a way that in approaching the contour from the left

the boundary value $R^+(t)$ of this function coincides with the previously chosen branch of $R(t)$. Then the following relation is valid:

$$R^-(t) = - R^+(t) = - R(t).$$

Consequently, applying the Sokhotski formulae we obtain

$$\Phi^+(t) = \frac{1}{R(t)} \left[\frac{1}{2} R(t) g(t) + \frac{1}{2\pi i} \int_L \frac{R(\tau) g(\tau)}{\tau - t} d\tau + P_{m-1}(t) \right],$$

$$\Phi^-(t) = - \frac{1}{R(t)} \left[- \frac{1}{2} R(t) g(t) + \frac{1}{2\pi i} \int_L \frac{R(\tau) g(\tau)}{\tau - t} d\tau + P_{m-1}(t) \right],$$

whence

$$\varphi(t) = - \frac{1}{R(t)} \frac{1}{\pi} \int_L \frac{R(\tau) f(\tau)}{\tau - t} d\tau + \frac{P_{m-1}(t)}{R(t)} \qquad (42.29)$$

$(2 P_{m-1}$ is again denoted by $P_{m-1})$.

It is readily observed that the second term of the right-hand side of formula (42.29) represents the general solution of the homogeneous equation (42.26) $(f(t) \equiv 0)$.

An analogous reasoning yields the following results.

2. *Solution bounded at the ends* a_k *and unbounded at the ends* b_k

$$\varphi(t) = - \frac{R_a(t)}{R_b(t)} \frac{1}{\pi} \int_L \frac{R_b(\tau)}{R_a(\tau)} \frac{f(\tau)}{\tau - t} d\tau. \qquad (42.30)$$

3. *Solution bounded at all the ends*

$$\varphi(t) = - R(t) \frac{1}{\pi} \int_L \frac{f(\tau)}{R(\tau)} \frac{d\tau}{\tau - t}, \qquad (42.31)$$

under the condition that the free term $f(t)$ satisfies the conditions

$$\int_L \frac{f(\tau)}{R(\tau)} \tau^{j-1} d\tau = 0 \quad (j = 1, 2, \ldots, m). \qquad (42.32)$$

Relying on formula (42.24) the reader can easily write down the solution in the general case.

We now quote the solution of a case which is of importance in applications—when the curve L is a section of the real axis $a \leqq t \leqq b$, representing the function $R(t)$ in such a way that only real values appear in our considerations.

1. Solution unbounded at both ends:

$$\varphi(t) = -\frac{1}{\sqrt{[(t-a)(b-t)]}}\left[\frac{1}{\pi}\int_a^b \frac{\sqrt{[(\tau-a)(b-\tau)]}f(\tau)}{\tau-t}d\tau + a_0\right].$$

(42.29′)

2. Solution bounded at the end a and unbounded at b:

$$\varphi(t) = -\sqrt{\left(\frac{t-a}{b-t}\right)}\frac{1}{\pi}\int_a^b \sqrt{\left(\frac{b-\tau}{\tau-a}\right)}\frac{f(\tau)}{\tau-t}d\tau.$$

(42.30′)

3. Solution bounded at both ends:

$$\varphi(t) = -\sqrt{[(t-a)(b-t)]}\frac{1}{\pi}\int_a^b \frac{f(\tau)}{\sqrt{[(\tau-a)(b-\tau)]}}\frac{d\tau}{\tau-t},$$

(42.31′)

under the condition

$$\int_a^b \frac{f(\tau)\,d\tau}{\sqrt{[(\tau-a)(b-\tau)]}} = 0.$$

(42.32′)

Many papers have been devoted to the solution of this problem by various particular methods. The references are given in the book of N. I. Muskhelishvili [17*].

§ 43. Direct solution of the Riemann problem

In the case of the Riemann problem with discontinuous coefficients and an open contour, besides the method given in the preceding subsections and based on reducing it to a problem with continuous coefficients solved before, another method is known which consists of a direct application of the method of solution for the problem with continuous coefficients. It is more convenient here to begin by considering the case of an open contour.

43.1. The problem for an open contour

We start from the homogeneous problem

$$\Phi^+(t) = G(t)\,\Phi^-(t). \tag{43.1}$$

A reasoning analogous to that of § 14.3 proves that the function†

$$\Phi(z) = e^{\Gamma(z)}, \quad \Gamma(z) = \frac{1}{2\pi i} \int_L \frac{\ln G(\tau)}{\tau - z}\, d\tau \tag{43.2}$$

satisfies boundary condition (43.1) everywhere except possibly at the ends. Let us examine the behaviour of $\Phi(z)$ at the ends. On the basis of formulae (8.1) and (8.2)

$$e^{\Gamma(z)} = (z - a_k)^{-\frac{\ln G(a_k)}{2\pi i}} e^{\Gamma_k(z)} = (z - a_k)^{-\frac{\theta_k}{2\pi} + i\frac{\ln \varrho_k}{2\pi}} e^{\Gamma_k(z)}, \tag{43.3}$$

$$e^{\Gamma(z)} = (z - b_k)^{\frac{\ln G(b_k)}{2\pi i}} e^{\Gamma_k^*(z)} = (z - b_k)^{\frac{\theta_k + \Delta_k}{2\pi} - i\frac{\ln \varrho_k^*}{2\pi}} e^{\Gamma_k^*(z)}, \tag{43.4}$$

where $\Gamma_k(z)$ and $\Gamma_k^*(z)$ are functions bounded in the vicinities of the points a_k and b_k, respectively. We shall take the value of the argument of $G(t)$ at the initial points a_k according to (42.5) or (42.6) and we introduce the integers $\varkappa_k$ which satisfy conditions (42.8) or (42.9). Consider the function

$$X(z) = \prod_{k=1}^{m} (z - b_k)^{-\varkappa_k} e^{\Gamma(z)}. \tag{43.5}$$

It is evident that it also satisfies the boundary condition and belongs to the given class of solutions on the ends.

The function $X(z)$ is called *the canonical function of the problem*. Since $\Gamma(\infty) = 0$ it is evident that the order of $X(z)$ at infinity is given by the relation

$$\varkappa = \sum_{k=1}^{m} \varkappa_k. \tag{43.6}$$

The number $\varkappa$ is the index of the problem. Now the solution is carried out in the usual way. Setting $G(t) = X^+(t)/X^-(t)$ we write the boundary condition in the form

$$\frac{\Phi^+(t)}{X^+(t)} = \frac{\Phi^-(t)}{X^-(t)}, \tag{43.7}$$

whence, for $\varkappa \geq 0$

$$\Phi(z) = X(z)\,P_\varkappa(z). \tag{43.8}$$

† arg $G(t)$ is, as before, regarded as varying continuously on all arcs.

If it is required that $\Phi(\infty) = 0$ then $P_\varkappa$ is to be replaced by $P_{\varkappa-1}$. By the usual method we find the solution of the non-homogeneous problem (for $\Phi(\infty) = 0$)

$$\Phi(z) = X(z)[\Psi(z) + P_{\varkappa-1}(z)], \qquad (43.9)$$

where

$$\Psi(z) = \frac{1}{2\pi i} \int_L \frac{g(\tau)}{X^+(\tau)} \frac{d\tau}{\tau - z}; \qquad (43.10)$$

for $\varkappa < 0$ the conditions of solubility

$$\int_L \frac{g(\tau)}{X^+(\tau)} \tau^{j-1} d\tau = 0 \qquad (j = 1, 2, \ldots, -\varkappa) \qquad (43.11)$$

must be satisfied.

43.2. The problem with discontinuous coefficients

Dividing the contour L into sections between the points of discontinuity of the coefficient $G(t)$ and regarding these sections as independent curves, we can solve the problem under consideration as the problem for an open contour. The distinction of the problem consists in the fact that the ends of the curves† coincide in pairs. We shall now outline the way of altering the solution of § 43.1 in order to deduce the solution of our problem.

We select a point of discontinuity t_1 to be the initial point and we define here $\arg G(t_1 + 0) = \theta$ by any concrete branch. By a continuous variation, $\arg G(t_2 - 0) = \theta_1 + \Delta$ is determined. The quantity $\arg G(t_2 + 0) = \theta_2$ is defined in such a way that‡

$$-2\pi < \arg G(t_2 - 0) - \arg G(t_2 + 0) < 0, \qquad (43.12)$$

if solutions unbounded at t_2 are required, and

$$0 < \arg G(t_2 - 0) - \arg G(t_2 + 0) < 2\pi, \qquad (43.13)$$

if we require solutions bounded at this point.

Similarly we proceed at all remaining points of discontinuity $(t_3, \ldots, t_n)$. At the point t_1 we define the integer $\varkappa$ which satisfies

---

† The term "the ends" will hereafter be applied not only to the ends of the curve but also to the points of discontinuity.

‡ It is assumed that t_2 is not a point of automatic boundedness. If it does belong to the latter class, the sign " = " is to be taken in the left-hand part of formula (43.13).

one of the conditions

$$-2\pi < \arg G(t_1 - 0) - \arg G(t_1 + 0) - 2\varkappa\pi < 0, \quad (43.14)$$

$$0 < \arg G(t_1 - 0) - \arg G(t_1 + 0) - 2\varkappa\pi < 2\pi, \quad (43.15)$$

depending on whether the solution is to be unbounded or bounded at the point t_1; the number $\varkappa$ is the index of the problem. Then

$$X(z) = (z - t_1)^{-\varkappa} e^{\Gamma(z)}, \quad \Gamma(z) = \frac{1}{2\pi i} \int_L \frac{\ln G(\tau)}{\tau - z} d\tau \quad (43.16)$$

is the canonical function of the homogeneous problem.

The solution of the non-homogeneous problem is given by formulae (43.9)–(43.11).

43.3. Examples

EXAMPLE 1. As the first example of application of the method described in this section we consider a problem encountered in solving a problem of mechanics.

Suppose that the contour L is the section $(0, l)$ of the real axis. Consider the boundary value problem

$$\Phi^+(t) = \frac{a - i(a_0 + a_1 t)}{a + i(a_0 + a_1 t)} \Phi^-(t) + g(t), \quad (43.17)$$

where a, a_0, a_1 are some prescribed real constants and $g(t)$ is a known function which satisfies the Hölder condition. We seek the solution which vanishes at infinity.

Since $|G(t)| = 1$ we have

$$\ln G(t) = 2i \arg[a - i(a_0 + a_1 t)] = -2i \arctan \frac{a_0 + a_1 t}{a} = 2i\theta. \quad (43.18)$$

The points with coordinates $a - i(a_0 + a_1 t)$ are geometrically represented by the section of the straight line parallel to the y-axis between the points $\zeta_1 = a - ia_0$ and $\zeta_2 = a - i(a_0 + a_1 l)$; hence $\varDelta = \{\theta\}_L < \pi$.

For definiteness we take $a < 0$. The following four cases may occur.

(1) ζ_1, ζ_2 lie both in quadrant I;
(2) ζ_1, ζ_2 lie both in quadrant IV;
(3) ζ_1 lies in quadrant I and ζ_2 in quadrant IV;
(4) ζ_1 lies in quadrant IV and ζ_2 in quadrant I.

For definiteness we seek the solution which is bounded at the origin of the coordinate system and unbounded at the other end. Let us consider in detail the fourth case (Fig. 16).

In accordance with (42.5) for $2\theta_1$ we have to take the branch which satisfies the condition $-2\pi < 2\theta_1 < 0$, the latter inequality being determined only by the chosen class of solutions. Then $0 < 2\theta_2 = 2(\theta_1 + \varDelta) < 2\pi$. By virtue of formula (42.9)

$$\varkappa = \left[\frac{\theta_1 + \varDelta}{\pi} \right] + 1 = 1.$$

15*

The canonical solution has the form

$$X(z) = (z - l)^{-1} e^{\Gamma(z)}, \quad \Gamma(z) = -\frac{1}{\pi} \int_0^l \frac{\arctan \dfrac{a_0 + a_1 \tau}{a}}{\tau - z} \, d\tau, \qquad (43.19)$$

and the general solution is given by the expression

$$\Phi(z) = X(z) [\Psi(z) + c_0], \qquad (43.20)$$

where

$$\Psi(z) = \frac{1}{2\pi i} \int_0^l \frac{g(\tau)}{X^+(\tau)} \frac{d\tau}{\tau - z}$$

and c_0 is an arbitrary constant.

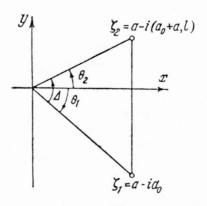

FIG. 16.

Without going into details for the other cases, we indicate the basic properties.

1st case.

$$-2\pi < 2\theta_2 = 2(\theta_1 + \Delta) < 0; \quad -1 < \frac{\theta_1 + \Delta}{\pi} < 0, \quad \varkappa = \left[\frac{\theta_1 + \Delta}{\pi} \right] + 1 = 0$$

2nd case.

$$-2\pi < 2\theta_2 = 2(\theta_1 + \Delta) < 0, \quad \varkappa = 0.$$

3rd case.

$$-4\pi < 2\theta_2 = 2(\theta_1 + \Delta) < -2\pi, \quad \varkappa = -1.$$

In all these cases the homogeneous problem is insoluble and the non-homogeneous is uniquely soluble; in the last case one additional condition is to be satisfied.

The reader is advised to investigate independently the solution in the remaining three possible classes of functions.

The solution described above will be used in § 37.4 in solving singular integral equations.

Let us make some remarks concerning the solution of the problems. From the viewpoint of practice when the solution has to be obtained in a definite form, the main difficulty is to compute the required integrals. From the theoretical viewpoint (the determination of the number of solutions) the problem consists, as the reader has found out from the example considered above, in the appropriate choice of the initial value of the argument (depending on the class of solutions) and in determining its final value from the initial by a continuous variation.

EXAMPLE 2. In order to become better acquainted with the technique of solving the problems, we consider one more example in which the data is chosen in such a way that the solution is given in terms of elementary functions.

Let L be the unit circle and L_1 and L_2 its right and left semi-circles, respectively. Let us solve the Riemann boundary value problem

$$\Phi^+(t) = G(t)\, \Phi^-(t) + \frac{(t^2+1)(4t+i)}{t^2+a^2}\left(\frac{t-i}{t+i}\right)^{\frac{it}{2}} \quad (|a| < 1), \qquad (43.21)$$

where

$$G(t) = \begin{cases} e^{\pi t} & \text{on } L_1, \\ e^{2\pi t} & \text{on } L_2. \end{cases}$$

The many-valued function $[(t-i)/(t+i)]^{it/2}$ is determined by a definite branch.

We first seek the solution in the most general class of functions havinig integrable singularities at both points of discontinuity $\pm i$.

Making use of formulae (43.16) we now find the canonical function of the homogeneous problem

$$\Gamma(z) = \frac{1}{2\pi i}\int_L \frac{\ln G(\tau)}{\tau - z}\, d\tau = \frac{\pi}{2\pi i}\int_{L_1}\frac{\tau}{\tau - z}\, d\tau + \frac{2\pi}{2\pi i}\int_{L_2}\frac{\tau}{\tau - z}\, d\tau$$

$$= \frac{\pi}{2\pi i}\int_L \frac{\tau}{\tau - z}\, d\tau + \frac{1}{2i}\int_i^{-i}\frac{\tau}{\tau - z}\, d\tau$$

$$= \frac{1}{2i}\int_L \frac{\tau}{\tau - z}\, d\tau - 1 + \frac{z}{2i}\ln\frac{z+i}{z-i}.$$

Applying to the integral the Cauchy formula for $z \in D^+$ and the Cauchy theorem for $z \in D^-$ we obtain

$$\Gamma^+(z) = \pi z - 1 + \frac{z}{2i}\ln\frac{z+i}{z-i},$$

$$\Gamma^-(z) = -1 + \frac{z}{2i}\ln\frac{z+i}{z-i}.$$

Hence

$$X^+(z) = e^{\pi z - 1}\left(\frac{z-i}{z+i}\right)^{\frac{iz}{2}},$$

$$X^-(z) = e^{-1}\left(\frac{z-i}{z+i}\right)^{\frac{iz}{2}}.$$

It can easily be verified by a direct substitution of the values $\pm i$ that at these points $X(z)$ behaves as $(z-i)^{-\frac{1}{2}}$, $(z+i)^{-\frac{1}{2}}$, i.e. it belongs to the considered class. Consequently the additional factors $(z \pm i)^{\varkappa}$ are not necessary, i.e. the index of the problem $\varkappa = 0$. For the canonical function and the solution of the problem itself to be fully defined we must indicate the manner of prescribing the many-valued function $\omega(z) = [(z-i)/(z+i)]^{iz/2}$, which is equivalent to prescribing the cut connecting the branch points $\pm i$ and the branch. Inserting X into the homogeneous boundary conditions we have

$$\omega^+(t) = \omega^-(t) \quad \text{on} \quad L_1, \quad \omega^+(t) = e^{\pi t}\omega^-(t) \quad \text{on} \quad L_2.$$

This means that the cut should be made along the left semi-circle L_2 and the value $\omega^-(t)$ on the right (exterior) side is deduced from the value $\omega^+(t)$ on the left (interior) side by multiplying by $e^{-\pi t}$. If we describe a path around one of the branch points (say i) exactly the factor $e^{2\pi i t/2} = e^{-\pi t}$ arises. We shall hereafter denote the limiting values on L from the inside by the same expression $[(t-i)/(t+i)]^{it/2}$; the values of $\omega^-(t)$ on L_1 are the same, while the values of $\omega^-(t)$ on L_2 are to be multiplied by $e^{-t\pi}$. The above reasoning is very important in the applications of boundary value problems to solving singular integral equations (see § 50.2).

The branch of $\omega(z)$ is chosen in such a way that $\omega^+(t)$ is identical with the corresponding function entering $g(t)$.

Let us now find $\Psi(z)$. According to formula (43.10)

$$\Psi(z) = \frac{1}{2\pi i}\int_L \frac{g(\tau)\,d\tau}{X^+(\tau)(\tau - z)} = \frac{e}{2\pi i}\int_L \frac{(\tau^2+1)(4\tau+i)\,e^{-\pi\tau}}{\tau^2+a^2}\,\frac{d\tau}{\tau - z}.$$

Making use of the theory of residues we obtain after some transformations

$$\Psi^+(z) = e\,\frac{1-a^2}{z^2+a^2}\left[\frac{iz-4a^2}{a}\sin a\pi - (i+4z)\left(\cos a\pi - e^{-\pi z}\frac{z^2+1}{1-a^2}\right)\right]$$

$$\Psi^-(z) = e\,\frac{1-a^2}{z^2-a^2}\left[\frac{iz-4a^2}{a}\sin a\pi - (i+4z)\cos a\pi\right].$$

By virtue of formula (43.9) in which we take $P_{\varkappa-1}$ instead of $P_\varkappa$ we have finally

$$\left.\begin{aligned}\Phi^+(z) &= \left(\frac{z-i}{z+i}\right)^{\frac{iz}{2}} e^{\pi z-1}[\Psi^+(z) + c],\\[2mm]\Phi^-(z) &= e^{-1}\left(\frac{z-i}{z+i}\right)^{\frac{iz}{2}}[\Psi^-(z) + c].\end{aligned}\right\} \qquad (43.22)$$

We advise the reader to solve the problem independently, assuming that an integrable infinity is admitted at the point i, the solution being bounded in the remaining part of the plane.

For reference we note the formulae giving the solution

$$
\left.
\begin{aligned}
\Phi^+(z) &= e^{\pi z}\left(\frac{z-i}{z+i}\right)^{\frac{iz}{2}}\frac{z+i}{z^2+a^2}\left[\sin a\pi\left(\frac{z}{a}-4az+3ai\right)+\right.\\
&\quad + \left.\cos a\pi(3iz+4a^2-1)+e^{-\pi z}(4z^2-3iz+1)\right]\\
\Phi^-(z) &= \left(\frac{z-i}{z+i}\right)^{\frac{iz}{2}}\frac{z+i}{z^2+a^2}\left[\sin a\pi\left(\frac{z}{a}-4az+3ai\right)+\right.\\
&\quad + \left.\cos a\pi(3iz+4a^2-1)\right]
\end{aligned}
\right\}
\tag{43.23}
$$

43.4. Concluding remarks

Making use of the methods of § 16 we can easily derive the solution for the case when the contour is composed of any number of closed or open non-intersecting curves, each of them containing an arbitrary finite number of points of discontinuity of the first kind. We shall however not dwell on this problem, since in the next section a slight modification of the method will enable us to present the solution for an even more general case in which an intersecting contour is admitted.

We now make one more remark concerning the possible generalizations of the solution described in § 42 and § 43.

It is well known to the reader that the general solution of the non-homogeneous problem is a sum of a particular solution of the non-homogeneous problem and the general solution of the homogeneous problem, and in selecting a second particular solution the constants entering the general solution are altered in a certain way. Constructing the general solution of the non-homogeneous problem, both parts of it were taken in the same class. This was expressed by the fact that the particular solution of the non-homogeneous problem contained the same canonical function as the solution of the homogeneous problem

$$
\Phi(z) = X(z)\frac{1}{2\pi i}\int_L \frac{g(\tau)}{X^+(\tau)}\frac{d\tau}{\tau-z} + X(z)P_\varkappa(z).
$$

Making use of the arbitrariness in the particular solution we may, in determining the general solution in any class of functions, take the particular solution of the non-homogeneous problem in another narrower class.

For instance, the general solution of the problem of § 42.2 belonging to the most general class of functions unbounded at all ends, can be taken in the form

$$
\Phi(z) = \frac{R_a(z)}{R_b(z)}\frac{1}{2\pi i}\int_L \frac{R_b(\tau)}{R_a(\tau)}\frac{d\tau}{\tau-z} + \frac{P_{m-1}(z)}{R(z)}.
$$

Here the particular solution is taken in the class of functions bounded at the ends a_k. The formula of inversion of the singular integral (§ 42.3) is altered correspondingly.

To conclude the section let us compare the two methods described in §§ 41–43.

In solving actual problems both methods are equivalent and the reader should employ that which appeals to him more. On general theoretical considerations concerning the problems to be solved, the second method (§ 43) is more convenient, since its formulae contain only known quantities and therefore are simpler than the corresponding formulae of § 42. However, the first method has the advantage over the second in that it is independent of the presence of a closed form of the solution; hence, it can be applied also in the cases when such a solution is unknown, for instance in generalized problems (see Problem 15 at he end of this chapter) and in the case of many unknown functions.

§ 44. Riemann problem for a complicated contour

44.1. Formulation of the problem

Let the contour L consist of a finite number of simple smooth closed or open curves, situated in an arbitrary way on the plane. The curves may have an arbitrary finite number of points of intersection or tangency and any number of them may be located inside others. Under these very general assumptions concerning the contour it may happen that it no longer divides the plane into the domains D^+ and D^- and a new formulation must be given to the Riemann problem. First, let us generalize the definition given in § 1 and let us agree to call a function sectionally analytic if it is analytic in every connected part of the plane which does not contain points of the contour. At the points of the contour there exist continuous limiting values from the left and from the right†, except at a finite number of points at which the limiting values either have infinities of an integrable order or remain bounded but depend on the manner of approaching the contour.

The Riemann boundary value problem can be stated as follows.

To determine a sectionally analytic function $\Phi(z)$ which satisfies on the contour L the boundary condition

$$\Phi^+(t) = G(t)\,\Phi^-(t) + g(t), \tag{44.1}$$

where $G(t)$, $g(t)$ are functions satisfying the Hölder condition everywhere, except at a finite number of points where they may have discontinuities of the first kind‡, and $G(t) \neq 0$.

When we speak of satisfying the boundary condition, the ends and points of intersection are excluded.

---

† It is assumed that a direction is established on all curves.
‡ The class of $g(t)$ can be extended (see the end of § 41.5).

The above formulated problem which is at first sight very complicated (the first methods of its solution were indeed very complicated), can be solved comparatively simply if we make use of the results of the solution of the problem with discontinuous coefficients, and introduce a small modification in the solution of the problem for the case of one closed curve with continuous coefficients. We now proceed to consider this modification.

44.2. A new method of solution of the Riemann problem for a closed curve

Suppose that the contour consists of one closed curve and $G(t)$ is a continuous function. Let us apply to the solution of the Riemann problem the method by which we solved in § 43.1 the same problem for an open contour.

Let us select on the contour a point t_1 and consider it to be the initial point. The closed contour is regarded as a special case of the open one, when the two ends coincide. Consider the function $e^{\Gamma(z)}$ where

$$\Gamma(z) = \frac{1}{2\pi i} \int_L \frac{\ln G(\tau)}{\tau - z} d\tau. \tag{44.2}$$

If $\varkappa = \operatorname{Ind} G(t) \neq 0$ the function $\ln G(t)$ has a discontinuity at t_1 and we have at this point the relation

$$\ln G(t_1 - 0) - \ln G(t_1 + 0) = 2\pi i \varkappa. \tag{44.3}$$

According to formula (8.3) we have in the vicinity of the point t_1

$$\Gamma(z) = \varkappa \ln(z - t_1) + \Gamma_0(z), \tag{44.4}$$

where $\Gamma_0(z)$ is a bounded function which on approaching t_1 tends to a definite limit. Hence

$$e^{\Gamma(z)} = (z - t_1)^{\varkappa} e^{\Gamma_0(z)}. \tag{44.5}$$

By the canonical function we understand the same function as in § 14.4. Then it can easily be verified that the canonical function of the considered Riemann problem has the form

$$X(z) = (z - t_1)^{-\varkappa} e^{\Gamma(z)}. \tag{44.6}$$

If $G(t)$ has points of discontinuity, then taking for t_1 the first such point and applying the same reasoning as in §43.2 we find that formula (44.6) yields the canonical solution in this case as well.

On the basis of the above constructed canonical function the general solution of the homogeneous and non-homogeneous problems may be deduced in the usual way (§§ 14.4, 14.5, 43.1); therefore there is no necessity for a special exposition.

44.3. The general case

We proceed to the general case stated in § 44.1, making for the time being the simplifying assumption that every point may serve as the end of not more than one curve of the contour. We begin by considering the principal problem—the construction of the canonical function.

DEFINITION. *By the canonical function of the Riemann problem we understand the sectionally analytic function* X(z) *which satisfies the homogeneous boundary condition*

$$X^+(t) = G(t) X^-(t)$$

and has the following properties.

1. *It has zero order everywhere in the finite part of the plane, except possibly at the ends*†.

2. *Near the ends it belongs to a given class* (*it is bounded or has an infinity of an integrable order*).

3. *At infinity it has the highest possible order* (*equal to the index of the problem in the considered class*).

Let us construct the canonical function for every simple curve $L_k (k = 1, 2, \ldots, m)$ of which the contour L is composed, observing the following rules.

1°. If L_k is an open contour we shall construct for it the canonical function in accordance with formulae of § 43.1, i.e.

$$X_k(z) = (z - b_k)^{-\varkappa_k} e^{\Gamma_k(z)}, \quad \Gamma_k(z) = \frac{1}{2\pi i} \int_{L_k} \frac{\ln G(\tau)}{\tau - z} d\tau. \quad (44.7)$$

2°. If L_j is a closed contour we select as the initial point a point b_j which is a point of discontinuity if there exist such points on the contour L_j, or an arbitrary point if $G(t)$ is continuous on L_j. The canonical function for the contour L_j is constructed in accordance

† Let us recall that "the ends" contain also the points of discontinuity (see § 43.2).

with the rules of § 43.3 or § 44.2, which leads to the formulae

$$X_j(z) = (z - b_j)^{-\varkappa_j} e^{\Gamma_j(z)}, \quad \Gamma_j(z) = \frac{1}{2\pi i} \int\limits_{L_j} \frac{\ln G(\tau)}{\tau - z} \, d\tau; \quad (44.8)$$

$\varkappa_k$ and $\varkappa_j$ in (44.7) and (44.8) are the indices of the coefficient $G(t)$ with respect to the curves L_k, L_j in the considered class of functions. The quantity

$$\varkappa = \sum_{k=1}^{m} \varkappa_k \quad (44.9)$$

will be called *the index of the Riemann problem for the complex contour in the considered class.*

We now prove that the canonical function of the Riemann problem for the whole complex contour $L = L_1 + L_2 \ldots + L_m$ is the product of the canonical functions for all simple contours constituting L:

$$X(z) = \prod_{k=1}^{m} X_k(z). \quad (44.10)$$

We begin the proof by showing that the last function satisfies the boundary condition. Let t be a point of a curve L_j, distinct from the ends and the points of intersection of the curves L_k. Substituting into the boundary condition the function (44.10) and taking into account that for every $k \neq j \, X_k^+(t) = X_k^-(t)$ we arrive after simplification at the relation

$$X_j^+(t) = G(t) X_j^-(t),$$

which is identically satisfied in accordance with the definition of the canonical function $X_j(z)$.

The first property of the canonical solution holds in view of formulae (44.7), (44.8). The second property also holds, since by definition all canonical functions for the simple curves belong to the considered class.

To establish the third property let us compute the order of $X(z)$ at infinity.

Inserting into formula (44.10) expressions (44.7), (44.8) and taking into account that $\Gamma_j(\infty) = 0$ for every curve L_j (closed or open) we find that the expansion in the vicinity of infinity starts from the term $(1/z)^{\varkappa}$. This completes the proof.

It can easily be established by reasoning similar to that at the end of § 41.1, that the canonical function is determined to within a constant term.

15 a*

If we normalize the canonical function by means of the condition

$$\lim_{z \to \infty} z^{\varkappa} X(z) = 1,$$

then formula (44.10) determines it uniquely.

Taking into account formulae (44.7), (44.8), (44.10) we can represent the canonical function in the form

$$X(z) = \prod_{j=1}^{m} (z - b_k)^{-\varkappa_j} e^{\Gamma(z)}, \quad \Gamma(z) = \frac{1}{2\pi i} \int_L \frac{\ln G(\tau)}{\tau - z} d\tau, \quad (44.11)$$

where b_k is the final point of the curve if the latter is open and the point taken for the beginning (and the end) of the contour if it is closed.

After the construction of the canonical function the general solutions of the homogeneous and non-homogeneous problems in a given class are found by a method which has already been used many times. Representing the coefficient of the Riemann problem in the form of the ratio of the canonical functions

$$G(t) = \frac{X^+(t)}{X^-(t)}$$

and reducing the boundary condition to the form

$$\frac{\Phi^+(t)}{X^+(t)} - \frac{\Phi^-(t)}{X^-(t)} = \frac{g(t)}{X^+(t)},$$

we arrive at the problem of the determination of an analytic function in terms of a given jump under the condition that the required function $\Phi(z)/X(z)$ has at infinity the order $-\varkappa$. Hence, by the usual considerations we obtain the general solution

$$\Phi(z) = X(z) [\Psi(z) + P_\varkappa(z)], \quad (44.12)$$

where

$$\Psi(z) = \frac{1}{2\pi i} \int_L \frac{g(\tau)}{X^+(\tau)} \frac{d\tau}{\tau - z}, \quad (44.13)$$

and $P_\varkappa(z)$ is a polynomial of degree $\varkappa$. When $\varkappa < 0$ the following $-\varkappa - 1$ conditions of solubility have to be satisfied:

$$\int_L \frac{g(\tau)}{X^+(\tau)} \tau^{j-1} d\tau = 0 \qquad (j = 1, 2, \ldots, -\varkappa - 1); \quad (44.14)$$

moreover, we set $P_\varkappa \equiv 0$.

The results have the same form as in the particular cases examined before (§ 14.5, 16.2, 43.1).

44.4. The case of coincidence of the ends

We shall briefly consider the nature of the case excluded so far when more than one end of the simple curves coincides at one point. It follows from formula (44.10) that the order of the canonical function $X(z)$ for the complex contour is equal to the sum of the orders of the canonical functions $X_k(z)$ for the simple contours constituting the complex contour. Thus, if at a given point t_0 several ends meet, then the order at t_0 of the canonical function is equal to the sum of the orders of the canonical functions for all simple curves having an end at the considered point. It follows that the orders of the latter canonical functions at the point t_0 cannot be arbitrary, but must be chosen in such a way that the resultant order at the point t_0 determines a canonical function belonging to the class given at the considered point. Furthermore, we have to observe the general requirement of the maximum of the order at infinity†. These conditions still leave a certain arbitrariness in the choice of the canonical functions for the simple contours ending at the point t_0; this arbitrariness, however, has no influence on the form of the canonical function for the whole contour.

Let us illustrate the above by an example.

Suppose that at a given point t_0 there meet three ends denoted by the numbers 1, 2, 3 and suppose that for these ends

$$e^{\Gamma_1(z)} = (z - t_0)^{\frac{1}{3}} e^{\Gamma_1^*(z)}, \quad e^{\Gamma_2(z)} = (z - t_0)^{\frac{2}{3}} e^{\Gamma_2^*(z)},$$
$$e^{\Gamma_3(z)} = (z - t_0)^{\frac{2}{3}} e^{\Gamma_3^*(z)},$$

where $\Gamma_k^*(z)$ are functions bounded in the vicinity of the point t_0. For definiteness we assume that the unknown canonical function is bounded at t_0.

It is readily observed that the following three possibilities arise for the exponents $\varkappa_1, \varkappa_2, \varkappa_3$:

$$(0, 0, -2), \quad (0, -1, -1), \quad (-1, 0, -1).$$

Disregarding all other possible factors we obtain for the canonical functions X_1, X_2, X_3 the admissible expressions

$$(e^{\Gamma_1(z)}, e^{\Gamma_2(z)}, (z - t_0)^{-2} e^{\Gamma_3(z)}),$$
$$(e^{\Gamma_1(z)}, (z - t_0)^{-1} e^{\Gamma_2(z)}, (z - t_0)^{-1} e^{\Gamma_3(z)}),$$
$$((z - t_0)^{-1} e^{\Gamma_1(z)}, e^{\Gamma(z)}, (z - t_0)^{-1} e^{\Gamma_3(z)}).$$

† This makes it impossible to increase the order of the zero at a finite point above the maximum possible.

In all three cases we have for the canonical function $X(z)$ the same expression, namely

$$(z - t_0)^{-2} e^{\Gamma_1(z)+\Gamma_2(z)+\Gamma_3(z)}.$$

§ 45. Exceptional cases and the general concept of index

45.1. Introductory remarks

In the theory of boundary value problems for analytic functions the concept of the index is of primary importance. It constitutes the fundamental quantitative characteristic of the set of solutions of the considered boundary value problem, influences the qualitative nature of the solution and is closely connected with the classes of functions admissible for solutions.

The concept of the index is defined in every problem independently. For instance the index of the Riemann problem (§ 12.1) was defined as the index of the coefficient $G(t)$ of the boundary condition with respect to the contour L, i.e. as the integer equal to the increment of the argument of $G(t)$ in describing the contour L, divided by 2π. It is necessary for this definition that the considered contour be closed and the function $G(t)$ be continuous and non-vanishing. A violation of any of these conditions renders the concept of the index meaningless and leads to the necessity of new definitions of the index in considering the relevant problems. The latter definitions are very diverse, suitable only in the cases under consideration, and they depend on the method of solution of the problem. As an example we may recall the definitions of the index of the Riemann problems with discontinuous coefficients (§§ 41.4, 43.2) and with open contour (§ 42.1). The index of this problem when the coefficient $G(t)$ had zeros or infinities (§ 15) was not defined at all. The investigation was carried out using the index of the function $G_1(t)$ deduced from $G(t)$ by separating the singularities and using the number of these singularities.

It is desirable to construct the concept of the index in such a way that it would be applicable in all cases of the boundary value problem under consideration. This can be done in the following way. First, the open contour is reduced to the case of a closed contour and discontinuous coefficients; next the contour L is somewhat deformed near the zeros and the points of discontinuity of the coefficients of the boundary condition in such a way that the new contour does not pass any more through these critical points which are now located either inside or outside L_0. The coefficients of the boundary condition on the deformed sections of the contour L are redefined in a continuous way, so that the solutions of the new "reduced" boundary value problem yield the solutions of the original problem in passing to the limit $L_0 \to L$.

The deduced new boundary value problem with continuous coefficients and closed contour is a type of a standard normal case. The index of this problem is taken as the index of the original problem.

It is readily observed that the index so defined depends essentially on the distribution of the critical points with respect to the contour L_0. Therefore the choice of the manner of deformation of the contour L into L_0 is to be determined by the class of admissible functions.

Let us briefly consider the realization of the outlined method by examining the Riemann problem when the coefficient $G(t)$ vanishes on the contour and has infinities of integral orders.

45.2. The homogeneous Riemann problem

Thus, we seek a sectionally analytic function $\Phi(z)$ which satisfies on a smooth closed contour the boundary condition

$$\Phi^+(t) = G(t)\,\Phi^-(t), \tag{45.1}$$

where $G(t)$ has at the points α_k zeros of orders m_k $(k = 1, 2, \ldots, \mu)$ and at the points β_j poles† of orders p_j $(j = 1, 2, \ldots, \nu)$; besides, it satisfies the Hölder condition everywhere on L except at the points β_j.

In solving the problem we shall extend the class of admissible functions, by a sectionally analytic function $\Phi(z)$ meaning a function analytic in D^+ and D^- and continuously continuable on L everywhere, except possibly at a finite number of points at which infinity of one of the limiting values $\Phi^+(t)$ or $\Phi^-(t)$ is possible—the other remaining bounded. Observe this is the only case in this book in which we admit non-integrable boundary values.

The admissible functions will be divided into classes, referring a sectionally analytic function $\Phi(z)$ to a class $H(\alpha_1, \alpha_2, \ldots, \alpha_r; \beta_1, \beta_2, \ldots, \beta_s)$ or, briefly, $H_{r,s}$ if near the points $\alpha_1, \alpha_1, \ldots, \alpha_r$, $1 \le r \le \mu$, $\beta_1, \beta_2, \ldots, \beta_s$, $1 \le s \le \nu$ one of the functions $\Phi^+(z)$ or $\Phi^-(z)$ can be unbounded while the other is bounded. The points near which an infinity of one of the functions $\Phi^+(z)$ or $\Phi^-(z)$ is admitted will be called the points determining the class of solutions.

The solution will be sought in the class $H_{r,s}$. Let us note that the most narrow class is the class of bounded functions $H_0(r = 0, s = 0)$. The solution of problem (45.1) in this class was already deduced in § 15.1.

We carry out the following constructions. We describe from the critical points α and β_j, circles T_{α_k}, T_{β_j} of sufficiently small radius ε. The arcs of the contour L cu by these circles are now replaced by arcs of the circles themselves located inside or outside L, in accordance with the following rules.

(1) If the zero of the function $G(t)$, the point α_k, determines the class of solutions $H_{r,s}$ then we take the arc of T_{α_k} located in D^- (Fig. 15); in the opposite case we take the arc located in D^+.

(2) For the poles at β_j we proceed conversely, i.e. if β_j determines the class of solutions $H_{r,s}$ then we take the arc of T_{β_j} belonging to D^+ and in the opposite case the arc in D^-.

Thus, every class of solutions $H_{r,s}$ is associated with a fully defined contour L_0 which is called the reduced contour of problem (45.1). This fact can be written in the form

$$L_0 = L' + \sum_{k=1}^{\mu} T_{\alpha_k} + \sum_{j=1}^{\nu} T_{\beta_j}, \tag{45.2}$$

† The term "pole" is here used in the same sense as in § 15.1.

L' denoting the remaining part of the contour L. Let us represent the coefficient $G(t)$ of the boundary condition (45.1) in the form

$$G(t) = \frac{\prod\limits_{k=1}^{\mu} (t - \alpha_k)^{m_k}}{\prod\limits_{j=1}^{\nu} (t - \beta_j)^{p_j}} \, G_1(t), \qquad (45.3)$$

where $G_1(t)$ is a function which satisfies the Hölder condition and does not vanish anywhere.

We proceed to determine on L_0 the reduced coefficient $G_0(t)$. Suppose that c is any of the critical points α_k or β_j and T_c the corresponding arc of the circle constituting a part of the reduced contour L_0; t'_c and t''_c are the points of intersection of the latter arc with the contour L (Fig. 17).

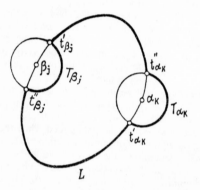

FIG. 17.

Consider the family of linear fractional functions $F(z)$ defined by the relation

$$\frac{F(z) - G_1(t'_c)}{F(z) - G_1(t''_c)} = A \, \frac{z - t'_c}{z - t''_c}, \qquad (45.4)$$

where A is an arbitrary constant. This family maps the circle T_c onto an elliptic bundle of circles passing through the points $G_1(t'_c)$ and $G_1(t''_c)$. Obviously, taking the radius ε of the circle T_c sufficiently small, by the suitable choice of the constant A we can select from the family (45.4) a function $F_c(z)$ which does not vanish on the arc T_c and the argument of which has an increment equal to the increment of the argument of the function $G_1(t)$ on the arc $\widehat{t'_c t''_c}$ of the contour L:

$$[\arg F_c(t)]_{T_c} = [\arg G_1(t)]_{\widehat{t'_c t''_c}}. \qquad (45.5)$$

Having constructed such functions for all critical points of the problem we define a function $G_0(t)$ on the reduced contour L_0, setting

$$G_0(t) = \begin{cases} \dfrac{\prod\limits_{k=1}^{\mu} (t-\alpha_k)^{m_k}}{\prod\limits_{j=1}^{\nu} (t-\beta_j)^{p_j}}\, G_1(t), & \text{if } t \in L', \\[3em] \dfrac{\prod\limits_{k=1}^{\mu} (t-\alpha_k)^{m_k}}{\prod\limits_{j=1}^{\nu} (t-\beta_j)^{p_j}}\, F_{\alpha_j}(t), & \text{if } t \in T_{\alpha_j}, \\[3em] \dfrac{\prod\limits_{k=1}^{\mu} (t-\alpha_k)^{m_k}}{\prod\limits_{j=1}^{\nu} (t-\beta_j)^{p_j}}\, F_{\beta_j}(t), & \text{if } t \in T_{\beta_j}, \end{cases} \qquad (45.6)$$

or, more briefly,

$$G_0(t) = \frac{\prod\limits_{k=1}^{\mu} (t-\alpha_k)^{m_k}}{\prod\limits_{j=1}^{\nu} (t-\beta_j)^{p_j}}\, G_{01}(t), \qquad (45.7)$$

the meaning of $G_{01}(t)$ being clear from the formulae themselves. It follows that the function $G_{01}(t)$ satisfies on L_0 the Hölder condition and does not vanish there.

We have thus constructed the reduced problem

$$\Phi_0^+(t) = G_0(t)\,\Phi_0^-(t) \quad \text{on} \quad L_0. \qquad (45.8)$$

The index of the reduced problem (45.8), i.e. the index of the coefficient $G_0(t)$ with respect to the contour L_0 will be called the index of the considered problem with respect to the class $H_{r,s}$. Taking into account the structure of the reduced contour L_0 and condition (45.5) it is easy to compute it. Denoting $\varkappa_1 = \text{Ind } G_1(t)$ we have

$$\varkappa = \text{Ind } G_0(t) = \varkappa_1 + \sum_{k=1}^{r} m_k - \sum_{j=s+1}^{\nu} p_j. \qquad (45.9)$$

In particular the index with respect to the class of bounded functions H_0 is the following:

$$\varkappa = \varkappa_1 - \sum_{j=1}^{\nu} p_j. \qquad (45.10)$$

Let $X_0(z)$ be the canonical function of the reduced problem (45.6). It is given by formulae (14.12), (14.9) of § 14.4

$$\left. \begin{aligned} X_0^+(z) &= e^{\Gamma_0^+(z)}, \quad X_0^-(z) = \frac{1}{z^\varkappa}\, e^{\Gamma_0^-(z)}, \\ \Gamma_0(z) &= \frac{1}{2\pi i} \int\limits_{L_0} \frac{\log\,[\tau^{-\varkappa}\, G_0(\tau)]}{\tau - z}\, d\tau. \end{aligned} \right\} \qquad (45.11)$$

It remains to pass to the limit $L_0 \to L$, i.e. to let $\varepsilon \to 0$ and to establish that the function

$$X(z) = \lim_{\varepsilon \to 0} X_0(z)$$

plays the role of the canonical function for the considered problem (45.1). It is convenient to transform first the expression for $\Gamma_0(z)$. Making use of formula (45.9) we represent the function of zero index $\tau^{-\varkappa} G_0(\tau)$ in the form

$$\tau^{-\varkappa} G(\tau) = \prod_{k=1}^{r} \left(1 - \frac{\alpha_k}{\tau}\right)^{m_k} \prod_{k=r+1}^{\mu} (\tau - \alpha_k)^{m_k} \times$$

$$\times \prod_{j=1}^{s} (\tau - \beta_j)^{-p_j} \prod_{j=s+1}^{v} \left(1 - \frac{\beta_j}{\tau}\right)^{-p_j} [\tau^{-\varkappa_1} G_{01}(\tau)],$$

the index of every factor being zero. In view of the structure of the contour L_0 and the preceding formula it is easy to derive for $\Gamma_0(z)$ the expression

$$\Gamma_0^+(z) = \ln \frac{\prod\limits_{k=r+1}^{\mu} (z - \alpha_k)^{m_k}}{\prod\limits_{j=1}^{s} (z - \beta_j)^{p_j}} + \frac{1}{2\pi i} \int\limits_{L_0} \frac{\ln[\tau^{-\varkappa_1} G_{01}(\tau)]}{\tau - z} \, d\tau,$$

$$\Gamma_0^-(z) = \ln \frac{\prod\limits_{j=s+1}^{v} \left(\dfrac{z - \beta_j}{z}\right)^{p_j}}{\prod\limits_{k=1}^{r} \left(\dfrac{z - \alpha_k}{z}\right)^{m_k}} + \frac{1}{2\pi i} \int\limits_{L_0} \frac{\ln[\tau^{-\varkappa_1} G_{01}(\tau)]}{\tau - z} \, d\tau.$$

Let us split the contour of integration L_0 into parts in accordance with (45.2) and pass to the limit $\varepsilon \to 0$. After the necessary transformations which we do not quote here, we obtain

$$\Gamma^+(z) = \lim_{\varepsilon \to 0} \Gamma_0^+(z) = \ln \frac{\prod\limits_{k=r+1}^{\mu} (z - \alpha_k)^{m_k}}{\prod\limits_{j=1}^{s} (z - \beta_j)^{p_j}} + \Gamma_1^+(z),$$

$$\Gamma^-(z) = \lim_{\varepsilon \to 0} \Gamma_0^-(z) = \ln \frac{\prod\limits_{j=s+1}^{v} \left(\dfrac{z - \beta_j}{z}\right)^{p_j}}{\prod\limits_{k=1}^{r} \left(\dfrac{z - \alpha_k}{z}\right)^{m_k}} + \Gamma_1^-(z),$$

where

$$\Gamma_1(z) = \frac{1}{2\pi i} \int\limits_{L} \frac{\ln[\tau^{-\varkappa_1} G_1(\tau)]}{\tau - z} \, d\tau.$$

We now obtain from formulae (45.11)

$$\left. \begin{aligned} X^+(z) &= \frac{\prod\limits_{k=r+1}^{\mu} (z - \alpha_k)^{m_k}}{\prod\limits_{j=1}^{s} (z - \beta_j)^{p_j}} X_1^+(z), \\[2em] X^-(z) &= \frac{\prod\limits_{j=s+1}^{v} (z - \beta_j)^{p_j}}{\prod\limits_{k-1}^{r} (z - \alpha_k)^{m_k}} X_1^-(z). \end{aligned} \right\} \tag{45.12}$$

$X_1(z)$ being given by the relations

$$X_1^+(z) = e^{\Gamma_1^+(z)}, \quad X_1^-(z) = \frac{1}{z^{\varkappa_1}} e^{\Gamma_1^-(z)};$$

it represents the canonical function of the homogeneous Riemann problem with the coefficient $G_1(t)$ and the contour L.

The function $X(z)$ has the properties of the canonical function. It can easily be verified that it satisfies the boundary condition (45.1) and its order at infinity is equal to the index of the problem $\varkappa = \varkappa_1 + \sum\limits_{k=1}^{r} m_k - \sum\limits_{j=s+1}^{\nu} p_j$; it is also evident that it belongs to the class $H_{r,s}$.

The general solution of the problem in the class $H_{r,s}$ can be obtained by the usual methods. It has the form

$$\Phi(z) = X(z)\, P_\varkappa(z), \tag{45.13}$$

where $P_\varkappa(z)$ is a polynomial of degree not exceeding $\varkappa$, with arbitrary coefficients. It follows from (45.13) that the problem has a solution when $\varkappa \geqq 0$ and then the number of linearly independent solutions is $\varkappa + 1$.

The greatest number of solutions of the problem occurs in the class $H_{\mu,\nu}$ in which infinities of the solution near every critical point are admitted; this follows from the formula (45.9) for the index. The smallest number of solutions occurs in the class of bounded functions H_0. In the latter case the canonical function has the form

$$X^+(z) = \prod_{k=1}^{\mu}(z - \alpha_k)^{m_k} X_1^+(z), \quad X^-(z) = \prod_{j=1}^{\nu}(z - \beta_j)^{p_j} X_1^-(z).$$

Formula (45.10) shows that an introduction into the coefficient $G(t)$ of binomials of the form $(t - \alpha)^m$ creating zeros of the coefficient, does not influence the number of solutions in the class H_0, while a creation of poles diminishes this number by the order of the pole. The problem is soluble if the sum of the orders of the poles does not exceed the index $\varkappa_1$ of the continuous part $G_1(t)$ of the coefficient $G(t)$. These results are identical with those of § 15.1.

The solution of the non-homogeneous problem can be deduced with the help of a reasoning analogous to that of §§ 14.5, 15.2.

The exposition of this section follows the paper of L. A. Chikin (1).

The assumption that the zeros and poles are of integal order is immaterial. The method is also applicable to more general cases. An investigation assuming that the zeros and poles are any power order has been published by the author (F. D. Gakhov [3]). The still more general case when

$$G(t) = \Pi\,(t - c_k)^{r_k} \ln^{s_k}(t - c_k)\, G_1(t), \tag{45.14}$$

has been investigated by Khvedelidze [11].

45.3. Exceptional points at contour ends

We briefly consider the case when the exceptional point, where the coefficient of the Riemann problem either vanishes or becomes infinite, occurs at the end point of the contour. Let

$$\Phi^+(t) = G(t)\,\Phi^-(t), \quad G(t) = (t - a)^r G_1(t), \tag{45.15}$$

r being any real number. To be specific we assume a to be the contour's initial point and, for the sake of simplicity, we assume this to be the only exceptional point.

The solution according to (43.2) is

$$\Phi(z) = e^{\Gamma(z)}, \quad \Gamma(z) = \frac{1}{2\pi i} \int_L \frac{\ln G(\tau)}{\tau - z} d\tau,$$

$$\Gamma(z) = \frac{r}{2\pi i} \int_L \frac{\ln(\tau - a)}{\tau - z} d\tau + \frac{1}{2\pi i} \int_L \frac{\ln G_1(\tau)}{\tau - z} d\tau.$$

It is required to investigate the behaviour of the solution at the exceptional point a. Using (8.25) and (8.10) we have:

$$\Gamma(z) = -\frac{r}{4\pi i} \ln^2(z - a) + \left[\frac{r}{2} - \frac{\ln G_1(a)}{2\pi i} \right] \ln(z - a) + \Gamma^*(z), \quad (45.16)$$

where $\Gamma^*(z)$ is bounded in the vicinity of a and has a definite limit at a. We set:

$$\theta(z) = \arg(z - a), \quad \alpha + i\beta = -\frac{\ln G_1(a)}{2\pi i}.$$

After simple transformations we obtain

$$\ln^2(z - a) = \ln^2 |z - a| + \theta^2 + 2i\theta \ln(z - a).$$

Substituting the latter expression into (45.16), we get

$$\Gamma(z) = \left[-\frac{r}{2\pi} \theta(z) + \frac{1}{2} r + \alpha \right] \ln(z - a) - \beta\theta +$$

$$+ i \frac{r}{4\pi} [\ln |z - a| + \theta^2(z)] + \Gamma^*(z).$$

Hence

where

$$\Phi(z) = (z - a)^{\psi(z)} \Phi^*(z),$$

$$\psi(z) = -\frac{r}{2\pi} \theta(z) + \frac{1}{2} r + \alpha,$$

whilst $\Phi^*(z)$ is a bounded function in the neighbourhood of a.

In deriving the formulae (8.25) we regarded the contour L as a cut for $\ln(z - a)$ and $\theta^+ = \theta$, $\theta^- = \theta - 2\pi$; therefore, for all t on the contour, including $t = a$, we have

$$\psi^+(z) - \psi^-(z) = r.$$

If z traverses point a in the positive direction, then $\psi(z)$ decreases monotonically if $r > 0$, but increases monotonically if $r < 0$. The solution at point a therefore has no definite order. The order depends on the approach path to a. On each arc departing from a the order is constant. This is the most distinctive feature of the case under consideration.

As regards the inhomogeneous problem, we only point out that the expression $g(t)/X^+(t)$ in the solution (43.10) is of variable order at point a. For solubility it is required that $g(t) = (t - a)^s g_1(t)$, where $r - s \leqq 0$ for solutions which are bounded at a, and $r - s < 1$ for unbounded solutions. These results are taken from the paper by Melnik (2).

In another paper (3), Melnik uses the Cauchy integral $\int\limits_L \dfrac{\ln\ln(\tau - a)}{\tau - z}\, d\tau$ to solve the boundary value problem if

$$G(t) = \prod (t - c_k)^{r_k} \ln(t - c_k)^{s_k}\, G_1(t).$$

Here the points c_k may be either end points of the contour or discontinuities where one of the limit values vanishes. The results are essentially the same as above.

§ 46. Hilbert boundary value problem with discontinuous coefficients

Consider the Hilbert boundary value problem (§ 27.1) assuming that the functions $a(s)$, $b(s)$, $c(s)$ in the boundary condition

$$a(s)\, u(s) + b(s)\, v(s) = c(s) \tag{46.1}$$

may have discontinuities of the first kind at a finite number of points of the contour.

This problem can be solved in many ways. For instance we can generalize the solution of the Hilbert problem given in Chapter IV to the case of discontinuous coefficients. This generalization can be carried out in two ways, as was done for the Riemann problem (§§ 41, 43) (see Problems 7, 8 at the end of this chapter). The second way consists in making use of the reducibility of the Hilbert problem to the Riemann problem, established in § 30. Since we wish to use the solution of the Riemann problem with discontinuous coefficients deduced above we choose the second way.

We can therefore solve then Hilbert problem for a circle or semicircle. The general case can be reduced to this case by means of conformal mapping.

Prior to proceeding to the next subsection the reader should refresh in his memory the contents of §§ 30.1, 30.2.

46.1. Hilbert problem for the semi-plane

On the basis of the considerations of § 30.2, the solution of the Hilbert problem (46.1) for the upper semi-plane is in the considered case given by the solution $\Phi^+(z)$ of the Riemann problem

$$\Phi^+(t) = -\frac{a(t) + ib(t)}{a(t) - ib(t)}\, \Phi^-(t) + \frac{2c(t)}{a(t) - ib(t)}, \tag{46.2}$$

which satisfies the condition

$$\Phi^-(z) = \overline{\Phi^+(\bar z)} = \overline{\Phi}^+(z) \tag{46.3}$$

(for notation see § 18.1).

We avail ourselves of the solution of Riemann's problem for a semi-circle (see § 14.2), adopting the form used in § 30.1:

$$\Phi^+(z) = X^+(z)\left[\Psi^+(z) + P_{2\varkappa}\left(\frac{z-i}{z+i}\right)\right], \quad \varkappa \geqq 0, \qquad (46.4)$$

$$\Phi^+(z) = X^+(z)\left[\Psi^+(z) + C\right], \quad \varkappa < 0, \qquad (46.5)$$

$$X(z) = e^{\Gamma(z)}, \quad \Gamma(z) = \frac{1}{2\pi i}\int\limits_{-\infty}^{+\infty} \ln\left[\left(\frac{t-i}{t+i}\right)^{-2\varkappa}\frac{a+ib}{a-ib}\right]\frac{dt}{t-z}, \qquad (46.6)$$

$$\Psi(z) = \frac{1}{\pi i}\int \frac{c(t)}{(a-ib)\,X^+(t)}\frac{dt}{t-z} \quad (\varkappa = \operatorname{Ind}(a+ib)), \qquad (46.7)$$

it being necessary to put $C = -\Psi^-(-i)$ if $\varkappa < 0$, and also to satisfy the solubility condition if $\varkappa < -1$:

$$\int\limits_{-\infty}^{+\infty} \frac{c(t)}{(a-ib)\,X^+(t)}\frac{dt}{(t+i)^k} = 0 \quad (k = 1, 2, \ldots, -\varkappa). \qquad (46.8)$$

From § 30.1 the arbitrary constants of the polynomial $P_{2\varkappa}$ can be chosen so that condition (46.3) is satisfied.

In the solution of practical problems it is usually more convenient to employ the method of analytic continuation by which formulae (46.4)–(46.8) were obtained, rather than use these formulae themselves. The procedure for practical problems is as follows.

1. Reduce Riemann's problem (46.2) to the problem with the continuous coefficient by the method of § 41.

2. Solve the reduced problem by analytic continuation.

3. Select arbitrary constants (for a negative index the solubility conditions) such that (46.3) is satisfied.

The method is illustrated by an example in § 46.2. Serious practical difficulties arise in selecting the branches for the solution of many-valued functions. The problem is solved in the same way for a circular domain.

Note, finally, that the first edition of this book gave another method of solving this problem based on § 44.2. As pointed out by Muskheli-shvili ([17*], p. 142) one encounters integrals which are divergent and so, generally speaking, the method does not yield finite solutions.

It follows that if the coefficients of the polynomial $P_\varkappa(z)$ are real $(\bar{P}_\varkappa(z) = P_\varkappa(z))$ then formula (46.6) defines a function satisfying condition (46.8).

For the solution of the Hilbert problem we employed the method of solution of the Riemann problem, presented in § 44.2. In solving practical problems we may also use the method of § 42. If then all the integrals turn out to be real, the solution of the Riemann problem is also a solution of the corresponding Hilbert problem. This follows directly from the fact that for real functions (and only for real functions) the relation $\bar{\Phi}(z) = \Phi(z)$ is satisfied.

46.2. Example

To illustrate the methods described above we consider a concrete example.

It is required to find a function $F(z) = u + iv$ which is analytic in the lower semi-plane, in accordance with the boundary condition on the real axis

$$u = \sigma(x) \quad (-1 < x < 1),$$
$$u_0 u - v_0 v = 0 \quad (x < -1, \quad x > 1),$$

where u_0, v_0 are constants satisfying the condition $u_0^2 + v_0^2 = 1$ and $\sigma(x)$ is a known function.

This problem is a particular case of the Hilbert boundary value problem with discontinuous coefficients

$$\mathrm{Re}\,\{[a(x) + ib(x)]\,F(x)\} = c(x)$$

for
$$a(x) = 1, \quad b(x) = 0, \quad c(x) = \sigma(x) \quad (-1 < x < 1),$$
$$a(x) = u_0, \quad b(x) = v_0, \quad c(x) = 0 \quad (x < -1, \quad x > 1).$$

The corresponding Riemann problem has the form

$$\Phi^+(x) = G(x)\,\Phi^-(x) + g(x),$$

$$G(x) = \begin{cases} -1, \\ -e^{2i\varphi}, \end{cases} \quad g(x) = \begin{cases} 2\sigma(x) & (-1 < x < 1), \\ 0 & (x < -1, \quad x > 1), \end{cases}$$

where $e^{i\varphi} = u_0 + iv_0$ and by a change of sign if necessary in the boundary condition we can ensure that the condition $0 \leqq \varphi \leqq \pi$ is satisfied.

The boundary value problem under consideration is equivalent to the last Riemann problem in the sense that every function $\Phi^-(z)$ satisfying the condition $\bar{\Phi}^-(z) = \Phi^+(z)$ is its solution.

The Riemann problem can be solved by the method of § 41. Thus we have

$$\frac{G(-1-0)}{G(-1+0)} = e^{2i\varphi}, \quad \frac{G(1-0)}{G(1+0)} = e^{-2i\varphi},$$

whence
$$\gamma_1 = \frac{\varphi}{\pi} - \varkappa_1, \quad \gamma_2 = -\frac{\varphi}{\pi} - \varkappa_2.$$

The integers $\varkappa_1$, $\varkappa_2$ are chosen in a way depending on the required class of solutions. If we seek a solution bounded at infinity and having at the ends -1, 1 infinities of an integrable order, then we have to take $\varkappa_1 = 1$, $\varkappa_2 = 0$.

The considered case differs from that investigated in § 42 by the fact that now the point at infinity belongs to the contour. The domains D^+ and D^- are on the same footing with respect to the point at infinity and therefore the fundamental auxiliary functions eliminating the discontinuity of the coefficient are to be taken for both functions Φ^+ and Φ^- in the form of the function $\omega^-(z)$ (see formula (41.7)). Selecting for z_0 the points $\pm i$ we have

$$\Phi^+(z) = \left(\frac{z+1}{z+i}\right)^{\frac{\varphi}{\pi}-1} \left(\frac{z-1}{z+i}\right)^{-\frac{\varphi}{\pi}} \Phi_1^+(z),$$

$$\Phi^-(z) = \left(\frac{z+1}{z-i}\right)^{\frac{\varphi}{\pi}-1} \left(\frac{z-1}{z-i}\right)^{-\frac{\varphi}{\pi}} \Phi_1^-(z).$$

Then we arrive at the Riemann problem

$$\Phi_1^+(x) = G_1(x)\,\Phi_1^-(x) + g_1(x).$$

In accordance with the general theory the coefficient

$$G_1(x) = \left(\frac{x-i}{x+i}\right)^{1-\frac{\varphi}{\pi}} \left(\frac{x-i}{x+i}\right)^{\frac{\varphi}{\pi}} G(x)$$

should be a continuous function. To elucidate this, consider in more detail the way of eliminating the discontinuity.

In the range $(-1, 1)$ $G(x) = -1$ and hence $G_1(x) = (x - i)/(x + i)$. In passing through the points -1 or $+1$, $G(x)$ acquires the value $-e^{2i\varphi}$. In the product $[(x-i)/(x+i)]^{1-\frac{\varphi}{\pi}} [(x-i)/(x+i)]^{\frac{\varphi}{\pi}}$, in accordance with the definition of the fundamental function, the first factor is discontinuous at the point -1 and the second at $+1$. In passing through the points $-1, 1$ from the left to the right these functions acquire the factors

$$e^{2\pi i\left(1-\frac{\varphi}{\pi}\right)} = e^{-2i\varphi}, \quad e^{2\pi i\frac{\varphi}{\pi}} = e^{2i\varphi},$$

respectively. It now follows at once that on the entire real axis

$$G_1(x) = -\frac{x-i}{x+i}.$$

Writing the boundary condition of the Riemann problem in the form

$$(x+i)\,\Phi_1^+(x) = -(x-i)\,\Phi_1^-(x) + g(x)(x+1)^{1-\frac{\varphi}{\pi}}(x-1)^{\frac{\varphi}{\pi}},$$

we easily obtain

$$(z+i)\,\Phi_1^+(z) = \Psi^+(z) + c_0 + c_1 z,$$

$$-(z-i)\,\Phi_1^-(z) = \Psi^-(z) + c_0 + c_1 z,$$

where

$$\Psi(z) = \frac{(-1)^{\frac{\varphi}{\pi}}}{\pi i} \int_{-1}^{1} \frac{\sigma(t)(t+1)^{1-\frac{\varphi}{\pi}}(1-t)^{\frac{\varphi}{\pi}}}{t-z} \, dt,$$

and c_0, c_1 are arbitrary constants. Hence

$$\Phi^+(z) = \left(\frac{z+1}{z+i}\right)^{\frac{\varphi}{\pi}-1}\left(\frac{z-1}{z+i}\right)^{-\frac{\varphi}{\pi}} \frac{\Psi^+(z)+c_0+c_1 z}{z+i}$$

$$= (z+1)^{\frac{\varphi}{\pi}-1}(z-1)^{-\frac{\varphi}{\pi}}[\Psi^+(z)+c_0+c_1 z].$$

$$\Phi^-(z) = \left(\frac{z+1}{z-i}\right)^{\frac{\varphi}{\pi}-1}(-1)\left(\frac{z-1}{z-i}\right)^{-\frac{\varphi}{\pi}} \frac{\Psi^-(z)+c_0+c_1 z}{z-i}$$

$$= -(z+1)^{\frac{\varphi}{\pi}-1}(z-1)^{-\frac{\varphi}{\pi}}[\Psi^-(z)+c_0+c_1 z].$$

For the many-valued function $f(z) = (z+1)^{\varphi/\pi-1}(z-1)^{-\varphi/\pi}$ to be fully defined, we now indicate the manner of making the cut connecting the branch points -1, $+1$. Inserting the solution of the homogeneous problem ($\Psi^+ = \Psi^- = 0$) into the

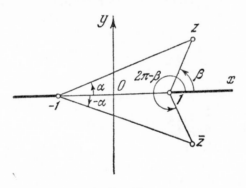

FIG. 18.

homogeneous boundary condition ($\sigma = 0$) we find that on the section $(-1, +1)$ of the axis the relation $f^+(x) = f^-(x)$ holds, i.e. on this section the function is continued analytically. Consequently, the cut should be made along the sections of the real axis $(-\infty, -1)$, $(1, \infty)$.

Finally let us show how to select the quantity $(-1)^{\varphi/\pi}$ and the arbitrary constants c_0, c_1 in order that the relation $\bar{\Phi}^-(z) = \overline{\Phi^-(\bar{z})} = \Phi^+(z)$ is valid. To compute $\overline{f(z)}$ we set $z+1 = r_1 e^{i\alpha_1}$, $z-1 = r_2 e^{i\alpha_2}$. Then (Fig. 18) $\bar{z}+1 = r_1 e^{-i\alpha_1}$, $\bar{z}-1 = r_2 e^{i(2\pi-\alpha_2)}$.

Consequently,

$$f(\bar{z}) = r_1^{\frac{\varphi}{\pi}-1} r_2^{-\frac{\varphi}{\pi}} e^{-i\alpha_1\left(\frac{\varphi}{\pi}-1\right)} e^{-i(2\pi-\alpha_2)\frac{\varphi}{\pi}},$$

$$\overline{f(\bar{z})} = r_1^{\frac{\varphi}{\pi}-1} r_2^{-\frac{\varphi}{\pi}} e^{i\alpha_1\left(\frac{\varphi}{\pi}-1\right)} e^{i(2\pi-\alpha_2)\frac{\varphi}{\pi}} = e^{2i\varphi} f(z).$$

Further, setting $(-1)^{\varphi/\pi} = e^{i\varphi}$ we obtain

$$\overline{\Psi}^-(z) = -\frac{e^{i\varphi}}{\pi i} \int_{-1}^{1} \frac{\sigma(t)(t+1)^{1-\frac{\varphi}{\pi}}(1-t)^{\frac{\varphi}{\pi}}}{t-z} dt = -e^{-2i\varphi}\Psi^+(z).$$

Inserting all quantities into the relation $\overline{\Phi}^-(z) = \Phi^+(z)$ we have

$$-(\overline{c}_0 + \overline{c}_1 z) e^{2i\varphi} = c_0 + c_1 z.$$

It follows from the relations $\overline{c}_k e^{2i\varphi} = -c_k$ $(k = 0, 1)$ that $ic_k e^{-i\varphi} = \overline{(ic_k e^{-i\varphi})}$ is a real quantity. Hence

$$c_0 = -iR_0 e^{i\varphi}, \quad c_1 = -iR_1 e^{i\varphi},$$

where R_0 and R_1 are real constants.

Hence we obtain the final form of the solution of the original problem

where
$$F(z) = \Phi^-(z) = i(1+z)^{\frac{\varphi}{\pi}-1}(1-z)^{-\frac{\varphi}{\pi}}[\Psi_1(z) + R_0 + R_1 z],$$

$$\Psi_1(z) = \frac{1}{\pi} \int_{-1}^{1} \frac{(1+t)^{1-\frac{\varphi}{\pi}}(1-t)^{\frac{\varphi}{\pi}}\sigma(t)}{t-z} dt.$$

46.3. Mixed boundary value problem for analytic functions

As an application of the above theory let us consider one particular example of the Hilbert boundary value problem with discontinuous coefficients, namely the so-called mixed boundary value problem for analytic functions, which is encountered in numerous applications. We state the problem.

Suppose that a closed contour L is divided by the points $a_1, b_1, \ldots, a_m, b_m$ into $2m$ parts. The union of arcs (a_k, b_k) will be denoted by L' while the union of arcs (b_k, a_{k+1}) by L'' $(k = 1, 2, \ldots, m; a_{m+1} = a_1)$.

It is required to determine a function analytic in D^+

$$F(z) = u + iv, \tag{46.9}$$

which satisfies the boundary conditions

$$u = f(s) \quad \text{on} \quad L', \quad v = h(s) \quad \text{on} \quad L'', \tag{46.10}$$

where $f(s)$, $h(s)$ are known functions satisfying the Hölder condition. At the points a_k, b_k, $F(z)$ should belong to a prescribed class.

We consider the case when the contour L is the real axis. Basically any problem can be reduced to this one by means of conformal mapping of the domain D^+ onto the upper semi-plane. It is then necessary to assume that $F(z)$ is bounded at infinity.

We assume that the point at infinity does not coincide with the points of division a_k, b_k. Consequently, L' consists of m finite sections (a_k, b_k) and L'' of $m-1$ finite sections and two infinite sections (b_m, ∞), $(-\infty, a_1)$.

The boundary conditions (46.10) can be written as one boundary condition of the Hilbert problem on the whole contour L:

$$a(s)\,u(s) + b(s)\,v(s) = c(s) \quad \text{on} \quad L, \tag{46.11}$$

where we have to set

$$a = 1, \quad b = 0, \quad c = f(s) \quad \text{on} \quad L';$$
$$a = 0, \quad b = 1, \quad c = h(s) \quad \text{on} \quad L''.$$

Construct the appropriate Riemann boundary value problem

$$\Phi^+(t) = G(t)\,\Phi^-(t) + g(t). \tag{46.12}$$

According to formula (46.2)

$$G(t) = -\frac{a+ib}{a-ib} = -1, \quad g(t) = \frac{2c}{a-ib} = 2f(t) \quad \text{on} \quad L',$$
$$G(t) = +1, \quad g(t) = 2ih(t) \quad \text{on} \quad L''.$$

Thus we have arrived at the boundary value problem (42.12′) investigated in § 42.2 and we may make use of the results deduced there. We shall quote the solution for the three most important classes; the derived formulae were given by M. V. Keldysh and L. I. Sedov (1).

1. *Solution unbounded at all the ends.*

$$\Phi(z) = \frac{1}{R(z)}\left[\frac{1}{2\pi i}\int_L \frac{R(\tau)\,g(\tau)}{\tau - z}\,d\tau + P_m(z)\right], \tag{46.13}$$

$$R(z) = 1\Big/\left(\prod_{k=1}^m (z - a_k)(z - b_k)\right).$$

If the additional condition $\Phi(\infty) = 0$ is given, $P_m(z)$ has to be replaced by $P_{m-1}(z)$. Next formulae will be written with this condition being taken into account.

2. *Solution bounded at the ends a_k and unbounded at the ends b_k.*

$$\Phi(z) = \frac{R_a(z)}{R_b(z)} \frac{1}{2\pi i} \int_L \frac{R_b(\tau)}{R_a(\tau)} \frac{g(\tau)}{\tau - z} d\tau, \qquad (46.14)$$

$$R_a(z) = \sqrt{\left(\prod_{k=1}^{m} (z - a_k) \right)}, \quad R_b(z) = \sqrt{\left(\prod_{k=1}^{m} (z - b_k) \right)}.$$

3. *Solution bounded at all the ends.*

$$\Phi(z) = R(z) \frac{1}{2\pi i} \int_L \frac{g(\tau)}{R(\tau)} \frac{d\tau}{\tau - z}. \qquad (46.15)$$

In this case the conditions of solubility

$$\int_L \frac{g(\tau)}{R(\tau)} \tau^{j-1} d\tau = 0 \quad (j = 1, 2, \ldots, m), \qquad (46.16)$$

have to be satisfied. It can easily be proved that the above formulae yielding the solution of the Riemann problem (46.12) are at the same time representing the solution of the original Hilbert problem, provided the coefficients of the polynomial $P_m(z)$ are real. To prove this, according to the remark at the end of the preceding subsection, it is necessary to show that all integrals entering these formulae are real. This fact follows from the following argument.

If $\tau \in L'$ then there is an odd number of negative binomials $(\tau - a_k)$, $(\tau - b_k)$ and therefore $R(\tau)$ and $R_b(\tau)/R_a(\tau)$ are imaginary. But $g(\tau) = 2f(\tau)$ is here a real function; consequently, in view of the presence of the factor $1/2\pi i$, the whole integral is real.

If $\tau \in L''$ there is an even number of negative binomials, $R(\tau)$ and $R_b(\tau)/R_a(\tau)$ are real and $g(\tau)/2\pi i = h(\tau)/\pi$ is also a real function. Hence the integral is again real.

It is readily observed that the function $P_m(z)/R(z)$ is the general solution of the homogeneous problem in the most general class of functions unbounded at all ends. Formula (42.24) yields the solution in any class when boundedness at an arbitrary p ends is required and integrable infinities at the remaining $2m - p$ ends are admitted. When $p < m$ the problem has $m - p$ linearly independent solutions and when $p > m$ there exists a unique solution, $p - m$ conditions of solubility (42.25) being satisfied.

46.4. The Dirichlet problem and its modifications for the plane with slits

Let us exclude from the complex plane the sections (a_k, b_k) $(k = 1, 2, \ldots, m)$ of the real axis and denote by D the remaining domain. By the boundary of the domain we understand the excluded sections (cuts) described twice, the upper edge being traversed in the direction of the positive axis while the lower edge is traversed the opposite direction. The fundamental problem consists in *the determination of a function harmonic in D in terms of its value on the edges of the cut.*

Because of its nature the problem is the Dirichlet problem for a multiply-connected domain in the particular case when all contours bounding the domain are reduced to sections of the real axis traversed twice. The particular case considered differs substantially from that examined in § 36 for a multiply-connected domain by the fact that now there are no points on the plane different from the points of $D + L$. Let us recall that the difficulties in solving the Dirichlet problem for a multiply-connected domain are due to the possibility of the occurence of many-valued functions. In § 36 these difficulties were met by an explicit isolation of the many-valued part $\ln(z - z_k)$ which is possible only when there are points present which do not belong to $D + L$. Consequently the above formulated problem cannot be solved by means of the method of § 36. The solution will be derived on the basis of other considerations consisting in investigating boundary value problems with discontinuous coefficients. It will be shown that the results are also essentially different from those of § 36.

We begin by considering the problem which is of independent interest and is auxiliary to the solution of the Dirichlet problem.

It is required to determine an analytic and single-valued function $\Phi(z) = u + iv$ in D, in terms of the values of its real part described on the cuts

$$u^+ = f^+(t), \quad u^- = f^-(t) \quad \text{on} \quad L, \tag{46.17}$$

where $f^+(t)$, $f^-(t)$ are known functions satisfying the Hölder condition. We assume that the boundary values are continuous functions; then the following conditions should be satisfied:

$$f^+(a_k) = f^-(a_k), \quad f^+(b_k) = f^-(b_k) \quad (k = 1, 2, \ldots, m). \tag{46.18}$$

We also consider, besides the unknown function $\Phi(z)$, the function $\bar{\Phi}(z) = \overline{\Phi(\bar{z})}$ which has already been encountered in § 30.2 and § 46.1. It was established before that if $\Phi(z)$ is analytic in D then $\overline{\Phi}(z)$ is also analytic in this domain. When z approaches the real axis, then $\Phi(z)$ and $\overline{\Phi}(z)$ tend to their limiting values on the axis from different semi-planes, these limiting values being complex conjugate.

We shall reduce the above problem to those considered before, making use of a device of N. I. Mushelishvili†. Consider the identity

$$\Phi(z) = \tfrac{1}{2}[\Phi(z) + \overline{\Phi}(z)] + \tfrac{1}{2}[\Phi(z) - \overline{\Phi}(z)] = \Psi(z) + \Omega(z) \quad (46.19)$$

and two functions analytic in D

$$\Psi(z) = \tfrac{1}{2}[\Phi(z) + \overline{\Phi}(z)], \quad \Omega(z) = \tfrac{1}{2}[\Phi(z) - \overline{\Phi}(z)]. \quad (46.20)$$

We proceed to investigate their behaviour on the real axis. On the sections (a_k, b_k) we have

$$\left.\begin{aligned} \operatorname{Re}\Psi(t) &= \tfrac{1}{2}[\operatorname{Re}\Phi(t) + \operatorname{Re}\overline{\Phi}(t)] = \tfrac{1}{2}[f^+(t) + f^-(t)], \\ \operatorname{Re}\Omega(t) &= \tfrac{1}{2}[f^+(t) - f^-(t)]. \end{aligned}\right\} \quad (46.21)$$

Denote by L' the remaining part of the axis (excluding the sections (a_k, b_k)). On L' both functions are analytically continuable through the axis and have the properties

$$\Psi(t) = \tfrac{1}{2}[\Phi(t) + \overline{\Phi}(t)] = \operatorname{Re}\Phi(t),$$

$$\Omega(t) = \tfrac{1}{2}[\Phi(t) - \overline{\Phi}(t)] = i\operatorname{Im}\Phi(t),$$

which is equivalent to the condition

$$\operatorname{Im}\Psi(t) = 0, \quad \operatorname{Re}\Omega(t) = 0 \quad \text{on} \quad L'.$$

Comparing the last relations with relations (46.21) we find that the functions $\Psi(z), \Omega(z)$ satisfy on the whole axis the following boundary conditions:

$$\operatorname{Re}\Psi(t) = \tfrac{1}{2}[f^+(t) + f^-(t)] \text{ on } L, \quad \operatorname{Im}\Psi(t) = 0 \text{ on } L', \quad (46.22)$$

$$\operatorname{Re}\Omega(t) = \tfrac{1}{2}[f^+(t) - f^-(t)] \text{ on } L, \quad \operatorname{Re}\Omega(t) = 0 \text{ on } L'. \quad (46.23)$$

Consequently, the determination of the function $\Psi(z)$ has been reduced to the solution of the mixed boundary value problem for analytic functions examined in the preceding subsection. For the function $\Omega(z)$ the problem is reduced to the determination of a function

† N. I. Muskhelishvili makes use of this device not for reduction to the mixed problem (the latter he solves later) but for a direct solution of the problem by the method of § 30.2.

analytic in the semi-plane in terms of the value of its real part prescribed on the x-axis, under the condition of analytic continuability into the complementary semi-plane. In view of the conditions of continuity (46.18) the boundary value or $\mathrm{Re}\,\Omega(z)$ on the axis is a continuous function.

The function $\Psi(z)$ is given by formulae of § 46.3 where we have to set

$$g(t) = \tfrac{1}{2}[f^+(t) + f^-(t)] \quad \text{on} \quad L, \quad g(t) = 0 \quad \text{on} \quad L'.$$

The determination of the function $\Omega(z)$ is carried out by applying the Schwarz operator: it is known that the latter in the case of the real axis is identical with the Cauchy type integral.

Thus, we have

$$\Omega(z) = \frac{1}{2\pi i} \int_L \frac{f^+(\tau) - f^-(\tau)}{\tau - z}\,d\tau. \tag{46.24}$$

Let us write the solution of the original problem in the class of functions bounded on some ends $c_1, c_2, \ldots, c_p$ and having integrable infinities at the remaining $2m - p$ ends, with the additional condition $\Phi(\infty) = 0$. According to formulae (42.24) and (46.24)

$$\Phi(z) = \frac{R_1(z)}{R_2(z)}\left[\frac{1}{4\pi i}\int_L \frac{R_2(\tau)}{R_1(\tau)}\,\frac{f^+(\tau) + f^-(\tau)}{\tau - z}\,d\tau + P_{m-p-1}(z)\right] +$$

$$+ \frac{1}{2\pi i}\int_L \frac{f^+(\tau) - f^-(\tau)}{\tau - z}\,d\tau, \tag{46.25}$$

where

$$R_1(z) = \sqrt{\Big/\!\Big(\prod_{k=1}^{p}(z - c_k)\Big)}, \quad R_2(z) = \sqrt{\Big/\!\Big(\prod_{k=p+1}^{2m}(z - c_k)\Big)}.$$

In the case when $p \geqq m$ we have to set $P_{m-p-1}(z) \equiv 0$ and add (when $p > m$) $p - m$ conditions of solubility

$$\int_L \frac{R_2(\tau)}{R_1(\tau)}\,[f^+(\tau) + f^-(\tau)]\,\tau^{j-1}\,d\tau = 0$$

$$(j = 1, 2, \ldots, p - m). \tag{46.26}$$

The reader can easily write down for himself the solutions for all the more important classes, making use of formulae (46.13) – (46.16).

If we discard the condition of continuity (46.18), then in view of the properties of the Cauchy type integral (§ 8.1) the second integral in formula (46.25) yields a function with a logarithmic singularity. Consequently, in this case bounded solutions do not exist. If the solution of the mixed problem is taken in the class of functions bounded at a given end, then the whole solution has a logarithmic singularity. Solutions of this nature are called by N. I. Muskhelishvili *almost bounded*. A singularity of this nature is generally speaking peculiar to the solution of the Dirichlet problem with discontinuous boundary values.

A comparison of the derived results with those of § 36 indicates the following essential difference. For the multiply-connected domain in the case of contours not reducing to slits, the solution of the considered problem in the class of single-valued functions is possible only under a special choice of the prescribed boundary values (the conditions of single-valuedness (36.20)). For the same problem in the case of contours reducing to cuts, the solution is possible for arbitrary boundary values, if only solutions having at the ends integrable infinities are admitted. In this case it is even possible to solve the homogeneous problem. The reasons for such a difference are not entirely clear.

Remark. Formulae (46.25) remain valid also in the case when, at the ends at which integrable infinities are admitted, the functions $f^+(t), f^-(t)$ have infinities of an integrable order (see § 43.4).

We proceed to solve the Dirichlet problem, i.e. to determine a function $u(x, y)$ harmonic and bounded in D which satisfies the boundary condition

$$u^+ = f^+(t), \quad u^- = f^-(t) \quad \text{on} \quad L, \tag{46.27}$$

where $f^+(t)$, $f^-(t)$ are known functions satisfying the condition of continuity (46.18). For the method of solution applied below it is required that the derivatives satisfy the Hölder condition.

We shall apply the method announced by M. V. Keldysh and L. I. Sedov (1). Denote by $\Phi(z) = u + iv$ the analytic function the real part of which is the required harmonic function. In general this function is many-valued. It is however easy to establish that its derivative

$$\Phi'(z) = \frac{\partial u}{\partial x} + i \frac{\partial v}{\partial x}$$

is a single-valued function. In fact, in view of the single-valuedness of u the imaginary part v acquires constant increments in describing any path enclosing any of the sections (a_k, b_k). It follows that the derivative returns to its original value on describing an arbitrary path in D, i.e. it is a single-valued function.

We apply the results obtained in the present subsection to determine $\Phi'(z)$, taking into account that the latter has at infinity a zero of order not lower than two.

In the most general class of functions having integrable infinities at all ends (and having at infinity a zero of order two) the solution has the form

$$\Phi'(z) = \frac{1}{R(z)} \frac{1}{2\pi i} \int_L \frac{R(\tau)[f'^+(\tau) + f'^-(\tau)]}{\tau - z} d\tau +$$

$$+ \frac{1}{2\pi i} \int_L \frac{f'^+(\tau) - f'^-(\tau)}{\tau - z} d\tau + \frac{c_0 + c_1 z + \cdots + c_{m-2} z^{m-2}}{R(z)}. \quad (46.28)$$

It is readily observed that all terms of the right-hand side have at infinity zeros of order two. For the first and third terms it is obvious. As concerns the second term it can be verified by integration by parts, taking into account that the integrated part vanishes in view of the continuity of the functions $f^+(t), f^-(t)$.

The function $\Phi(z)$ is given by the formula

$$\Phi(z) = \int_0^z \Phi'(z) \, dz + C, \quad (46.29)$$

where C is a real constant.

To determine the real constants $c_0, c_1, \ldots, c_{m-2}$, C we make use of the relations $\operatorname{Re} \Phi(a_k) = f(a_k)$. Setting in (46.29) $z = a_k$ $(k = 1, 2, \ldots, m)$ and integrating along a path not intersecting L (for instance along the left edge of the x-axis) we arrive at m equations

$$\operatorname{Re} \int_0^{a_k} \Phi'(z) \, dz + C = f^+(a_k). \quad (46.30)$$

To prove the solubility of the latter system, consider the corresponding homogeneous system. This is obtained by setting in the boundary value problem $f^+(t) = f^-(t) \equiv 0$. Thus we have

$$\operatorname{Re} \int_0^{a_k} \frac{c_0 + c_1 \tau + \cdots + c_{m-2} \tau^{m-2}}{R(\tau)} d\tau + C = 0 \quad (k = 1, 2, \ldots, m).$$

$$(46.31)$$

BVP 16

By successive subtraction we obtain the system of $m - 1$ equations

$$\operatorname{Re} \int_{a_k}^{a_{k+1}} = 0 \quad (k = 1, 2, \ldots, m - 1).$$

Taking into account that in the sum $\int_{a_k}^{b_k} + \int_{b_k}^{a_{k+1}}$ the first term has a purely imaginary integrand, we obtain the system

$$\int_{b_k}^{a_{k+1}} \frac{c_0 + c_1 \tau + \cdots + c_{m-2} \tau^{m-2}}{R(\tau)} d\tau = 0 \quad (k = 1, 2, \ldots, m - 1). \quad (46.32)$$

Since the denominator has a constant sign, to satisfy the relations the numerator should change its sign at least once on every section (b_k, a_{k+1}). Consequently, the polynomial of degree $m - 2$ changes its sign at least $m - 1$ times. Such a polynomial, however, vanishes identically. Now it is obvious that also $C = 0$.

Thus, the homogeneous system (46.31) corresponding to the system (46.30) has only zero solutions. Consequently, the non-homogeneous system has a unique solution for an arbitrary right-hand side. This implies the solubility of the Dirichlet problem itself. We have at the same time deduced an algorithm for deriving the solution.

Observe that in the book of N. I. Muskhelishvili [17*] another method of solving the Dirichlet problem is presented; it does not require the assumption of differentiability of the boundary values. This method is based on a solution of one more auxiliary problem (see Problem 11 at the end of this chapter).

§ 47. The dominant equation for open contours

47.1. Basic concepts and notation

We start the exposition from the simplest case of an equation with continuous coefficients, the contour of which is composed of non-intersecting open curves. A generalization to the case of a complex contour composed of an arbitrary finite number of arbitrary located closed and open curves (see § 44) is almost automatic, assuming that the coefficients of the equation have discontinuities of the first kind at a finite number of points.

We shall hereafter use the notation and terminology for integral equations of Chapter III, in particular in § 20.1, and for the Riemann problem first of all in §§ 42, 43.

Thus, suppose that the contour $L = L_1 + L_2 + \cdots + L_m$ consists of m open non-intersecting smooth curves. As in § 20.1 the singular operator and the singular integral equation will be written in the form

$$K\varphi \equiv a(t)\,\varphi(t) + \frac{1}{\pi i} \int_L \frac{M(t, \tau)}{\tau - t}\,\varphi(\tau)\,d\tau = f(t) \qquad (47.1)$$

or

$$K\varphi \equiv K^\circ \varphi + k\varphi$$

$$\equiv a(t)\,\varphi(t) + \frac{b(t)}{\pi i} \int_L \frac{\varphi(t)}{\tau - t}\,d\tau + \int_L k(t, \tau)\,\varphi(\tau)\,d\tau = f(t). \qquad (47.2)$$

The functions $a(t)$, $b(t) = M(t, t)$ are assumed to satisfy the Hölder condition. As distinct from § 20.1 we require that the function

$$k(t, \tau) = \frac{1}{\pi i} \cdot \frac{M(t, \tau) - M(t, t)}{\tau - t}$$

satisfies the Hölder condition with respect to both variables. To this end it is sufficient that the kernel $M(t, \tau)$ satisfies the Hölder condition with respect to the variable t and has a partial derivative with respect to τ which satisfies the Hölder condition with respect to both variables.

The solution $\varphi(t)$ of the equation will be sought in the same classes as to which the boundary values of the solutions of the Riemann problem belong, i.e. *in the classes of functions which satisfy the Hölder condition everywhere inside the curve, and are bounded or have an integrable singularity at the ends* (the last requirement should be specified in every actual problem). For the free term $f(t)$ we may take a function of the same class as that in which the free term $g(t)$ of the Riemann problem in § 41.5 was taken. To begin with, for the sake of simplicity of formulation we assume that the function $f(t)$ satisfies the Hölder condition.

The reason for strengthening the condition imposed upon the regular kernel $k(t, \tau)$, as compared with § 20.1, is due to the fact that we have to ensure the integrability of the product $k(t, \tau)\,\varphi(\tau)$ on those ends at which unbounded solutions are admitted.

As before, let us define the adjoint operator

$$K'\psi \equiv a(t)\,\psi(t) - \frac{1}{\pi i} \int_L \frac{M(\tau,\,t)}{\tau - t}\,\psi(\tau)\,d\tau, \tag{47.3}$$

$$K'\psi \equiv K^\circ\psi + k'\psi$$

$$\equiv a(t)\,\psi(t) - \frac{1}{\pi i} \int_L \frac{b(\tau)\,\psi(\tau)}{\tau - t}\,d\tau + \int_L k'(t,\,\tau)\,\psi(\tau)\,d\tau, \tag{47.4}$$

where

$$b(\tau) = M(\tau,\,\tau), \quad k'(t,\,\tau) = \frac{1}{\pi i} \frac{M(\tau,\,t) - M(\tau,\,\tau)}{\tau - t}.$$

The equation $K'\psi = 0$ is homogeneous, adjoint to the considered equation.

47.2. Solution of the dominant equation

We begin by solving the dominant equation

$$K^\circ\varphi \equiv a(t)\,\varphi(t) + \frac{b(t)}{\pi i} \int_L \frac{\varphi(\tau)}{\tau - t}\,d\tau = f(t). \tag{47.5}$$

Introducing the analytic function

$$\Phi(z) = \frac{1}{2\pi i} \int_L \frac{\varphi(\tau)}{\tau - z}\,d\tau$$

and proceeding as in § 21.1 we arrive at the Riemann boundary value problem for the contour L,

$$\Phi^+(t) = G(t)\,\Phi^-(t) + g(t), \tag{47.6}$$

$$G(t) = \frac{a(t) - b(t)}{a(t) + b(t)}, \quad g(t) = \frac{f(t)}{a(t) + b(t)}, \tag{47.7}$$

with the additional condition $\Phi(\infty) = 0$.

We shall consider the normal (non-exceptional) case when $G(t)$ does not vanish anywhere and is bounded; for the original equation this corresponds to the condition $a(t) \pm b(t) \neq 0$.

To simplify the following formulae let us divide the equation (47.5) throughout by $\sqrt{[a^2(t) - b^2(t)]}$; we assume therefore that the coefficients of this equation satisfy the condition

$$a^2(t) - b^2(t) = 1. \tag{47.8}$$

In view of the necessity to divide the solutions of the singular integral equation into classes let us at once carry out a classification of the solutions of the Riemann problem.

We shall uniformly denote all ends a_k and b_k by the letters $c_1, c_2, \ldots, c_{2m}$. Suppose that $c_1, c_2, \ldots, c_q$ are not ends of automatic boundedness (non-singular); the remaining ends $c_{q+1}, \ldots, c_{2m}$ are assumed to be singular i.e. the ends of automatic boundedness. Let us divide all ends into three groups

$$c_1, c_2, \ldots, c_p; \quad c_{p+1}, \ldots, c_q; \quad c_{q+1}, \ldots, c_{2m}. \tag{47.9}$$

A sectionally analytic function is said *to belong to class* $H(c_1, c_2, \ldots, c_p)$ if it constitutes a solution of the Riemann boundary value problem (47.6), satisfies the Hölder condition everywhere on L except the ends, has on the ends of group I integrable infinities and is bounded on all the remaining ends. The points $c_1, c_2, \ldots, c_p$ at which the solution may have an infinity will be called *the ends defining the class* †.

The class of solutions bounded at all the ends (the narrowest class) will be denoted by H_0.

The above classification will be used not only for the solutions of Riemann problems, but generally for all sectionally analytic functions. Let us also introduce the concept of *adjoint* classes. If in the above definition of the class we exchange the points of groups I and II, we obtain the class adjoint to the considered one. Thus, if $H(c_1, c_2, \ldots, c_p)$ is the considered class, then $H'(c_1, c_2, \ldots, c_p) = H(c_{p+1}, \ldots, c_q)$ is the adjoint class.

† The classification introduced is somewhat different from that of N. I. Muskhelishvili [17*]. The latter author considers as the fundamental the most general class of functions having integrable infinities at all non-singular ends; the remaining classes are deduced by narrowing this class. Consequently, in this case the points defining the class are all non-singular ends at which the boundedness of the solution is required. We, however, consider as the fundamental the least general class of solutions which are bounded at all the ends. The remaining classes are deduced by extending the fundamental one, admitting integrable infinities at some ends; the latter ends define the class. The alteration of the classification performed above is necessary to attain unification with the classification in the exceptional cases, when $G(t)$ may have zeros or poles (see § 45). In that case, naturally the points at which the solution has an infinity have been regarded not as ordinary but as exceptional.

The solution of the dominant equation (47.5) is obtained by means of the Sokhotski formula

$$\varphi(t) = \Phi^+(t) - \Phi^-(t). \tag{47.10}$$

We work from the formulae of § 43.1, which give the solution of the Riemann problem with open contour. Reasoning analogous to that in § 21.2 leads to the formula

$$\varphi(t) = \frac{1}{2}\left[1 + \frac{1}{G(t)}\right]g(t) +$$

$$+ X^+(t)\left[1 - \frac{1}{G(t)}\right]\left[\frac{1}{2\pi i}\int\limits_L \frac{g(\tau)}{X^+(\tau)} \frac{d\tau}{\tau - t} - \frac{1}{2}P_{\varkappa-1}(t)\right]. \tag{47.11}$$

This expression indicates that the class of solutions of the integral equation is identical with the class of solutions of the corresponding Riemann problem. (Let us recall that by assumption the free term $f(t)$ belongs to the same class as the required solution.)

Introducing into the expression $X^+(t) = \prod\limits_{k=1}^{m}(t - b_k)^{-\varkappa_k}[\sqrt{G(t)}]\,e^{\Gamma(t)}$ the values of $G(t)$, $g(t)$ from formulae (47.7) and taking into account condition (47.8) we transform formula (47.11) to the form

$$\varphi(t) = a(t)f(t) - \frac{b(t)Z(t)}{\pi i}\int\limits_L \frac{f(\tau)}{Z(t)} \frac{d\tau}{\tau - t} + b(t)Z(t)P_{\varkappa-1}(t), \tag{47.12}$$

where

$$Z(t) = [a(t) + b(t)]\,X^+(t) = [a(t) - b(t)]\,X^-(t) = \prod\limits_{k=1}^{m}(t - b_k)^{-\varkappa_k}e^{\Gamma(t)},$$

$$\Gamma(t) = \frac{1}{2\pi i}\int\limits_L \frac{\ln G(\tau)}{\tau - t}\,d\tau, \quad \varkappa = \sum\limits_{k=1}^{m}\varkappa_k. \tag{47.13}$$

If $\varkappa \leqq 0$ we have to set $P_{\varkappa-1}(t) \equiv 0$. For $\varkappa < 0$ the solution in the considered class exists if and only if the conditions of solubility

$$\int\limits_L \frac{f(\tau)}{Z(\tau)}\tau^{j-1}\,d\tau = 0 \quad (j = 1, 2, \ldots, -\varkappa). \tag{47.14}$$

are satisfied. The last term of formula (47.12) constitutes the general solution of the homogeneous dominant equation $K^\circ\varphi = 0$.

Introducing as in § 21.2 the operator R

$$Rf \equiv a(t)f(t) - \frac{b(t)Z(t)}{\pi i} \int\limits_L \frac{f(\tau)}{Z(\tau)} \frac{d\tau}{\tau - t}, \tag{47.15}$$

we can write the general solution of the dominant equation in the form

$$\varphi(t) = Rf + \sum_{k=1}^{\varkappa} C_k \varphi_k(t), \tag{47.16}$$

where $\varphi_k(t) = b(t)\,Z(t)\,t^{k-1}$ are eigenfunctions of the operator K°, i.e. they are linearly independent solutions of the homogeneous dominant equation $K^\circ \varphi = 0$.

We proceed to investigate the problem of classes of solutions. For the latter we introduce a classification analogous to the above classification for the solutions of the Riemann problem. The difference consists only in the fact that now we deal not with sectionally analytic functions defined in the entire plane, but with functions prescribed only on the contour.

DEFINITION. *A solution $\varphi(t)$ of the singular integral equation, which satisfies everywhere on L, except possibly its ends, the Hölder condition, has at the points of group I (47.9) an integrable infinity and is bounded near all the remaining ends, is said to belong to the class* $h(c_1, c_2, \ldots, c_p)$.

The points $c_1, c_2, \ldots, c_p$ at which the solution has an infinity are assumed to define the class.

A solution of the integral equation belonging to class $h(c_{p+1}, \ldots, c_q)$ will be called a solution of class adjoint to the class $h(c_1, c_2, \ldots, c_p)$. It is readily observed that the limiting values of sectionally analytic functions of class $H(c_1, c_2, \ldots, c_p)$ are functions of class $h(c_1, c_2, \ldots, c_p)$. The class of solutions bounded at all the ends (the narrowest class) will be denoted by h_0. The above classification will be applied not only to the solutions of a singular integral equation but generally to arbitrary functions prescribed on the contour.

In view of formula (47.13) and the properties of the canonical function (formulae (43.3)–(43.5)) the function $Z(t)$ (N. I. Muskhelishvili [17*] called it *the canonical function of the integral equation* (47.5) *of considered class*) has the form

$$Z(t) = \omega_0(t) \sum_{k=1}^{2m} (t - c_k)^{\gamma_k} \quad (\gamma_k = \alpha_k + i\beta_k), \tag{47.17}$$

where $\omega_0(t)$ satisfies the Hölder condition and it does not vanish anywhere, and $\mathrm{Re}\gamma_k$ is defined by the conditions

$$\left.\begin{array}{llll} -1 < \alpha_k < 0 & \text{for} & k = 1, 2, \ldots, p, \\ 0 > \alpha_k < 1 & \text{for} & k = p + 1, \ldots, q, \\ \alpha_k = 0 & \text{for} & k = q + 1, \ldots, 2m. \end{array}\right\} \qquad (47.18)$$

It follows from the expression for the function $Z(t)$ and formula (47.11) that if a solution of the Riemann problem (47.6) is taken in class $H(c_1, \ldots, c_p)$, then the solution of the integral equation (46.5) belongs to class $h(c_1, \ldots, c_p)$.

On the ends $c_{q+1}, \ldots, c_{2m}$ the solution vanishes and it is easy to see that it satisfies the Hölder condition up to the ends. On the singular ends $c_{q+1}, \ldots, c_{2m}$ the solution is bounded but it has no definite limit and, consequently, it is discontinuous. It is readily observed that class h_0 coincides with the class of functions satisfying the Hölder condition on the whole contour, if and only if all ends are non-singular.

The index $\varkappa$ of the Riemann problem (47.6) in the considered class will be called *the index of the integral equation itself*. The properties of the dominant singular equation can be stated as in § 21.2 where the solution and the index have to be regarded as belonging to the same class. We shall not repeat the relevant satements. As in the case of the Riemann problem the admittance of unboundness of a solution near a non-singular end increases the number of solutions of the integral equation by unity. The equation has the greatest number of solutions in class $h(c_1, c_2, \ldots, c_q)$ and the smallest number in the class h_0. The difference between the numbers of solutions in these classes is equal to the number q of the non-singular ends (the number of the conditions of solubility is taken as the number of solutions with the negative sign).

It can easily be verified that R transforms any function $f(t)$ of class $h(c_1, c_2, \ldots, c_p)$ into a function $\varphi(t)$ of the same class; in doing so the properties of the function $Z(t)$ have to be taken into account. This fact justifies the condition imposed upon the free term $f(t)$ in the beginning of this subsection.

Remark 1. It was already indicated in § 43.4 that in determining the general solution of the Riemann boundary value problem in a class $H(c_1, \ldots, c_p)$ we can take the particular solution of the homogeneous problem entering the formula as the general solution, in an arbitrary less general class $H(c_1, \ldots, c_r)$ $(r < p)$. Let us denote the canonical function of the Riemann problem in class $H(c_1, \ldots, c_r)$ by $X(z)$ and in class $H(c_1, \ldots, c_r)$ by $X_1(z)$. The canonical functions of the dominant singular integral equation associated with it in accordance with formula (47.13), will be denoted by $Z(t)$ and $Z_1(t)$.

Then the general solution of the dominant equation in class $h(c_1, \ldots, c_p)$ can be written in the form

$$\varphi(t) = a(t)f(t) - \frac{b(t)Z_1(t)}{\pi i} \int_L \frac{f(\tau)}{Z_1(\tau)} \frac{d\tau}{\tau - t} + b(t)Z(t)P_{\varkappa-1}(t). \quad (47.11')$$

Remark 2. If for the free term $f(t)$ we take functions of more general classes, as was done for $g(t)$ at the end of § 41.5, then the solution of the considered equation can possess additional singularities in accordance with the singularities of the free term. The nature of these singularities was considered in sufficient detail at the end of § 41.5, we shall not therefore dwell on this point here.

47.3. Solution of the equation adjoint to the dominant equation

Consider the equation

$$K^{o\prime}\psi \equiv a(t)\psi(t) - \frac{1}{\pi i} \int_L \frac{b(\tau)\psi(\tau)}{\tau - t} d\tau = f(t), \quad (47.19)$$

adjoint to the dominant equation (47.5).

Introducing the sectionally analytic function (see § 21.3)

$$\Omega(z) = \frac{1}{2\pi i} \int_L \frac{b(\tau)\psi(\tau)}{\tau - z} d\tau, \quad (47.20)$$

we arrive at the boundary value problem

$$\Omega^+(t) = \frac{1}{G(t)} \Omega^-(t) + \frac{b(t)f(t)}{a(t) - b(t)}, \quad (47.21)$$

where $G(t)$, as in the preceding subsection, is given by formula (47.7). The homogeneous problem $\Omega^+(t) = [1/G(t)]\Omega^-(t)$ is called the adjoint to the problem $\Phi^+(t) = G(t)\Phi^-(t)$. If $X(z)$ is the canonical function of the Riemann problem (47.6), then evidently $X'(z) = 1/X(z)$ is the canonical function of problem (47.21). Also, if $X(z)$ belongs to a class $H(c_1, \ldots, c_p)$, then $X'(z)$ belongs to the adjoint class H'. If the order at infinity of the canonical function $X(z)$ (equal to the index of the problem in the class $H(c_1, \ldots, c_p)$) is $\varkappa$, then obviously, the order of $X'(z) = 1/X(z)$ is $-\varkappa$. Consequently, the index $\varkappa'$ of the adjoint problem in the adjoint class $H(c_{p+1}, \ldots, c_q)$ of the solutions is equal to the index $\varkappa$ of the problem taken with opposite sign.

The general solution of the boundary value problem (47.21) in the class $H(c_{p+1}, \ldots, c_q)$ can be written in the form

$$\Omega(z) = \frac{[X^+(z)]^{-1}}{2\pi i} \int_L \frac{X^+(\tau)\, b(\tau) f(\tau)}{a(\tau) - b(\tau)} \frac{d\tau}{\tau - z} + [X(z)]^{-1} P_{-\varkappa-1}(z), \quad (47.22)$$

where $P_{-\varkappa-1}$ is a polynomial of degree $-\varkappa - 1$ with arbitrary coefficients.

From the solution (47.22) of the Riemann problem, in accordance with the Sokhotski formulae

$$b(t)\,\psi(t) = \Omega^+(t) - \Omega^-(t) \tag{47.23}$$

we obtain as before the solution of the adjoint equation (47.19)

$$\psi(t) = a(t)f(t) + \frac{1}{\pi i Z(t)} \int_L \frac{Z(\tau)\, b(\tau) f(\tau)}{\tau - t}\, d\tau + \frac{1}{Z(t)} P_{-\varkappa-1}(t), \tag{47.24}$$

which obviously belongs to the adjoint class $h(c_{p+1}, \ldots, c_q)$.

If $-\varkappa \leqq 0$ we have to set $P_{-\varkappa-1}(t) \equiv 0$ and if $-\varkappa < 0$ the conditions of solubility

$$\int_L Z(\tau)\, b(\tau) f(\tau)\, \tau^{j-1}\, d\tau = 0 \quad (j = 1, 2, \ldots, \varkappa), \tag{47.25}$$

have to be satisfied. Formula (47.24) indicates that the functions $\psi_j(t) = [1/Z(t)] t^{j-1}$ $(j = 1, \ldots, -\varkappa)$ are linearly independent solutions of the adjoint equation (47.19). Consequently, the conditions of solubility (47.14) of the considered dominant equation (47.5) can be written in the form

$$\int_L f(\tau)\, \psi_j(\tau)\, d\tau = 0 \quad (j = 1, 2, \ldots, -\varkappa). \tag{47.26}$$

This is the second Noether theorem (see § 23.2), proved here for the particular case of the dominant equation. Similarly to the other Noether theorems, this theorem will be proved in § 48.3 for the general case of the complete equation.

47.4. Examples

1. For the first example let us examine the dominant singular integral equation of the first kind.

Set in equation (47.5) $a(t) \equiv 0$. In view of the condition $a(t) \pm b(t) \neq 0$, $b(t)$ does not vanish anywhere on L. Dividing by $b(t)$ the equation can be written in the form

$$\frac{1}{\pi i} \int_L \frac{\varphi(\tau)}{\tau - t} \, d\tau = f(t).$$

Consequently, the solution of the integral equation of first kind is the inversion of the Cauchy type integral. This problem was solved in § 42.3. In this case all ends are non-singular ($q = 2m$). In the class $h(c_1, \ldots, c_p)$ the problem has for $p \geqq m$, $p - m$ linearly independent solutions; for $p < m$ it has a solution only if $m - p$ additional conditions of solubility are satisfied. We shall not write the relevant formulae again.

2. Consider the integral equation encountered in solving a mechanical problem

$$a\varphi(t) + \frac{a_0 + a_1 t}{\pi} \int_0^l \frac{\varphi(\tau)}{\tau - t} \, d\tau = b_0 + b_1 t. \tag{47.27}$$

The contour L is here the section of the real axis $(0, l)$; a, a_0, a_1, b_0, b_1 are real constants.

By the usual method we reduce the problem to the Riemann boundary value problem

$$\Phi^+(t) = G(t) \Phi^-(t) + g(t);$$

$$G(t) = \frac{a - i(a_0 + a_1 t)}{a + i(a_0 + a_1 t)}; \quad g(t) = \frac{b_0 + b_1 t}{a + i(a_0 + a_1 t)}.$$

We have arrived at the boundary value problem examined in § 43.3. The solution of the original equation (47.27) is obtained with the help of the Sokhotski formula

$$\varphi(t) = \Phi^+(t) - \Phi^-(t).$$

Making use of formula (47.11) and the solution (43.20) let us write down the solution in class $h(l)$ (i.e. bounded at the origin of the coordinate system and having an integrable infinity at the end of the interval) for the fourth case (Fig. 16)

$$\varphi(t) = a(t) f(t) - \frac{b(t) Z(t)}{\pi i} \int_0^l \frac{f(\tau)}{Z(\tau)} \frac{d\tau}{\tau - t} + b(t) Z(t) P_{\varkappa - 1}(t).$$

We have to set here

$$\varkappa = 1, \quad a(t) = \frac{a}{\sqrt{[a^2 + (a_0 + a_1 t)^2]}}, \quad b(t) = i \frac{a_0 + a_1 t}{\sqrt{[a^2 + (a_0 + a_1 t)^2]}},$$

$$f(t) = \frac{b_0 + b_1 t}{\sqrt{[a^2 + (a_0 + a_1 t)^2]}}, \quad Z(t) = (t - l)^{-1} e^{\Gamma(t)},$$

$$\Gamma(t) = -\frac{1}{\pi} \int_0^l \frac{\arctan\left(\dfrac{a_0}{a} + \dfrac{a_1}{a} \tau\right)}{\tau - t} \, d\tau, \quad P_{\varkappa - 1}(t) = c_0 = \text{const.}$$

16 a*

The homogeneous equation has one linearly independent solution $b(t)Z(t)$ in the considered case.

The reader is advised to consider the remaining cases (1, 2, 3) independently.

§ 48. Complete equation for open contours

The theory of the complete integral equation (47.1) in the considered case consists of two parts, similar to those for a closed contour.

1. Regularization, including an elaboration of the methods of regularization and an examination of the connection between the solutions of the original singular equation and the deduced regular equations (the problems of equivalent regularization).

2. Basic properties of the singular equation (the Noether theorems).

The methods of investigation will be the same as in the case of closed contours (Chapter III). We shall hereafter abbreviate the description, omitting the reasoning when it is completely analogous to that of Chapter III.

We shall observe that the difference between the considered case and the case of closed contours consists in the fact that the product of two integrable functions $\varphi(t), \psi(t)$ may turn out to be non-integrable. To ensure the integrability of the product we have to take care of the fact that both functions have no infinities at the same end. Recalling the properties of the adjoint classes introduced in the preceding section we easily infer that a product of functions belonging to adjoint classes is always integrable.

The original part of the theory described below consists in establishing in which class (see § 47) the corresponding functions should be taken.

48.1. Regularization by solving the dominant equation

The basic information about the regularization of the complete singular integral equation was presented in §§ 22 and 24, the concepts and notations of which will constantly be used here.

In the case under consideration all three methods of regularization for the closed contour, are valid; the regularization from the left, from the right and regulariziation by solving the dominant equation. The most simple for the investigation of the behaviour of the appearing kernels and the admissible classes of solutions is the last method, we shall therefore begin by describing it.

Let us write equation (47.1) in the form

$$a(t)\,\varphi(t) + \frac{b(t)}{\pi i}\int_L \frac{\varphi(t)}{\tau - t}\,d\tau = f(t) - \int_L k(t, \tau)\,\varphi(\tau)\,d\tau \quad (48.1)$$

or, symbolically,

$$K^\circ\varphi = f - k\varphi \qquad (48.1')$$

and let us solve it as the dominant equation, regarding the right-hand side as a known function. The solution will be sought in a class

$$h(c_1, c_2, \ldots, c_p) = h.$$

For simplicity we assume that $f(t)$ satisfies the Hölder condition. If, as it was assumed in § 47.1, $k(t,\tau)$ satisfies the Hölder condition with respect to both variables and $\varphi(t)$ belongs to class h, then it can easily be proved that $k\varphi$ satisfies the Hölder condition and, consequently, the whole right-hand side of relation (48.1) satisfies this condition.

Applying formula (47.11) we obtain (see § 24.6)

$$\varphi(t) + \int_L K(t, \tau)\,\varphi(\tau)\,d\tau = f_1(t), \qquad (48.2)$$

where

$$\left.\begin{aligned}
K(t, \tau) &\equiv R_t k(t, \tau) \\
&= a(t)\,k(t, \tau) - \frac{b(t)\,Z(t)}{\pi i}\int_L \frac{k(\tau_1, \tau)}{Z(\tau_1)}\,\frac{d\tau_1}{\tau_1 - t}, \\
f_1(t) &= a(t)f(t) - \\
&\quad - \frac{b(t)\,Z(t)}{\pi i}\int_L \frac{f(\tau)}{Z(\tau)}\,\frac{d\tau}{\tau - t} + b(t)\,Z(t)\,P_{\varkappa-1}(t),
\end{aligned}\right\} \quad (48.3)$$

$$Z(t) = [a(t) + b(t)]\,X^+(t) = [a(t) - b(t)]\,X^-(t) = \prod_{k=1}^{2m}(t - c_k)^{-\varkappa_k}\,e^{\Gamma(t)},$$

$$\Gamma(t) = \frac{1}{2\pi i}\int_L \frac{\ln G(\tau)}{\tau - t}\,d\tau.$$

If $\varkappa \leqq 0$ we have to set $P_{\varkappa-1} \equiv 0$ and when $\varkappa < 0$ the functional relations

$$\int_L \left[\int_L \frac{k(t, \tau)}{Z(t)}\,t^{j-1}\,dt\right]\varphi(\tau)\,d\tau = \int_L t^{j-1}\frac{f(t)}{Z(t)}\,dt \qquad (48.4)$$

$$(j = 1, 2, \ldots, -\varkappa),$$

should be satisfied. These relations can also be written in the form

$$\int_L \varrho_j(\tau)\,\varphi(\tau)\,d\tau = f_j, \tag{48.4'}$$

where

$$\varrho_j(\tau) = \int_L \frac{k(t,\tau)}{Z(t)}\,t^{j-1}\,dt, \quad f_j = \int_L t^{j-1}\frac{f(t)}{Z(t)}\,dt. \tag{48.5}$$

If the considered equation (48.1) is soluble then conditions (48.4) are satisfied.

48.2. Investigation of the regularized equation

We proceed to investigate the kernel $K(t,\tau)$ and the free term $f_1(t)$ of equation (48.2) obtained by the regularization of the original singular equation (48.1). Under the assumptions with respect to the given functions $k(t,\tau)$, $f(t)$ made in the preceding subsection, it is evident that these functions satisfy the Hölder condition at all points of the contour distinct from its ends. Let us consider the behaviour on the ends.

According to formula (47.17)

$$Z(t) = \omega_0(t)\prod_{k=1}^{2m}(t-c_k)^{\gamma_k}; \quad \gamma_k = \alpha_k + i\beta_k, \tag{48.6}$$

$$-1 < \alpha_k < 0 \quad \text{for} \quad k = 1, 2, \ldots, p,$$

$$0 < \alpha_k < 1 \quad \text{for} \quad k = p+1, \ldots, q,$$

$$\alpha_k = 0 \quad \text{for} \quad k = q+1, \ldots, 2m,$$

$\omega_0(t)$ is a function satisfying the Hölder condition which does not vanish anywhere.

It can easily be derived from formulae (48.3) that the functions $K(t,\tau)$ and $f_1(t)$ behave at the ends $c_1, c_2, \ldots, c_p$ as the function $Z(t)$, i.e. they have infinities of order α_k. On the remaining ends they are bounded. Thus, the kernel $K(t,\tau)$ and the free term $f_1(t)$ which possess at the ends $c_1, \ldots, c_p$ a fixed infinity, do not satisfy the conditions imposed upon these functions in the Fredholm theory†. However, by a change of the variables for φ and the argument of integration τ,

---

† Only in the case of a fixed infinity of order smaller than unity, i.e. of the form $(\tau - t)^{-\alpha}$ ($0 < \alpha < 1$) this order is diminished by iteration, and there exists a number m such that the mth iterated kernel is bounded.

it is easy to eliminate the fixed infinities and to reduce the equation to a form in which the kernel and the free term obey all conditions of the Fredholm theory.

Denote

$$T(t) = \prod_{k=1}^{p} (t - c_k)^{\gamma_k} \tag{48.7}$$

$$\varphi(t) = T(t)\, \varphi_0(t), \tag{48.8}$$

and make the substitution

$$\varphi_0(t) + \int_L \frac{K(t, \tau)\, T(\tau)}{T(t)}\, \varphi_0(\tau)\, d\tau = \frac{f_1(t)}{T(t)}.$$

Now the fixed infinities with respect to the variable t are absent, but the same infinities with respect to the variable τ have appeared. To eliminate the latter we perform the change of the independent variables

$$\tau_1 = \int_{a_k}^{\tau} T(\tau)\, d\tau, \quad t_1 = \int_{a_k}^{t} T(t)\, dt \quad \text{on} \quad L_k \ (k = 1, 2, \ldots, m). \tag{48.9}$$

Denoting again $\varphi_0[t(t_1)]$ by $\varphi_0(t_1)$ we arrive at the equation

$$\varphi_0(t_1) + \int_{\Gamma} K_1(t_1, \tau_1)\, \varphi_0(\tau_1)\, d\tau_1 = f_2(t_1), \tag{48.10}$$

in which the new kernel $K_1(t, \tau)$ and the free term $f_2(t)$ are bounded on the new contour of integration Γ.

Observe that since in general the inversion of formulae (48.9) is not unique, the new contour Γ can turn out to be self-intersecting which is of no essential importance. Having found all solutions of the Fredholm equation (48.10), with the help of formulae (48.8) we obtain the solutions of equation (48.2) belonging to the class $h(c_1, \ldots, c_p)$.

Since the solutions of equations (48.2) and (48.10) are connected by the simple relation (48.8), it follows that although equation (48.2) is not exactly a Fredholm equation, all Fredholm theorems turn out to be valid for it. The only difference is that its solutions are not bounded as it follows from the Fredholm theory; these solutions belong to class $h(c_1, \ldots, c_p)$.

Consequently, we may state that as a result of regularization we reduced a singular equation to a Fredholm equation.

If we take the homogeneous equation

$$\psi(t) + \int_L K(\tau, t)\,\psi(\tau)\,d\tau = 0, \tag{48.11}$$

adjoint to (48.2) and perform a change of variables analogous to (48.8), (48.9) it can readily be established that we arrive at the Fredholm equation

$$\psi_0(t_1) + \int_\Gamma K_1(\tau_1, t_1)\,\psi_0(\tau_1)\,d\tau_1 = 0, \tag{48.12}$$

adjoint to equation (48.10).

48.3. Other methods of regularization. Equivalent regularization

It has already been indicated that the two other methods of regularization based on a product of two singular operators are also applicable in this case. However, if we take the regularizing operator (formula (22.16)) in an arbitrary way without taking care of some special conditions, then as a rule the kernel of the regularized equation has on the ends fixed singularities the order of which depends not only on the selected class of solutions and the behaviour of the coefficients of the considered equation on the ends, but also on the Hölder exponent of these coefficients. This results in serious difficulties in investigating the allowable classes of solutions. These difficulties may be avoided by a special choice of the regularizing operator. One of the operators which are suitable for regularization, is the operator R which appeared as the solving operator in the solution of the non-homogeneous dominant equation.

First we prove that the operator

$$R\omega \equiv a(t)\,\omega(t) - \frac{b(t)\,Z(t)}{\pi i}\int_L \frac{\omega(\tau)}{Z(\tau)}\,\frac{d\tau}{\tau - t} \tag{48.13}$$

is regularizing for the operator

$$K\varphi \equiv K^\circ\varphi + k\varphi \equiv$$

$$\equiv a(t)\,\varphi(t) + \frac{b(t)}{\pi i}\int_L \frac{\varphi(\tau)}{\tau - t}\,d\tau + \int_L k(t, \tau)\,\varphi(\tau)\,d\tau. \tag{48.14}$$

We already know (formula (47.16)) that if $K^\circ\varphi = f$, then

$$\varphi = Rf + \sum_{k=1}^{\varkappa} C_k\varphi_k, \quad \text{where} \quad \varphi_k = b(t)\,Z(t)\,t^{k-1}$$

are the eigenfunctions of the operator K°. This implies the identity

$$K^\circ Rf \equiv f. \qquad (48.15)$$

Consequently, $K^\circ R \equiv E$ where E is the unit operator. Hence also $RK^\circ \equiv E$. It follows that the operator R is regularizing from the left and from the right for the dominant operator K° and consequently also for the complete operator K. The kernel of the regularized equation is simple. Its behaviour on the ends has been investigated in the preceding subsection.

Having in mind applications of the above methods to an investigation of the problems of equivalent regularization and to a derivation of the basic properties (Noether theorems), let us consider the problem of the eigenfunctions of the operator R constructed for the solution of the equation $K^\circ \varphi = f$ in class $h(c_1, \ldots, c_p)$.

Consider the equation

$$R\omega \equiv a(t)\,\omega(t) - \frac{b(t)Z(t)}{\pi i} \int_L \frac{\omega(\tau)}{Z(\tau)} \frac{d\tau}{\tau - t} = f(t). \qquad (48.16)$$

Introducing the new function $\psi(t) = \omega(t)/Z(t)$ we obtain the dominant equation

$$a(t)\,\psi(t) - \frac{b(t)}{\pi i} \int_L \frac{\psi(t)}{\tau - t}\, d\tau = \frac{f(t)}{Z(t)}.$$

If $f(t)$ belongs to class $h(c_1, \ldots, c_p)$ then it is easy to see that $f(t)/Z(t)$ belongs to class $h(c_{p+1}, \ldots, c_q)$. The index of the last equation in class $h(c_{p+1}, \ldots, c_q)$ is $-\varkappa$. Solving by the ordinary method the last equation in class $h(c_{p+1}, \ldots, c_q)$ we obtain the solution of the original equation (48.16) in class $h(c_1, \ldots, c_p)$ in accordance with the formula $\omega(t) = Z(t)\,\psi(t)$. This solution has the form

$$\omega(t) = K^\circ f + b(t)\, P_{-\varkappa - 1}(t). \qquad (48.17)$$

If $\varkappa \geqq 0$ we have to set $P_{-\varkappa - 1}(t) \equiv 0$ and when $\varkappa > 0$ the $\varkappa$ conditions of solubility

$$\int_L f(t)\, t^{j-1}\, dt = 0 \quad (j = 1, 2, \ldots, \varkappa), \qquad (48.18)$$

should be satisfied. Thus, we have obtained the following result.

For $\varkappa \geqq 0$ the homogeneous equation $R\omega = 0$ is insoluble. The non-homogeneous equation for $\varkappa = 0$ is absolutely and uniquely soluble *and for $\varkappa > 0$ soluble (uniquely) only if $\varkappa$ conditions* (48.18) *are satisfied.*

For $\varkappa < 0$ the homogeneous equation has $\varkappa$ linearly independent solutions. The non-homogeneous equation is absolutely soluble and its solutions depends on $\varkappa$ arbitrary constants.

Making use of the deduced properties of the operator R we can proceed to the problem of equivalent regularization in the sense of the definition of § 24.1.

If the index of the operator K in a prescribed class is non-negative ($\varkappa \geqq 0$) we apply regularization from the left. The equation

$$RK\varphi = Rf \tag{48.19}$$

is a Fredholm equation equivalent to the original singular equation (48.14).

If $\varkappa \leqq 0$ we apply regularization from the right. Introducing a new function $\omega(t)$ by the substitution

$$\varphi = R\omega \tag{48.20}$$

we arrive at the equation

$$KR\omega = f, \tag{48.21}$$

which is equivalent to the original singular equation, in the sense that in the considered class both equations (48.14) and (48.21) are simultaneously soluble or insoluble, and, in the case of solubility, to every solution of equation (48.21) there corresponds in accordance with formula (48.20) a definite solution of the original equation (48.14), and conversely, to every solution $\varphi(t)$ of the original equation there correspond, in accordance with the formula

$$\omega(t) = K^{\circ}\varphi + b(t) P_{-\varkappa-1}(t), \tag{48.22}$$

solutions of equation (48.14).

Thus, the problem of equivalent regularization in the generalized sense is solved for an open contour in exactly the same way as for a closed one.

48.4. Basic properties of the singular equation

All preceding results lead the reader to the conclusion that the basic properties of the singular equation are expressed by the so-called Noether theorems, essentially the same as in the case of a closed contour. Their derivation can be carried out in the same way as for a closed contour in § 23.

First, we easily establish that if the functions $\varphi(t)$ and $\psi(t)$ are such that $\psi K \varphi$ and $\varphi K' \psi$ are integrable, in particular if φ and ψ belong to adjoint classes, then the following identity holds:

$$\int_L \psi K \varphi \, dt = \int_L \varphi K' \psi \, dt. \tag{48.23}$$

Next, the same reasoning as in § 23.2 leads to the basic properties of the equation of the considered type, which by analogy to the theorems deduced in the simplest case by Noether (see Historical Notes to Chapters III and VI) are sometimes called the Noether theorems. Without repeating the derivations we state the theorems.

THEOREM 1. *The number of linearly independent solutions (of an arbitrary class) of the homogeneous equation $K\varphi = 0$ is finite.*

THEOREM 2. *The necessary and sufficient condition of solubility in a given class h of the non-homogeneous equation $K\varphi = f$ consists in satisfying the relations*

$$\int_L f(t) \psi(t) \, dt = 0, \tag{48.24}$$

where $\psi(t)$ is an arbitrary solution of the adjoint class h', of the adjoint homogeneous equation $K'\psi = 0$.

THEOREM 3. *If n is the number of linearly independent solutions of class h of the homogeneous equation $K\varphi = 0$, $\varkappa$ is the index of the equation in this class and n' is the number of linearly independent solutions of the adjoint class h' of the adjoint homogeneous equation $K'\psi = 0$, then*

$$n - n' = \varkappa. \tag{48.25}$$

Thus, the difference between the above theorems and the corresponding theorems of § 23.2 consists only in the fact that now the solutions of the given equation $K\varphi = 0$ and the adjoint equation $K'\psi = 0$ have necessarily to be taken in adjoint classes.

The theory of equivalent regularization from the left, described in §§ 24.4 and 24.5 for a closed contour, has not so far been investigated for the case under consideration. The problem whether the theorem of § 24.5 can automatically be extended to the case of open contours or whether some difficulties will be encountered, is not clear.

§ 49. The general case

49.1. Equation on a complex contour and with discontinuous coefficients

We already know from §§ 41–46 that the theory of the Riemann boundary value problem in the general case of an arbitrary complex contour and discontinuous coefficients, in principle differs little from the case of simple open contours. Taking into account the close connection between the singular integral equations and the Riemann boundary value problem, the same should be expected for the equations under consideration.

Consider the dominant equation

$$K°\varphi \equiv a(t)\,\varphi(t) + \frac{b(t)}{\pi i} \int_L \frac{\varphi(\tau)}{\tau - t}\, d\tau = f(t) \qquad (49.1)$$

under the assumption that the contour L is composed of an arbitrary finite number of closed or open and arbitrarily located curves. The coefficients $a(t)$, $b(t)$ are assumed to satisfy the Hölder condition everywhere on the contour, except at a finite number of points $t_1, \ldots, t_l$ where they have discontinuities of first kind. All ends of the curves and the points of discontinuity will uniformly be called "the ends" (see § 43.2) and denoted by $c_1, c_2, \ldots, c_r$. For simplicity we assume that the ends do not coincide with the points of self-intersection of the contour. As in all analogous cases we require that the equation is to be satisfied everywhere except at the ends and the points of self-intersection. We also suppose that the condition $a^2(t) - b^2(t) = 1$ is satisfied.

Let $c_1, c_2, \ldots, c_q$ be the non-singular and $c_{q+1}, \ldots, c_r$ the singular ends. Exactly as in § 47.2 we introduce the adjoint classes $h(c_1, c_2, \ldots, c_p) = h$ and $h(c_{p+1}, \ldots, c_q) = h'$. We seek the solution of the equation in a class h. The free term $f(t)$ can in the general case be assumed to belong to the same class h.

A closed form of the solution of equation (49.1) can be obtained by the same method as in the simpler cases considered earlier (§§ 21.2, 47.2). Introducing the analytic function

$$\Phi(z) = \frac{1}{2\pi i} \int_L \frac{\varphi(\tau)}{\tau - z}\, d\tau, \qquad (49.2)$$

we arrive at the Riemann boundary value problem

$$\Phi^+(t) = G(t)\,\Phi^-(t) + g(t), \tag{49.3}$$

$$G(t) = \frac{a(t) - b(t)}{a(t) + b(t)}, \quad g(t) = \frac{f(t)}{a(t) + b(t)} \tag{49.4}$$

under the assumption of § 44 and under the condition $\Phi(\infty) = 0$.

Taking, in accordance with formulae (44.11) – (44.13), the solution of this problem in a class H, by means of the Sokhotski formula

$$\varphi(t) = \Phi^+(t) - \Phi^-(t) \tag{49.5}$$

we are led to the solution of the original equation (49.1). It is given by the formulae

$$\varphi(t) = a(t)f(t) - \frac{b(t)Z(t)}{\pi i}\int_L \frac{f(t)}{Z(\tau)}\,\frac{d\tau}{\tau - t} + b(t)\,Z(t)\,P_{\varkappa-1}(t), \tag{49.6}$$

$$Z(t) = [a(t) + b(t)]\,X^+(t) = [a(t) - b(t)]\,X^-(t)$$

$$= \prod_{k=1}^{r}(t - c_k)^{-\varkappa_k}\,e^{\Gamma(t)}, \quad \Gamma(z) = \frac{1}{2\pi i}\int_L \frac{\ln G(\tau)}{\tau - z}\,d\tau, \tag{49.7}$$

the numbers $\varkappa_k$ being chosen on each end in accordance with the selected class of solutions.

If $\varkappa \leqq 0$ we have to set $P_{\varkappa-1}(t) \equiv 0$; for $\varkappa < 0$ the conditions of solubility

$$\int_L \frac{f(\tau)}{Z(\tau)}\,\tau^{j-1}\,d\tau = 0 \quad (j = 1, 2, \ldots, -\varkappa), \tag{49.8}$$

should be satisfied. In analogous way we solve the equation

$$K^{\circ\prime}\psi \equiv a(t)\,\psi(t) - \frac{1}{\pi i}\int_L \frac{b(\tau)\,\psi(\tau)}{\tau - t}\,d\tau = f(t),$$

adjoint to the dominant equation.

In investigating the complete equation we encounter one peculiarity which makes the case of discontinuous coefficients different from that of open contours. The way of seperating out the dominant part, consisting in making use of the identity

$$\int \frac{M(t,\tau)\,\varphi(\tau)}{\tau - t}\,d\tau = M(t,t)\int \frac{\varphi(\tau)}{\tau - t}\,d\tau + \int \frac{M(t,\tau) - M(t,t)}{\tau - t}\,\varphi(\tau)\,d\tau,$$

cannot be applied in the case of a discontinuous function M, since near the point of discontinuity the integrand has a singularity of the form $(\tau - t)^{-1}$. Hence, the complete singular integral equation should at once be considered with the dominant part separated out

$$\mathrm{K}\varphi = \mathrm{K}^\circ\varphi + k\varphi \equiv a(t)\,\varphi(t) + \frac{b(t)}{\pi i}\int\limits_L \frac{\varphi(\tau)}{\tau - t}\,d\tau +$$

$$+ \int\limits_L k(t,\,\tau)\,\varphi(\tau)\,d\tau = f(t). \quad (49.9)$$

On the same ground the adjoint equation

$$\mathrm{K}'\psi \equiv a(t)\,\psi(t) - \frac{1}{\pi i}\int\limits_L \frac{b(\tau)\,\psi(\tau)}{\tau - t}\,d\tau + \int\limits_L k(\tau,\,t)\psi(\tau)\,d\tau = f(t) \quad (49.10)$$

should be regarded as an independent type of equation, not reducible to the former equation.

We shall not dwell here on the theory of the complete equation. We indicate only that it can be constructed for the case under consideration, analogously to the case of simple open contours investigated in the preceding section.

49.2. Example

Let L be the unit circle and L_1 and L_2 its right and left semi-circles, respectively.

Let us solve the dominant singular equation

$$a(t)\varphi(t) + \frac{b(t)}{\pi i}\int\limits_L \frac{\varphi(\tau)}{\tau - t}\,dt = \frac{4(t^2 + 1)(4t + i)}{e^{\pi t}(4t^2 + 1)}\left(\frac{t - i}{t + i}\right)^{\frac{it}{2}},$$

where

$$a(t) = \tfrac{1}{2}(e^{-\pi t} + 1), \quad b(t) = \tfrac{1}{2}(e^{-\pi t} - 1) \quad \text{on} \quad L_1,$$

$$a(t) = \cosh \pi t, \quad b(t) = -\sinh \pi t \quad \text{on} \quad L_2,$$

and the many-valued function $[(t - i)/(t + i)]^{it/2}$ is defined by a definite branch in the following class of functions:

    (a) $h(i, -i)$;   (b) $h(i)$.

Having calculated by means of formulae (49.4) the coefficients $G(t)$, $g(t)$ we arrive at the Riemann boundary value problem

$$\Phi^+(t) = G(t)\,\Phi^-(t) + \frac{4(t^2 + 1)(4t + i)}{4t^2 + 1}\left(\frac{t - i}{t + i}\right)^{\frac{it}{2}},$$

where

$$G(t) = e^{\pi t} \quad \text{on} \quad L_1, \quad G(t) = e^{2\pi t} \quad \text{on} \quad L_2.$$

This is the boundary value problem examined in § 43.3 (Example 2), for $a = \tfrac{1}{2}$. We can make use of the solution presented there, imposing upon it the additional condition $\Phi^-(\infty) = 0$.

(a) To obtain the required solution of class $h(i, -i)$ we have to set in formula (43.22) $a = \frac{1}{2}$, $c = 0$. Then we obtain

$$\Phi^+(z) = \left(\frac{z-i}{z+i}\right)^{\frac{iz}{2}} \frac{6(iz-1)e^{\pi z} + 4(z^2+1)(4z+i)}{4z^2+1},$$

$$\Phi^-(z) = \left(\frac{z-i}{z+i}\right)^{\frac{iz}{2}} \frac{6(iz-1)}{4z^2+1}.$$

Taking into account the rules given in the solution of the boundary value problem, concerning the determination of the limiting values, we have

$$\Phi^+(t) = \left(\frac{t-i}{t+i}\right)^{\frac{it}{2}} \frac{6(it-1)e^{\pi t} + 4(t^2+1)(4t+i)}{4t^2+1} \quad \text{on } L,$$

$$\Phi^-(t) = \left(\frac{t-i}{t+i}\right)^{\frac{it}{2}} \frac{6(it-1)}{4t^2+1} \quad \text{on } L_1,$$

$$\Phi^-(t) = \left(\frac{t-i}{t+i}\right)^{\frac{it}{2}} \frac{6(it-1)}{4t^2+1} e^{-\pi t} \quad \text{on } L_2.$$

Hence

$$\varphi(t) = \left(\frac{t-i}{t+i}\right)^{\frac{it}{2}} \frac{6(it-1)(e^{\pi t}-1) + 4(t^2+1)(4t+i)}{4t^2+1} \quad \text{on } L_1,$$

$$\varphi(t) = \left(\frac{t-i}{t+i}\right)^{\frac{it}{2}} \frac{12(it-1)\sinh \pi t + 4(t^2+1)(4t+i)}{4t^2+1} \quad \text{on } L_2.$$

(b) In class $h(i)$ the index of the equation $\varkappa = -1$ and therefore the equation is soluble only if the condition

$$\int_L \frac{(\tau - i)(4\tau + i)e^{-\pi \tau}}{4\tau^2 + 1} \, d\tau = 0$$

is satisfied. This relation is easily shown to be satisfied. As before we obtain

$$\Phi^+(t) = e^{\pi t}(t+i)\left(\frac{t-i}{t+i}\right)^{\frac{it}{2}} \frac{6i + 4e^{-\pi t}(4t^2 - 3it + 1)}{4t^2+1} \quad \text{on } L,$$

$$\Phi^-(t) = 6i\left(\frac{t-i}{t+i}\right)^{\frac{it}{2}} \frac{t+i}{4t^2+1} \quad \text{on } L_1,$$

$$\Phi^-(t) = 6ie^{-\pi t}\left(\frac{t-i}{t+i}\right)^{\frac{it}{2}} \frac{t+i}{4t^2+1} \quad \text{on } L_2,$$

$$\varphi(t) = \frac{t+i}{4i^2+1}\left(\frac{t-i}{t+i}\right)^{\frac{it}{2}} [6i(e^{\pi t}-1) + 4(4t^2 - 3it + 1)] \quad \text{on } L_1,$$

$$\varphi(t) = \frac{t+i}{4t^2+1}\left(\frac{t-i}{i+i}\right)^{\frac{it}{2}} [12i\sinh \pi t + 4(4t^2 - 3it + 1)] \quad \text{on } L_2.$$

49.3. Exceptional cases

In § 45 a generalization of the formulation of the Riemann problem was indicated, in which the general theory contains also the exceptional cases when the coefficient $G(t)$ has on the contour zeros and poles; also, the class of allowable functions was extended. For solutions, sectionally analytic functions were admitted, one of the limiting values of which had infinities at a finite number of points.

This formulation of the Riemann problem can be used for the appropriate generalization of the singular integral equations, in order to include in the general theory the exceptional cases, when $a^2(t) - b^2(t) = 0$ at isolated points of the contour.

The limiting values of solutions of the Riemann problem are then expressed by integrals taken not in the sense of their Cauchy principal values, but by the so-called integrals in the sense of Cauchy–Hadamard, by which the finite parts of diverging integrals are understood. The equations with Cauchy–Hadamard integrals yield as a particular case the equations with Cauchy kernel, including their exceptional cases.

We shall not deal here with this problem, since it would require introducing a great number of auxiliary concepts, and we refer the reader to the paper of L. A. Chikin (1).

49.4. Approximate methods

The development in recent years of approximate methods of solving singular integral equations (see § 21.5) has been extended, though to a lesser extent, to closed contour equations. For this purpose Ivanov (3) used the method of moments (see § 21.5). For solving the equation

$$\frac{1}{2\pi} \int_{-1}^{1} \frac{\varphi(t)}{t - x} dx + \frac{1}{2\pi} \int_{-1}^{1} K(x, t)\,\varphi(t)\,dt = f(x)$$

Kalandiya (1) used an interpolation polynomial for an approximate singular integral, and Gauss's quadrature formula for the regular integral. The problem reduces to the solution of a set of linear algebraic equations in the values of the required function at the nodes of interpolation.

Reference should also be made to the work of Pykhteyev (1) who has elaborated approximate methods of evaluating two singular integrals

$$\frac{1}{\pi} \int_{-1}^{1} \frac{f(t)}{t - x} dt, \quad \frac{\sqrt{(1 - x^2)}}{\pi} \int_{-1}^{1} \frac{f(t)}{\sqrt{(1 - t^2)}} \frac{dt}{t - x},$$

which occur in various mechanics problems in connexion with the inversion of the Cauchy integral.

§ 50. Historical notes

The problem of solving the Riemann problem with discontinuous coefficients arose simultaneously with the formulation of the problem itself. Riemann [22] (see Historical notes of Chapter II) stated the problem from the very beginning as a problem with discontinuous coefficients. The first solutions of the problem in such a formulation were obtained by Hilbert [9] in 1905 and Plemelj (2) in 1907, in the beginning of the study of boundary value problems. The Riemann problem for an open contour is first encountered in a paper of Carleman (1) in 1922 (see Historical notes of Chapters II and III).

A complete solution of the Riemann problem with open contours was given in 1941 independently by two authors: N. I. Muskhelishvili (1) by the method given in § 43 and F. D. Gakhov (2) by the method of §§ 41, 42 †. In the same paper F. D. Gakhov also presented a solution of the problem with discontinuous coefficients. Applying the method of N. I. Muskhelishvili, D. A. Kveselava (5) solved this problem somewhat later. The papers of Hilbert and Plemelj for the solution of Gakhov and the paper of Carleman for Muskhelishvili's solution, are the direct predecessors of these concluding papers.

Hilbert, Plemelj and Carleman who were acquainted with the method of solution of problems with discontinuous coefficients and open contours did not deduce complete solutions of them only because at that time there did not exist any effective solution of this problem in the simplest case of a closed contour and continuous coefficients. It is characteristic that N. I. Mushelishvili, F. D. Gakhov and D. A. Kveselava solved problems with discontinuous coefficients and open contours directly, reducing one to the other, while earlier every problem was being solved independently by its own methods and the papers containing the solutions of these problems were separated by a period of 15 years.

F. D. Gakhov [4] presented further applications of his method to the theory of the Riemann boundary value problem with many unknown functions. By using functions of matrices it was possible to obtain the fundamental functions eliminating the discontinuity in the same form as in the case of one unknown function, the only difference being that the exponent λ is here a matrix. The same method was applied by L. I. Chibrikova (1) to derive the solution of the generalized Riemann problem with discontinuous coefficients and open contours (see Problems 15, 16 at the end of this chapter).

The method of N. I. Muskhelishvili was also applied to the solution of the Riemann problem for complex contours. The first solution of this problem announced by Trjizinsky (1) was every cumbersome. D. A. Kveselava (3) proposed to take the canonical solution of the problem in the form of a product of the canonical solutions for the simple contours, which led to a considerable simplification of the solution. However, there still remained complications due to the difference

---

† In this connection it should be observed that a rigorous justification of this method became possible only on the basis of the formulae on the behaviour of a Cauchy type integral at the points of discontinuity of the density, which were derived by N. I. Muskhelishvili (see §§ 8.1–8.4).

in the way of construction of the canonical solutions for closed and open contours. F. D. Gakhov and L. I. Chibrikova (1) proposed, for the derivation of the canonical solution in the case of a closed contour, to apply the method used in the solution of the problem for an open contour (see § 44.2); thus, the problem was reduced to the solution for the case of discontinuous coefficients (the discontinuity occurs not for the coefficient itself but for its logarithm). The application of this device simplified finally the solution for the case of a complex contour and this solution acquired the same form as the solution for a multiply-connected domain.

A particular case of the Hilbert boundary value problem with discontinuous coefficients—the mixed boundary value problem for analytic functions—was investigated by Volterra as early as in 1883. From then on this problem has been solved by many authors by means of many particular methods, the basis of which is constituted by the property of the function $\sqrt{[\pi(z - a_k)(z - b_k)]}$ which passes from real to purely imaginary values (and conversely), in passing through the points a_k, b_k of the real axis. Almost complete results were derived by Signorini (1); a full solution was presented by M. V. Keldysh and L. I. Sedov (1). The same methods were employed for solving other problems, for instance the Dirichlet problem for the plane with slits.

The general method of reduction of the mixed boundary value problem to the Riemann problem with discontinuous coefficients was first used by F. D. Gakhov in 1941 (1) as an illustration of the proved reducibility of the Hilbert boundary value problem to the Riemann problem. The basis of the method of reduction of the Hilbert problem to the Riemann problem with discontinuous coefficients was given by N. I. Mushelishvili [17*]. The latter work contains also applications of the method to solving many particular cases of the considered problem (the Dirichlet problem for the plane with slits and similar problems).

The theory of singular integral equations with open contours treated in § 48 and in §§ 49.1 and 49.2, was established in 1942 by N. I. Muskhelishvili and D. A. Kveselava in the paper (1). In the further development an essential role was played by the paper of D. A. Kveselava (5) (in 1944). The main value of this work consists in the investigation of the application of the operator R (§ 49.3) as the regularizing operator from the left and from the right; on this basis a simple construction of the theory of singular equation was performed, which we presented in § 49.3. The singular equation in the case of discontinuous coefficients was examined by F. D. Gakhov [3], the method of investigation being the same as in the paper of N. I. Muskhelishvili and D. A. Kveselava; in view of the close connection between the cases of open contours and discontinuous coefficients the last paper does not contain any essentially original results.

From the earlier papers on this topic we mention that of Carleman (1), which has already been quoted in Chapters II–III, VI. He announced the solution of the non-homogeneous dominant equation on the segment $(0, 1)$ in the class of functions bounded at one end and having an infinity at the other end. The solution in this class is unique. There is no mention of the possibility of other classes of solutions in Carleman's paper.

Problems on Chapter VI

1. Prove that the solution of the Riemann problem for an open contour is independent of the form of the curve supplementing the considered contour to a closed one (F. D. Gakhov [3]).

2. Suppose that L is the unit circle and L_1 and L_2 are its upper and lower semicircles, respectively.

Solve for this contour L the Riemann boundary value problem

$$\Phi^+(t) = G(t)\,\Phi^-(t) + (t-1)^{\frac{2\pi-5}{\pi}}\,(t+1)^{\frac{5-\pi}{\pi}}\,\frac{e^{4t}}{t^2+a^2} \qquad (|a|<1),$$

where

$$G(t) = e^{14i} \quad \text{on} \quad L_1, \qquad G(t) = e^{4i} \quad \text{on} \quad L_2,$$

and the many-valued function $[(t-1)/(t+1)]^{5/\pi}$ is defined by one of its branches. Consider the classes of functions having an integrable infinity at the points

(a) $t_1 = 1$, $t_2 = -1$; (b) $t_1 = 1$; (c) $t_1 = -1$,

and bounded at all remaining points of the plane z.

(d) Solve the problem in the class of everywhere bounded functions.

(e) Prove that in the class of functions bounded everywhere and vanishing at infinity the problem has a solution only for special values of the parameter a; find these values of the parameter and derive the solution for this case.

Answer.

(a) $\Phi^+(z) = \dfrac{e^{4i}}{z+1}\left(\dfrac{z+1}{z-1}\right)^{\frac{5-\pi}{\pi}}\left[\dfrac{a^2+1}{a}\,\dfrac{z\sin 4a + a\cos 4a}{z^2+a^2}\,e^{-4i} + \right.$

$$\left. + \dfrac{(z^2-1)\,e^{4(z-i)}}{z^2+a^2} + c_0 + c_1 z\right],$$

$$\Phi^-(z) = \dfrac{1}{z+1}\left(\dfrac{z+1}{z-1}\right)^{\frac{5-\pi}{\pi}}\left[\dfrac{a^2+1}{a}\,\dfrac{z\sin 4a + a\cos 4a}{z^2+a^2}\,e^{-4i} + c_0 + c_1 z\right];$$

(b) $\Phi^+(z) = e^{4i}\left(\dfrac{z+1}{z-1}\right)^{\frac{5-\pi}{\pi}}\left[\dfrac{(z-1)\,e^{4z-4i}}{z^2+a^2} + \right.$

$$\left. + \dfrac{a(1-z)\cos 4a + (a^2+z)\sin 4a}{a(z^2+a^2)}\,e^{-4i} + c\right],$$

$$\Phi^-(z) = \left(\dfrac{z+1}{z-1}\right)^{\frac{5-\pi}{\pi}}\left[\dfrac{a(1-z)\cos 4a + (a^2-z)\sin 4a}{a(z^2+a^2)}\,e^{-4i} + c\right];$$

(c) $\Phi^+(z) = e^{4i}\left(\dfrac{z+1}{z-1}\right)^{\frac{5-2\pi}{\pi}}\left[\dfrac{(z+1)(e^{4z}-\cos 4a)}{z^2+a^2}\,e^{-4i} + \right.$

$$\left. + \dfrac{(a^2-z)\sin 4a}{a(z^2+a^2)}\,e^{-4i} + c\right]$$

$$\Phi^-(z) = \left(\dfrac{z+1}{z-1}\right)^{\frac{5-2\pi}{\pi}}\left[\dfrac{(a^2-z)\sin 4a - a(z+1)\cos 4a}{a(z^2+a^2)}\,e^{-4i} + c\right];$$

(d) $\Phi^+(z) = (z-1)\left(\dfrac{z+1}{z-1}\right)^{\frac{5-\pi}{\pi}} \dfrac{a\,e^{4z} - (z\sin 4a + a\cos 4a)}{a\,(z^2 + a^2)}$,

$\Phi^-(z) = (1-z)\left(\dfrac{z+1}{z-1}\right)^{\frac{5-\pi}{\pi}} \dfrac{z\sin 4a + a\cos 4a}{a\,(a^2 + z^2)}\,e^{-4i}$;

(e) $a = \pm\dfrac{\pi}{4}$, $\Phi^+(z) = (z-1)\left(\dfrac{z+1}{z-1}\right)^{\frac{5-\pi}{\pi}} \dfrac{e^{4z}+1}{z^2 + \dfrac{\pi^2}{16}}$

$$\Phi^+(z) = \left(\dfrac{z+1}{z-1}\right)^{\frac{5-\pi}{\pi}} \dfrac{(z-1)\,e^{-4i}}{z^2 + \dfrac{\pi^2}{16}}.$$

By $[(z+1)/(z-1)]^{5/\pi}$ we understand the function analytic in the plane cut along L_1, the limiting values of which from the inside of the circle coincide with the corresponding function entering the boundary condition.

3. Let L, L_1, L_2 denote the same contours as in the preceding problem. Solve the Riemann boundary value problem

$$\Phi^+(t) = G(t)\,\Phi^-(t) + \frac{1}{25}\left(\frac{t+1}{t-1}\right)^{\frac{4t}{\pi}} \frac{t^2-1}{t^2-a^2} \quad (|a| < 1),$$

$$G(t) = \begin{cases} e^{5it} & \text{on} \quad L_1, \\ e^{-3it} & \text{on} \quad L_1, \end{cases}$$

in the most general class of functions having an infinity of an integrable order at both points of the discontinuity.

Answer. The index of the problem in the considered class is $\varkappa = -2$, and the problem therefore is in general insoluble. The solution exists only when $a = +\pi/5$ and is then given by the formulae

$$\Phi^+(z) = (z^2-1)\left(\frac{z+1}{z-1}\right)^{\frac{4z}{\pi}} \frac{e^{5iz}+1}{25\,z^2 - \pi^2},$$

$$\Phi^-(z) = \left(\frac{z+1}{z-1}\right)^{\frac{4z}{\pi}} \frac{z^2-1}{25\,z^2 - \pi^2}.$$

By $[(z+1)/(z-1)]^{4z/\pi}$ we understand the function analytic in the plane cut along L_2, the limiting values of which from the inside of the unit circle coincide with the function $[(t+1)/(t-1)]^{4t/\pi}$ entering the boundary condition.

4. Suppose that L is a smooth closed contour and $G(t)$ is a non-vanishing function which satisfies the Hölder condition everywhere, except at the points $t_1, t_2, \ldots, t_l$ where it has discontinuities of first kind. We define the index of $G(t)$ as follows:

$$\varkappa = \frac{1}{2\pi}\,\{\arg G(t)\}_L,$$

and on the sections of L between the points of discontinuity the argument varies continuously, while in passing through the point of discontinuity it satisfies one of the conditions

$$0 \leqq \arg G(t_k - 0) - \arg G(t_k + 0) < 2\pi,$$
$$-2\pi < \arg G(t_k - 0) - \arg G(t_k + 0) < 0.$$

Prove that the index defined in this way coincides with the index defined in § 41.4 (N. I. Muskhelishvili [17*], § 85).

5*. Suppose that L is a contour composed of m open curves and $f(t)$ is a known function satisfying the Hölder condition $P_{m-1}(t)$ being a polynomial of degree $(m-1)$, the coefficients of which are not prescribed beforehand. Prove that the Cauchy type integral

$$\frac{1}{\pi i} \int_L \frac{\varphi(\tau)}{\tau - t} \, d\tau = f(t) + P_{m-1}(t)$$

may be inverted in the class of functions bounded at all ends and that then the polynomial $P_{m-1}(t)$ is uniquely determined (N. I. Muskhelishvili [17*], § 89).

6*. Suppose that L and $f(t)$ are the same as in the preceding problem and c_k are constants not prescribed beforehand. Prove that the Cauchy type integral

$$\frac{1}{\pi i} \int_L \frac{\varphi(\tau)}{\tau - t} \, d\tau = f(t) + c_k \quad t \in L_k \ (k = 1, 2, \ldots, m)$$

may be inverted in the class of bounded functions and that all constants c_k are then uniquely determined (N. I. Muskhelishvili [17*], § 90).

7*. Let there be given on a closed contour L a real function $f(s)$ which is continuous on L, except at the points $t_1, t_2, \ldots, t_l$ where it has discontinuities of the first kind. Denote the limiting values of $f(s)$ at the points of discontinuity by $f(s_k - 0) = a_k$, $f(s_k + 0) = b_k$. Prove that the solution of the Dirichlet problem

$$u = f(s)$$

for the domain D bounded by the contour L can be represented in the form

$$u(x, y) = v(x, y) + \sum_{k=1}^{l} \frac{a_k - b_k}{\pi} \arctan \frac{y - \beta_k}{x - \alpha_k} \quad (t_k = \alpha_k + i\beta_k),$$

where $v(x, y)$ is the solution of the Dirichlet problem with continuous coefficients.
Prove that in the class of functions having order of growth below the logarithmic $\left(\dfrac{u(x, y)}{\ln r_i} \to 0 \text{ where } r_i \text{ is the distance from the point of discontinuity } t_i \text{ to an arbitrary point of the domain}\right)$ this solution is unique.

8*. Solve the Hilbert problem with discontinuous coefficients by the method of § 28.2 (by means of determination of the regularizing factor). Make use of the results of the preceding problem (E. K. Stolyarova (1)).

9*. Suppose that a domain D is symmetric with respect to the x-axis and the required function $\Phi(z)$ is obtained by a symmetry continuation from the upper semi-plane into the lower one. The real part of $\Phi(z)$ is prescribed on the part of the contour of D which lies in the upper semi-plane, while on the symmetric part of the contour its imaginary part is known. Prove that this mixed boundary value problem is absolutely soluble in the class of bounded functions (A. V. Bitsadze (1)).

Hint. First map the domain D onto the unit circle.

10*. Suppose that D is the plane with the cuts (a_k, b_k) along the x-axis (§ 46.4). Determine a function $\Phi(z) = u + iv$ analytic and bounded in the domain D, if on the edges of the cut the following data is given:

$$u^+ = f^+(t) + c_k, \quad u^- = f^-(t) + c_k \quad \text{on} \quad (a_k, b_k) \quad (k = 1, 2, \ldots, m).$$

Here f^+, f^- are known functions which satisfy the Hölder condition, c_k are constants not prescribed beforehand. Prove that the problem is absolutely and uniquely soluble and that all constants c_k are then uniquely determined (N. I. Muskhelishvili [17*], § 91).

11*. On the basis of the results of the preceding problem solve the Dirichlet problem for the plane with cuts along the real axis (§ 46.3), without assuming the differentiability of the boundary values (N. I. Muskhelishvili [17*], § 91).

Hint. Consider the function $U(x, y) = u(x, y) + \sum\limits_{k=1}^{m} a_k \omega_k(z)$ where

$$\omega_k(z) = \frac{1}{b_k - a_k} [(z - b_k) \ln (z - b_k) - (z - a_k) \ln (z - a_k)].$$

12. Determine the function $w = f(z)$ which maps the lower semi-plane $y < 0$ onto a semi-plane bounded by the straight line $u = kv$ with a cut along a vertical section of length C made from the origin of the coordinate system, under the condition that the interval $|x| < 1$ corresponds to the edges of the cut and the intervals $|x| > 1$ to the sections of the straight line $u = kv$.

Answer.

$$f(z) = \pm i C e^{i\varphi} (z + 1)^{\frac{\varphi}{\pi}} (z - 1)^{1 - \frac{\varphi}{\pi}}.$$

C is a real constant and $\varphi = \arctan k$. The sign in the formula depends on whether the cut is directed upwards or downwards.

Hint. Make use of the fact that the required function should constitute the solution of the Hilbert boundary value problem from the example of § 46.4, when $\sigma = 0$, in the class of functions bounded at the points $z = -1, z = 1$ and behaving as z when $z \to \infty$.

Making use of the geometric properties of the Zhukovski function $W = \frac{1}{2}[z + (1/z)]$ and its logarithm, solve the following two problems.

13. Determine the function $f(z) = u + iv$ analytic in the lower semi-plane, bounded at the points $z = -1, z = 1$ and behaving as z when $z \to \infty$, in accordance with the boundary condition.

$$\text{Im } f(x) = 0 \quad (x < -1 \quad \text{and} \quad x > 1),$$
$$|f(x)| = 1 \quad (-1 < x < 1).$$

14. Determine the function $f(z) = u + iv$ analytic in the lower semi-plane, bounded at the points $z = -1, z = 1$ and behaving as $\ln z$ when $z \to \infty$, in accordance with the boundary condition

$$\operatorname{Im} f(x) = 0 \quad (x < -1),$$
$$\operatorname{Im} f(x) = \pi \quad (x < 1),$$
$$\operatorname{Re} f(x) = 0 \quad (-1 < x < 1).$$

14*. Consider the boundary value problem

$$\Phi^+[a(t)] = G(t)\,\Phi^-(t),$$

all conditions of § 17 being satisfied, except that we now admit discontinuities of first kind at the points $t_1, t_2, \ldots, t_l$ for the coefficient $G(t)$.
Prove that the substitution

$$\Phi^+(z) = \Pi\,[z - \alpha(t_k)]^{\lambda_k}\,\Phi_1^+(z),$$
$$\Phi^-(z) = \Pi\left(\frac{z - t_k}{z - z_0}\right)^{\lambda_k}\Phi^-(z) \quad (z_0 \in D^+)$$

reduces this boundary value problem to a boundary value problem of the same kind with the continuous coefficient

$$G_1(t) = G(t)\,\Pi\left(\frac{t - t_k}{\alpha(t) - \alpha(t_k)}\right)^{\lambda_k}(t - z_0)^{-\lambda_k}$$

(L. I. Chibrikova (1)).

16*. Prove that the preceding boundary value problem, in the case when the contour L is open and $\alpha(t)$ maps the contour onto itself preserving the direction, the problem may be reduced by the method of § 42.1 to the problem

$$\Phi^+[\alpha_1(t)] = G_1(t)\,\Phi^-(t)$$

for a closed contour with a discontinuous coefficient $G_1(t)$ and a continuous function $\alpha_1(t)$ (L. I. Chibrikova (1)).

17. Consider the generalized singular integral equation

$$a(t)\,\varphi[\alpha(t)] + b(t)\,\varphi(t) + \frac{a(t)}{\pi i}\int_L \frac{\varphi(\tau)}{\tau - \alpha(t)}\,d\tau - \frac{b(t)}{\pi i}\int_L \frac{\varphi(\tau)}{\tau - t}\,d\tau = f(t)$$

(see Problem 10 to Chapter III and Problem 15 to Chapter VI), where L is a closed contour; $a(t)$, $b(t)$ have discontinuities of first kind at a finite number of points. Reduce this equation to the generalized Riemann boundary value problem and investigate it (L. I. Chibrikova (1)).

18. Carry out the same for the case when L is an open contour.

19. Suppose that L is the unit circle and L_1 and L_2 are its upper and lower semi-circles, respectively.

Solve the dominant singular equation

$$a(t)\,\varphi(t) + \frac{b(t)}{\pi i}\int_L \frac{\varphi(\tau)}{\tau - t}\,d\tau = 16\,(t-1)^{\frac{2\pi-5}{\pi}}\,(t+1)^{\frac{5-\pi}{\pi}}\,\frac{e^{4t-7}}{16t^2 + \pi^2}\,,$$

where

$$\begin{cases} a(t) = \cos 7, \\[2mm] b(t) = -i\sin 7 \quad \text{on} \quad L_1; \end{cases} \qquad \begin{cases} a(t) = \dfrac{e^{-7i} - e^{-3i}}{2}, \\[2mm] b(t) = \dfrac{e^{-7i} - e^{-3i}}{2} \quad \text{on} \quad L_2 \end{cases}$$

in the class of functions

(a) $h(-1,1)$; (b) $h(1)$; (c) $h(-1)$; (d) h_0.

Hint. Make use of the results of Problem 2 above

Answer.

(a) $\varphi(t) = \dfrac{16}{t+1}\left(\dfrac{t+1}{t-1}\right)^{\frac{5-\pi}{\pi}}\left[\dfrac{(t^2-1)\,e^{4t}}{16t^2+\pi^2} - \dfrac{16+\pi^2}{16\,(16t^2+\pi^2)}\,(1-e^{-14i}) + c\right]$ on L_1,

$\qquad \varphi(t) = \dfrac{16}{t+1}\left(\dfrac{t+1}{t-1}\right)^{\frac{5-\pi}{\pi}}\left[\dfrac{(t^2-1)\,e^{4t}}{16t^2+\pi^2} - \dfrac{16+\pi^2}{16\,(16t^2+\pi^2)}\,(1-e^{-4i}) + c^*\right]$ on L_2,

where

$$c^* = \frac{\sin 2}{\sin 7}\,e^{5i}c;$$

(b) $\varphi(t) = 16\left(\dfrac{t+1}{t-1}\right)^{\frac{5-\pi}{\pi}}\dfrac{(t-1)\,(e^{4t} - e^{-14i} + 1)}{16t^2+\pi^2}$ on L_1,

$\qquad \varphi(t) = 16\left(\dfrac{t+1}{t-1}\right)^{\frac{5-\pi}{\pi}}\dfrac{(t-1)\,(e^{4t} - e^{-4i} + 1)}{16t^2+\pi^2}$ on L_2;

(c) $\varphi(t) = 16\left(\dfrac{t+1}{t-1}\right)^{\frac{5-2\pi}{\pi}}\dfrac{(t+1)\,(e^{4t} - e^{-14i} + 1)}{16t^2+\pi^2}$ on L_1,

$\qquad \varphi(t) = 16\left(\dfrac{t+1}{t-1}\right)^{\frac{5-2\pi}{\pi}}\dfrac{(t+1)\,(e^{4t} - e^{-4i} + 1)}{16t^2+\pi^2}$ on L_2;

(d) $\varphi(t) = 16\left(\dfrac{t+1}{t-1}\right)^{\frac{5-\pi}{\pi}}\dfrac{(t-1)\,(e^{4t} - e^{-14i} + 1)}{16t^2+\pi^2}$ on L_1,

$\qquad \varphi(t) = 16\left(\dfrac{t+1}{t-1}\right)^{\frac{5-\pi}{\pi}}\dfrac{(t-1)\,(e^{4t} - e^{-4i} + 1)}{16t^2+\pi^2}$ on L_2.

20. Prove that the integro-differential equation for the wing of a plane

$$\frac{\Gamma(t)}{B(t)} - \frac{1}{\pi} \int_{-a}^{a} \frac{\Gamma'(\tau)}{\tau - t} \, d\tau = f(t)$$

can be reduced to the Fredholm integral equation

$$\Gamma(t) + \int_{-a}^{a} K(t, \tau) \, \Gamma(\tau) \, d\tau = F(t) + c_1 + c_2 \arcsin \frac{t}{a} \, ,$$

where

$$K(t, \tau) = \frac{1}{2\pi B(\tau)} \ln \frac{a^2 - t\tau + \sqrt{[(a^2 - t^2)(a^2 - \tau^2)]}}{a^2 - t\tau - \sqrt{[(a^2 - t^2)(a^2 - \tau^2)]}} \, ,$$

$$F(t) = \int_{-a}^{a} B(\tau) \, K(t, \tau) \, f(\tau) \, d\tau,$$

c_1, c_2 are arbitrary constants.

Hint. Make use of the formula of inversion of the Cauchy type integral in the class of functions unbounded on both ends, and integrate the derived equation with respect to t.

CHAPTER VII

INTEGRAL EQUATIONS SOLUBLE IN CLOSED FORM

AN EXACT solution of integral equations usually requires the use of infinite processes (the method of successive approximations, infinite sets of linear algebraic equations etc.). However, for some types of integral equations the solution can be obtained by a finite number of quadratures of the defined function, i.e. by integral equations which are soluble in closed form. It is characteristic of such equations that their principal variable enters the kernel as an analytic function. Analytic continuation in the complex plane is used as the method of solution. By introducing an auxiliary analytic function the equation can be reduced to the Riemann boundary value problem and its solution can thus be obtained in closed form.

An example of the method is the solution given in § 21 of the dominant singular integral equation with a Cauchy kernel. The analytic continuation in the complex plane is effected by a Cauchy integral for which the required solution of the equation acts as the density. Sokhotski's formulae for the limit values of the integral enabled the equation to be reduced to the Riemann boundary value problem. A similar method with appropriate modifications enables solutions of some other types of equations to be obtained in closed form. In §§ 51, 52 we consider integral equations with kernels of automorphic type, their principal part being a Cauchy kernel. In these sections use will be made of Sokhotski's formulae. In the final section we solve equations with kernels of an essentially new type (logarithmic kernels and kernels with a weak power singularity). For the functions effecting the analytic continuation in the complex plane we derive formulae for the limit values (counterparts of Sokhotski's formulae) by which the equations are reduced to the Riemann boundary value problem.

§ 51. Equations with automorphic kernels and a finite group

The solution in a closed form in Chapters III and VI was obtained only for dominant (characteristic) equations. For the whole set of complete equations, however, only the possibility of regularization has been established, i.e. the possibility of reducing the equation to a Fredholm equation. Nevertheless, the method by which the dominant equation is solved, if appropriately generalized, makes it possible to derive a closed form of the solution also for some special types of the complete equation. The present section is devoted to the determination of such types.

To solve the complete equation which will be written in the form

$$a(t)\,\varphi(t) + \frac{b(t)}{\pi i} \int\limits_{L_0} \left[\frac{1}{\tau - t} + K(t, \tau) \right] \varphi(\tau)\,d\tau = f(t), \quad (51.1)$$

we shall introduce, as was done for the dominant equation, the function

$$\Phi(z) = \frac{1}{2\pi i} \int\limits_{L_0} \left[\frac{1}{\tau - z} + K(z, \tau) \right] \varphi(\tau)\,d\tau. \quad (51.2)$$

In order that this function be analytic and single-valued, it is first of all necessary that the kernel $K(t, \tau)$ be single-valued and analytic with respect to the variable t. This is the first essential limitation of the applicability of the considered method. We assume that it is satisfied. The essential difference between the function defined by integral (51.2) and the function defined by a Cauchy type integral consists in the fact that in the former there may be other lines of singularities besides L_0, which are the singular lines of the function $K(z, \tau)$, under the condition that the variable τ describes the integration contour L_0. In order that the considered method be effective, it is necessary that for all possible lines of discontinuity of function (51.2), the singular equation (51.1) enables us to construct the boundary conditions of the Riemann problem. The types of singular integral equations for which a closed form of solution will be given below, are selected on the basis of the above general principles. Their main characteristic is the automorphicity of the kernel.

For problems of this kind it will be necessary to go beyond the scope of the fundamentals of the theory of complex variable which have been sufficient throughout the book so for, and to make use of more advanced topics, namely the theory of automorphic functions.

51.1. Some results of the theory of finite groups of linear fractional transformations and of automorphic functions

For future use it is sufficient to know only the simplest facts of the theory of automorphic functions which we shall present here, for the reader's convenience. The details may be found in the book of L. R. Ford [1] or in Chapter VII of the book of V. V. Golubyev, *Analytic theory of differential equations*.

The set of all linear fractional transformations constitutes a group. In general this group is infinite. We shall here consider only finite groups of linear fractional transformations.

Let there be given the transformations $\omega_1(z), \ldots, \omega_m(z)$. If we construct a set containing these transformations, the inverse transformations and products constructed from them in all possible ways, then this set constitutes a group. This group is called the group *generated by the transformations* $\omega_1(z), \ldots, \omega_m(z)$, and the latter are called the *fundamental transformations of the group*. The groups generated by one transformation *cyclic*. A knowledge of cyclic groups is important, since every group contains cyclic subgroups.

Let us find the form of the coefficients $\alpha, \beta, \gamma, \delta$ of the transformation

$$\omega_1(z) = \frac{\alpha z + \beta}{\gamma z + \delta}, \tag{51.3}$$

so that the latter generates a finite cyclic group. Assume that the determinant of the transformation ω_1 is unity,

$$\alpha\delta - \beta\gamma = 1. \tag{51.4}$$

This can be achieved by dividing the coefficients of the transformation by $\sqrt{(\alpha\delta - \beta\gamma)}$.

Since the cyclic group generated by the transformation ω_1 contains all consecutive iterations

$$\omega_2 = \omega_1(\omega_1), \quad \omega_3 = \omega_1(\omega_1(\omega_1)), \ldots,$$

in order that this group be finite, it is necessary and sufficient that one of these iterations is identical with z, for instance

$$\omega_n(z) \equiv z.$$

Let us derive an expression for the nth iteration $\omega_n(z)$. To this end we make use of the following representation of the linear fractional function (Ford [1], p. 24):

$$\frac{\omega_1(z) - \xi_1}{\omega_1(z) - \xi_2} = K \frac{z - \xi_1}{z - \xi_2}, \tag{51.5}$$

where ξ_1, ξ_2 are the fixed points of the transformation $\omega_1(z)$ and K is a number not equal to unity

$$K = \frac{\alpha - \gamma \xi_1}{\alpha - \gamma \xi_2}, \tag{51.6}$$

which satisfies the following conditions, provided conditions (51.4) are satisfied

$$K + \frac{1}{K} = (\alpha + \delta)^2 - 2. \tag{51.7}$$

From (51.5) we obtain

$$\frac{\omega_n(z) - \xi_1}{\omega_n(z) - \xi_2} = K^n \frac{z - \xi_1}{z - \xi_2},$$

whence

$$\omega_n(z) = \frac{(K^n \xi_2 - \xi_1) z + (1 - K^n) \xi_1 \xi_2}{(K^n - 1) z + \xi_2 - K^n \xi_1}.$$

It follows from the last expression that $\omega_n(z) \equiv z$ if and only if $K^n = 1$, i.e. when

$$K = e^{\frac{2\pi i k}{n}} \quad (k = 1, 2, \ldots, n - 1).$$

Substituting this expression for K into (51.7) we obtain

$$(\alpha + \delta)^2 - 2 = e^{\frac{2\pi i k}{n}} + e^{-\frac{2\pi i k}{n}} = 2 \cos \frac{2\pi k}{n}.$$

Hence

$$(\alpha + \delta)^2 = 4 \cos^2 \frac{\pi k}{n}$$

and, consequently,

$$\alpha + \delta = \pm 2 \cos \frac{\pi k}{n} \quad (k = 1, 2, \ldots, n - 1). \tag{51.8}$$

This formula will essentially be used later. Relations (51.4) and (51.8) are the necessary and sufficient conditions for the cyclic group generated by the substitution (51.3) to have n members.

As examples let us consider the cases when $n = 2$ and $n = 3$.

(1) $n = 2$. $k = 1$; $\cos \frac{\pi k}{n} = \cos \frac{\pi}{2} = 0$.

Condition (51.8) yields $\delta = -\alpha$ and the substitution

$$\omega(z) = \frac{\alpha z + \beta}{\gamma z - \alpha},$$

for $-\alpha^2 - \beta\gamma = 1$ is the general form of the substitution which is identical with its inverse, $\omega[\omega(z)] = z$.

(2) $n = 3$. $k = 1, 2$; $\cos\dfrac{\pi}{3} = \dfrac{1}{2}$; $\cos\dfrac{2\pi}{3} = -\dfrac{1}{2}$.

Conditions (51.8) take the form $\alpha + \delta = \pm 1$, and we obtain two two-parametric families of substitutions

$$\omega(z) = \frac{\alpha z + \beta}{\gamma z + 1 - \alpha}, \quad \alpha(1 - \alpha) - \beta\gamma = 1,$$

$$\omega(z) = \frac{\alpha z + \beta}{\gamma z - 1 - \alpha}, \quad \alpha(1 + \alpha) + \beta\gamma = -1,$$

which satisfy the condition $\omega_3(z) \equiv z$.

We now present the definition of automorphic functions and we proceed to establish some of their properties. The functions invariant with respect to a group of linear fractional transformations are called *automorphic functions*.

The rational automorphic functions constitute the only class of single-valued functions automorphic with respect to finite groups.

Points or figures obtained one from the other by means of the substitutions of the group are called *equivalent*. An automorphic function takes the same value at equivalent points. The domain which does not contain two distinct equivalent points but does contain points equivalent to any point of the plane with respect to the group under consideration, is called the *fundamental domain of the group* (accordingly, the fundamental domain of the automorphic function).

If a group does not contain integral linear transformations and

$$\omega_k(z) = \frac{\alpha_k z + \beta_k}{\gamma_k z + \delta_k} \quad (k = 1, 2, \ldots, m)$$

are the fundamental transformations of the group, then the fundamental domain is the exterior of the so-called isometric circles of the fundamental transformations, the equations of which are

$$|\gamma_k z + \delta_k| = 1.$$

The automorphic functions have in the fundamental domain the same number of zeros and poles and, moreover, any of their values is taken the same number of times. It follows from this property that if an automorphic function has no poles in the fundamental domain, then it is identically constant.

A special role in the theory of automorphic functions is played by the so-called *fundamental function of the group*. This is the automorphic function

$$F(z) = \sum_{k=0}^{n-1} \frac{1}{\omega_k(z) - a}, \tag{51.9}$$

where a is some number, taking in every fundamental domain any of its values once.

If the function

$$F(z) = \sum_{k=0}^{n-1} \omega_k(z) \tag{51.10}$$

is not an identical constant, then it is the fundamental automorphic function having a pole at infinity.

We shall later need an automorphic function which has a pole of a certain order $\varkappa$ at infinity. It can be expressed as a simple polynomial $P_\varkappa(F)$ in terms of function (51.10). In the general case it can be expressed in terms of function (51.9) in the form

$$\frac{P_\varkappa(F)}{[F(z) - F(z_\infty)]^\varkappa}, \tag{51.11}$$

where z_∞ is the point at infinity or any point equivalent to it.

51.2. Reduction of a complete singular integral equation to a boundary value problem

Suppose that L_0 is a smooth curve which for simplicity will hereafter be regarded as closed, and

$$\omega_0(t) \equiv t, \quad \omega_1(t), \ldots, \omega_{n-1}(t)$$

are linear fractional functions constituting a group.

Consider the equation

$$a(t)\,\varphi(t) + \frac{b(t)}{\pi i} \int_{L_0} \sum_{k=0}^{n-1} \left[\frac{1}{\tau - \omega_k(t)} - \frac{1}{\tau - \omega_k(\infty)} \right] \varphi(\tau)\,d\tau = f(t), \tag{51.12}$$

where $a(t)$, $b(t)$, $f(t)$ are known functions which satisfy the Hölder condition, and $a^2(t) - b^2(t) = 1$.

Introduce the function

$$\Phi(z) = \frac{1}{2\pi i} \int_{L_0} \sum_{k=0}^{n-1} \left[\frac{1}{\tau - \omega_k(z)} - \frac{1}{\tau - \omega_k(\infty)} \right] \varphi(\tau) d\tau, \quad (51.13)$$

which satisfies the condition $\Phi(\infty) = 0$. It is analytic everywhere, except on the curve L_0 and the curves L_k the equations of which are

$$\tau - \omega_k(z) = 0 \quad \text{or} \quad z = \omega_k^{-1}(\tau), \quad \tau \in L_0 \quad (k = 1, 2, \ldots, n-1).$$

Let us examine the behaviour of $\Phi(z)$ on the lines L_k. As z crosses the curve L_0, $\omega_k^{-1}(z)$ crosses the curve L_k. Substituting in (51.3) $\omega_k^{-1}(z)$ for z we observe that all terms of the sum vary continuously, except for the $(k+1)$th which constitutes a Cauchy type integral with the line of discontinuities L_0. Assume that all functions $\omega_k^{-1}(z)$ map the curve L_0 onto the curves L_k preserving the direction and that all curves L_k either have no common points or the number of points of intersection is finite. It is then readily observed that for any point of the curves L_k which is not a point of self-intersection, for $z \to \omega_k^{-1}(t) \in L_k$ $(t \in L_0)$ the following relation holds:

$$\Phi^{\pm} [\omega_k^{-1}(t)]$$

$$= \pm \tfrac{1}{2} \varphi(t) + \frac{1}{2\pi i} \int_{L_0} \sum_{j=0}^{n-1} \left[\frac{1}{\tau - \omega_j [\omega_k^{-1}(t)]} - \frac{1}{\tau - \omega_j(\infty)} \right] \varphi(\tau) d\tau.$$

Since $\omega_k(z)$ constitute a group, the sum in the right-hand side of the last relation differs only in the order of the terms from the sum

$$\sum_{k=0}^{n-1} \left[\frac{1}{\tau - \omega_k(t)} - \frac{1}{\tau - \omega_k(\infty)} \right].$$

Hence, on the contours L_k the formulae analogous to the Sokhotski formulae are valid

$$\left. \begin{aligned} \varphi(t) &= \Phi^+ [\omega_k^{-1}(t)] - \Phi^- [\omega_k^{-1}(t)], \\ \frac{1}{\pi i} \int_{L_0} \sum_{k=0}^{n-1} \left[\frac{1}{\tau - \omega_k(t)} - \frac{1}{\tau - \omega_k(\infty)} \right] \varphi(\tau) d\tau & \\ &= \Phi^+ [\omega_k^{-1}(t)] + \Phi^- [\omega_k^{-1}(t)] \\ (k = 0, 1, \ldots, n-1). & \end{aligned} \right\} \quad (51.14)$$

Introducing these expressions into the general equation (51.12) we find that the function $\Phi(z)$ satisfies the boundary conditions

$$\Phi^+ [\omega_k^{-1}(t)] = \frac{a(t) - b(t)}{a(t) + b(t)} \Phi^- [\omega_k^{-1}(t)] + \frac{f(t)}{a(t) + b(t)},$$

which can also be written in the form

$$\Phi^+(t) = G\left[\omega_k(t)\right]\Phi^-(t) + g\left[\omega_k(t)\right] \quad (k = 0, 1, \ldots, n-1), \quad (51.15)$$

where the usual notations have been introduced

$$G(t) = \frac{a(t) - b(t)}{a(t) + b(t)}, \quad g(t) = \frac{f(t)}{a(t) + b(t)}. \quad (51.16)$$

The form of the boundary conditions resembles the familiar Riemann boundary value problem. However, to avoid mistakes let us emphasize that it is not exactly the Riemann problem. The difference consists in the fact that we seek a function *having a definite analytic representation* (51.13). Hence, in solving problems we have to take care of the prescribed analytic nature of the function, besides satisfying the boundary conditions on the lines L_k.

Let us examine in detail the function $\Phi(z)$. Replacing in relation (51.13) z by $\omega_j(z)$ and taking into account that the $\omega_j(z)$ constitute a group we find that $\Phi(z)$ satisfies the condition

$$\Phi\left[\omega_j(z)\right] = \Phi(z), \quad (51.17)$$

i.e. it remains invariant with respect to the substitutions of the group.

Consequently, the required function $\Phi(z)$ is an *automorphic function with the prescribed group of transformations* $\omega_0(z), \omega_1(z), \ldots, \omega_{n-1}(z)$.

Thus, the solution of the considered singular equation (51.12) is reduced to the boundary value problem in which it is required to determine a sectionally analytic automorphic function $\Phi(z)$ having n lines of discontinuity L_n on which its limiting values satisfy the boundary conditions (51.15) and the additional condition $\Phi(\infty) = 0$.

We observe that the boundary conditions (51.15) for the curves $L_1, \ldots, L_{n-1}$ could be obtained from the boundary condition for the curve L_0 by means of relation (51.17).

Remark. If among the transformations of the group there are transformations which map the curve L_0, changing the direction, then the reasoning should be somewhat altered.

Suppose that the transformations of the group $\omega_{k_1}^{-1}(z), \omega_{k_2}^{-1}(z), \ldots,$ map the contour L_0, changing the direction. Then for the curves $L_{k_j} \ (j = 1, 2, \ldots)$ the functions $\Phi^+(z), \Phi^-(z)$ are exchanged and the boundary condition (51.15) should be replaced by the condition

$$\Phi^+(t) = \frac{1}{G\left[\omega_{kj}(t)\right]}\Phi^-(t) - \frac{f\left[\omega_{kj}(t)\right]}{a\left[\omega_{kj}(t)\right] - b\left[\omega_{kj}(t)\right]} \quad \text{on} \quad L_{kj} \quad (51.15')$$

$$(j = 1, 2, 3, \ldots).$$

51.3. Solution of the boundary value problem

We now proceed to solve the boundary value problem (51.15) to which we have reduced the solution of the singular equation (51.12). The contour $L = L_0 + L_1 + \cdots + L_{n-1}$ on which the boundary condition is given, is in general self-intersecting. To solve the boundary value problem we make use of the method of § 44. As we agreed before, the contour L_0 is regarded as closed. Then all contours L_k equivalent to L_0 are also closed.

As was established in § 44.3 the canonical function $X(z)$ for the union of contours $L_0, L_1, \ldots, L_{n-1}$ can be constructed as the product of the canonical functions $X_0(z), X_1(z), \ldots, X_{n-1}(z)$ for each of the contours separately. However, in the considered case there is no necessity to construct a canonical function for each contour L_k, for if $X_0(z)$ is the canonical function for the contour L_0

$$X_0^+(t) = G(t)\,X_0^-(t),$$

then replacing in this relation $t \in L_0$ by $\omega_k(t)$ $(t \in L_n)$ we have

$$X_0^+[\omega_k(t)] = G[\omega_k(t)]\,X_0^-[\omega_k(t)] \quad (k = 0, 1, \ldots, n-1). \quad (51.18)$$

This implies that $X_k(z) = X_0[\omega_k(z)]$.

Thus, the canonical functions of the boundary value problem for the contours L_k equivalent to the contour L_0 are functions equivalent to the canonical function $X_0(z)$ for the fundamental contour L_0.

According to § 44.2

$$X_0(z) = (z - t_0)^{-\varkappa}\, e^{\Gamma_0(z)},$$

where t_0 is a point of the contour L_0 distinct from the points of intersection, $\varkappa$ is the index of the function $G(t)$,

$$\Gamma_0(z) = \frac{1}{2\pi i} \int\limits_{L_0} \frac{\ln G(\tau)}{\tau - z}\, d\tau. \quad (51.19)$$

Consequently, the canonical function for the whole contour L has the form

$$X(z) = \prod_{k=0}^{n-1} X_0[\omega_k(z)] = \prod_{k=0}^{n-1} [\omega_k(z) - t_0]^{-\varkappa}\, e^{\sum\limits_{k=0}^{n-1} \Gamma_0[\omega_k(z)]}. \quad (51.20)$$

It does not vanish anywhere, except possibly at the point at infinity or points equivalent to it, where it has order $\varkappa$, and it is invariant with respect to all transformations of the group

$$X(z) = X[\omega_k(z)] \quad (k = 1, 2, \ldots, n-1). \quad (51.21)$$

Let us now find the general solution of the homogeneous problem

$$\Phi^+(t) = G[\omega_k(t)] \, \Phi^-(t), \quad t \in L_k \quad (k = 0, 1, \ldots, n-1), \quad (51.22)$$

which satisfies the condition $\Phi(\infty) = 0$.

Inserting into the last boundary condition the expression for $G[\omega_k(t)]$ from the relation

$$X^+(t) = G[\omega_k(t)] \, X^-(t),$$

we obtain

$$\frac{\Phi^+(t)}{X^+(t)} = \frac{\Phi^-(t)}{X^-(t)}, \quad t \in L_k \quad (k = 0, 1, \ldots, n-1).$$

These relations indicate that the function $\Phi(z)/X(z)$ is analytic in the entire plane except at the point at infinity and the points equivalent to the latter, where it can have poles of order $\varkappa - 1$. Moreover, it follows from (51.17) and (51.21) that it remains invariant with respect to all substitutions $\omega_k(z)$ of the group. Consequently, it is a rational automorphic function which has in the fundamental domain a pole of order $\varkappa - 1$[†], and such a function has the form (see formula (51.11))

$$\frac{\Phi(z)}{X(z)} = \frac{P_{\varkappa-1}(F)}{[F(z) - F(z_\infty)]^{\varkappa-1}},$$

where $F(z)$ is the fundamental function of group (51.9), $P_{\varkappa-1}$ is an arbitrary polynomial of degree not exceeding $\varkappa - 1$ and z_∞ denotes the point at infinity or one of the points equivalent to it.

Thus, in the case of a positive index $\varkappa > 0$ the homogeneous problem (51.22) has $\varkappa$ linearly independent solutions given by the formula

$$\Phi(z) = X(z) \frac{P_{\varkappa-1}(F)}{[F(z) - F(z_\infty)]^{\varkappa-1}}. \quad (51.23)$$

For $\varkappa \leqq 0$ the homogeneous problem has no solutions.

If it is possible to take the function (51.10) as the fundamental automorphic function, then the solution takes a simpler form, namely

$$\Phi(z) = X(z) \, P_{\varkappa-1}(F). \quad (51.24)$$

[†] This reasoning shows the difference between the considered problem and the ordinary Riemann boundary value problem. If we solved our problem without taking into account relation (51.13) and relation (51.17) following from it, then applying as usual the generalized Liouville theorem, we would find that the ratio $\Phi(z)/X(z)$ was a polynomial of degree $n(\varkappa - 1)$.

17 a*

We now proceed to the solution of the non-homogeneous problem (51.15). By the usual method we reduce the boundary condition to the form

$$\frac{\Phi^+(t)}{X^+(t)} - \frac{\Phi^-(t)}{X^-(t)} = \frac{g[\omega_k(t)]}{X^+(t)}, \quad t \in L_k \quad (k = 0, 1, \ldots, n - 1).$$

This is the problem of determination of an analytic (in our case also automorphic) function in terms of its jump. Making use of formula (51.14) it can easily be verified that the sectionally analytic function

$$\Psi(z) = \frac{1}{2\pi i} \sum_{k=0}^{n-1} \int_{L_0} \frac{g(\tau)}{X^+(\tau)} \frac{d\tau}{\tau - \omega_k(z)} \tag{51.25}$$

remains invariant with respect to all transformations of the group and it satisfies the last boundary condition. Consequently, the function

$$X(z) \frac{1}{2\pi i} \sum_{k=0}^{n-1} \int_{L_0} \frac{g(\tau)}{X^+(\tau)} \frac{d\tau}{\tau - \omega_k(z)}$$

is a particular solution of the non-homogeneous problem (51.15).

Adding the general solution of the homogeneous problem we obtain the general solution of the non-homogeneous problem (51.15)

$$\Phi(z) = X(z) \left\{ \Psi(z) + \frac{P_{\varkappa-1}(F)}{[F(z) - F(z_\infty)]^{\varkappa-1}} \right\}, \tag{51.26}$$

where $X(z)$ is the canonical solution given by formula (51.24) and $\Psi(z)$ is determined by formula (51.25).

For $\varkappa < 0$ we have to set $P_{\varkappa-1} \equiv 0$ and, moreover, we have to require that $\Psi(z)$ has at infinity a zero of order $-\varkappa + 1$. The expansion of the latter function in the vicinity of infinity has the form

$$\Psi(z) = -\frac{1}{2\pi i} \sum_{j=1}^{\infty} \left[\sum_{k=0}^{n-1} \int_{L_0} \frac{\omega_k'(\tau) \, \omega_k^{j-1}(\tau) \, g(\tau)}{X^+(\tau)} \, d\tau \right] z^{-j}.$$

Equating to zero the coefficients of z^{-j} $(j = 1, 2, \ldots, -\varkappa)$ we obtain the conditions of solubility of the problem

$$\int_{L_0} \frac{g(\tau)}{X^+(\tau)} \sum_{k=0}^{n-1} \omega_k'(\tau) \, \omega_k^{j-1}(\tau) \, d\tau = 0 \quad (j = 1, 2, \ldots, -\varkappa). \tag{51.27}$$

Thus, similarly to the case of the ordinary Riemann problem we have the following results.

The homogeneous boundary value problem (51.22) *has, for* $\varkappa > 0$, $\varkappa$ *linearly independent solutions, while, for* $\varkappa \leqq 0$, *it is insoluble.*

The non-homogeneous boundary value problem (51.15) *is, for* $\varkappa \geqq 0$, *absolutely soluble. For* $\varkappa < 0$, *it is soluble only if* $-\varkappa$ *conditions* (51.27) *are satisfied.*

51.4. Solution of the integral equation

Making use of the solution (51.26) of the boundary value problem (51.15), in accordance with the Sokhotski formula

$$\varphi(t) = \Phi^+(t) - \Phi^-(t) \tag{51.28}$$

we can find the solution of the original integral equation (51.12). Reasoning analogous to that of §§ 21.2 and 47.2 leads to the relation

$$\varphi(t) = \frac{1}{2}\left[1 + \frac{1}{G(t)}\right]g(t) +$$

$$+ X^+(t)\left[1 - \frac{1}{G(t)}\right]\left\{\Psi(t) - \frac{1}{2}\frac{P_{\varkappa-1}(F)}{[F(t) - F(z_\infty)]^{\varkappa-1}}\right\}.$$

Introducing here the values of $X(t)$ and $\Psi(t)$ from formulae (51.20), (51.25) and $G(t)$, $g(t)$ from formulae (51.16) and taking into account the assumed condition $a^2(t) - b^2(t) = 1$ we obtain

$$\varphi(t) = a(t)f(t) - b(t)Z(t)\sum_{k=0}^{n-1}\frac{1}{\pi i}\int_{L_0}\frac{f(\tau)}{Z(\tau)}\frac{d\tau}{\tau - \omega_k(t)} +$$

$$+ b(t)Z(t)\frac{P_{\varkappa-1}(F)}{[F(t) - F(z_\infty)]^{\varkappa-1}}, \tag{51.29}$$

where

$$Z(t) = [a(t) + b(t)]X^+(t) = [a(t) - b(t)]X^-(t)$$

$$= \prod_{k=0}^{n-1}[\omega_k(t) - t_0]^{-\varkappa}e^{\sum_{k=0}^{n-1}\Gamma_0[\omega_k(t)]}, \tag{51.30}$$

$$\Gamma_0(t) = \frac{1}{2\pi i}\int_L\frac{\ln G(\tau)}{\tau - t}d\tau, \quad G(t) = \frac{a(t) - b(t)}{a(t) + b(t)}.$$

If $\varkappa \leqq 0$ we have to set $P_{\varkappa-1} \equiv 0$, and for $\varkappa < 0$ the following conditions of solubility occur:

$$\int_{L_0}\left[\sum_{k=0}^{n-1}\omega_k'(\tau)\omega_k^{l-1}(\tau)\right]\frac{f(\tau)}{Z(\tau)}d\tau = 0 \quad (j = 1, 2, \ldots, -\varkappa). \tag{51.31}$$

Thus, for the integral equation we have results analogous to those obtained in the preceding subsection for the corresponding boundary value problem.

51.5. Example

Let us solve the singular integral equation

$$\frac{1}{2}(t^p + t^{-p})\,\varphi(t) + \frac{t^{-p} - t^p}{2\pi i} \int_{|\tau|=1} \left(\frac{1}{\tau - t} + \frac{1}{\tau - \alpha + t}\right) \varphi(\tau)\,d\tau = t^{-p}f(t)$$

$$(0 < \alpha < 2,\ p \text{ an integer}).$$

We introduce a sectionally analytic function automorphic with respect to the group $\omega_0(z) \equiv z$, $\omega_1(z) \equiv \alpha - z$,

$$\Phi(z) = \frac{1}{2\pi i} \int_{|\tau|=1} \left(\frac{1}{\tau - z} + \frac{1}{\tau + z - \alpha}\right) \varphi(\tau)\,d\tau.$$

It has two lines of discontinuity: the unit circle L_0 and the circle L_1, $|\tau - \alpha| = 1$ obtained from L_0 by means of the mapping $\omega_1(z) = \alpha - z$.

The boundary conditions on the contour $L = L_0 + L_1$ have the form

$$\Phi^+(t) = t^{2p}\,\Phi^-(t) + f(t) \quad \text{on} \quad L_0,$$

$$\Phi^+(t) = (\alpha - t)^{2p}\,\Phi^-(t) + f(\alpha - t) \quad \text{on} \quad L_1.$$

The canonical functions for the simple contours composing L are given by the expressions

$$X_0^+(z) = 1, \quad X_0^-(z) = z^{-2p}, \quad X_1^+(z) = 1, \quad X_1^-(z) = (\alpha - z)^{-2p}.$$

The following reasoning is an illustration not only of the method described in this section, but also generally of the method of solution of boundary value problems in the case of intersecting contours. In view of the overlapping of the domains bounded by the simple contours composing L, these problems possess peculiar difficulties. We shall therefore describe the solution in greater detail than it has been customary so far.

Each curve L_k divides the plane into two complementary domains D_k^+, D_k^- ($k = 0, 1$). The whole contour L divides the plane into four connected domains which may conveniently be denoted as the products of the simple domains by the intersection of which they have been obtained (Fig. 19).

In each domain, on the basis of the formula $X(z) = \prod\limits_{k=0}^{1} X_k(z)$ the canonical function is given by the expressions

$$X(z) = \begin{cases} X_0^+ X_1^+ = 1 & \text{in} \quad D_0^+ D_1^+, \\ X_0^+ X_1^- = (\alpha - z)^{-2p} & \text{in} \quad D_0^+ D_1^-, \\ X_0^- X_1^+ = z^{-2p} & \text{in} \quad D_0^- D_1^+, \\ X_0^- X_1^- = z^{-2p} (\alpha - z)^{-2p} & \text{in} \quad D_0^- D_1^-. \end{cases}$$

$\Phi(z)$ has at infinity the order 2 and $X(z)$ the order $4p$. The ratio $\Phi(z)/X(z)$ has the order $-(4p - 2)$. Taking into account that the

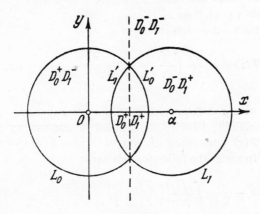

Fig. 19.

point at infinity is a fixed point of the group and hence it belongs to both fundamental domains (these are semi-planes bounded by the straight line passing through the point $z = \alpha/2$), this order should be divided by 2.

For the fundamental automorphic function of the group we take the function

$$F(z) = \frac{1}{z} + \frac{1}{\alpha - z} = \frac{\alpha}{z(\alpha - z)},$$

which has a simple pole at the origin†.

The general solution of the homogeneous problem for $p > 0$ has the form

$$\frac{P_{2p-1}(F)}{[F(z) - F(\infty)]^{2p-1}} = c_0 + \frac{c_1}{F(z)} + \cdots + \frac{c_{2p-1}}{F^{2p-1}(z)}.$$

† The simplest function $z + \omega(z)$ cannot be employed, since it reduces to a constant.

The general solution of the boundary value problem can be written in the form

$$\Phi(z) = \frac{X(z)}{2\pi i} \int\limits_{|\tau|=1} \frac{f(\tau)}{X^+(\tau)} \left[\frac{1}{\tau - z} + \frac{1}{\tau - \alpha + z} \right] d\tau + c_0 +$$

$$+ c_1 z(\alpha - z) + \cdots + c_{2p-1} z^{2p-1}(\alpha - z)^{2p-1}$$

(the constants $c_k \alpha^k$ are here replaced by c_k).

In computing the integral we have to remember that for $X^+(\tau)$ various expressions are taken on different sections of the curve L (1 on L_0' and $(\alpha - \tau)^{-2p}$ on L_0'')†.

If $p \leqq 0$ the solution has the form

$$\Phi(z) = \frac{X(z)}{2\pi i} \int\limits_{L_0} \frac{f(\tau)}{X^+(\tau)} \frac{2\tau - \alpha}{\tau^2 - z^2 - \alpha(\tau - z)} d\tau,$$

and in the case $p < 0$, according to the general theory $-2p$ conditions of solubility have to be satisfied. These can be derived by expanding $\Phi(z)$ into series of powers $F(z) = \alpha/[z(\alpha - z)]$.

It is easy to verify the following relations:

$$\Phi(z) = -\frac{1}{2\pi i} \int\limits_{L_0} \frac{f(\tau)}{X^+(\tau)} \frac{F(z)}{F(\tau)} \frac{F'(\tau) d\tau}{F(\tau) - F(z)} = \sum_{j=1}^{\infty} c_j F^j(z),$$

$$c_j = -\frac{1}{2\pi i} \int\limits_{L_0} \frac{f(\tau)}{X^+(\tau)} \frac{F'(\tau)}{F^{j+1}(\tau)} d\tau.$$

For the solubility of the problem it is necessary and sufficient that

$$\int\limits_{L_0} \frac{f(\tau)}{X^+(\tau)} (2\tau - \alpha) [\tau(\alpha - \tau)]^{j-1} d\tau = 0 \quad (j = 1, 2, \ldots, -2p).$$

Let us make use of the solution of the boundary value problem to solve the original integral equation. We have

$$Z(t) = [\alpha(t) + b(t)] X^+(t) = t^{-p} X^+(t).$$

---

† It seems that we are confronted with an integral having a discontinuous density; however, this is not so. If we introduce $X(z)$ into the integrand and consider the density $[X(z)/X^+(\tau)]f(\tau)$, then it is readily observed that it behaves as a continuous function in passing through the ends of the sections L_0', L_0''.

The general solution of the integral equation can be written in the form

$$\varphi(t) = \frac{1}{2}(t^{-2p} + 1)f(t) - \frac{(t^{-2p} - 1)X^+(t)}{2\pi i} \times$$

$$\times \int_{|\tau|=1} \frac{f(\tau)}{X^+(\tau)} \frac{2\tau - \alpha}{\tau^2 - t^2 - \alpha(\tau - t)} d\tau + \frac{1}{2}(t^{-2p} - 1)X^+(t) \times$$

$$\times [c_0 + c_1 t(\alpha - t) + \cdots + c_{2p-1} t^{2p-1}(\alpha - t)^{2p-1}].$$

If $p < 0$ the following $-2p$ conditions of solubility should be satisfied:

$$\int_{|\tau|=1} \frac{f(\tau)}{X^+(\tau)}(2\tau - \alpha)[\tau(\alpha - \tau)]^{j-1} d\tau = 0 \quad (j = 1, 2, \ldots, -2p).$$

51.6. The case when the auxiliary analytic function does not vanish at infinity

In the kernels of the examined integral equations there appear additional terms of the form $1/[\tau - \omega_k(\infty)]$. Hence, the introduced auxiliary functions $\Phi(z)$, similarly to the Cauchy type integrals, possessed the property $\Phi(\infty) = 0$. This enabled us to write down on the basis of the generalized Liouville theorem the general solution in which all terms of the polynomial $P_{\varkappa-1}(F)$ constitute linearly independent solutions of the homogeneous equation. If the above additional terms be eliminated from the kernel, then there arise difficulties with the explicit determination of linearly independent solutions. Let us briefly consider this case.

Thus, suppose that the integral equation has the form

$$a(t)\varphi(t) + \frac{b(t)}{\pi i}\int_L \sum_{k=0}^{n-1} \frac{1}{\tau - \omega_k(t)}\varphi(\tau) d\tau = f(t). \quad (51.12')$$

The corresponding auxiliary analytic function has the form

$$\Phi(z) = \frac{1}{2\pi i}\int_L \sum_{k=0}^{n-1} \frac{1}{\tau - \omega_k(z)}\varphi(\tau) d\tau. \quad (51.13')$$

We have to consider separately the following two essentially distinct cases.

$1°$. Integral transformations of the group. In this case it follows from the properties of automorphic functions that the function $\Phi(z)$

has at infinity one simple pole in every fundamental domain. Hence, it can easily be deduced† that formula (51.23) also in the considered case yields the general solution containing $\varkappa$ linearly independent solutions.

2°. Fractional transformations of the group. In this case $\sum\limits_{k=0}^{n-1} \omega_k(z) \neq$ const and $P(z) = \sum\limits_{k=0}^{n-1} \omega_k(z)$ can be taken for the fundamental function of the group.

If we expand $\Phi(z)$ as series in the vicinity of infinity, the free term of this expansion has the form

$$\Phi(\infty) = \frac{1}{2\pi i} \sum_{k=0}^{n-1} \int\limits_L \frac{\varphi(\tau)\,d\tau}{\tau - \omega_k(\infty)}. \tag{51.32}$$

If the substitution into the last integral of the general solution of the integral equation leads to the identical vanishing of the former, for arbitrary values of the parameters, then the considered case does not differ at all from that examined above, and the general solution again is given by formula (51.23).

Assume now that integral (51.32) does not identically vanish. Then $\Phi(\infty) \neq 0$ and in the solutions of the boundary value problem and the integral equation the polynomial $P_{\varkappa-1}$ has to be replaced by $P_{\varkappa}$. Thus, the general solution contains $\varkappa + 1$ parameters. However, not all are independent. It turns out that there exists one linear relation between them; we now proceed to derive it.

Preserving all previous notation let us write down the solutions of the boundary value problem and the singular integral equation

$$\Phi(z) = X(z)\left[\Psi(z) + P_{\varkappa}(F)\right], \tag{51.33}$$

$$\varphi(t) = a(t)f(t) - b(t)Z(t)\sum_{k=0}^{n-1}\frac{1}{\pi i}\int\limits_L \frac{f(\tau)}{Z(\tau)}\frac{d\tau}{\tau - \omega_k(t)} + b(t)Z(t)P_{\varkappa}(F). \tag{51.34}$$

To obtain the required linear relation let us calculate the value of $\Phi(\infty)$ in two ways. First, from solution (51.33) of the boundary value problem, next from the definition of $\Phi(z)$ (formula (51.32)), substituting into the latter $\varphi(t)$ from the solution (51.34) of the singular equation.

† It has to be taken into account here that the point at infinity is a fixed point of the group (see the reasoning in the solution of the example (51.5)).

Taking into account that the canonical function has at infinity a zero of order $\varkappa$ and $F^\varkappa = [z + \sum_{k=1}^{n-1} \omega_k(z)]^\varkappa$ has a pole of the same order, and denoting by c_k the coefficients of the polynomial $P_\varkappa$, we obtain from (51.33)

$$\Phi(\infty) = \lim_{z \to \infty} [X(z) c_\varkappa F^\varkappa(z)] = \alpha c_\varkappa, \qquad (51.35)$$

where

$$\alpha = \lim_{z \to \infty} [z^\varkappa X(z)].$$

Comparing (51.32) and (51.35) we arrive at the relation

$$c_\varkappa \left[\alpha - \frac{1}{2\pi i} \sum_{k=0}^{n-1} \int_L \frac{b(\tau) Z(\tau) F^\varkappa(\tau)}{\tau - \omega_k(\infty)} d\tau \right]$$

$$= \frac{1}{2\pi i} \sum_{k=0}^{n-1} \int_L \frac{d\tau}{\tau - \omega_k(\infty)} \left[a(\tau) f(\tau) - \frac{b(\tau) Z(\tau)}{\pi i} \times \right.$$

$$\left. \times \sum_{k=0}^{n-1} \int_L \frac{f(\tau_1)}{Z(\tau_1)} \frac{d\tau_1}{[\tau_1 - \omega_j(\tau)]} + b(\tau) Z(\tau) P_{\varkappa-1}(F) \right]. \quad (51.36)$$

If the coefficient of $c_\varkappa$ does not vanish, the last relation makes it possible to express the parameter $c_\varkappa$ in terms of the remaining parameters. Introducing the determined value of $c_\varkappa$ into formula (51.34) we obtain the general solution containing $\varkappa$ linearly independent parameters.

Suppose now that the coefficient of $c_\varkappa$ vanishes. We shall prove that relation (51.36) cannot identically be satisfied and, consequently, one of the parameters $c_0, c_1, \ldots, c_{\varkappa-1}$ can be expressed by the other parameters. In fact, under the above assumption, relation (51.36) can be written in the form

$$\frac{1}{2\pi i} \sum_{k=0}^{n-1} \int_L \frac{\varphi(\tau)}{\tau - \omega_k(\infty)} d\tau = 0.$$

If we assume that this relation is identically satisfied, this means that $\Phi(\infty) = 0$ which contradicts the original assumption. Thus, in all cases the general solution contains $\varkappa$ linearly independent parameters.

The above investigation follows the paper of I. A. Paradoksova, which has not yet been published.

51.7. Additional remarks

We have investigated here only the simplest type of complete singular equation soluble in a closed form. The paper of F. D. Gakhov and L. I. Chibrikova (1), according to which the description of the present section is carried out, contains the solution of a number of other types which can be solved by analogous methods. These equations may be divided into two groups: the first contains equations soluble as here in § 51.4, with the help of automorphic functions (see Problem 2), the second contains equations the solution of which is carried out by means of more complicated functions, namely functions acquiring some factors under a transformation of the group. The latter group contains the familiar equations employed by Tricomi [27] and A. V. Bitsadze (1), in solving boundary value problems for equations of mixed type (see Problems 4 and 5).

The reader who wants to become better acquainted with the problem considered is advised to read the above mentioned paper of F. D. Gakhov and L. I. Chibrikova. However, he should be warned that this paper contains one essential inaccuracy. The kernels of the integral equations are there taken without the terms $1/[\tau - \omega_k(\infty)]$ and, consequently, the corresponding automorphic functions do not, generally speaking, vanish at infinity. The investigation of the preceding section of the relation between the $\varkappa + 1$ parameters entering the general solution, has not been carried out there. Therefore, the number of linearly independent solutions given in the paper is greater than the true number by unity.

We make two more remarks.

1*. If L_0 is an open curve, then all equivalent curves L_k are also open. Here the usual problems of the classes of solutions arise. If the ends of the curves L_k do not coincide, the problem of allowable classes is solved in exactly the same way as in the case of the dominant equation on a complex contour (§ 44.3). If, now, the ends coincide, the singularities may overlap and the solution constructed for every simple curve in the class of integrable functions may be non-integrable. In this case, therefore, an additional investigation is necessary (see § 44.4).

2*. In order not to complicate the problem we assumed that the contour L_0 on which the equation was considered, is simple. No complications arise if L_0 is a union of closed and open curves possessing a finite number of points of intersection.

The next section is devoted to generalizations in the case of infinite groups.

§ 52. Continuation. The case of an infinite group

It is natural to make an attempt to generalize the formulation of the problem to the case of an infinite group. An investigation proves that this generalization would be difficult on the basis of the previous formulae, by passing to the limit. It is simpler to formulate the problem for the infinite group anew, in such a way that the case considered above is a particular case of the new formulation. We begin by formulating the boundary value problem. For simplicity we confine ourselves to the case of a closed contour.

52.1. The Riemann boundary value problem

Suppose that D is one of the domains composed wholly of the ordinary points of a functional (Fuchsian or elementary) group of linear fractional transformations

$$\omega_0(z) \equiv z, \quad \omega_1(z), \quad \omega_2(z), \ldots,$$

and let L_0 be a smooth closed curve located wholly in the domain D. We denote by $L_k (k = 1, 2, \ldots)$ the curves equivalent to L_0, i.e. the curves into which the curve L_0 is transformed under the transformations of the group. We assume that the curves L_k do not overlap, although they may intersect.

We shall call a function $\Phi(z)$ defined in D, a sectionally analytic automorphic function, if it has the following properties.

1. It is automorphic with respect to the substitutions of the group

$$\Phi[\omega_k(z)] = \Phi(z).$$

2. It is analytic at all points of the domain D, except at the points of the contour $L = L_0 + L_1 + \ldots$.

3. It is continuously continuable through all points of the contour L, except for the points of self-intersection near which it is bounded.

The Riemann problem consists now in determination of a sectionally analytic automorphic function $\Phi(z)$ which satisfies on the curve L the boundary condition

$$\Phi^+(t) = G(t)\,\Phi^-(t) + g(t), \tag{52.1}$$

where $G(t)$, $g(t)$ are functions prescribed on L_0, which satisfy the Hölder condition, and, moreover, $G(t) \neq 0$.

The curves L_k are also the lines of discontinuity for the function $\Phi(z)$. As in the preceding section it would be easy to construct the boundary conditions for them. This is not necessary, however, since the solution which will later be derived taking into account the boundary condition (52.1) and the automorphicity of the unknown function, will automatically satisfy the boundary conditions on the whole contour L.

We now present the solution of the above formulated problem. As in the ordinary case (§ 14) the basis of the solution is the solution of the simplest problem of the jump

$$\Phi^+(t) - \Phi^-(t) = g(t). \tag{52.2}$$

The ordinary Cauchy integral does not yield its solution since the requirement of automorphicity is not satisfied. We construct the counterpart of the Cauchy integral for automorphic functions.

Let $F(z)$ be the simple automorphic function of the group having one simple pole z_0 in the fundamental domain R, and $g(t)$ a defined function on L_0 satisfying Hölder's condition. We construct the integral

$$\Phi(z) = \frac{1}{2\pi i} \int_{L_0} g(\tau)\, \frac{F'(\tau)}{F(\tau) - F(z)}\, d\tau. \tag{52.3}$$

It defines a sectionally analytic automorphic function having a line of discontinuity L_0 and vanishing at z_0 (and at all equivalent points). The kernel of the integral may be represented as

$$\frac{F'(\tau)}{F(\tau) - F(z)} = \frac{1}{\tau - z} + \Omega\,(\tau, z),$$

where Ω is a continuous function on L_0. Hence it is easily deduced that formulae similar to Sokhotski's ((4.9) and (4.10)) hold for the limit values of $\Omega(z)$ on L_0:

$$\Phi^+(t) - \Phi^-(t) = g(t),$$

$$\Phi^+(t) + \Phi^-(t) = \frac{1}{\pi i} \int\limits_{L_0} g(\tau)\,\frac{F'(\tau)}{F(\tau) - F(z)}\,d\tau. \tag{52.4}$$

The integral (52.3) therefore yields the solution of the jump problem (52.2) in the class of sectionally analytic automorphic functions which vanish at z_0. The solution of the problem with zero jump and zero at z_0 vanishes identically by virtue of the theorem of analytic continuability and the properties of automorphic functions. Hence the solution of (52.3) is unique in the class under consideration.

By the canonical function $X(z)$ of problem (52.1) we understand an automorphic function which satisfies the homogeneous boundary condition

$$X^+(t) = G(t)\,X^-(t)$$

and has zero order everywhere in the fundamental domain R, except at the point z_0 where it has the maximum possible order equal to the index $\varkappa$ of $G(t)$. It can easily be shown that the canonical function is given by the formula

$$X(z) = [F(z) - F(t_0)]^{-\varkappa}\,e^{\Gamma(z)}, \tag{52.5}$$

to within a constant term. Here $F(z)$ is the simple automorphic function introduced above with the simple pole at z_0

$$\Gamma(z) = \frac{1}{2\pi i} \int\limits_{L_0} \ln G(\tau)\,\frac{F'(\tau)}{F(\tau) - F(z)}\,d\tau; \tag{52.6}$$

t_0 is the initial point of the path of integration in this integral (cf. § 44.2).

The usual reasoning leads to the general solution of problem (52.1)

$$\Phi(z) = X(z)\,[\Psi(z) + P_\varkappa(F)], \tag{52.7}$$

where

$$\Psi(z) = \frac{1}{2\pi i} \int\limits_{L_0} \frac{g(\tau)}{X^+(\tau)}\,\frac{F'(\tau)}{F(\tau) - F(z)}\,d\tau, \tag{52.8}$$

$X(z)$ is given by formula (52.5) and $P_\varkappa$ is an arbitrary polynomial of degree $\varkappa$.

If the required solution be subjected to the condition $\Phi(z_0) = 0$ (this case is encountered in solving the singular equation), then $P_\varkappa$ has to be replaced by $P_{\varkappa-1}$.

If in the last case $\varkappa \le 0$, then we have to set $P_{\varkappa-1} \equiv 0$, and for $\varkappa < 0$ the following conditions of solvability have to be satisfied:

$$\int\limits_{L_0} \frac{g(\tau)}{X^+(\tau)}\,[F(\tau)]^{j-1}\,F'(\tau)\,d\tau = 0 \quad (j = 1, 2, \ldots, -\varkappa). \tag{52.9}$$

The latter is obtained by expanding (52.8) as a series in powers of $F(z)$ and equating the first $-\varkappa$ coefficients of the expansion to zero. The formulation and the solution of the problem for the case of open contours presents no difficulties.

52.2. The singular integral equation

Consider the singular integral equation

$$a(t)\varphi(t) + \frac{b(t)}{\pi i} \int_{L_0} \frac{F'(\tau)}{F(\tau) - F(t)} \varphi(\tau)\, d\tau = f(t), \tag{52.10}$$

where, as before, the known functions $a(t), b(t), f(t)$ satisfy the Hölder condition and $L_0, F(t)$ have the same meaning as in the preceding subsection.

Introducing the sectionally analytic automorphic function

$$\Phi(z) = \frac{1}{2\pi i} \int_{L_0} \frac{F'(\tau)}{F(\tau) - F(z)} \varphi(\tau)\, d\tau, \tag{52.11}$$

we arrive as before at the Riemann boundary value problem

$$\Phi^+(t) = G(t)\,\Phi^-(t) + g(t), \tag{52.12}$$

$$G(t) = \frac{a(t) - b(t)}{a(t) + b(t)}, \quad g(t) = \frac{f(t)}{a(t) + b(t)}, \tag{53.13}$$

examined in the preceding subsection. In view of the property of the fundamental automorphic function $F(z)$ (simple pole at z_0) the function $\Phi(z)$ should satisfy the additional condition $\Phi(z_0) = 0$.

Making use of the solution of the boundary value problem (42.12) derived in the preceding subsection, in accordance with the Sokhotski formula,

$$\varphi(t) = \Phi^+(t) - \Phi^-(t)$$

we obtain the solution of the integral equation (52.10)

$$\varphi(t) = a(t)f(t) - \frac{b(t) Z(t)}{\pi i} \int_{L_0} \frac{f(\tau)}{Z(\tau)} \frac{F'(\tau)}{F(\tau) - F(t)}\, d\tau + b(t) Z(t) P_{\varkappa-1}(F), \tag{52.14}$$

where

$$Z(t) = [a(t) + b(t)]\, X^+(t) = [a(t) - b(t)]\, X^-(t) = [F(t) - F(t_0)]^{-\varkappa}\, e^{\Gamma(t)}. \tag{52.15}$$

If $\varkappa \le 0$ we have to set $P_{\varkappa-1} \equiv 0$. For $\varkappa < 0$ the conditions of solubility

$$\int_{L_0} \frac{f(\tau)}{Z(\tau)} [F(\tau)]^{j-1} F'(\tau)\, d\tau = 0 \quad (j = 1, 2, \ldots, -\varkappa), \tag{52.16}$$

should be satisfied.

The reasoning behind the latter formulae is analogous to that in § 51.4.

52.3. Some applications

It seems at first sight that automorphic functions are a very special class which is only encountered in exceptional cases. This is not so at all. In fact, ordinary periodic and double-periodic functions are atomorphic. Consequently such useful theoretical and practical functions as exponential, trigonometric and hyperbolic functions belong to this class; then there are elliptic functions—the most useful of the higher transcendental functions. Various periodic and double-periodic processes which arise in nature may be reduced to such functions. Hence the theory has numerous applications. A few applications, mainly mathematical, will now be discussed.

$1°$. *Riemann problem for periodic functions.* It is required to find the sectionally analytic periodic function $\Phi(z)$ with period 2π which has limit values on a smooth closed contour L located within the periodic strip, which satisfy the boundary condition:

$$\Phi^+(t) = G(t)\,\Phi^-(t) + g(t), \qquad (52.17)$$

where $G(t)$, $g(t)$, as above, satisfy Hölder's condition and $G(t) \neq 0$.

The group of transformations is generated by the substitution $\omega_1(z) = z + 2\pi$. The fundamental domain is a vertical strip 2π in width, e.g. the strip $0 \leq \operatorname{Re} z \leq 2\pi$. The fundamental automorphic function is $e^{iz} = e^{-y} e^{ix}$. It vanishes at the upper end of the strip, but has a simple pole at the lower end†. (The point z_0 is the lower end.)

According to (52.3), the solution of the jump problem

$$\Phi_0^+(t) - \Phi_0^-(t) = g(t)$$

is given by a function

$$\Phi_1(z) = \frac{1}{2\pi} \int_L g(\tau) \frac{e^{i\tau}\, d\tau}{e^{i\tau} - e^{iz}} = \frac{1}{4\pi i} \int_L g(\tau) \left[\cot\frac{\tau - z}{2} + i \right] d\tau, \quad (52.18)$$

which has a first-order zero at the lower end of the strip. The canonical function, according to (52.5) and (52.6), is given by the

---

† For the integral function e^{iz} the infinitely remote point is essentially singular and as $y \to \infty$ it has a pole of infinitely great order. But an infinitely remote point belongs to all fundamental domains. The order of the pole must therefore be distributed between all of them. From the general property of fundamental automorphic functions whereby each assumes its value once, it follows that at the lower end of each strip the order of the pole must be set equal to unity. The same is also true of the order of the zero at the upper end.

expression

$$X(z) = (e^{iz} - e^{it_0}) e^{\Gamma(z)},$$

$$\Gamma(z) = \frac{1}{4\pi} \int_L \ln G(\tau) \cot\frac{\tau - z}{2} d\tau. \qquad (52.19)$$

The general solution, according to (52.7) and (12.8), is

$$\Phi(z) = X(z)\left\{\int \frac{g(\tau)}{X^+(\tau)} \left[\cot\frac{\tau - z}{2} + i\right] d\tau + P_\varkappa(e^{iz})\right\}. \qquad (52.20)$$

If a solution is required which vanishes at the lower end of the strip, then $P_\varkappa$ has to be replaced by $P_{\varkappa-1}$. In this case it is necessary to put $P_{\varkappa+1} \equiv 0$ if $\varkappa \leq 0$; if $\varkappa < 0$ the following conditions must be fulfilled

$$\int \frac{g(\tau)}{X^+(\tau)} e^{i\tau(k-1)} d\tau = 0, \quad k = 1, 2, \ldots, -\varkappa. \qquad (52.21)$$

2°. *The Hilbert problem for periodic functions in a semi-plane.* Find the function $F(z) = u + iv$ which is analytic in the upper half-plane and has a period 2π, satisfying the boundary condition

$$a(x) u(x) + b(x) v(x) = c(x). \qquad (52.22)$$

By the method of § 30.2 we reduce this to the condition of the Riemann problem:

$$\Phi^+(x) = -\frac{a(X) + ib(X)}{a(X) - ib(X)} \Phi^-(x) + \frac{c(X)}{a(X) - ib(X)}. \qquad (52.23)$$

In the sectionally analytic function $\Phi^+(z) = F(z)$, $\Phi^-(z) = \bar{F}(z)$, which by virtue of the latter relations must satisfy the condition

$$\Phi^-(z) = \overline{\Phi^+}(z); \qquad (52.24)$$

the boundary condition (52.23) may be assumed to be defined on the interval $[0, 2\pi]$. We use the formula of the foregoing section. Setting

$$\arg(a + ib) = \arctan\frac{b(t)}{a(t)} = \theta(t),$$

we have

$$\Gamma(z) = \frac{1}{4\pi} \int_0^{2\pi} \ln\left[-\frac{a + ib}{a - ib}\right] \cot\frac{\tau - z}{2} d\tau$$

$$= \frac{1}{4\pi} \int_0^{2\pi} [2\theta(\tau) + \pi] \cot\frac{\tau - z}{2} d\tau.$$

Here the initial point of integration is $t_0 = 0$. Introducing the constant multiplier C into the canonical function (52.19) and selecting the latter in accordance with the condition $\overline{X(z)} = X(z)$, we have

$$X(z) = \left(\sin\frac{z}{2}\right)^{-2\varkappa} e^{\Gamma(z)}.$$

Having regard to (52.24), the solution is

$$F(z) = X(z)\left[\frac{1}{2\pi i}\int\limits_0^{2\pi} \frac{c(\tau)}{X^+(\tau)\,[a(\tau) - ib(\tau)]}\cot\frac{\tau - z}{2}\,d\tau + \right.$$

$$\left. + \sum_{k=0}^{\varkappa}(\alpha_k\cos kz + \beta_k\sin kz)\right]. \qquad (52.25)$$

If $\varkappa < 0$ the last sum in the solution is discarded. Moreover, the following condition must be satisfied

$$\int\frac{c(\tau)\cos k\tau}{X^+(\tau)\,[a(\tau) - ib(\tau)]}\,d\tau = \int\frac{c(\tau)\sin k\tau}{X^+(\tau)\,[a(\tau) - ib(\tau)]}\,d\tau = 0$$

$$(k = 0, 1, \ldots, -\varkappa - 1).$$

3°. *The mixed boundary value problem of periodic analytic functions in a semi-plane.* Consider the mixed problem solved in § 43.3 and find the analytic function $F(z) = u + iv$ in the upper semi-plane which satisfies the boundary condition

$$u = f(t)\ \text{ on }\ L' = \sum_{k=1}^m a_k b_k,\quad v = h(t)\ \text{ on }\ L'' = \sum_{k=1}^m b_k a_{k+1}$$

$$(a_{m+1} = a_1) \qquad (52.26)$$

on condition that the defined intervals $a_k b_k$ and $b_k a_{k+1}$ $(k = 1, 2, \ldots, m)$ occur on $[0, \pi]$ and that the required function is continued periodically with period π.

Proceeding in the same way as in § 43.3, we arrive at the Riemann boundary value problem

$$\Phi^+(t) = G(t)\,\Phi^-(t) + g(t),$$

where

$$G(t) = -1,\quad g(t) = 2f(t)\ \text{ on }\ L',$$

$$G(t) = 1,\quad g(t) = 2h(t)\ \text{ on }\ L'',$$

for the sectionally analytic periodic function $\Phi(z)$ with period π which satisfies the condition $\overline{\Phi(z)} = \Phi(z)$.

For the case under consideration the fundamental function is e^{2iz}. We have

$$\Gamma(z) = \frac{1}{2\pi i} \int_0^\pi \ln G(\tau) \cot(\tau - z)\, d\tau$$

$$= \frac{1}{2\pi i} \sum_{k=1}^m \int_{a_k}^{b_k} \pi i \cot(\tau - z)\, d\tau = \frac{1}{2} \ln \prod_{k=1}^m \frac{\sin(z - b_k)}{\sin(z - a_k)}.$$

In the class of functions which are bounded at b_k and integrable at a_k, the canonical function is

$$X(z) = e^{\Gamma(z)} = \sqrt{\left[\prod_{k=1}^m \frac{\sin(z - b_k)}{\sin(z - a_k)}\right]}.$$

The solution of the problem in the particular class yields the formula

$$\Phi(z) = \frac{1}{\pi} \sqrt{\left[\prod_{k=1}^m \frac{\sin(z - b_k)}{\sin(z - a_k)}\right]} \times$$

$$\times \left[\sum_{k=1}^m \int_{a_k}^{b_k} \sqrt{\left[\prod_{k=1}^m \frac{\sin(\tau - a_k)}{\sin(\tau - b_k)}\right]} f(\tau) \cot(\tau - z) + \right.$$

$$\left. + \sum_{k=1}^n \int_{b_k}^{a_k+1} \sqrt{\left[\prod_{k=1}^m \frac{\sin(\tau - a_k)}{\sin(\tau - b_k)}\right]} h(\tau) \cot(\tau - z)\, d\tau + a_0 \right], \quad (52.27)$$

where a_0 is an arbitrary real constant.

The solution in other classes may be obtained in the same way.

$4°$. *Determination of the flow past a periodic lattice.* This is an application from hydromechanics. Suppose that in the z plane we have a lattice with its profile formed by p rows of sections for which $y = m\pi$ and $a_k < x < b_k$ ($k = 1, 2, \ldots, p$; $m = 0, \pm 1, \pm 2, \ldots$) and q rows of sections for which $y = \frac{2n + 1}{2}\pi$, $c_j < x < d_j$ ($j = 1, 2, \ldots, q$; $n = 0$, $\pm 1, \pm 2, \ldots$), and that a potential stream of an incompressible fluid flows past it.

Determination of the function of the stream velocities merely entails determination of the function $\Phi(z) = u - iv$ satisfying the following conditions:

(a) $\Phi(z)$ is a sectionally analytic periodic function with the period πi and the line of discontinuities

$$L = \sum_{k=1}^{p}(a_k, b_k) + \sum_{j=1}^{q}\left(c_j + \frac{\pi i}{2}, \ d_j + \frac{\pi i}{2}\right);$$

(b) $\Phi(+\infty) = 0$, $\Phi(-\infty)$ is finite;

(c) at points a_k, $c_j + \dfrac{\pi i}{2}$ $F(z)$ is finite, and at points b_k, $d_j + \dfrac{\pi i}{2}$ it is integrable;

(d) on L the imaginary part of $\Phi(z)$ must assume the following prescribed values when approaching L from above or below

$$v^+ = f^+(t), \quad v^- = f^-(t) \quad \text{on} \ \ L, \tag{52.28}$$

the values of $f^+(t)$ and $f^-(t)$ here coinciding at the end-point.

This is the Dirichlet problem of § 46.4 for a plane having slits and the additional condition of periodicity except for a slight difference in that the imaginary part v is defined instead of the real part u.

Introducing, as in § 46.4, the functions

$$\Psi(z) = \frac{1}{2}[\Phi(z) + \overline{\Phi(\bar{z})}], \quad \Omega(z) = \frac{1}{2}[\Phi(z) - \overline{\Phi(\bar{z})}],$$

we arrive at the mixed problem in Ψ, and at the problem of determining the analytic (periodic) function from the imaginary part $\operatorname{Im}\Omega(t) = f^+(t) - f^-(t)$ on L for Ω.

The fundamental function is e^{2z}. The function Ω is found by the formula

$$\Omega(z) = \frac{1}{2\pi i}\int_L \frac{2ie^{2\tau}[f^-(\tau) - f^+(\tau)]\,d\tau}{e^{2\tau} - e^{2z}}$$

$$= \frac{1}{2\pi}\int_L [f^-(\tau) - f^+(\tau)]\,[\coth(\tau - z) + 1]\,d\tau. \tag{52.29}$$

For the function $\Psi(z)$, we have

$$\Gamma(z) = \frac{1}{2\pi i}\int_L \ln(-1)\coth(\tau - z)\,d\tau$$

$$= \frac{1}{2}\ln\left[\prod_{k=1}^{p}\frac{\sinh(z - a_k)}{\sinh(z - b_k)}\prod_{j=1}^{q}\frac{\cosh(z - c_j)}{\cosh(z - d_j)}\right].$$

Hence the canonical function in the required class of functions is

$$X(z) = \sqrt{\left[\prod_{k=1}^{p} \frac{\sinh(z - a_k)}{\sinh(z - b_k)} \prod_{j=1}^{q} \frac{\cosh(z - c_j)}{\cosh(z - d_j)}\right]}.$$

The solution of the mixed problem is found by the formula

$$\Psi(z) = \frac{-X(z)}{2\pi} \int_L \frac{f^+(\tau) + f^-(\tau)}{X^+(\tau)} [\coth(\tau - z) + 1] \, d\tau. \quad (52.30)$$

The solution of the considered problem is defined as the sum of the function

$$\Phi(z) = \Psi(z) + \Omega(z). \quad (52.31)$$

We will now illustrate the use of elliptic functions by an example.

5°. *The Riemann problem for an even elliptic function.* For a double-periodic function with periods ω_1, ω_2 one of the fundamental domains is a parallelogram with the vertices 0, ω_1, $\omega_2 + \omega_1$, ω_2. The fundamental automorphic function is the Weierstrass function $\wp(z)$. At the vertices of the parallelogram of periods it has poles of second order. One of the equivalent vertices, for example, $z = 0$, is taken as z_0. The contour L is assumed to be simple, smooth and closed and located within the paralellogram. The Riemann problem is completely solved in accordance with § 52.1.

The solution of the jump problem is found by the formula

$$\Phi(z) = \frac{1}{2\pi i} \int_L g(\tau) \frac{\wp'(\tau) d\tau}{\wp(\tau) - \wp(z)}.$$

The canonical function is given by the expression

$$X(z) = [\wp(z) - \wp(t_0)]^{-\varkappa} e^{\Gamma(z)},$$

$$\Gamma(z) = \frac{1}{2\pi i} \int \ln G(\tau) \frac{\wp'(\tau) d\tau}{\wp(\tau) - \wp(z)},$$

whilst the general solution for $\varkappa \le 0$ is written as

$$\Phi(z) = \frac{X(z)}{2\pi i} \left\{ \int \frac{g(\tau)}{X^+(\tau)} \frac{\wp'(\tau) d\tau}{\wp(\tau) - \wp(z)} + P_\varkappa[\wp(z)] \right\}. \quad (52.32)$$

If $\varkappa < 0$, it is necessary to put $P_\varkappa = 0$; for $\varkappa < 1$ the following solubility conditions must be fulfilled

$$\int_L \frac{g(\tau)}{X^+(\tau)} \wp^{k-1}(\tau)\wp'(\tau) \, d\tau = 0, \quad k = 1, 2, \ldots, -\varkappa - 1.$$

The case of a doubly-periodic function has special features as compared with the cases of automorphic functions considered previously; in fact, a fundamental automorphic function with one simple pole in the fundamental domain is no longer present. Hence an arbitrary double-periodic function is expressed rationally not by one, but by two doubly-periodic functions ($\wp$ and $\wp'$). For this reason the solution of the Riemann problem is not quite the same for the cases of even, odd and arbitrary doubly-periodic functions.

We confine ourselves to the even function case. The solution for the other two cases is given by Chibrikova (2) from whose paper all this subsection is taken.

Chibrikova (2) also gave a solution for a number of other problems by the use of periodic and double-periodic functions and, in particular, the solution of the problem of the flow past a doubly-periodic lattice; inversion formulae for a number of integrals with kernels formed from elliptic functions were also given (see Problems 10 and 11 at the end of the Chapter).

For simplicity a closed contour is taken throughout. No difficulties arise in considering any number of closed and open lines within the fundamental domain.

52.4. Integral equation with a non-fundamental automorphic function

Consider the integral equation

$$a(t)\varphi(t) + \frac{b(t)}{\pi i} \int_{L_0} \frac{H'(\tau)}{H(\tau) - H(t)} \varphi(t) \, dt = f(t), \qquad (52.33)$$

where, as distinct from (52.10), in § 52.2, the automorphic function $H(z)$ in the kernel assumes its value in the fundamental domain R not once, but ν times.

We introduce the sectional automorphic function

$$\Phi(z) = \frac{1}{2\pi i} \int \frac{H'(\tau)}{H(\tau) - H(z)} \varphi(\tau) d\tau. \qquad (52.34)$$

It vanishes at ν points $z_1, \ldots, z_2$ (the multiple zero is repeated as many times as its multiplicity), where $H(z)$ has poles. The denominator of the kernel $H(\tau) - H(z)$ vanishes ν times for each point $\tau \in L_0$.

The function $\Phi(z)$ therefore has ν lines of discontinuity in the fundamental domain, viz. L_0 and $(\nu - 1)$ curves $L_0^{(1)}, \ldots, L_0^{(\nu-1)}$, the equations of which are defined in implicit form by the solution of the equation

$$H(z) = H(\tau) \quad (\tau \in L_0). \tag{52.35}$$

Putting $\Psi(z)$ for the inverse function of $H(z)$, the equations of the contours $L_k^{(j)}$ $(j = 0, 1, \ldots, \nu - 1; \; k = 0, 1, 2 \ldots)$ are given by the branches $\Psi_k^{(j)}$ of the many-valued function Ψ:

$$z = \Psi_k^{(j)} [H(\tau)] \tag{52.36}$$

$$(\tau \in L_0; \quad j = 0, 1, \ldots, \nu - 1; \quad k = 0, 1, 2, \ldots).$$

First suppose that the branch points of the functions $\Psi_k^{(j)}(z)$ are not located within the contour L_0; all the contours L_0^j are also then closed. It is implied from the definition of $\Psi_k^{(j)}$ $(H[\Psi_k^{(j)}(t)] \equiv t)$ that

$$H(z) = H \{\Psi_k^{(j)}[H(z)]\} = H(z).$$

The function $H(z)$ is therefore automorphic not only in the prescribed group $\omega_k(z)$ of linear transformations, but also in the group† of (non-linear) transformations $\Psi_k^{(j)}[H(z)]$.

On L_0 formulae analogous to (52.4) hold good for the limit values of $\Phi(z)$:

$$\left. \begin{array}{c} \Phi^+(t) - \Phi^-(t) = \varphi(t), \\[2mm] \Phi^+(t) + \Phi^-(t) = \dfrac{1}{\pi i} \displaystyle\int\limits_{L_0} \dfrac{H'(\tau)}{H(\tau) - H(t)} \varphi(\tau) \, d\tau. \end{array} \right\} \tag{52.37}$$

By using them we arrive in the usual way at the boundary value problem on L_0:

$$\Phi^+(t) = G(t) \Phi^-(t) + g(t), \tag{52.38}$$

$$G = \frac{a - b}{a + b}, \quad g = \frac{f}{a + b}. \tag{52.39}$$

There is no need to form the boundary conditions for the rest of the lines of discontinuity $L_0^{(j)}$ within R (or for equivalent lines in other fundamental domains), since the solution of the boundary value problem (52.38), having regard to its automorphicity in relation to all

---

† The fact that the transformations Ψ_j form a complete group follows from the fact that they exhaust all the singular lines $L_0^{(j)}$ lying within the fundamental domain.

the transformations of the group, automatically satisfies the boundary conditions on all the lines of discontinuity.

The real difference between this and the earlier problems is that the required function $\Phi(z)$ is a function of $H(z)$ by virtue of (52.34).

In expressing the solution of the jump problem (partial solution of the inhomogeneous problem) by an integral of type (52.3) we satisfy this condition automatically. If, moreover, the solution of the homogeneous problem with positive index is taken as the polynomial of $H(z)$, all the conditions are satisfied. Bearing these circumstances in mind and proceeding in the same way as before, we get the solution of the boundary value problem and then obtain the solution of the initial equation (52.33) from (52.37). Assuming, as usual, that $a^2 - b^2 = 1$ we write the equation as

$$\varphi(t) = a(t) f(t) - b(t) Z(t) \frac{1}{\pi i} \int_{L_0} \frac{f(\tau)}{Z(\tau)} \frac{H'(\tau)}{H(\tau) - H(z)} d\tau +$$
$$+ b(t) Z(t) P_{\varkappa+1}(H), \qquad (52.40)$$

where

$$\left.\begin{array}{l} Z(t) = X^+(t) [a(t) + b(t)] = X^-(t) [a(t) - b(t)], \\ X(z) = [H(z) - H(t_0)]^{-\varkappa} e^{\Gamma(z)}, \\ \Gamma(z) = \dfrac{1}{2\pi i} \int_{L_0} \ln G(\tau) \dfrac{H'(\tau)}{H(\tau) - H(z)} d\tau, \end{array}\right\} \qquad (52.41)$$

and t_0 is the initial point of integration on L_0.

If the contour L_0 separates the branch points of the functions $\Psi_k^{(j)}$ so that some are within the contour and someout side, then the $L_0^{(j)}$ become open curves. But this does not complicate the solution in any way. As seen in Chapter VI, the main feature of the solution of the boundary value problems of integral equations for a closed contour is that solutions in different classes are possible, namely, bounded and unbounded at end-points. But in the case in question the open contours $L_0^{(j)}$ are equivalent to a closed contour L_0 on which the solution is continuous and therefore bounded. Hence, owing to the equivalence, only one class of bounded solutions is possible on $L_0^{(j)}$. The solution of the equation is given by (52.40) and (52.41) in this case also. As regards the case when L_0 is an open line (or even a set of open lines) the same may be said as in § 51.7 in 1*.

§ 52.5. Example

Consider the integral equation

$$\frac{1}{2}\left[(2t-1)^{\varkappa}+(t^2-4t+2)^{\varkappa}\right]\varphi(t)+$$

$$+\frac{1}{2\pi i}\left[(2t-1)^{\varkappa}-(t^2-4t+2)^{\varkappa}\right]\int_{L_0}\frac{2F(\tau)F'(\tau)}{F^2(\tau)-F^2(t)}\varphi(\tau)\,d\tau=f(t),$$

where L_0 is a circle $|2z-7|=1$, and $F(z)=2z^2/(2z-1)$ is the fundamental automorphic function of the group $\omega_0(z)=z$, $\omega_1(z)=z/(2z-1)$.

The fundamental domains of the group are the exterior and interior of the circle $|2z-1|=1$. We put R_0 for the exterior domain where L_0 is located, and R_1 for the interior. The automorphic function of kernel $H(z)=F^2(z)$ has second-order poles at an infinitely-remote point in R_0 and at the point $z=1/2$ in R_1.

The auxiliary sectionally automorphic function

$$\Phi(z)=\frac{1}{2\pi i}\int_{L_0}\frac{H'(\tau)}{H(\tau)-H(z)}\varphi(\tau)\,d\tau$$

has a second-order zero at an infinitely remote point in R_0. In each fundamental domain the function $\Phi(z)$ has two singular lines; their equations are:

in R_0: $z=\tau$, (L_0)

$$z=\Psi_0^{(1)}(\tau)=\frac{-\tau^2-\tau\sqrt{(\tau^2+2\tau-1)}}{2\tau-1}\quad(\tau\in L_0),\qquad (L_0^{(1)})$$

in R_1: $z=\dfrac{\tau}{2\tau-1}$ $(\tau\in L_0)$ (L_1)

$$z=\Psi_1^{(1)}(\tau)=\frac{-\tau^2+\tau\sqrt{(\tau^2+2\tau-1)}}{2\tau-1}\quad(\tau\in L_0).\qquad (L_1^{(1)})$$

The transformations Ψ_0^1, Ψ_0^1 are branches of the many-valued function $[\Psi(2z-2)+z^2]^2-z^2(z^2+2z-1)=0$ with the branch points $z=-1+\sqrt{2}$. These are located outside L_0; all the contours are therefore closed.

The boundary value problem on the contour L_0 is

$$\Phi^+(t)=\left(\frac{t^2-4t+2}{2t-1}\right)^{\varkappa}\Phi^-(t)+\frac{f(t)}{(2t-1)^{\varkappa}},\quad t\in L_0.$$

On the other three contours they may be obtained by a substitution of variables.

Since the number of lines of discontinuity of $\Phi(z)$ is finite, the canonical function can be obtained as the product of the canonical functions of each of the simple contours. For L_0 the canonical function is found by the method of § 14.6. We have:

$$X^+(z)=\frac{p_-(z)}{q_-(z)}=\left[\frac{z-(2-\sqrt{2})}{2z-1}\right]^{\varkappa},$$

$$X^-(z)=\frac{q_+(z)}{p_+(z)}=\frac{1}{z-(2+\sqrt{2})}.$$

For the other contours the canonical functions can be found by a substitution of the variables from consideration of the properties of automorphicity. (But it is simpler to use the method of § 14.6.) The contours L_0, $L_0^{(1)}$, L_1, $L_1^{(1)}$ divide the plane into the five domains D_0^+, $D_0^{(1)+}$, D_1^+, $D_1^{(1)+}$, $D^- = D_0^- \cdot D_0^{(1)} \cdot D_1^- \cdot D_1^{(1)-}$. For the whole set of domains the canonical function is

$$X(z) = \begin{cases} \left[\dfrac{1-2z}{z^2+4z-2} \right]^{\varkappa}, & z \in D_0^+,\ D_1^+, \\[3mm] \left[\dfrac{2z-1}{z^2-4z+2} \right]^{\varkappa}, & z \in D_0^{(1)+},\ D_1^{(1)+}, \\[3mm] \left[\dfrac{2z-1}{(z^2+4z-2)(z^2-4z+2)} \right]^{\varkappa}, & z \in D^-. \end{cases}$$

If $\varkappa > 0$ the canonical function has zeros of order $2\varkappa$ at the points $z = 1/2$, ∞. In the fundamental domain R_0 the function $\Phi(z)/X(z)$ has a pole of order $2\varkappa - 2$ at $z = \infty$ if $\varkappa > 0$. Accordingly, the solution of the boundary value problem is

$$\Phi(z) = \frac{X(z)}{2\pi i} \int\limits_{L_0} \frac{f(\tau)}{X^+(\tau)(2\tau - 1)^2} \frac{2F(\tau)\,F'(\tau)}{F^2(\tau) - F^2(t)}\, d\tau + X(z)\, P_{\varkappa-1}[F^2(z)].$$

The solution of the initial equation is

$$\varphi(t) = \left[1 + \left(\frac{2t-1}{t^2-4t+2} \right)^{\varkappa} \right] \frac{f(t)}{2(2t-1)^{\varkappa}} +$$

$$+ \left(\frac{1-2t}{t^2+4t-2} \right)^{\varkappa} \left[1 - \left(\frac{2t-1}{t^2-4t-2} \right) \right] \times$$

$$\times \frac{1}{2\pi i} \int \frac{f(\tau)}{X^+(\tau)(2\tau-1)^{\varkappa}} \frac{2F(\tau)\,F'(\tau)\,d\tau}{F^2(\tau) - F^2(t)} +$$

$$+ \left(\frac{1-2t}{t^2+4t-2} \right)^{\varkappa} \left[1 - \left(\frac{2t-1}{t^2-4t+2} \right)^{\varkappa} \right] P_{\varkappa-1}[F^2(t)].$$

If $\varkappa < 0$ it is necessary to put $P_{\varkappa-1} \equiv 0$. A solution exists if the following conditions are satisfied

$$\int \frac{f(\tau)}{X^+(\tau)(2\tau-1)^{\varkappa}} [F(\tau)]^{2k-1}\, F'(\tau)\, d\tau = 0 \quad (k = 1, 2, \ldots, -\varkappa).$$

§ 53. Some types of integral equations with power and logarithmic kernels

The feature of the kernels of all the integral equations which we have so far considered has been the presence of a term which becomes a first-order infinity when the arguments coincide. In this section we investigate equations with kernels which become an infinity of the order of a logarithm or power less than unity when the arguments coincide. In the nature of their infinity they are like kernels of the

Fredholm type, but owing to the additional feature (the changed form of the kernel on crossing the principal diagonal of the square in which they are defined) they make the equations singular.

Their method of solution (reduction to the Riemann boundary value problem by analytic continuation in the complex plane) also relates them to singular integral equations with a Cauchy kernel.

First we consider equations with a power kernel, which are a generalization of the classical Abel equation, historically the first of the integral equations to become known to mathematicians.

53.1. Abel's integral equation

A solution of the classical Abel equation is

$$A\varphi \equiv \int_\alpha^x \frac{\varphi(t)}{(x-t)^\mu}\, dt = g(x) \quad (0 < \mu < 1). \tag{53.1}$$

Here use is made of the integral-calculus formula

$$\int_\tau^x (x-t)^{\mu-1}(t-\tau)^{-\mu}\, dt = \frac{\pi}{\sin\mu\pi}; \tag{53.2}$$

to find it we carry out the substitution $t = \tau + s(x - \tau)$. As a result the integral becomes

$$\int_0^1 s^{-\mu}(1-s)^{\mu-1}\, ds = \frac{\Gamma(\mu)\,\Gamma(1-\mu)}{\Gamma(1)} = \frac{\pi}{\sin\mu\pi}. \tag{53.3}$$

Substituting in (53.1) τ for t and t for x, we multiply by $(x-t)^{\mu-1}\, dt$, integrate from α to x and make use of Dirichlet's formula for reversal of the order of integration:

$$\int_\alpha^x dt \int_\alpha^t F(x,t,\tau)\, d\tau = \int_\alpha^x d\tau \int_\tau^x F(x,t,\tau)\, dt.$$

We have

$$\int_\alpha^x \frac{g(t)\, dt}{(x-t)^{1-\mu}} = \int_\alpha^x \varphi(\tau)\, d\tau \int_\tau^x (x-t)^{\mu-1}(t-\tau)^{-\mu}\, d\tau$$

$$= \frac{\pi}{\sin\mu\pi} \int_\alpha^x \varphi(\tau)\, d\tau; \tag{53.4}$$

18*

hence

$$\varphi(x) = \frac{\sin \mu \pi}{\pi} \frac{d}{dx} \int\limits_{\alpha}^{x} \frac{g(t)\, dt}{(x - t)^{1-\mu}}. \tag{53.5}$$

If $g(x)$ is a differentiable function, after integrating by parts and then differentiating, the solution becomes

$$\varphi(x) = \frac{\sin \mu \pi}{\pi} \left[\frac{g(\alpha)}{(x - \alpha)^{1-\mu}} + \int\limits_{\alpha}^{x} \frac{g'(t)\, dt}{(x - t)^{1-\mu}} \right]. \tag{53.6}$$

The equivalence of (53.5) to the initial equation (53.1) after transformation is implied since the equation

$$B\psi \equiv \int\limits_{\alpha}^{x} \frac{\psi(t)\, dt}{(x - t)^{1-\mu}} = 0,$$

corresponding to the transforming operator, has no other solutions than a zero solution. This can be found by applying a transformation similar to that by which (53.4) was obtained from (53.1).

Suppose that (53.1) is defined on some section $\alpha \leqq x \leqq \beta$. Assuming that $g'(x)$ is continuous over the entire section including the endpoints, we find from (53.6) that if $g(\alpha) \neq 0$ the solution of the equation for $x = \alpha$ becomes an infinity of order $1 - \mu$. The foregoing reasoning points to the uniqueness of an equation in this class. Hereafter the solution is required in the class of functions of type

$$\varphi(x) = \frac{\varphi^*(x)}{(x - \alpha)^{1-\mu-\varepsilon}} \quad (\varepsilon > 0), \tag{53.7}$$

where $\varphi^*(x)$ is a function which satisfies Hölder's condition throughout the interval $[\alpha, \beta]$. We consider in which class the right-hand side $g(x)$ should now be taken.

We investigate the behaviour at point α. Substituting into equation (53.1) the expression

$$\varphi(t) = \frac{\varphi^*(t)}{(t - \alpha)^{1-\mu-\varepsilon}}$$

and also substituting $t = \alpha + s(x - \alpha)$, we obtain:

$$g(x) = (x - \alpha)^{\varepsilon} \int\limits_{0}^{1} \frac{\varphi^*[\alpha + s(x - \alpha)]}{s^{1-\mu-\varepsilon}(1 - s)^{\mu}}\, ds.$$

The integral is a bounded function. Hence $g(x)$ must have a zero of order ε at the initial point; similarly at β. The right-hand side therefore is

$$g(x) = (x - \alpha)^\varepsilon \, g^*(x) \quad (\varepsilon > 0). \tag{53.8}$$

With regard to $g^*(x)$ we assume that it has a derivative which satisfies Hölder's condition (the sufficient condition). Formula (53.6) shows that the solution $\varphi(x)$ for $g(x)$ in this class belongs to the class (53.7). For the same reasons as above the solution is also unique in this class.

53.2. Integral with a power kernel

We consider an analytic function represented by the integral

$$\Phi(z) = [(z - a)(\beta - z)]^{\frac{1}{2}\mu - \frac{1}{2}} \int_\alpha^\beta \frac{\varphi(t)\,dt}{(t - z)^\mu}. \tag{53.9}$$

Here the many-valued function $[(z - \alpha)(\beta - z)]^{\frac{1}{2}\mu - \frac{1}{2}}(t - z)^{-\mu}$ is defined in a plane cut along the section $\alpha \leq x \leq \beta$ by some branch. The density of the integral $\varphi(t)$ is taken from the class of functions of type

$$\varphi(x) = \frac{\varphi^*(x)}{[(x - \alpha)(\beta - x)]^{1 - \mu - \varepsilon}}. \tag{53.7'}$$

The integral (53.9) plays the same role in the solution of the integral equation in the next subsection as the Cauchy integral did in the solution of the dominant equation with a Cauchy kernel in § 21.1.

Clearly $\Phi(z)$ is an analytic function throughout the complex plane cut along the section $[\alpha, \beta]$. At any points of the cut distinct from the end-points the function $\Phi(z)$ satisfies Hölder's condition if $\varphi(t)$ belongs to the class (53.7'). In the vicinity of the point at infinity and the end-points α, β, we obtain from (53.9)

$$\Phi(z) = O\left(\frac{1}{z}\right), \tag{53.10a}$$

$$\Phi(z) = O\left[(z - \alpha)^{-\frac{1-\mu}{2}}\right], \tag{53.10b}$$

$$\Phi(z) = O\left[(\beta - z)^{-\frac{1-\mu}{2}}\right], \tag{53.10c}$$

where the symbol $O(x^\nu)$ denotes a quantity of order x^ν as $x \to 0$. For simplicity we put:

$$[(z - \alpha)(\beta - z)]^{\frac{1}{2}(1-\mu)} = R(z). \tag{53.11}$$

Then

$$\Phi(z) = \frac{1}{R(z)} \int\limits_{\alpha}^{\beta} \frac{\varphi(t)\,dt}{(t-z)^{\mu}}. \tag{5.9}$$

We put $\Phi^{\pm}(x)$ for the limit values of $\Phi(z)$ on the approaches to the cut from above and below respectively and write $\Phi(z)$ as

$$\Phi(z) = \frac{1}{R(z)} \left[\int\limits_{\alpha}^{x} \frac{\varphi(t)\,dt}{(t-z)^{\mu}} + \int\limits_{x}^{\beta} \frac{\varphi(t)\,dt}{(t-z)^{\mu}} \right]. \tag{53.12}$$

We express the limit values $\Phi^{\pm}(x)$ in terms of the values of the integrals on the right-hand side.

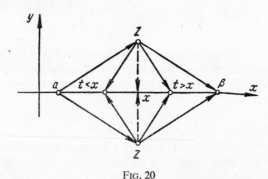

Fig. 20

On the upper approach to the cut we ascribe the argument 0 to numbers $z - \alpha$ and $z - \beta$ if $z \to x \in [\alpha, \beta]$, and the argument 0 to the number $t - z$ if $z \to x < t < \beta$, or the argument $-\pi$ if $z \to x > t > \alpha$ (see Fig. 20). By definition we have

$$R^{+}(x) = R(x) = [(x - \alpha)(\beta - x)]^{\frac{1}{2}(1-\mu)}.$$

Then from (53.12)

$$\Phi^{+}(x) = \frac{1}{R(x)} \left[e^{\mu\pi i} \int\limits_{\alpha}^{x} \frac{\varphi(t)\,dt}{(x-t)^{\mu}} + \int\limits_{x}^{\beta} \frac{\varphi(t)\,dt}{(t-x)^{\mu}} \right]. \tag{53.13}$$

If z tends to x on the lower approach, the arguments of the numbers $\beta - z$ and $t - z$ with $\beta > t > x$ tend to the same values as on the upper approach, whilst the arguments of $z - \alpha$ and $t - z$ tend to

2π and π respectively if $\alpha < t < x$, i.e. they increase by 2π:

$$R^-(x) = e^{2\pi i \frac{1}{2}(1-\mu)} R(x) = -e^{-\mu\pi i} R(x).$$

Hence

$$\Phi^-(x) = -\frac{1}{R(x)}\left[\int\limits_a^x \frac{\varphi(t)\,dt}{(x-t)^\mu} + e^{\mu\pi i}\int\limits_x^\beta \frac{\varphi(t)\,dt}{(t-x)^\mu}\right]. \qquad (53.13')$$

Solving relations (53.13) and (53.13') in the integrals, we obtain another pair of formulae:

$$\int\limits_\alpha^x \frac{\varphi(t)\,dt}{(x-t)^\mu} = \frac{e^{\mu\pi i}\Phi^+(x) + \Phi^-(x)}{e^{2\mu\pi i} - 1} R(x), \qquad (53.14)$$

$$\int\limits_x^\beta \frac{\varphi(t)\,dt}{(t-x)^\mu} = -\frac{\Phi^+(x) + e^{\mu\pi i}\Phi^-(x)}{e^{2\mu\pi i} - 1} R(x). \qquad (53.14')$$

The pairs of formulae (53.13), (53.14) for a function represented by integral (53.9) are counterparts of Sokhotski's formulae (4.8), (4.9) and (4.10) for a Cauchy integral.

53.3. The generalized Abel integral equation

Consider the integral equation

$$a(x)\int\limits_\alpha^x \frac{\varphi(t)\,dt}{(x-t)^\mu} + b(x)\int\limits_x^\beta \frac{\varphi(t)\,dt}{(t-x)^\mu} = f(x) \quad (0 < \mu < 1). \quad (53.15)$$

It is assumed that the functions $a(x)$, $b(x)$ defined on $[\alpha, \beta]$ do not vanish simultaneously and that they satisfy Hölder's condition. We take the right-hand side in the form

$$f(x) = [(x-a)(b-x)]^\varepsilon f^*(x) \quad (\varepsilon > 0), \qquad (53.16)$$

where $f^*(x)$ has a derivative in class H on $[\alpha, \beta]$. The solution is required in the class of functions

$$\varphi(x) = \frac{\varphi^*(x)}{[(x-a)(b-x)]^{1-\mu\varepsilon}} \quad (\varepsilon > 0), \qquad (53.17)$$

where $\varphi^*(x)$ satisfies Hölder's condition on $[\alpha, \beta]$.

For the solution we make use of the notion of analytic continuation in the complex plane (see § 21.1). We reintroduce the

18a*

analytic function represented in § 53.2 by an integral with a power kernel for which the required solution of the integrals acts as the density:

$$\Phi(z) = \frac{1}{R(z)} \int_a^\beta \frac{\varphi(t)\,dt}{(t-z)^\mu}.$$ (53.9)

Substituting into equation (53.15) the values of the integrals from formulae (53.14), we find a relation which holds on $[\alpha, \beta]$:

$$\Phi^+(x) = \frac{e^{\mu\pi i}\,b(x) - a(x)}{e^{\mu\pi i}\,a(x) - b(x)}\,\Phi^-(x) + \frac{(e^{2\mu\pi i} - 1)f(x)}{R(x)\,[e^{\mu\pi i}\,a(x) - b(x)]}.$$ (53.18)

This is the boundary condition of the Riemann problem with the contour $[\alpha, \beta]$. Owing to (53.9) its solution must satisfy conditions (53.10). By solving the boundary value problem we are able to obtain the solution of the initial integral equation (53.15) from formulae (53.14) by solving the ordinary Abel integral equation.

The solution of integral equation (53.15) in class (53.17) thus merely entails determining the solution of the Riemann boundary value problem (53.18) which satisfies conditions (53.10) and then solving Abel's equation (53.14) in the class of functions (53.17).

The proof of equivalence of these two problems is based on the following lemma (see § 21.1).

LEMMA. *If the limit values of a function $\Phi(z)$ which is analytic in a plane cut along $[\alpha, \beta]$ and which satisfies conditions* (53.10), *are associated with $\varphi(x)$ by relation* (53.14) *(or* (53.14')), *then the representation* (53.9) *holds for $\Phi(z)$.*

Proof. It is obvious that functions $\Phi(z)$ and $\Phi(x)$, associated by representation (53.9), satisfy relation (53.14). Suppose that some function $\Phi_1(z)$ exists other than (53.9) which also satisfies (53.14). Substituting into the latter Φ and Φ_1, we find by subtraction that their difference $\Phi_2(z)$ must satisfy the homogeneous boundary condition of the Riemann problem

$$\Phi_2^+(x) = -e^{-\mu\pi i}\Phi_2^-(\psi).$$ (53.19)

We find its solution in the class of functions with a permissible infinity of integrable order at both end-points † (see § 43.1) satisfying the condition (53.10a).

---

† In the other possible classes there is only the zero solution.

Here

$$G(t) = e^{(1-\mu)\pi i}, \quad \varkappa = 1, \quad P_{\varkappa-1}(z) = C = \text{const},$$

$$\Gamma(z) = \frac{1}{2\pi i} \int_a^\beta \frac{\ln G(t)}{t - z}\, dt = \frac{(1 - \mu)\pi i}{2\pi i} \ln\left(\frac{z - \beta}{z - \alpha}\right).$$

Hence from formula (43.5)

$$X(z) = (z - \beta)^{-1}\, e^{\Gamma(z)} = (z - \beta)^{-\frac{1}{2}(1+\mu)}\, (z - \alpha)^{-\frac{1}{2}(1-\mu)}.$$

The general solution of the problem $P_{\varkappa-1}(z)$, $X(z)$ is therefore written as

$$\Phi_2(z) = C(z = \alpha)^{-\frac{1}{2}(1-\mu)}\, (\beta - z)^{-\frac{1}{2}(1+\mu)}. \tag{53.20}$$

The condition (53.10c) can only be satisfied by putting $C = 0$. Consequently, $\Phi_2(z) = 0$, i.e. $\Phi_1(z) \equiv \Phi(z)$, as required.

We shall now establish the equivalence. If $\varphi(x)$ is a solution of equation (53.15) which belongs to class (53.17), then a function $\Phi(z)$ with the representation (53.9) must satisfy conditions (53.10) and, according to the lemma, it must be the solution of the boundary value problem (53.18).

Conversely, let $\Phi(z)$ be the solution of boundary value problem (53.18) satisfying conditions (53.10). The function $\varphi(t)$ belonging to class (53.17) is found from Abel' sequation (53.14) or (53.14'). From § 53.1 the latter is unconditionally and single-valuedly soluble. According to the lemma, $\varphi(t)$ and $\Phi(t)$ are associated by (53.9); therefore (53.14') also holds. We multiply the latter by $a(x)$ and $b(x)$ respectively and then add them. Having regard to the boundary condition (53.18), we arrive at equation (53.15) and the equivalence is established.

The homogeneous boundary value problem (53.18) corresponds to the homogeneous equation (53.15) ($f \equiv 0$). We will establish the relationship between the classes of their solutions.

Confining ourselves to the initial point α, we set

$$\frac{e^{\mu\pi i}\, b(x) - a(x)}{e^{\mu\pi i}\, a(x) - b(x)} = G(x).$$

Let $G(\alpha) = \varrho e^{i\theta}$. According to formulae (42.5) and (42.6), we obtain the solution in the class of bounded functions in α by selecting $\theta = \arg G(\alpha)$ within $0 < \theta < 2\pi$, whilst by taking it within $-2\pi < \theta \leqq 0$, we have unbounded solutions. If the least positive θ is seen to satisfy the condition $\theta \leqq (1 - \mu)\pi$, both possible classes of solutions of the boundary value problem are suitable for solving the equation.

If, however, $\theta > (1 - \mu)\pi$, the condition (53.10 a) makes it necessary to discard the unbounded solution and leave only the bounded solution.

The right-hand side of Abel's equation (53.14) belongs, because of the multiplier $R(x)$, to the class (53.8)† in every case, so that the solution of equation (53.15) is in the class (53.7′). Here unbounded solutions of the initial equation (53.15) always correspond, according to formula (53.6), to the unbounded solutions of the boundary value problem. To the bounded solutions of the boundary problem, either bounded or unbounded solutions of the equation may correspond, depending on the quantity $\theta = \arg G(\alpha)$.

Without dwelling on the details we now indicate how the classes of solutions of the inhomogeneous equation are related to the corresponding boundary value problem. The solution of the boundary value problem is expressed in terms of Cauchy integrals, whilst according to the results of § 8.4, they convert functions (the density of the integral) having an infinity of power order, into functions of the same class. The denominator of the free term of the boundary condition (53.18) contains the factor $(x - \alpha)^{\frac{1}{2}(1 - \mu)}$. It is therefore to be inferred that in order to obtain the $\Phi^{\pm}(x)$ which have an infinity of order $\frac{1}{2}(1 - \mu) - \varepsilon$ or less at point α (this ensures a right-hand side in class (53.8) for equation (53.14)), it is necessary to take $f(x)$ with zero of order ε if $x = \alpha$. For this it is sufficient to take $f(x)$ from class (53.16).

Note one special case. If $a(x) = b(x) = 1$, equation (53.15) becomes

$$\int\limits_{\alpha}^{\beta} \frac{\varphi(t)\,dt}{|x - t|^{\mu}} = f(x). \tag{53.21}$$

This is Abel's integral equation with constant limits. The corresponding boundary value problem (53.18) is the jump problem:

$$\Phi^{+}(x) - \Phi^{-}(x) = \frac{e^{\mu \pi i} + 1}{R(x)} f(x). \tag{53.22}$$

† Since $a(x)$, $b(x)$ have a derivative which satisfies Holder's condition, it is to be inferred that $g^{*}(x)$ belongs to the same class.

A unique solution is determined by the Cauchy integral:

$$\Phi(z) = \frac{e^{\mu\pi i} + 1}{2\pi i} \int\limits_\alpha^\beta \frac{f(t)}{R(t)} \frac{dt}{t - z}.$$

The conditions (53.11) are fulfilled by virtue of the properties of a Cauchy integral:

It remains to solve the ordinary Abel equation

$$\int\limits_\alpha^x \frac{\varphi(t)}{(x - t)^\alpha} dt = \frac{1}{2} f(x) - \frac{\cot\mu\pi}{2\pi} R(x) \int\limits_\alpha^\beta \frac{f(t)}{R(t)} \frac{dt}{t - x}. \quad (53.23)$$

For $f(x)$ from class (53.16), formula (53.6) gives a solution of the initial equation (53.21) which belongs to class (53.17).

53.4. Integral with a logarithmic kernel

We will consider the analytic function represented by the integral:

$$\Phi(z) = \psi(z) \int\limits_0^\infty [P(t - z) \ln(t - z) + Q(t - z)] \varphi(t) \, dt, \quad (53.24)$$

where $P(w)$, $Q(w)$ are integral analytic functions of their argument, $\varphi(t)$ is a function which satisfies Hölder's condition, and $\psi(z)$ is an analytic function which will be defined later. These functions are such that the integral is convergent for all finite z, and $\psi(z)$ is an analytic function selected so that $\Phi(z)$ tends to zero as the argument tends to infinity. The contour of integration is a singular line for $\Phi(z)$; it is analytic everywhere outside it. The function $\Phi(z)$ is therefore analytic in a plane cut along the real positive semi-axis. We find its limit values on the approaches of the cut.

On the upper approach we ascribe the argument $-\pi$ to the number $t - x$ if $t < x < \infty$, or the argument 0 if $t > x > 0$ (see Fig. 20). If z tends to x on the lower approach, the argument is increased by 2π if $t < x < \infty$, but it remains the same if $t > x > 0$.

Consequently,

$$\left.\begin{aligned} \ln(t - z) &\to \ln|x - t|, \quad z \to x \pm i0, \quad t > x > 0, \\ \ln(t - z) &\to \ln|x - t) \mp \pi i, \quad z \to x \pm i0, \quad t < x < \infty. \end{aligned}\right\} \quad (53.25)$$

P and Q have identical limit values $P(x - t)$, $Q(x - t)$ on both approaches, being continuous functions throughout the finite plane.

We substitute $\Phi^+(x)$, $\Phi^-(x)$ for the limit values of $\Phi(x)$ on the upper and lower approaches, respectively. Sub-dividing the integral with respect to the sequence of bounds $\int\limits_0^x + \int\limits_x^\infty$ and passing to the limit in each as $z \to x \pm i0$, we obtain the following formulae by the use of (53.25):

$$
\left.
\begin{aligned}
\frac{1}{\psi^+(x)} \Phi^+(x) &= \int\limits_0^\infty [P(t-x)\ln|x-t| + \\
&\quad + Q(t-x)]\varphi(t)\,dt - \pi i \int\limits_0^x P(t-x)\,\varphi(t)\,dt, \\
\frac{1}{\psi^-(x)} \Phi^-(x) &= \int\limits_0^\infty [P(t-x)\ln|x-t| + \\
&\quad + Q(t-x)]\varphi(t)\,dt + \pi i \int\limits_0^x P(t-x)\varphi(t)\,dt.
\end{aligned}
\right\} \quad (53.26)
$$

After inversion these yield another pair of formulae:

$$
\left.
\begin{aligned}
-2\pi i \int\limits_0^x P(t-x)\varphi(t)\,dt &= \frac{\Phi^+(x)}{\psi^+(x)} - \frac{\Phi^-(x)}{\psi^-(x)}, \\
2\int\limits_0^\infty [P(t-x)\ln|x-t| + Q(t-x)]\,\varphi(t)\,dt & \\
&= \frac{\Phi^+(x)}{\psi^+(x)} + \frac{\Phi^-(x)}{\psi^-(x)}.
\end{aligned}
\right\} \quad (53.27)
$$

The two pairs of formulae (53.26) and (53.27) play the same role for the integral (53.24) as Sokhotski's formulae (4.8)–(4.10), or formula (53.13) and (53.14), play for an integral with a power kernel.

If $\psi(z)$ is chosen appropriately, the limit values $\Phi^\pm$ are everywhere finite, including the end-points 0, ∞, and they satisfy Hölder's condition. The choice is made from consideration of the behaviour of the integral functions P and Q at infinity. In each particular case the question must be investigated independently with a study of the asymptotic properties of the kernel as $z \to \infty$.

Consider now the case of a finite integration interval $[\alpha, \beta]$. For simplicity we put $X(w) \equiv 1$. For single-valuedness of the function in the neighbourhood of the infinitely remote point, the form of the logarithmic kernel is slightly changed (see § 37.1).

We investigate the limit values of

$$\Phi(z) = \int\limits_{\alpha}^{\beta} \left[\ln\left(1 - \frac{\alpha - t}{\alpha - z}\right) + Q(t - z) \right] \varphi(t) \, dt \qquad (53.28)$$

on the approaches of the cut $[\alpha, \beta]$.

Putting

$$A = \int\limits_{\alpha}^{\beta} \varphi(t) \, dt, \qquad (53.29)$$

the relation (53.28) becomes

$$\Phi(z) = \int\limits_{\alpha}^{\beta} [\ln(t - z) + Q(t - z)] \varphi(t) - A \ln(\alpha - z).$$

For the same reasons as for an infinite integration interval, two pairs of formulae are again obtained

$$
\left.
\begin{aligned}
\Phi^+(x) = \int\limits_{\alpha}^{\beta} [\ln|x - t| + Q(t - x)] \varphi(t) \, dt - \\
- \pi i \int\limits_{0}^{x} \varphi(t) \, dt - A \ln(x - \alpha) + \pi i A, \\
\Phi^-(x) = \int\limits_{\alpha}^{\beta} [\ln|x - t| + Q(t - x)\varphi(t) \, dt + \\
+ \pi i \int\limits_{0}^{x} \varphi(t) \, dt - A \ln(x - \alpha) - \pi i A,
\end{aligned}
\right\} \qquad (53.30)
$$

$$
\left.
\begin{aligned}
- 2\pi i \int\limits_{\alpha}^{x} \varphi(t) \, dt = \Phi^+(x) - \Phi^-(x) - 2\pi i A, \\
2 \int\limits_{\alpha}^{\beta} [\ln|t - x| + Q(t - x)] \varphi(t) \, dt \\
= \Phi^+(x) + \Phi^-(x) + 2A \ln(x - \alpha).
\end{aligned}
\right\} \qquad (53.31)
$$

With $\Phi(z)$ represented by (53.28) they are counterparts of Sokhotski's formulae.

BVP 18b

53.5. Integral equations with logarithmic kernels

1. Consider the integral equation

$$b(x) \int\limits_0^\infty [P(t - x) \ln |t - x| + Q(t - x)] \varphi(t) \, dt -$$

$$- \pi i a(x) \int\limits_0^x P(t - x)\varphi(t) \, dt = \tfrac{1}{2}f(\psi), \qquad (53.32)$$

where $P(w)$, $Q(w)$ are defined integral functions, whilst $a(x)$, $b(x)$, $f(x)$ are functions, defined on the real semi-axis, which satisfy Hölder's condition.

We again introduce from § 53.4 a function analytic in a plane cut along the real positive semi-axis

$$\Phi(z) = \psi(z) \int\limits_0^\infty [P(t - z) \ln(t - z) + Q(t - z)] \varphi(t) \, dt. \qquad (53.24)$$

Inserting into equation (53.32) the values of the integrals from formulae (53.27), we get the boundary condition of the Riemann problem

$$\Phi^+(x) = \frac{a(x) - b(x)}{a(x) + b(x)} \frac{\psi^+(x)}{\psi^-(x)} \Phi^-(x) + \frac{f(x)\psi^+(x)}{a(x) + b(x)}. \qquad (53.33)$$

Assuming that the solution of the boundary value problem is found, the first formula in (53.27) yields the following integral equation for the required function $\varphi(t)$

$$2\pi i \int\limits_0^x P(t - x)\varphi(t) \, dt = \frac{\Phi^-(x)}{\psi^-(x)} - \frac{\Phi^+(x)}{\psi^+(x)}. \qquad (53.34)$$

This Volterra equation with a difference kernel belongs to the class of equations of the convolution kind. It may be solved in closed form by use of the single-sided Laplace transform.

The boundary value problem has to be solved in the class of functions that are bounded at the origin and which vanish at infinity. The most complicated problem in the solution is that of finding the type of the function $\psi(z)$. This cannot be done from general considerations and it is by no means practicable to do so for all functions P and Q. This limits the applicability of the method to solving integral equations of type (53.32).

The equivalence of the boundary value problem (53.33) to the initial integral equation (53.32) is established by similar reasoning to that in § 53.3. It consists in proving that if $\Phi(z)$ is obtained from the solution of the boundary value problem (53.33), whilst $\varphi(z)$ is found from (53.34), then they are related by representation (53.24). For the difference $\Phi_2(z)$ between two admissible solutions of (53.34) which vanish at infinity, the boundary condition is $\Phi_2^+(x)/\psi^+(x)$ $= \Phi_2^-(x)/\psi^-(x)$. If $\Phi(z)/\psi(z) \to 0$ as z tends to infinity in any direction, it follows that $\Phi(z) \equiv 0$, as required.

We illustrate this by an example.

EXAMPLE. Solve the integral equation

$$\int_0^\infty H_0^{(1)}[k\,|x - t|]\,\varphi(t)\,dt = e^{i\alpha x} \quad (x > 0), \tag{53.35}$$

where k and α are constants. The kernel here is the first Hankel function of zero kind, which, as is known†, has the following analytic representation:

$$H_0^{(1)}(w) = \frac{2i}{\pi}\,J_0(w)\ln w + Q(w),$$

where J_0 is a Bessel function of zero kind, and Q is an everywhere-convergent known power series. Here $P \equiv J_0$, $a = 0$ and $b = 1$.

The following asymptotic representation holds for Hankel functions.‡

$$H_0^{(1)}(w) \sim Aw^{-\frac{1}{2}}e^{iw} + O\left(\frac{1}{w}\right), \tag{53.36}$$

where $O(1/w)$ is a quantity which tends to zero of order $1/w$ as $w \to \infty$, and A is a definite constant.

Bearing these considerations in mind, it can be established without difficulty that the function $\psi(z)$ may be taken in the form $\psi(z) = z^{-1/2}\,e^{ikz}$. In fact, introducing the auxiliary analytic function by the formula

$$\Phi(z) = z^{-\frac{1}{2}}e^{ikz}\int_0^\infty \left[\frac{2i}{\pi}\,J_0\,k(t - z)\ln(t - z) + Q(t - z)\right]\varphi(t)\,dt, \tag{53.37}$$

from the asymptotic representation (53.36), we have:

$$\Phi(z) \sim z^{-\frac{1}{2}}e^{ikz}\int_0^\infty A\,\frac{e^{-ik(t-z)}\varphi(t)}{\sqrt{[k(t - z)]}}\,dt = Az^{-1}\int_0^\infty \frac{e^{it}\varphi(t)}{\sqrt{\left[k\left(\frac{t}{z} - 1\right)\right]}}\,dt = O\left(\frac{1}{z}\right)$$

(provided that the integral $\int_0^\infty e^{it}\,\varphi(t)\,dt$ converges).

† See, for example, R. O. Kuz'min, *Bessel functions* (Besselevy funktsii), ONTI, p. 44, 1935.

‡ *Ibid.*, p. 74.

18b*

If the limit value $\sqrt{z}$ on the upper approach of the cut is assumed to be $\sqrt{x}$, then on the lower approach it is $-\sqrt{x}$. Therefore $\psi^+(x) = -\psi^-(x) = x^{-1/2}e^{ikx}$ and hence, having regard to equation (53.25), the formulae (53.27) may be written as

$$(\sqrt{x})\, e^{-ikx}[\Phi^+(x) - \Phi^-(x)] = 2 \int_0^\infty H_0^{(1)}[k\,|x-t|]\,\varphi(t)\,dt = 2e^{i\alpha x},$$

$$4 \int_0^x J_0[k(t-x)]\,\varphi(t)\,dt = (\sqrt{x})\, e^{-ikx}[\Phi^+(x) + \Phi^-(x)].$$

It follows, from the first equation, that

$$\Phi(z) = \frac{1}{\pi i} \int_0^\infty \frac{e^{i(k+\alpha)t}}{\sqrt{t}}\,\frac{dt}{t-z}.$$

From the second equation, having regard to Sokhotski's formula, we have

$$\int_0^x J_0[k(x-t)]\,\varphi(t)\,dt = \frac{(\sqrt{x})\, e^{-ikx}}{2\pi i} \int_0^\infty \frac{e^{i(k+\alpha)t}}{\sqrt{t}}\,\frac{dt}{t-x}. \tag{53.38}$$

(Owing to the evenness of $J_0(w)$, a change in sign of the argument of the kernel does not alter the equation.)

The integral on the right-hand side can be evaluated as follows. We set $k + \alpha = \beta$, substitute variables $t = u^2$, differentiate the integral with respect to the parameter β and then integrate the resulting first-order linear differential equation with respect to β. As a result we have

$$\int_0^\infty \frac{e^{i\beta t}}{\sqrt{t}}\,\frac{dt}{t-x} = (i-1)\left(\sqrt{\frac{\pi}{2}}\right) e^{i\beta x} \int_0^\beta \frac{e^{-ix\tau}}{\sqrt{\tau}}\,d\tau.$$

Equation (53.38) may therefore be written as

$$\int_0^x J_0[k(x-t)]\,\varphi(t)\,dt = \frac{(1+i)}{2\sqrt{(2\pi)}}\,(\sqrt{x})\, e^{i\alpha x} \int_0^\beta \frac{e^{-ixt}}{\sqrt{\tau}}\,d\tau. \tag{53.39}$$

This latter equation belongs to equations of the convolution kind. It is solved by the use of Fourier or Laplace integral transforms. Omitting the process of solution, the final result is

$$\varphi(x) = \frac{\sqrt{(\alpha+k)}}{2i\sqrt{(2\pi)}} \int_{-i-\gamma\infty}^{i+\gamma\infty} \frac{\sqrt{(k+w)}}{\alpha+w}\,e^{-ixw}\,dw, \tag{53.40}$$

where γ is a positive number.

Note that the initial equation (53.35) is also an equation of the convolution kind; it may be solved most simply by a direct application of the Fourier integral transform. The solution is only given here to illustrate the method.

2. We briefly consider the equation with finite limits:

$$b(x) \int_\alpha^\beta [\ln |t - x| + Q(t - x)] \varphi(t) \, dt - \pi i a(x) \int_\alpha^x \varphi(t) \, dt = \tfrac{1}{2} f(\psi).$$
(53.41)

Introducing the analytic function (53.28) and using formulae (53.31), we arive at, the boundary value problem

$$\Phi^+(x) = \frac{a(x) - b(x)}{a(x) + b(x)} \Phi^-(x) + \frac{f(\psi) + 2\pi i A b(x) - 2 A a(x) \ln (x - \alpha)}{a(x) + b(x)}.$$
(53.42)

The required solution of the equation is obtained from the first formula of (53.31)

$$\varphi(x) = \frac{1}{2\pi i} [\Phi^-(x) - \Phi^+(x)]'.$$

It depends linearly on the constant A. The latter is found from the relation

$$A = \int_\alpha^\beta \varphi(t) \, dt.$$
(53.29)

The question of permissible classes will not be considered.

53.6. Various possible generalizations

We have solved in closed form singular integral equations with kernels of three types: (1) with a Cauchy kernel (§ 23.1, 51, 52), called below polar kernels; (2) with a power kernel (§ 53.3); and (3) with logarithmic kernels. In all cases use was made of the same method of analytic continuation in the complex plane by means of a function represented by an integral with appropriately selected kernels.

It is natural to enquire into the possiblity of solutions in closed form for types of equations with kernels which are combinations of the singularities of the cited three types. The singularities may be combined either as products, or as sums. No significant results have as yet been obtained in this direction, so we must confine ourselves to indicating possible lines of enquiry.

The starting point must be the consideration of formulae for the limit values of functions which analytically continue an equation defined on the contour in the complex plane. With the addition of singularities the formulae may be produced by adding the respective formulae for integrals with polar, power or logarithmic kernels (Sokhotski's formulae and their counterparts).

Let us illustrate these remarks, taking as an example the simplest combination when purely polar and logarithmic kernels are added in a kernel, without other complicating terms, and when a part of the real axis is the contour.

In this case the continuation into the complex plane is carried out by the formula

$$\Phi(z) = \frac{p}{2\pi i} \int_\alpha^\beta \frac{\varphi(t)}{t - z} dt + \frac{1}{2\pi i} \int_\alpha^\beta \varphi(t) \ln\left(1 - \frac{\alpha - t}{\alpha - z}\right) dt. \quad (53.43)$$

By adding Sokhotski's formulae (4.8)–(4.10) and their counterparts (53.30), (53.31), we obtain the following two pairs of formulae

$$\Phi(x) = \pm \frac{p}{2} \varphi(x) + \frac{p}{2\pi i} \int_\alpha^\beta \frac{\varphi(x)}{t - x} dt \pm \frac{1}{2} \int_\alpha^x \varphi(t) \, dt +$$

$$+ \frac{1}{2\pi i} \int_\alpha^\beta \ln |t - x| \, \varphi(t) \, dt -$$

$$- \frac{1}{2\pi i} [\ln(x - \alpha) \pm \pi i] \int_\alpha^\beta \varphi(t) \, dt, \quad (53.44)$$

$$\left.\begin{array}{l} \Phi^+(x) - \Phi^-(x) - A = p\varphi(x) - \displaystyle\int_\alpha^x \varphi(t) \, dt, \\[2mm] \Phi^+(x) + \Phi^-(x) + \dfrac{1}{\pi i} A \ln(x - \alpha) \\[2mm] = \dfrac{p}{\pi i} \displaystyle\int_\alpha^\beta \dfrac{\varphi(t) \, dt}{t - x} + \dfrac{1}{\pi i} \displaystyle\int_\alpha^\beta \ln |t - x| \, \varphi(t) \, dt, \end{array}\right\} \quad (53.45)$$

where
$$A = \int_\alpha^\beta \varphi(t) \, dt. \quad (53.29)$$

From consideration of the last pair of formulae, it is easily deduced that an integral equation of the type

$$a(x)\left[p\varphi(x) - \int_\alpha^x \varphi(t) \, dt \right] +$$

$$+ \frac{b(x)}{\pi i} \int_\alpha^\beta \left[\frac{p}{t - x} + \ln |t - x| \right] \varphi(t) \, dt + f(x), \quad (53.46)$$

where $a(x)$, $b(x)$, $f(x)$ are defined functions†, can be reduced to the Riemann boundary value problem

$$\Phi^+(x) = \frac{a(x) - b(x)}{a(x) + b(x)} \Phi^-(x) + \frac{f(x) + Aa(x) - \dfrac{1}{\pi i} Ab(x) \ln(x - \alpha)}{a(x) + b(x)}.$$
(53.47)

Suppose that the functions $\Phi^\pm(z)$ have been found from the solution of the boundary value problem. The solution of the initial equation can then be obtained by bringing in the first formula of (53.45). Setting therein

$$\int_\alpha^x \varphi(t)\, dt = u, \quad \frac{1}{p}[\Phi^+(x) - \Phi^-(x) - A] = g(x),$$

the first-order linear differential equation for u is

$$u' - \frac{1}{p} u = g(x);$$

by solving it

$$u = e^{\frac{x}{p}} \left[\int_\alpha^x g(x) e^{-\frac{x}{p}}\, dx + C \right]$$
(53.48)

the solution of the initial equation is found to be

$$\varphi(x) = u' = \frac{1}{p} e^{\frac{x}{p}} \left[\int_\alpha^x g(x) e^{-\frac{x}{p}}\, d\psi + C \right] + g(x).$$
(53.49)

The constants A and C are determined by solving a set of two linear equations formed as follows: one is formed by setting in (53.48) $x = \beta$ and having regard to the fact that $u(\beta) = A$ by virtue of (53.29), and the other is then obtained by placing the right-hand side of (53.49) in (53.29). Solving the equations we find that

$$C = 0, \quad A = e^{\frac{\beta}{p}} \int_\alpha^\beta g(x) e^{-\frac{x}{p}}\, dx.$$

† For a generalization of all following operations, it is sufficient to require that they satisfy Hölder's condition and the condition that $a(x) \pm b(x) \neq 0$.

Equation (53.43) may thus be solved in closed form. The qualitative side of the question is determined by the value of the index of the coefficient $(a(x) - b(x))/(a(x) + b(x))$ of the Riemann problem.

We now draw some general conclusions.

In the foregoing reasoning two circumstances were of real significance. Firstly—the possibility of expressing all the terms of the integral equation containing the required function in terms of the limit values of an auxiliary analytic function (reduction of the equation to the boundary value problem); secondly—the presence of a relationship between the required function $\varphi(x)$ and the limit values $\Phi^+(x)$ such that the solution for φ could be given in closed form (on condition that $\Phi^+(x)$ were already known). In the case then under consideration both conditions were fulfilled and this led to success.

With the other combinations of singularities (power-logarithmic, polar-power, etc.) no difficulty arises in writing out types of integral equations which are reducible to the boundary value problem, but in the other respect—finding a relationship which permits solution in a closed form for $\varphi(x)$—serious difficulties arise and we are still in no position to finalize the study.

Another possible generalization would be the introduction into the power kernel of a multiplier in the form of an integral function. The main obstacle is due to the difficulties in finding and using a function $\psi(z)$ which gives an estimate of the behaviour of the auxiliary analytic function $\Phi(z)$ at infinity. Investigations have so far been confined to the case when the asymptotic formulae give $\psi(z)$ as the product of an exponential and power function.

With other functions, e.g. trigonometric functions, it would be necessary to overcome the difficulties due to the possibility of their having an infinite set of zeros. A solution of boundary value problems in such classes of functions (apart from automorphic functions) has still to be realized.

A third conceivable generalization is to discard the difference nature of the argument of a kernel's additional terms (e.g. P and Q in the formulae of § 53.4). The main difficulty is to determine the nature of the singularities of $\Phi(z)$. In this case the latter are no longer all concentrated on the contour. Questions of this kind have still to be treated.

It should, however, be pointed out that the topics dealt with in this section have only recently attracted attention and interesting new results should occasion no surprise.

§ 54. Historical notes

The contents of § 51 follow the paper of F. D. Gakhov and L. I. Chibrikova which has already been mentioned.

From the early papers on this topic we first of all mention the paper of Gellerstedt (1) in 1936, and then the papers of S. G. Mikhlin [16], D. I. Sherman (2) (1948) and Bitsadze (1) (1953). In all these papers the method of solution of the dominant (characteristic) equation is applied to the solution of complete equations of a particular type.

The contents of § 52 follows the paper of L. I. Chibrikova (2). It should be observed that the work on the determination of types of singular integral equations soluble in closed form, has only recently begun and is still being carried out. New interesting results are to be expected.

In her later papers L. I. Chibrikova has given many new examples of the use of automorphic functions for solving various boundary value problems. Reference should be made to her papers (3), (4) in which various boundary value problems are solved for mixed (elliptic-hyperbolic) differential equations, and more particularly, to her paper (5) which contains a solution in closed form in terms of automorphic functions of the Hilbert boundary value problem for some polygons bounded by two circles.

The idea underlying the solution of equations with power and logarithmic kernels (§ 53) was put forward by Carleman (2) in 1922. This was essentially the same as the idea of analytic continuation in the complex plane which he used in solving the dominant equation with a Cauchy kernel. In the cited paper he solved Abel's equation with constant limits

$$\int\limits_0^1 \frac{\varphi(t)\, dt}{|t - x|^\mu} = f(x),$$

and also the following simple equation with a logarithmic kernel

$$\int\limits_0^1 \ln|x - t|\, \varphi(t)\, dt = f(x).$$

Carleman's auxiliary analytic functions had a branch point at infinity which caused a number of difficulties in the solution.

In 1958 Heins and MacCamy (1) showed that the method of analytic continuation in the complex plane enables the solution of an integral equation of the type

$$\int\limits_0^\infty [P(x - t)\ln|x - t| + Q(x - t)]\, \varphi(t)\, dt = f(x),$$

where P and Q are even integral functions, to be reduced to the simplest equation of the convolution type

$$\int\limits_0^x P(x - t)\, \varphi(t)\, dt = f(x).$$

The solution of Abel's generalized equation (53.35) in §§ 53.2, 53.3 was given by Sakalyuk (1). The contents of §§ 53.4–53.6 are partly taken from the present author's article (6), but some of the material has not previously been published.

A final reference should be made to two papers in which similar problems are investigated on the basis of quite different ideas.

Akhiyezer and Shchelrin (1) have obtained solutions for Abel's equation with constant limits and the equation with a logarithmic kernel as a special case from formulae for the inversion of integrals with hypergeometric kernels. M. G. Krein obtained solutions of these two equations as a special case using a method of solving integral equations which he discovered in investigations connected with the inverse Sturm–Liouville problem.

Problems on Chapter VII

1. Prove that the equation

$$\frac{1}{2}(1 + t^{\varkappa}) + \frac{1 - t^{\varkappa}}{2\pi i} \int\limits_{|\tau| = 1} \left(\frac{1}{\tau - t} + \frac{3t + 1}{3\tau t + \tau - t + 1} + \right.$$

$$\left. + \frac{3t - 1}{3\tau t - \tau + t + 1} - \frac{18\tau}{9\tau^2 - 1} \right) \varphi(\tau)\, d\tau = f(t)$$

belongs to the type of equations (51.12), and solve it.

Answer.

$$\varphi(t) = \frac{1}{2}(t^{-\varkappa} + 1) f(t) - \frac{t^{-\varkappa} - 1}{2\pi i} \int\limits_{|\tau| = 1} \left(\frac{1}{\tau - t} + \frac{3t + 1}{3\tau t + \tau - t + 1} + \right.$$

$$\left. + \frac{3t - 1}{3\tau t - \tau + t + 1} - \frac{18\tau}{9\tau^2 - 1} \right) f(\tau)\, d\tau + (t^{-\varkappa} - 1) P_{\varkappa - 1}(F),$$

$$F(z) = 9z \frac{1 - z^2}{1 - 9z^2} \quad (\varkappa \geqq 0).$$

2. By the same method solve the following equation on an open contour

$$\varphi(t) - \lambda \int\limits_0^1 \left(\frac{1}{\tau - t} + \frac{(1 + i)\, t - 1}{(1 + i)\, \tau t - \tau - t + 1} - \frac{2}{2\tau - 1 + i} \right) \varphi(\tau)\, d\tau = f(t).$$

3. Consider the singular integral equation

$$a(t)\, \varphi(t) + b(t) \sum_{k=0}^{n-1} \frac{H[\omega_k(t)]}{\pi i} \int\limits_L \left[\frac{1}{\tau - \omega_k(t)} - \frac{1}{\tau - \omega_k(\infty)} \right] \varphi(\tau)\, d\tau = f(t), \qquad (1)$$

where $\omega_0(z) \equiv z, \omega_1(z), \ldots, \omega_{n-1}(z)$ is the group of linear fractional transformations, $H(z)$ is a rational function having no poles on L_0 and $a^2 - b^2 = 1$.

Prove that with the help of the analytic automorphic function

$$\Phi(z) = \sum_{k=0}^{n-1} \frac{H[\omega^k(z)]}{2\pi i} \int\limits_{L_0} \left[\frac{1}{\tau - \omega_k(z)} - \frac{1}{\tau - \omega_k(\infty)} \right] \varphi(\tau)\, d\tau$$

the equation can be reduced to the boundary value problem

$$\Phi^+(t) = \frac{a[\omega_k(t)] - H[\omega_k(t)]\, b[\omega_k(t)]}{a[\omega_k(t)] + H[\omega_k(t)]\, b[\omega_k(t)]} \Phi^-(t) + \frac{H[\omega_k(t)]\, f[\omega_k(t)]}{a[\omega_k(t)] + H[\omega_k(t)]\, b[\omega_k(t)]}, \qquad (2)$$

$$t \in L_k \quad (k = 0, 1, \ldots, n - 1).$$

Hence derive the solution of equation (1) for $\varkappa \geqq -\nu + 1$ in the form

$$\varphi(t) = a(t) f(t) - b(t)\, Z(t) \sum_{k=0}^{n-1} \frac{H[\omega_k(t)]}{\pi i} \times$$

$$\times \int\limits_{L_0} \frac{f(\tau)}{Z(\tau)} \left[\frac{1}{\tau - \omega_k(t)} - \frac{1}{\tau - \omega_k(\infty)} \right] d\tau + \frac{b(t)\, Z(t)\, P_{\varkappa + \nu - 1}(F)}{[F(t) - F(z_\infty)]^{\varkappa - 1} \prod\limits_{j=1}^{\nu} [F(t) - F(z_j)]};$$

$Z(t) = [a(t) + H(t) b(t)]X^+(t)$, $X(t)$ is the canonical function of Problem 2, z_j are the poles of the function $H(z)$ (F. D. Gakhov and L. I. Chibrikova (1)).

4. Consider the singular equation

$$a(t)\,\varphi(t) + b(t) \sum_{k=0}^{n-1} \frac{p_k(t)}{\pi i} \int_{L_0} \left[\frac{1}{\tau - \omega_k(t)} - \frac{1}{\tau - \omega_k(\infty)} \right] \varphi(\tau)\,d\tau = f(t), \quad (1)$$

the functions $\omega_k(z)$ constituting a cyclic group; $p_k(z)$ are rational functions which satisfy the conditions

$$p_k(z) = \prod_{j=0}^{k-1} p_1[\omega_j(z)], \quad \prod_{j=0}^{n-1} p_1[\omega_j(z)] = 1.$$

Prove that with the help of the analytic function

$$\Phi(z) = \sum_{k=0}^{n-1} \frac{p_k(z)}{2\pi i} \int_{L_0} \left[\frac{1}{\tau - \omega_k(z)} - \frac{1}{\tau - \omega_k(\infty)} \right] \varphi(\tau)\,d\tau,$$

satisfying the condition

$$p_j(z)\,\Phi[\omega_j(z)] = \Phi(z) \quad (j = 1, 2, \ldots, n-1),$$

the equation is reduced to the boundary value problem

$$\Phi^+(t) = \frac{a[\omega_k(t)] - b[\omega_k(t)]}{a[\omega_k(t)] + b[\omega_k(t)]}\, \Phi^-(t) + \frac{p_k(t)\,f[\omega_k(t)]}{a[\omega_k(t)] + b[\omega_k(t)]},$$

$$t \in L_k \quad (k = 0, 1, \ldots, n-1).$$

Hence, derive the solution of equation (1) in the form

$$\varphi(t) = a(t)f(t) - b(t)\,Z(t) \sum_{k=0}^{n-1} \frac{p_k(t)}{\pi i} \times$$

$$\times \int_{L_0} \frac{f(\tau)}{Z(\tau)} \left[\frac{1}{\tau - \omega_k(t)} - \frac{1}{t - \omega_k(\infty)} \right] d\tau + b(t)\,Z(t)\,\theta(t),$$

where

$$\theta(z) = \theta_0(z) \frac{P_{\varkappa + \nu - 1}(F)}{[F(z) - F(z_\infty)]^{\varkappa - 1} \prod_{j=1}^{\nu} [F(z) - F(z_j)]}; \quad \theta_0(z) = \sum_{k=0}^{n-1} p_k(z);$$

$z_1, z_2, \ldots, z$ are the zeros of the function $\theta_0(z)$ (F. D. Gakhov and L. I. Chibrikova (1)).

5. Prove that the Tricomi equation [24]

$$\varphi(t) - \lambda \int_0^1 \left[\frac{1}{\tau - t} - \frac{1}{\tau + t - 2\tau t} \right] \varphi(\tau)\,d\tau = f(t)$$

belongs to the type of equations (1) of the preceding problem.

6. Reduce the solution of the equation

$$a(t)\,\varphi(t) + \frac{b(t)}{\pi i} \sum_{k=0}^{2n-1} (-1)^k \int_{L_0} \frac{\varphi(\tau)}{\omega_k(\tau) - t}\, d\tau = f(t), \tag{1}$$

where $\omega_k(t)$ constitute a cyclic group, to the boundary value problem

$$\Phi^+(t) = \frac{a[\omega_k(t)] - b[\omega_k(t)]}{a[\omega_k(t)] + b[\omega_k(t)]}\, \Phi^-(t) + \frac{(-1)^k \omega_k'(t) f[\omega_k(t)]}{a[\omega_k(t)] + b[\omega_k(t)]}$$

for the function

$$\Phi(z) = \sum_{k=0}^{2n-1} \frac{(-1)^k}{2\pi i} \int_{L_0} \frac{\varphi(\tau)}{\omega_k(\tau) - z}\, d\tau.$$

Hence, derive the solution of equation (1) in a closed form.
Prove that the equation

$$\varphi(t) = \frac{1}{\pi} \int_0^1 \left(\frac{1}{\varphi - t} + \frac{1 - 2\tau}{\tau + t - 2\tau t} \right) \varphi(\tau)\, d\tau,$$

employed by A. V. Bitsadze (1) in solving boundary value problems for equations of mixed type, belongs to the type of equations (1) (F. D. Gakhov and L. I. Chibrikova (1)).

7. Prove that by means of the expansion into simple fractions, in the case of finite groups the kernel $F'(\tau)/[F(\tau) - F(t)]$ is transformed into the expression

$$\sum_{k=0}^{n-1} \left[\frac{1}{\tau - \omega_k(t)} - \frac{1}{\tau - \omega_k(\infty)} \right].$$

8. Solve the integral equation

$$\frac{1}{2} [1 + (t-1)^\varkappa (3-t)^\varkappa]\, \varphi(t) +$$

$$+ \frac{1 - (t-1)^\varkappa (3-1)^\varkappa}{2\pi i} \int_{L_0} \frac{H'(\tau)}{H(\tau) - H(t)}\, \varphi(\tau)\, d\tau = f(t),$$

where L_0 is the circle $|z - 1| = 1/2$,

$$H = F(z) - \frac{1}{F(z)} = \frac{16 - z^2(4 - z^2)}{4z(4 - z)},$$

and $F(z)$ is the fundamental automorphic function of the group Γ:

$$\omega_0(z) \equiv z, \quad \omega_1(z) = 4 - z, \quad F(z) = \frac{1}{\omega_0(z)} + \frac{1}{\omega_1(z)}, \quad \varkappa > 0.$$

Answer.

$$\varphi(t) = \frac{(t-1)^\varkappa (3-t)^\varkappa + 1}{2(t-1)^\varkappa (3-t)^\varkappa}\, f(t) - \frac{1 - (t-1)^\varkappa (3-t)^\varkappa}{(t-1)^\varkappa (3-t)^\varkappa} \times$$

$$\times \left[\frac{(4-t)t}{3t^2 - 12t - 16} \right]^\varkappa \frac{1}{2\pi i} \int_{L_0} f(\tau) \left[\frac{3\tau^2 - 12\tau - 16}{(4-\tau)\tau} \right]^\varkappa \frac{H'(\tau)}{H(\tau) - H(t)}\, d\tau -$$

$$- \frac{1 - (t-1)^\varkappa (3-t)^\varkappa}{2(t-1)^\varkappa (3-t)^\varkappa} \left[\frac{(4-t)t}{3t^2 - 12t - 16} \right]^\varkappa P_{\varkappa - 1}[H(t)].$$

9. Show that the integral

$$\frac{1}{\pi i} \int\limits_{L_0} \frac{H'(\tau)}{H(\tau) - H(t)} \varphi(\tau) \, d\tau = f(t),$$

where $H(z) = \dfrac{1}{\cos z} - \cos z$ and L_0 is the circle $\left| z - \dfrac{\pi}{4} \right| = 1/2$, admits the inversion

$$\varphi(t) = \frac{1}{\pi i} \int\limits_{L_0} \frac{H'(\tau)}{H(\tau) - H(t)} f(\tau) \, d\tau.$$

Hint. $f(z) = \cos z$ is the fundamental automorphic function generated by the group of transformations $\omega_1(z) = z + 2\pi$, $\omega_2(z) = -z$.

10. Suppose that $F(z)$ is a single- or double-periodic function with period ω and having a zero of first order if $z = 0$. Show that for the integral

$$\frac{1}{\pi i} \int\limits_0^\omega \frac{F'(\tau - t)}{F(\tau - t)} \varphi(\tau) \, d\tau = f(t)$$

the inversion formula holds:

$$\varphi(t) = \frac{1}{\pi i} \int\limits_0^\omega \frac{F'(\tau - t)}{F(\tau - t)}) f(\tau) \, d\tau,$$

if the condition $\int\limits_0^\omega f(t) \, dt = 0$ is fulfilled.

Generalize for the case when the ratio F'/F is replaced by the ratio of the two periodic functions F_1 and F_2 with period ω which satisfy the condition $F_1(0) = 1$, $F_2(0) = 1$ (see L. I. Chibrikova (2), p. 107).

11. From consideration of the results of Problem 10, derive the following formulae for the Jacobian functions ϑ_1, sn, cn and dn:

(1) $\displaystyle \frac{1}{2\pi^2} \int\limits_0^{2\pi} \frac{\vartheta_1'\left(\dfrac{\tau - t}{2\pi}\right)}{\vartheta_1\left(\dfrac{\tau - t}{2}\right)} \varphi(\tau) \, d\tau = f(t),$ $\displaystyle \varphi(t) = -\frac{1}{2\pi^2} \int\limits_0^{2\pi} \frac{\vartheta_1'\left(\dfrac{\tau - t}{2\pi}\right)}{\vartheta_1\left(\dfrac{\tau - t}{2}\right)} f(\tau) \, d\tau;$

(2) $\displaystyle \frac{1}{\pi i} \int\limits_0^{4k} \frac{\operatorname{cn}(\tau - t) \operatorname{dn}(\tau - t)}{\operatorname{sn}(\tau - t)} \varphi(\tau) \, d\tau = f(t),$

$$\varphi(t) = \frac{1}{\pi i} \int\limits_0^{4k} \frac{\operatorname{cn}(\tau - t) \operatorname{dn}(\tau - t)}{\operatorname{sn}(\tau - t)} f(\tau) \, d\tau;$$

(3) $\displaystyle \frac{1}{\pi i} \int\limits_0^{4k} \frac{\operatorname{cn}(\tau - t)}{\operatorname{sn}(\tau - t)} \varphi(\tau) \, d\tau = f(t),$ $\displaystyle \varphi(t) = \frac{1}{\pi i} \int\limits_0^{4k} \frac{\operatorname{cn}(\tau - t)}{\operatorname{sn}(\tau - t)} f(\tau) \, d\tau;$

(4) $\dfrac{1}{\pi i} \displaystyle\int_0^{4k} \dfrac{\varphi(\tau)}{\operatorname{sn}(\tau - t)} \, d\tau = f(t), \quad \varphi(t) = \dfrac{1}{\pi i} \displaystyle\int_0^{4k} \dfrac{f(\tau)}{\operatorname{sn}(\tau - t)} \, d\tau;$

(5) $\dfrac{1}{\pi i} \displaystyle\int_0^{4k} \dfrac{\operatorname{dn}(\tau - t)}{\operatorname{sn}(\tau - t)} \, \varphi(\tau) \, d\tau = f(t), \quad \varphi(t) = \dfrac{1}{\pi i} \displaystyle\int_0^{4k} \dfrac{\operatorname{dn}(\tau - t)}{\operatorname{sn}(\tau - t)} f(\tau) \, d\tau;$

(6) $\dfrac{1}{\pi i} \displaystyle\int_0^{4k} \dfrac{\operatorname{cn}(\tau - t)}{\operatorname{sn}(\tau - t)\operatorname{dn}(\tau - t)} \varphi(\tau) \, d\tau = f(t),$

$$\varphi(t) = \dfrac{1}{\pi i} \int_0^{4k} \dfrac{\operatorname{cn}(\tau - t)}{\operatorname{sn}(\tau - t)\operatorname{dn}(\tau - t)} f(\tau) \, d\tau.$$

12. Find Hilbert's inversion formulae (6.6), (6.7) from the results of Problem 10 on the condition that

$$\int_0^{2\pi} u\,(\sigma) = \int_0^{2\pi} v\,(\sigma) \, d\sigma = 0.$$

13. Solve the Abel-type equation

$$\int_\alpha^x (x - t)^\mu \varphi(t) \, dt = f(x), \quad 0 < \mu < 1.$$

Answer.

$$\varphi(t) = \frac{\sin \mu\pi}{\mu\pi} \frac{d^2}{dx^2} \int_0^x \frac{f(t)}{(x - t)^\mu} \, dt.$$

14. Solve Schölmilch's integral equation

$$\frac{2}{\pi} \int_0^{\frac{\pi}{2}} \varphi(x \sin \theta) \, d\theta = f(x).$$

Answer.

$$\varphi(x) = f(0) + x \int_0^{\frac{\pi}{2}} f\,'(x \sin \psi) \, d\psi$$

(Whittaker and Watson, *A Course of Modern Analysis*, Pt. I, 2nd ed. Fizmatgiz, p. 323, 1962.)

REFERENCES

A. MANUALS, MONOGRAPHS AND REVIEWS

1. L. FORD, *Automorphic functions*, McGraw-Hill, New York (1929).
2. F. D. GAKHOV, Linear boundary value problems of the theory of complex variable functions. *Izv. Kazan. fiz. mat. obshch.*, X ser. 3, 39–79 (1938).
3. F. D. GAKHOV, Boundary value problems for analytic functions and singular integral equations. *Izv. Kazan. fiz. mat. obshch.*, XIV, ser. 3, 75–160 (1949).
4. F. D. GAKHOV, Riemann boundary value problem for a system of n pairs of functions. *Uspekhi matem. nauk*, VII, 50, No. 4, 3–54 (1952).
5. L. A. GALIN, *Contact problems of the theory of elasticity* (Kontaktnye zadachi teorii uprugosti). Gostekhizdat (1953).
6. G. M. GOLUZIN, *Geometric theory of complex variable functions* (Geometricheskaya teoriya funktsii kompleksnogo peremennogo). Gostekhizdat (1952).
7. E. GOURSAT, *Cours d'analyse mathématique*, 1. Paris (1923).
8. C. HERMITE, *Cours d'analyse*. Paris (1873).
9. D. HILBERT, *Grundzüge der Integralgleichungen*, 2. Aufl., Dritter Abschnitt. Leipzig–Berlin (1924).
10. Z. I. KHALILOV, *Linear equations in linear normed spaces* (Lineynye uraveniya v lineiynykh normirovannykh prostranstvakh). Izd. Akad. Nauk Azerb. SSR, Baku (1949).
11. B. V. KHVEDELIDZE, *Linear discontinuous boundary value problems of the theory of functions, singular integral equations and some applications* (Lineinye razryvnye granichnye zadachi teorii funktsii, singulyarnye integral'nye uravneniya i nekotorye ikh prilozheniya). Trud. Tbilsk. Mat. Inst., Akad. Nauk Gruz. SSR, XXIII, pp. 3–158 (1956).
12. V. D. KUPRADZE, *Some new remarks on the theory of integral singular equations.* Trudy Tbil. Univ., *42*, 1–23 (1951).
13. V. D. KUPRADZE, *Boundary value problems of the theory of vibrations and integral equations* (Granichnyye zadachi teorii kolebanii i integralnyye uravneniya). Gostekhizdat (1950).
14. M. A. LAVRENT'EV and B. V. SHABAT, *Methods of the theory of complex variable functions* (Metody teorii funktsii kompleksnogo peremennogo). Gostekhizdat (1951).
15. S. G. MIKHLIN, The plane problem of the theory of elasticity. *Trudy Seism. Inst. Akad. Nauk SSSR*, No. 65 (1935).
16. S. G. MIKHLIN, Singular integral equations. *Uspekhi matem. nauk*, III, No. 3 (25), 29–112 (1948).
17. N. I. MUSKHELISHVILI, *Singular integral equations* (Singulyarnyye integral'nyye uravneniya). Gostekhizdat (1946). English translation—Noordhoff, Groningen, Holland (1953).

17*. N. I. MUSKHELISHVILI, *Singular integral equations* (Singulyarnye integral'nye uravneniya), 2nd ed. Fizmatgiz (1962).

18. N. I. MUSKHELISHVILI, *Some basic problems of mathematical theory of elasticity* (Nekotorye osnovnye zadachi matematicheskoi teorii uprugosti). Izd. Akad. Nauk SSSR (1954). English translation—Noordhoff, Groningen, Holland (1953).

19. E. PICARD, *Leçons sur quelques types simples d'équations aux dérivées partielles.* Gauthier-Villars, Paris (1927).

20. H. POINCARÉ, *Leçons de Mécanique Céleste*, Dover, New York, vol. III, Ch. X.

21. I. I. PRIVALOV, *Boundary properties of analytic functions* (Granichnyye svoistva analiticheskikh funktsii). Gostekhizdat (1950).

22. B. RIEMANN, *Gesammelte Mathematische Werke.* Leipzig (1892).

23. V. I. SMIRNOV, *Course of higher mathematics* (Kurs vysschei matematiki), vol. 3, part II. Gostekhizdat (1948). English translation by Pergamon Press (1964).

24. I. SNEDDON, *Fourier transforms.* International series in pure and applied mathematics, McGraw-Hill, New York (1951).

25. YU. V. SOKHOTSKI, *On definite integrals and functions employed in expansions into series* (Ob opredelennykh integralakh i funtsiyakh upotrebljayemykh pri razlozhenii v ryady). St. Petersburg (1873).

26. E. TITCHMARSH, *Introduction to the theory of Fourier integrals.* Clarendon Press, Oxford (1937).

27. F. TRICOMI, Sullo equazioni lineari alle derivative parziali di 2° ordino di tipo misto. *Memorie della R. Academia Nazionale dei Linci*, ser. V, vol. XIV, fasc. VII (1923).

28. G. G. TUMASHEV and M. T. NUZHIN, Inverse boundary value problems. *Uch. zap. Kazan. Un.*, *115*, 6 (1955).

29. I. N. VEKUA, Systems of differential equations of elliptic type. *Matem. sb.*, 31/73; 2, 218–314 (1952).

30. I. N. VEKUA, *New methods of solution of elliptic equations* (Novye metody resheniya ellipticheskikh uravnenii). Gostekhizdat (1948).

30*. I. N. VEKUA, *Generalized analytic functions* (Obobshchennye analiticheskiye funktsii). Fizmatgiz (1959). English translation—Pergamon Press (1962).

31. N. P. VEKUA, *Systems of singular integral equations* (Sistemy singulyarnykh integralnykh uravnenii). Gostekhizdat (1950).

B. ORIGINAL PAPERS IN RUSSIAN

AKHIYEZER, N. I., and V. A. SCHHEBRINA, (1) On the inversion of some singular integrals. *Uch. zap. Karkov univ.*, XXV, 4, 191–8 (1957).

AKSENT'EV, L. A., (1) The sufficient conditions for the single-sheetedness of three inverse problems. *Uch. zap. Kazan. univ.*, *117*, No. 2, 32–5 (1957).

ALEKSEYEV, A. D., (1) On a singular integral equation on a contour of class *R*. *Dokl. Akad. Nauk SSSR*, *136*, No. 3, 525–8 (1961).
(2) Singular integral equation on a contour of class *R*. Trudy Tbil. matem. inst.

ANDRIANOV, S. N., (1) On the existence and number of the solution of the inverse boundary value problem in the theory of analytic functions. *Uch. zap. Kazan. univ.*, *113*, No. 10, 21–30 (1953).

BATYREV, A. V., (1) Approximate solution of the Riemann–Privalov problem. *Uspekhi matem. nauk, 11*, No. 5, 71–6 (1956).

BITSADZE, A. V., (1) On the problem of equations of mixed type. *Trudy matem. inst. im. V. A. Steklova, 61*, 1–58 (1953).

CHERSKII, YU. I., (1) The general singular equation and the convolution equation. *Matem. sborn., 41* (81), No. 3, 277–95 (1957).
(2) To the solution of the Riemann boundary value problem in a class of generalized functions. *Dokl. Akad. Nauk SSSR, 125*, No. 6, 500–503 (1959).

CHIBRIKOVA, L. I., (1) Special cases of generalized Riemann problem. *Uch. zap. Kazan. univ., 112*, No. 10, 129–54 (1952).
(2) On the Riemann boundary value problem for automorphic functions. *Uch. zap. Kazan. univ., 116*, No. 4, 59–109 (1956).
(3) To the solution of the Tricomi boundary value problem for the equation. $\frac{\partial^2 u}{\partial x^2} + \text{sgn } y \frac{\partial^2 u}{\partial y^2} = 0$. *Uch. zap. Kazan. univ., 117*, No. 9, 43–51 (1957).
(4) Effective solution of Hilbert's boundary value problem for some polygonals bounded by two circles. *Uch. zap. Kazan. univ., 117*, No. 2, 22–6 (1957).
(5) New method of solving one boundary value problem of the mixed type. *Uch. zap. Kazan. univ., 117*, No. 9, 44–47 (1957).

CHIBRIKOVA, L. I., and V. S. ROGOZHIN, (1) On the reduction of some boundary value problems to the generalized Riemann problem. *Uch. zap. Kazan. univ., 112*, No. 10, 123–27 (1952).

CHIKIN, L. A., (1) Special cases of the Riemann boundary value problem and singular integral equations. *Uch. zap. Kazan. univ., 113*, No. 10, 57–104 (1952).
(2) On the stability of the Riemann boundary value problem. *Dokl. Akad. Nauk SSSR, 111*, No. 1, 44–6 (1956).
(3) Stability of the Riemann boundary value problem. *Uch. zap. Rostov. univ., 43*, No. 6, 119–125 (1959).

FEL'D, YA. I., (1) On infinite sets of linear algebraic equations associated with problems in semi-infinite periodic structures. *Dokl. Akad. Nauk SSSR, 102*, No. 2, 257–60 (1955).

FOK, V. A., (1) On some integral equations of mathematical physics. *Dokl. Akad. Nauk SSSR, 37*, 147 (1942).
(2) On some integral equations of mathematical physics. *Matem. sborn., 14* (56), No. 1, 3–50 (1944).

GAGUA, M. B., (1) On an application of multiple Cauchy integrals. In the symposium *Investigations into contemporary problems in the theory of complex-variable functions* (Issledovaniya po sovremennym problemam teorii funktsii kompleksnogo peremennogo), 345–52. Fizmatgiz (1960).

GAKHOV, F. D., (1) On the Riemann boundary value problem. *Matem. sborn., 2* (44), No. 4, 673–83 (1937).
(2) Boundary value problems of the theory of analytic functions and singular integral equations. Doctor's Thesis, Tbilisi (1941).
(3) On inverse boundary value problems. *Dokl. Akad. Nauk SSSR, 86*, No. 4, 649–52 (1952).
(4) On inverse boundary value problems. *Uch. zap. Kazan. univ., 113*, No. 10, 9–20 (1953).

(5) On the inverse boundary value problem for a multiply-connected domain. *Uch. zap. Rostov. ped. inst.*, No. 3, 19–27 (1955).

(6) On new types of integral equations soluble in closed form. In the symposium *Problems in continuum mechanics* (Problemy mekhaniki sploshnoi sredy). *Akad. Nauk SSSR*, 102–114 (1961).

GAKHOV, F. D., and L. I. CHIBRIKOVA, (1) On some types of singular integral equations solvable in a closed form. *Matem. sborn.*, *35*, 395–436 (1954).

(2) On the Riemann boundary value problem for the case of intersecting contours. *Uch. zap. Kazan. univ.*, *113*, No. 10, 107–110 (1953).

GAKHOV, F. D., and E. G. KHASABOV, (1) On the Hilbert boundary value problem for a multiply connected domain. *Izv. vuzov. Matematika*, 2, No. 1, 12–21 (1958).

GAKHOV, F. D., and YU. M. KRIKUNOV, (1) Topological methods of the complex variable and their application to inverse boundary value problems. *Izv. Akad. Nauk SSSR*, ser. mat., 20, 206–240 (1956).

GAKHOV, F. D., and I. M. MELNIK, (1) Singular points of the contour in the inverse boundary value problem of the theory of analytic functions. *Ukr. mat. zhur.*, *11*, No. 1, 25–37 (1959).

GANIN, M. P., (1) Boundary value problems of the theory of polyharmonic functions. *Uch. zap. Kazan. univ.*, *111*, No. 10, 9–13 (1951).

(2) Boundary value problems for polyanalytic functions. *Dokl. Akad. Nauk SSSR*, *80*, No. 3, 313–16 (1951).

(3) Equivalent regularizing operator for a system of singular integral equations *Dokl. Akad. Nauk SSSR*, *79*, No. 3, 385–87 (1951).

(4) On a general boundary value problem for analytic functions. *Dokl. Akad. Nauk SSSR*, *79*, No. 6, 921–24 (1951).

(5) On a general boundary value problem for analytic functions. Thesis, Kazan (1952).

IVANOV, V. V., (1) Approximate solution of singular integral equations. *Dokl. Akad. Nauk SSSR*, *110*, No. 1, 15–18 (1956).

(2) On the application of the method of moments and the mixed method to the approximate solution of singular integral equations. *Dokl. Akad. Nauk SSSR*, *114*, No. 5, 945–8 (1957).

(3) Approximate solution of singular integral equations in the case of open contours of integration. *Dokl. Akad. Nauk SSSR*, *111*, No. 5, 933–6 (1956).

(4) Some properties of Cauchy integrals and their applications. *Dokl. Akad. Nauk SSSR*, *121*, No. 5, 793–4 (1958).

KAKICHEV, V. A., Boundary properties of the Cauchy integral of many variables. *Uch. zap. shakt. ped. inst.*, *2*, No. 6, 25–90 (1959).

KALANDIYA, A. I., (1) On the approximate solution of one class of singular integral equations. *Dokl. Akad. Nauk SSSR*, *125*, No. 4, 715–18 (1959).

KELDYSH, M. V., (1) Conformal mappings of multiply-connected domains onto canonical domains. *Uspekhi matem. nauk*, *6*, 90–119 (1939).

KELDYSH, M. V., and L. I. SEDOV, (1) Effective solution of some boundary value problems for harmonic functions. *Dokl. Akad. Nauk SSSR*, *16*, No. 1, 7–10 (1937).

KHASABOV, E. G., Hilbert's generalized boundary value problem. *Uch. zap. Rostov-on-Don univ.*, *46*, No. 7, 257–78 (1959).

KHVEDELIDZE, B. V., (1) On the Poincaré boundary value problem in the theory of logarithmic potential for a multiply-connected domain. *Soobshch. Akad. Nauk Gruz. SSR*, *2*, No. 7, 10, 571–8 and 865–72 (1941).

(2) Notes on my paper *Linear boundary value problems. Soobshch. Akad. Nauk Gruz. SSR*, XXI, No. 2, 129–30 (1958).

KRASNOVIDOVA, I. S., and V. S. ROGOZHIN, (1) The sufficient conditions for single-sheetedness of the solution of the inverse boundary value problem. *Uspekhi matem. nauk*, VIII, No. 1 (53), 151–3 (1953).

KREIN, M. G., (1) Integral equations on a semi-line with the kernel depending on the argument difference. *Uspekhi matem. nauk*, XIII, No. 5 (83), 413–16 (1958).

KRIKUNOV, YU. M., (1) On the solution of the generalized Riemann boundary value problem and a linear singular integro-differential equation. *Uch. zap. Kazan. univ., 112*, No. 10, 191–9 (1952).

(2) On the solution of the generalized Riemann problem. *Dokl. Akad. Nauk SSSR, 86*, No. 2, 269–72 (1952).

(3) Generalized Riemann boundary value problem and a linear singular integro-differential equation. *Uch. zap. Kazan. univ., 116*, No. 4, 3–30 (1956).

KUPRADZE, V. D., (1) Theory of integral equations with integral in the sense of Cauchy principal value. *Soobshch. Akad. Nauk Gruz. SSR, 2*, No. 7, 587–96 (1941).

(2) On the problem of equivalence in the theory of singular integral equations. *Soobshch. Akad. Nauk Gruz. SSR, 2*, No. 9, 793–8 (1941).

(3) On the theory of integral equations with integral in the sense of the Cauchy principal value. *Izv. Akad. Nauk SSSR*, ser. matem., 5 (1941).

KVESELAVA, D. A., (1) Solution of a boundary value problem of the theory of functions. *Dokl. Akad. Nauk SSSR, 53*, No. 8, 683–6 (1946).

(2) Some boundary value problems of the theory of functions. *Trudy Tbil. matem. inst., 16*, 39–80 (1948).

(3) Hilbert boundary value problem and singular integral equations in the case of intersecting contours. *Trudy Tbil. matem. inst.*, 17, 1–27 (1949).

(4) Riemann–Hilbert problem for a multiply-connected domain. *Soobshch. Akad. Nauk Gruz. SSSR, 6*, No. 8, 581–90 (1945) (in Georgian).

(5) Singular integral equations with discontinuous coefficients. *Trudy Tbil. matem. inst., 13*, 1–27 (1944) (in Georgian, with an extensive Russian summary).

MAGNARADZE, L. G., (1) On a generalization of the Plemelj–Privalov theorem. *Soobshch. Akad. Nauk Gruz. SSR, 8*, No. 8 (1947).

(2) Theory of a class of linear singular integro-differential equations. *Soobshch. Akad. Nauk Gruz. SSR, 6*, No. 2, 103–110 (1943).

MANDZHAVIDZE, F. G., and B. V. KHVEDELIDZE, On the Riemann–Privalov problem in continuous coefficients. *Dokl. Akad. Nauk SSSR, 123*, No. 5, 791–4 (1958).

MELNIK, I. M., (1) Limiting values of an analytic function represented by a curvilinear integral. *Soobshch. Akad. Nauk Gruz. SSR, 17*, No. 8, 681–6 (1956).

(2) An exceptional case of the Riemann boundary value problem. *Trudy Tbil. matem. inst., 24*, 149–162 (1957).

(3) Behaviour of a Cauchy inetgral near points of discontinuity of density and the singular case of the Riemann boundary value problem. *Uch. zap. Rostov. univ., 43*, No. 6, 57–71 (1959).

MIKHAILOV, L. G., (1) Limiting values of an analytic function represented by a curvilinear integral. *Uch. zap. Tadzh. univ.*, *1*, 32–5 (1952).

(2) A boundary value problem of Riemann type for a system of differential equations of first order of elliptic type. *Uch. zap. Tadzh. univ.*, *10*, 32–79 (1957).

MIKHLIN, S. G., (1) The problem of equivalence in the theory of singular integral equations. *Matem. sborn.*, *3* (45), 121–41 (1938).

(2) On a class of singular integral equations. *Dokl. Akad. Nauk SSSR*, *24*, No. 4, 315–7 (1938).

MUSKHELISHVILI, N. I., (1) An application of the Cauchy type integral to a class of singular integral equations. *Trudy Tbil. matem. inst.*, X, 1–43 (1941).

NATALEVICH, V. K., (1) Non-linear singular integral equations and non-linear boundary value problems of the theory of analytic functions. *Uch. zap. Kazan. univ.*, *112*, No. 10, 155–90 (1952).

NUZHIN, M. T., (1) On some inverse boundary value problems and their application in predetermining the cross-sectional shape of twisted bars. *Uch. zap. Kazan. univ.*, *109*, No. 1 (1949).

PARADOKSOVA, I. A., (1) On one integral equation with an automorphic kernel. *Dokl. Akad. Nauk SSSR*, *125*, No. 3, 496–9 (1959).

(2) On one singular integral equation associated with a cyclic group of linear fractional transformations. *Izv. vyssh. uch. zav.*, *Mat.*, *5*, 136–43 (1960).

(3) On the number of solution of one complete singular integral equation. *Uch. zap. Rostov. ped. inst.*, No. 5, 51–8 (1960).

PRIVALOV, N. I., (1) On a boundary value prolbem. *Matem. sborn.*, *41*, 4, 519–26 (1950).

PYKHTEYEV, G. N., On the evaluation of some singular integrals with Cauchy kernels. *Prikl. mat. i mekh.*, *23*, 1074–82 (1959).

ROGOZHIN, V. S., (1) Some boundary value problems for the polyharmonic equation. *Uch. zap. Kazan. univ.*, *110*, No. 3, 71–94 (1950).

(2) Finding the shape of a body from a specified pulsed impact pressure. *Prikl. mat. i mekh.*, *23*, No. 3, 589–91 (1959).

(3) A new integral representation of a sectionally-analytic function and its application. *Dokl. Akad. Nauk SSSR*, *135*, No. 4, 791–3 (1960).

(4) On the number of solutions of the exterior inverse boundary value problem. *Uch. zap. Rostov. univ.*, *46*, No. 7, 155–8 (1959).

(5) The sufficient conditions for single-sheetedness of the solution of inverse boundary value problems. *Prikl. mat. i mekh.*, *22*, No. 6, 804–807 (1958).

RODIN, YU. A., (1) On the solubility conditions of Riemann and Hilbert boundary value problems on Riemann surfaces. *Dokl. Akad. Nauk SSSR*, *129*, No. 6, 1234–7 (1959).

SAKALYUK, K. D., (1) Abel's generalized integral equation. *Dokl. Akad. Nauk SSSR*, *131*, No. 4, 748–51 (1960).

SHERMAN, D. I., (1) On the general theory of potential. *Izv. Akad. Nauk SSSR*, ser. matem., *10*, 121–34 (1946).

(2) On the methods of solution of some singular integral equations. *Prikl. matem. i mekh.*, *12*, 423–53 (1948).

(3) On a case of regularization of singular equations. *Prikl. matem. i mekh.*, *15*, 75–83 (1951).

(4) On some problems of the theory of potential. *Prikl. matem. i mekh.*, *9*, 479–88 (1945).

(5) On the connection of the basic problem of the theory of elasticity with a special case of the Poincaré problem. *Prikl. matem. i mekh.*, *17*, 685–92 (1953).

SIMONENKO, I. B., (1) The Riemann boundary value problem with continuous coefficients. *Dokl. Akad. Nauk SSSR, 124*, No. 2, 278–81 (1959).

SOBOLEV, S. L., (1) On a limit problem in the theory of logarithmic potential and its application to the reflection of plane elastic waves. *Trudy Seism. inst.*, No. 11, 1–9 (1930).

STOLYAROVA, Z. K., (1) General linear boundary value problem for equations of elliptic type. *Uch. zap. Kazan. univ., 111*, No. 8, 149–60 (1951).

USMANOV, N. K., (1) Boundary value problems for partial differential equations of the first order of elliptic type. *Trudy inst. fiz. i matem. Latv. SSR*, No. 1, 41–100 (1950).

(2) On the boundary value problems of functions satisfying a system of differential equation. *Trudy inst. fiz. i matem. Latv. SSR*, No. 2, 59–100 (1950).

VEKUA, I. N., (1) On singular linear integral equations. *Dokl. Akad. Nauk SSSR, 26*, No. 8, 335–8 (1940).

(2) Integral equations with the singular Cauchy integral. *Trudy Tbil. matem. inst.*, X, 45–72 (1941).

(3) On the theory of singular integral equations. *Soobshch. Akad. Nauk Gruz. SSR*, III, No. 9, 869–76 (1942).

(4) On one linear Riemann boundary value problem. *Trudy Tbil. matem. inst.*, XI, 109–139 (1942).

(5) On some properties of the solutions of sets of elliptic equations. *Dokl. Akad. Nauk SSSR, 98*, No. 2, 181–4 (1954).

ZERAGIYA, P. K., (1) On the solution of polyharmonic equations. *Trudy Tbil. matem. inst.*, VIII, 135–63 (1940).

C. REFERENCES IN NON-RUSSIAN LANGUAGES

BERTRAND, G., (1) Le probléme de Dirichlet et le potential de simple couche. *Bull. des Sciences Math.*, *47*, 2-e ser. (1923).

BITTNER, L., (1) Plemelische Randwertformeln für mehrfache Cauchy-Integral. *J. angew. Math. und Mech.*, *39*, No. 9–11, 347–9 (1959).

CARLEMAN, T., (1) Sur la résolution de certaines équations intégrales. *Arkiv för Matem. och Physik, 16*, No. 26 (1932).

(2) Abelsche Integralgleichung mit konstanten Grenzen *Mat. Z.*, Bd. 15, 111–20 (1922).

DEMTCHENKO, (1) *Problèmes mixtes harmoniques.* Gauthier-Villars, Paris (1933).

GELLERSTEDT, S., (1) Quelques problèmes mixtes. *Arkiv för Math. Astr. och Physik, 26*, No. 3 (1936).

HARNACK, A., (1) Beiträge zur Theorie des Cauchy'schen Integral. *Berichte d. k. Sächs. Ges. d. Wiss., Math.-Phys. Classe, 37*, 379–98 (1885).

HASEMAN, C., (1) *Anwendung der Theorie der Integralgleichung auf einige Randwertaufgaben.* Göttingen (1907).

HEINS, A. E., and R. C. MacCAMY, (1) A function-theoretic solution of certain integral equations. *Q. J. Math.*, *9*, 34, 132–43 (1958).

HILBERT, D., (1) Über eine Anwendung der Integralgleichungen. Verhandl. des III. Internat. Mathematiker Kongresses, Heidelberg (1904).

NOETHER, F., (1) Über eine Klasse singulärer Integralgleichungen. *Mathem. Ann.*, *92*, 42–63 (1921).

PLEMELJ, I., (1) Ein Ergänzungssatz ... *Monatshefte für Math. u. Phys.*, *19*, 205–210 (1908).

(2) Riemannsche Funktionenscharen. *Ibid.*, 211–245.

RIABOUHINSKY, (1) Sur la détermination d'une surface, d'après les données qu'elle porte. *C. R. Acad. sci.*, 18 (1929).

SIGNORINI, (1) Sopra un problema ol contoro nella teoria delle funzioni di variable complessa. *Annali di matematica* (3), 25 (1916).

TRJIZINSKY, W., (1) Singular integral equations. *Trans. Amer. math. Soc.*, *60*, No. 2 (1946).

VOLTERRA, V., (1) Sopra alkune condizioni caratteristische per functioni di variable complessa. *Annali di matematica* (2), 11 (1883).

WIENER, N., and E. HOPF, (1) Über eine Klasse singulärer Integralgleichungen. *Sitz. Berliner Akad. Wiss.*, 696–706 (1931).

INDEX

Made in Great Britain